James P.M. Syvitski David C. Burrell Jens M. Skei

Fjords

Processes and Products

With 216 Figures

Springer-Verlag
New York Berlin Heidelberg London Paris Tokyo

James P.M. Syvitski
Geological Survey of Canada, Bedford Institute of Oceanography,
Dartmouth, Nova Scotia B2Y 4A2, Canada

David C. Burrell
Institute of Marine Science, University of Alaska,
Fairbanks, Alaska 99701, U.S.A.

Jens M. Skei
Norwegian Institute for Water Research,
Oslo 3, Norway

Library of Congress Cataloging in Publication Data
Syvitski, James P. M.
Fjords : processes and products.
Bibliography: p.
Includes indexes.
1. Fjords. 2. Environmental protection.
I. Burrell, David C. II. Skei, Jens M. III. Title.
GB454.F5S98 1986 551.46'09 86-13951

Media conversion by David Seham Associates, Metuchen, New Jersey.
Printed and bound by Quinn-Woodbine, Woodbine, New Jersey.
Printed in the United States of America.

9 8 7 6 5 4 3 2 1

ISBN 0-387-96342-1 Springer-Verlag New York Berlin Heidelberg
ISBN 3-540-96342-1 Springer-Verlag Berlin Heidelberg New York

To our families

Patrick, Estelle, Tonya, and Alexander Syvitski
Helen, Kari, Michael, and Elizabeth Burrell
Anne-Grete, Kari, and Fiona Skei

Preface

Fjords are both an interface and a buffer between glaciated continents and the oceans. They exhibit a very wide range in environmental conditions, both in dynamics and geography. Some are truly wonders of the world with their dizzying mountain slopes rising sharply from the ocean edge. Others represent some of the harshest conditions on earth, with hurricane winds, extremes in temperature, and catastrophic earth and ice movements.

Fjords are unique estuaries and represent a large portion of the earth's coastal zone. Yet they are not very well known, given the increasing population and food pressures, and their present industrial and strategic importance. Temperate-zone estuaries have had many more years of intense study, with multiyear data available. Most fjords have not been impacted by man but, if history repeats itself, that condition will not last long. Fjords present some unique environmental problems, such as their usually slow flushing time, a feature common to many silled environments. Thus there is presently a need for management guidelines, which can only be based on a thorough knowledge of the way fjords work.

Fjords are, in many respects, perfect natural oceanographic and geologic laboratories. Source inputs are easily identified and their resulting gradients are well developed. Throughout this book, we emphasize the potential of modeling processes in fjords, with comparisons to other estuary, lake, shelf and slope, and open ocean environments.

Although primarily concerned with geologic aspects, this is the first book to integrate physical, chemical, and biological aspects and their interaction in fjords. Our text attempts to include state-of-the-art summaries and an extensive reference list on fjord geomorphology, geophysics, glaciology, hydrology, sedimentology, geotechnical engineering, biogeochemistry, biotic influences, and environmental case studies. Over 1100 references, from industry, government, and university scientists, are provided to the reader.

The book is principally addressed to advanced students, research professionals, and environmental scientists in the earth science and oceanographic community. Each chapter is linked with the others, but also stands on its own, thus allowing a reader to skip chapters where necessary. The book is loosely divided into three sections. The first two chapters form the introduction to the book and paint the background in a descriptive manner. Chapter 1 defines the fjord environment exemplified with comparisons to other environmental settings and describes the "why" behind past and present fjord research. Chapter 2 provides a more detailed description of the various fjord provenances, especially in regard to their glacial and postglacial history.

The second of the book's sectors expands on the introduction in a more quantitative manner, detailing the main operative processes and their products. Chapter 3 describes the terrestrial input parameters, how they are affected by the local drainage basin characteristics, climate, glacial and sea-level history, and how these parameters control the developments of fjord-head deltas. Chapter 4 details fjord circulation and mixing processes, flushing events and ice influences, on the deposition or erosion of sediment. Chapter 5 concerns itself with the mass-physical properties of the seafloor and slope stability problems and processes common to many fjords. Chapter 6 reviews the biota-sediment interactions, including a detailed description of the pelagic and littoral processes (production, grazing, and sedimentation), and the benthic environment (community structure, production, infauna and epifauna, stress effects). Chapter 7, on biogeochemistry, involves both subaqueous and subsurface diagenetic reactions and the partitioning of chemical species, in both oxic and anoxic environments and the redox zone in between.

The last sector of the book deals with the implications and applications of the preceding theoretical analysis to environmental management and to future studies in fjords. Chapter 8 provides 12 case histories chosen to cover the wide variety of environmental problems that face fjord managers and other environmentalists. The last chapter (9) lists, by discipline, some of the major gaps in our knowledge base on fjords with proposals for their solution.

The three authors of this book have principally worked in fjords throughout their careers and are presently involved in the management of some very large research programs on fjords. Our access to a large pool of unpublished data has inevitably led to the inclusion of some primary data. Diagrams or facts, unreferenced, are mainly from this pool. Much of the book has, however, come from the published literature. We would like to thank all the authors and publishing houses that have allowed us the use of their information and diagrams.

We particularly wish to thank all those people who have assisted us in the writing of this book. Drs. G.P. Glasby and H. Holtedahl kindly provided valuable information on New Zealand and Norwegian fjords, respectively. Our noteworthy chapter reviewers include J.T. Andrews, G.S.Boulton, R. Gilbert, J. Gray, G.E. Farrow, D.L. Forbes, F.J. Hein, P.R. Hill, P. Mudie, T.H. Pearson, D.J.W. Piper, C.T. Schafer, D.C.C. Sego, S.F. Sugai, R.B. Taylor, G. Vilks, G. Hong, S. Henrichs, K. Moran, and C.I. Winiecki. In addition, many colleagues at the fjord sections of the Norwegian Institute for Water Research, the Atlantic Geoscience Centre, and the Institute of Marine Science (University of Alaska) are acknowledged for their direct or indirect support of this project. Our most sincere thanks are extended to the Bedford Institute of Oceanography Word Processing Unit (D. Anderson) and drafting facility (Art Cosgrove and Harvey Slade), and those production assistants at the Atlantic Geoscience Centre (A. Simms, D. Beattie, W. LeBlanc). Finally we would like to thank our associates at Springer-Verlag for their understanding, co-operation, and great patience.

James P.M. Syvitski
David C. Burrell
Jens M. Skei

Contents

Part 1 Introduction

1

Fjords and Their Study

1.1 Definition, Distribution, and History

The usual scientific definition of a fjord, and the one adopted in this book, is a deep, high-latitude estuary which has been (or is presently being) excavated or modified by land-based ice. The term derives from the Old Norse "fjorthr," which is close to the present-day Icelandic usage. Throughout this book we use the modern Norwegian spelling "fjord," except that the anglicized "fiord" is retained in proper names. There are also many other designators in a variety of local languages, such as the Celtic "loch" or "lough." In Nordic usage, "fjord" is a generic name for a wide variety of marine inlets (and, formerly, bodies of fresh water), including, in southern Scandinavia, a number of shallow, temperate-zone estuaries somewhat removed from the types of fjords generally referenced here. Fairbridge (1968) has advocated the Swedish name "fjärd" for this latter variety. The distinction is a useful one, but the term itself has not been generally accepted into the scientific literature.

All estuaries are ephemeral geologic features, but fjords are the youngest of all—products of the general retreat of ice and sea-level fluctuations that have occurred since the last glacial maximum some 17,000 years BP. As a class, fjords are therefore immature, non-steady-state systems, evolving and changing over relatively short time scales. Columbia Glacier in Prince William Sound, Alaska, has recently started to retreat from the terminal moraine locality shown in Figure 1.1. By the year 2000 it is expected that a new fjord some 25 to 40 km in length and more than 400 m deep will occupy the valley shown. This is an extreme example, but all silled fjords, whether presently associated with glaciers or not, are immature estuaries and hence sites of net sediment accumulation. As scientifically defined, fjords are predominantly features of mountainous regions which presently,

FIGURE 1.1. The birth of a fjord: Columbia Glacier (Prince William Sound, Alaska) located at the terminal moraine sill in 1976. The glacier is now retreating rapidly, and within 30 to 50 years Columbia Bay is expected to be a new fjord some 40 km long with water depths in excess of 300 m. (Photo courtesy L.R. Mayo, U.S. Geological Survey, Fairbanks, Alaska.)

or in the recent past, have supported ice fields feeding valley glaciers. Hence fjords sensu stricto do not occur in flat polar terrain (as along the Alaska-Canada Beaufort Sea coast). Otherwise fjords have a worldwide distribution at high latitudes in both hemispheres (Fig. 1.2). The principal fjord provinces occur along the coasts of North and South America (above 42° latitude); the Kerguelen Islands, South Georgia, the Russian and Canadian arctic archipelagos, Svalbard and other high latitude islands; the southwest coast of New Zealand's South Island; Antarctica; Iceland and Greenland; and northern Europe, including the British Isles above around 56°N. Since fjords are at least partially ice scoured, the typical configuration (Fig. 1.3) is a long, narrow, deep, and steep-sided inlet, which is frequently branched and sinuous, but may be remarkably straight in whole or in part where the ice has followed major fault zones (see below). In these high-latitude regions, vertical gradients have been accentuated by isostatic uplift (Fairbridge, 1980). As a class, fjords are the deepest of all estuaries.

Fjords usually, but not inevitably, contain one or more submarine sills (Fig. 1.4). The internal

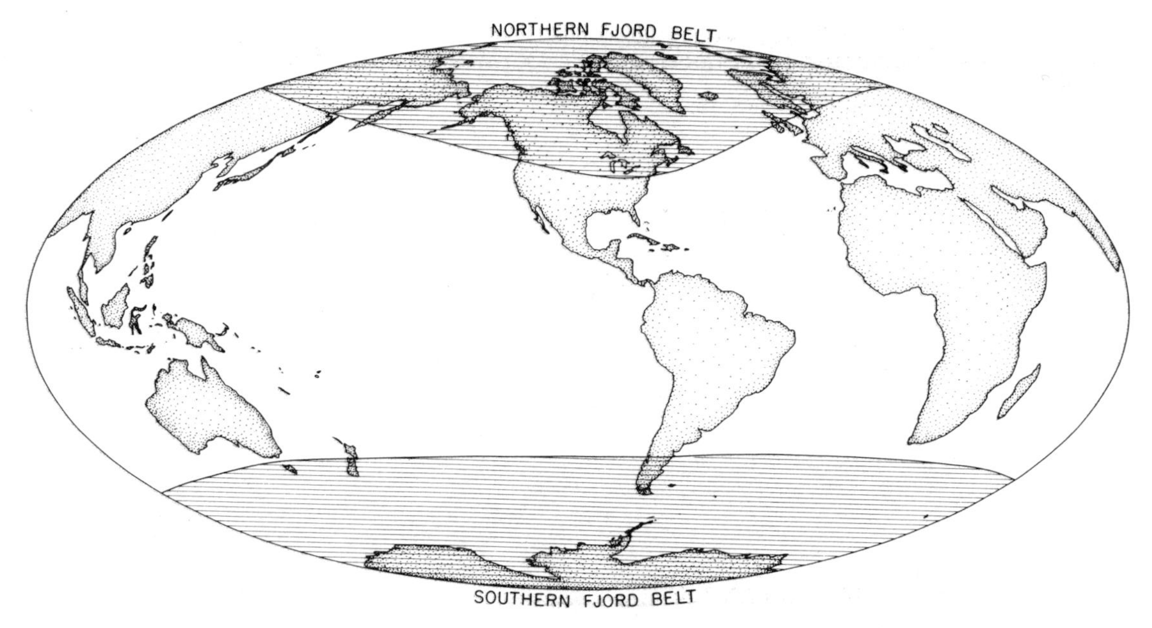

FIGURE 1.2. Generalized worldwide distribution of fjords.

basins defined by these sills are characteristic features of fjords, which determine many of their distinctive physical and biogeochemical characteristics. Silled types of fjords are therefore emphasized in the following chapters. In Norway, a "poll" is a landlocked inlet, and this term has now been incorporated into the scientific literature (Matthews and Heimdal, 1980; Wassmann, 1984) to describe a variety of fjord having an extremely shallow sill located within the near-surface circulation zone, and correspondingly distinctive physical and chemical properties.

When the permanent ice fields retreated during postglacial times, northern hemisphere fjords were selected by early man as desirable habitation sites, possibly by 7000 years BP (Brattegard, 1980). In historical times Viking navigators sailed from the Norwegian fjords to explore, exploit, and settle similar environments right across the north Atlantic: the Faeroe Islands and Iceland at the beginning and end of the ninth century, Greenland and North America some 100 to 150 years later. These exploits, well known from the rich legacy of Saga literature, illustrate the early importance of fjords and their environs as ports and farmsteads (and point up the incredible Viking navigation skills). The fjords of Greenland and Baffin Island were specifically documented following the voyages of John Davis in 1585, and Robert Bylot and William Baffin (in 1616) in search of the Northwest Passage. Today their named McBeth and Cambridge Fiords stand beside the neighboring Itirbilung and Inugsuin inlets, contrasting native Inuit and northern European cultures.

European exploration of the northwest coast of America was delayed until after the profound revolution in scientific thought and method that occurred in and around the 17th century. The great English explorer-navigators of the 18th century, for example, were thoroughly imbued with the spirit and objectives of the nascent Royal Society, which published a "Catalogue of Directions" for "seamen and other far travellers" (partially authored by Hooke) in 1665–66. The first cartographic surveys of the fjord coasts of British Columbia and Alaska were executed in a thoroughly scientific manner (see Burrell, 1980), and many useful natural history observations and collections were made. However, from the description of Howe Sound, British Columbia, in June 1792 (Vancouver, 1798):

> . . . a fresh southerly gale, attended with dark gloomy weather added greatly to the dreary prospect of the surrounding country. The low fertile shores we had been accustomed to see, . . . here no longer existed:

FIGURE 1.3. Aerial photograph of McBeth Fiord, Baffin Island (1949: NAPL T219L105).

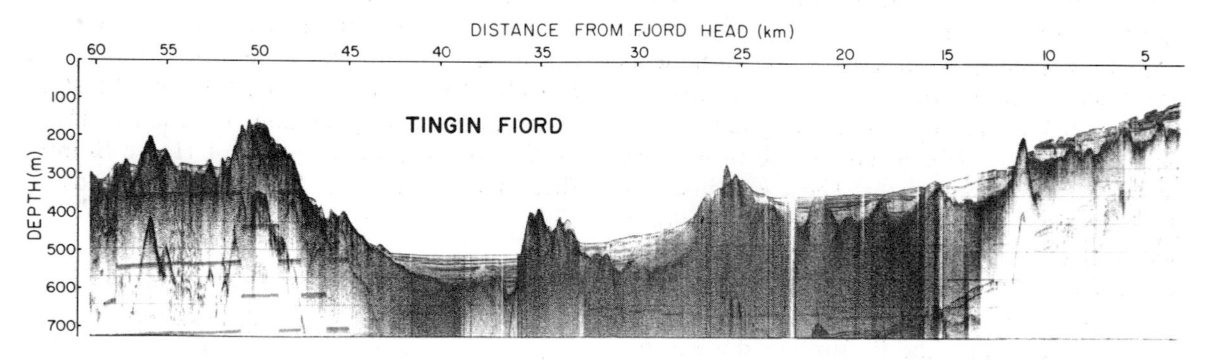

FIGURE 1.4. Longitudinal seismic profile of Tingin Fiord, Baffin Island, showing sediment infilling.

their place was now occupied by the base of a stupendous snowy barrier, thinly wooded and rising from the sea abruptly to the clouds; from whose frigid summit, the dissolving snow in foaming torrents rushed down the sides and chasms of its rugged surface, exhibiting altogether a sublime, though gloomy spectacle, which animated nature seemed to have deserted . . .

it is apparent that these inlets, now a spectacular tourist attraction (Fig. 1.1), were of scant aesthetic interest in this pre-Romantic age. The northwest America surveys of Cook (Cook and King, 1784) and Vancouver (Vancouver, 1798) took place during the "Little Ice-Age" period when the glaciers were much farther advanced than at present. In fact, portions of Figure 4.26 detailing the Neoglacial ice advance in Glacier Bay, Alaska, are based on their observations. At that time, the huge fjord complex that presently comprises Glacier Bay National Monument was buried beneath ice and resembled an ice shelf more than a fjord coast.

The existence of sills at the mouths of fjords appears to have been first documented by Captain Cook in 1774 during his voyage through Tierra del Fuego, and it seems likely from his descriptions of anchorages that Vancouver was also well aware of this physiographic feature. Later, Darwin similarly associated sills with fjords when he suggested (Darwin, 1846) that the best anchorage for a ship was at the mouth of a fjord where the water was shallower. As noted above, sills are now regarded as a characteristic but not essential prerequisite of fjord-estuaries. As areas surrounding fjords have become more populated and ship traffic has increased, very detailed bathymetric information has become available. Today, the most comprehensive bathymetric surveys within fjords are morphometric analyses of submarine features (e.g., Hay et al., 1983a).

Some of the earliest studies of fjords focused on origins. Their characteristic geomorphology generated a great deal of speculation, ranging from a purely tectonic genesis to fluvial or glacially modified mountain valleys. In the early 19th century, Esmark (1824) concluded that Norwegian fjords were the result of glacial excavation. This glacial genesis concept was later disputed by Gregory who advocated a causal connection between Tertiary dislocations and the development of fjords. Gregory (1913) categorically stated that all fjord systems of the world owed their distinctive structure to earth movements, a theory also supported by de Geer (1910). Several geomorphological "schools" took up the challenge, including Johnson (1915), who argued most strongly against the tectonic theory. The accepted view today is one of accommodating compromise: fjords owe their origin to fluvial action along fault lines, with subsequent and dominant excavation by glaciers following the path of least resistance (primarily the proto-river valley). The fact that fjords are found only within coastal terrain once dominated by Quaternary ice sheets underscores the essential influence of glacial processes (Holtedahl, 1967).

1.2 Environmental Setting and Study

Except for some polar inlets during the winter, all fjords are estuaries. Major freshwater inflow is likely to be at the head, and, as is shown in Chapter 4, the brackish water typically flows toward the mouth as a surface plume. As with all estuaries, fjords are therefore transition regions between the land and open oceans, regions of strong physical and chemical gradients where fresh and salt waters mix and react. Where "fjord-type" stratified circulation is well developed, marked physical and biogeochemical gradients—spatial changes in properties—may be apparent down-inlet from head to mouth (Fig. 1.5). Gradients are also characteristic of the sediments. The coarser, fluvially transported sediment deposited in the upper parts of fjords gradually becomes finer seaward (Fig. 1.6). Similar lithologic gradients occur near sills as a result of winnowing (Gilbert, 1978; Gade and Edwards, 1980). Many environmental chemical pollutants have relatively short residence times under estuarine conditions and, if released into fjords, may also exhibit strong concentration gradients within the water column, biota, and sediments. High concentrations occur near the pollution source, decreasing laterally as a result of dilution and reaction (Fig. 1.7).

Most importantly, fjords of the type predominantly considered in this volume—steep, deep, and characteristically silled inlets—encompass a number of distinctive oceanographic environ-

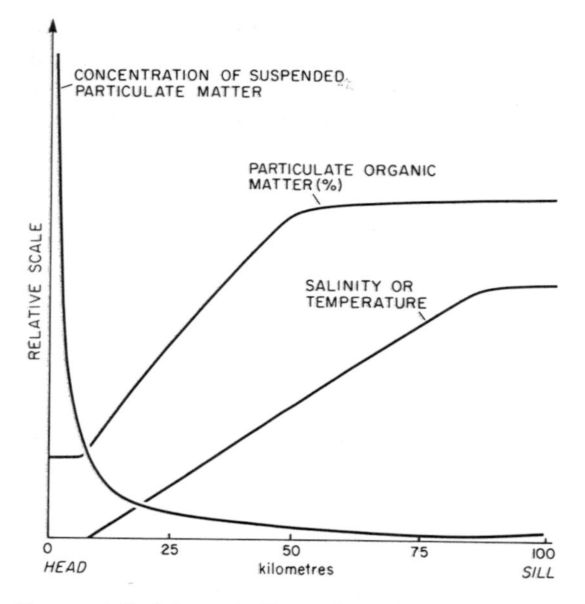

FIGURE 1.5. Schematic illustration of gradients within a typical British Columbia fjord during the summer, showing the longitudinal distributions of surface temperature and salinity, the concentration of suspended particulate matter, and relative percentage of suspended organic detritus.

ments, which make them particularly exciting sites for estuarine research. The near-surface "estuarine zone," basically common to all estuaries, is underlain by marine water which, in silled fjords, may be physically restrained in basin enclosures. Such coastal-zone, mini-ocean basins offer unique opportunities for studying terrestrial input into quasi-closed marine systems. Often the circulation above and below the top of the sill is poorly coupled, and, in deep fjords, processes and reactions within the basins may be spatially and temporally separated from those occurring in the upper-zone estuarine environment. The resultant pronounced vertical hydrographic gradients in these deep fjords influence both biota and sediments. For instance, some fjords show strong physical-chemical gradients, ranging from fully oxygenated water masses at the surface to totally isolated anoxic regions at depth (Fig. 1.8).

The concept of fjords as mini-oceans where inputs and outputs can be more easily measured

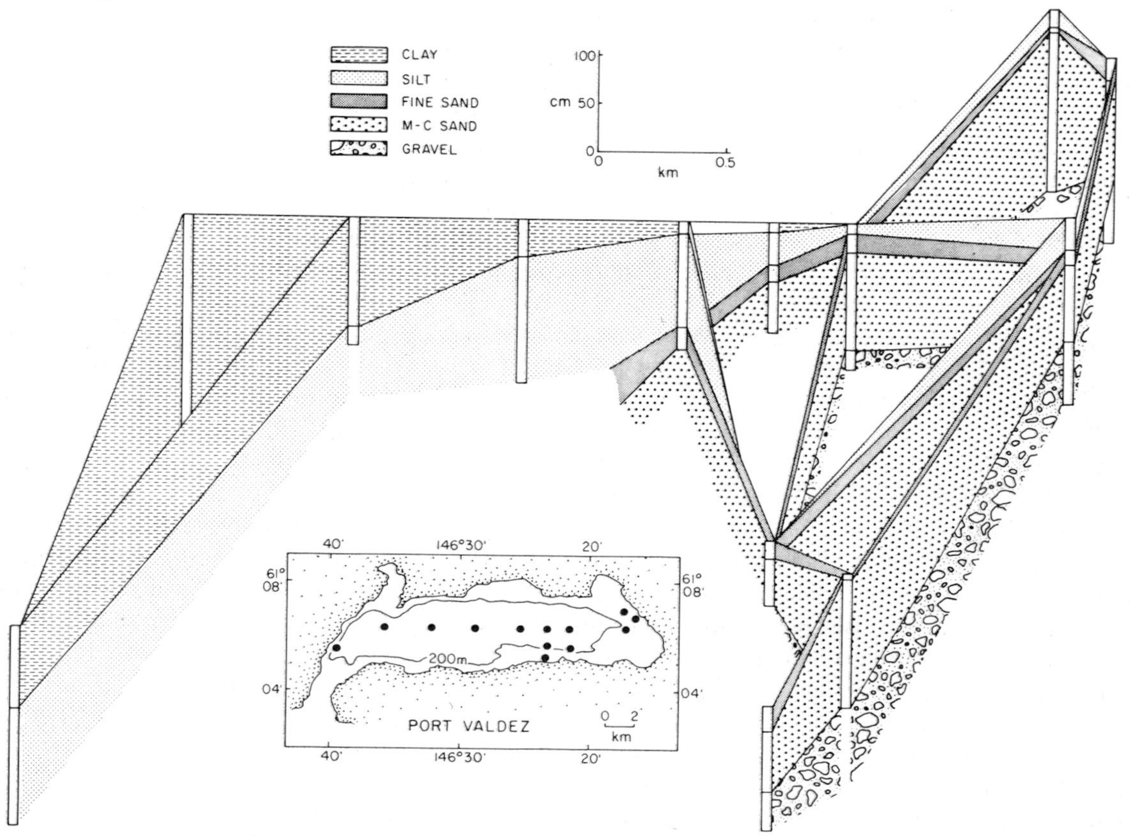

FIGURE 1.6. Three-dimensional "fence" diagram showing gradients in grain size of the near-surface sediments from head to mouth within Port Valdez, Prince William Sound, Alaska (after Sharma, 1979). (The coarse-grained sediment at the head of the fjord is predominantly the result of slumping caused by the major 1964 earthquake.)

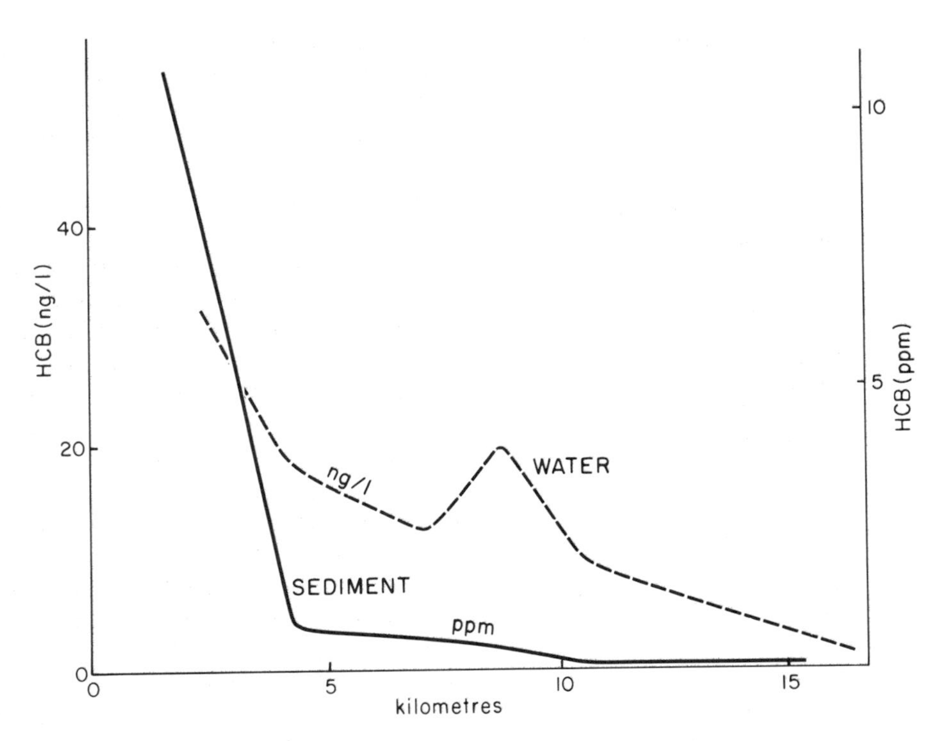

FIGURE 1.7. Gradient of a pollutant (HCB) in water and sediments of Frierfjord, Norway, away from the source (Skei, 1981a).

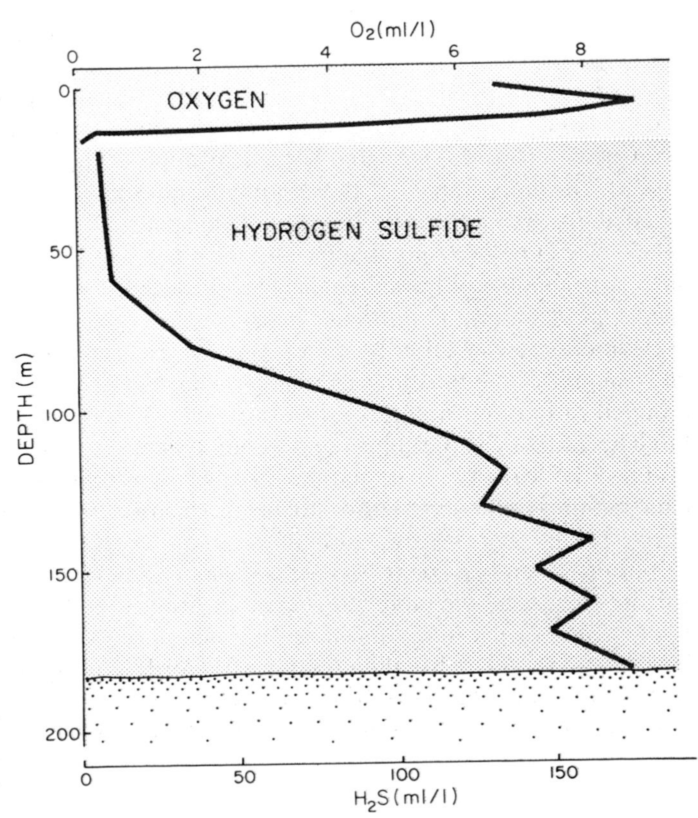

FIGURE 1.8. The vertical distribution of oxygen and hydrogen sulfide in Framvaren, a permanently anoxic Norwegian fjord (after Skei, 1983c).

TABLE 1.1. Parameters affecting fjord circulation, sedimentation, biogeochemistry, and biota.

A. Glacial	- Relative sea-level history - Wet vs. cold based glaciers - Floating vs. tidewater vs. hinterland glaciers - Style and rates of glacier advances and retreats - Basal shear stress
B. Fluvial	- Transport rates of bed load, suspended and dissolved loads - Runoff characteristics (e.g., jökolhlaup events) - Paraglacial history - Stratification and turbidity
C. Climatic	- Glacier movement including iceberg production - Sea-ice conditions - Thermal stratification - Wind events (waves, upwelling, aeolian transport) - Terrestrial and marine biomass production
D. Geographic	- Fetch length - Fjord dimensions (e.g., basin & sill depths, width, volume) - Relative sea-level history - Tidal characteristics - Coriolis effect - Flushing dynamics
E. Geotechnical	- Frequency and size of slope failures - Mass transport process - Seiches and tsunami waves

and modeled than in the larger and more remote open ocean systems has great practical application. The importance of these studies relies on an ability to parameterize fjords and elucidate details of end-member conditions. Because there are a large number of fjords spanning a very wide spectrum of natural conditions, there is an extensive choice available to match specific requirements. The controlling parameters can be grouped in five interrelated categories: glacial, fluvial, climatic, geographic, and geotechnical (Table 1.1). Most of these parameters are variable (some predictable, some stochastic), with scalar harmonics on the order of seconds to centuries. Even historical "constants" such as sill depth can be variable on a geologic time scale. Some of the many possible end-member fjords include: (1) high sedimentation rate fjords with particulate material settling through a density-stabilized water column; (2) low sedimentation rate fjords where detritus settles through water that is only intermittently stratified; (3) well-mixed fjords where circulation is dominated by tide and wave action; (4) polar fjords subject to periglacial processes on land, and influenced by the presence of sea ice through at least part of the year; (5) glacial fjords where glacier ice tongues (tidewater or floating) control circulation and sedimentation dynamics; (6) fjords subject to semi-continuous subaqueous slope failures and sediment gravity flows; and (7) fjords containing anoxic basin waters and marked temporal and spatial biogeochemical gradients.

Table 1.2 provides a few interesting statistics that are germane to the concept of fjords as mini-ocean systems. The total volume of water within all fjords is approximately the same as that for freshwater lakes, or for saline lakes and inland seas. As a percent of the world hydrosphere, however, the amount is rather insignificant (much less than 1%). Similarly, the volume of unconsolidated sediment in fjords is very small compared with the volume of sedimentary rocks within the lithosphere, or even the volume of continental margin sediments. Nevertheless, fjords have acted as efficient sediment traps in recent geologic times, retaining perhaps one-quarter of the fluvial sediment effluxed from the land over the last 100,000 years. Fjords exhibit a very wide range of sedimentation rates at the present time (Chapter 4), from the highest recorded natural marine values, to rates approaching those characteristic of deep-sea bains.

1.3 The Past, Present, and Future of Fjord Research

Systematic study of physical processes was first begun in Norwegian fjords over 100 years ago. By the early years of this century the fundamental modes of fjord circulation were well understood (Ekman, 1875; Murray, 1886; Gran, 1900; Helland-Hansen, 1906; Grund, 1909; Gaarder, 1916; see also historical review by Gade, 1970). Investigations have been extended to fjords in other parts of the world only in comparatively recent times. At the present time, probably the most extensive synoptic and time-series data sets are for Norwegian and British Columbian inlets (e.g., Saelen, 1967; Pickard,

TABLE 1.2. Fjords in perspective.

(a) Total water volume	
FJORDS	(1.4) 10^{14} m^3
Freshwater lakes	(1.3) 10^{14} m^3
Saline lakes and inland seas	(1.0) 10^{14} m^3
(b) Uplift	
FJORDS (isostasy) (i) average (decay curve)	10 mm yr^{-1}
(ii) extreme (Glacier Bay)	40 mm yr^{-1}
Tectonic (mountainbuilding) (i) average (pulse)	3–10 mm yr^{-1}
(ii) extreme (Japan)	70 mm yr^{-1}
(c) Erosion	
FJORD areas	0.1–3 mm yr^{-1}
Oceania areas (e.g., Taiwan)	1 mm yr^{-1}
World (average)	0.06 mm yr^{-1}
(d) Accumulation	
FJORDS	0.1–9000 mm yr^{-1}
Mouths of large rivers	100–500 mm yr^{-1}
Gulf of Mexico (Mississippi region)	0.2 mm yr^{-1}
Molasse Basin (Swiss Alps)	0.1–0.4 mm yr^{-1}
California Continental Borderland	0.1–4.5 mm yr^{-1}
Vema Fracture Zone	1.2 mm yr^{-1}
Deep Sea basins	0.001–0.1 mm yr^{-1}
(e) Volume of sediment in FJORDS as compared with	
Total volume of sedimentary rocks on earth	0.03%
Total sediment volume on the continental margins	0.1%
Volume of marine sediment deposited within the last 100,000 years	24%

1975). Comprehensive reviews of the present status of fjord physical oceanography have been given by Gade and Edwards (1980) and Farmer and Freeland (1983).

Numerical models of estuarine circulation abound (e.g., Wang and Kravitz, 1980), and there are numerous examples detailing the effects of flow through sill constrictions (e.g., see Farmer and Freeland, 1983). However, bottom-water renewal has only been specifically addressed using these techniques comparatively recently (e.g., Niebauer, 1980). Because of the complex vertical density structure in deep fjords, simplified two-layer models are most common (Klinck et al., 1981; Hamilton et al., 1985). Gravitational circulation in the near-surface "estuarine" zone has been modeled by Winter (1973).

Following the debate on the formation of fjords, geologic research since the 1950s has concentrated on sediment transport and deposition processes. Many of those fjords that are presently experiencing glaciomarine sedimentation have exceptionally high rates of hinterland erosion and sedimentation, comparable or superior to the monsoonal environments of Oceania. Many fjords have also experienced exceptional rates of uplift, comparable to the extreme diastrophic earth movements of New Zealand or Japan (Table 1.2). Geophysicists find fjords ideal sites to measure parameters indicative of postglacial isostatic rebound and sea-level fluctuations. To the geotechnical engineer, fjords offer models of nearly every form of subaqueous slope failure, release mechanisms, mass sediment transport, and gravity flow deposits.

Attempts at modeling these natural processes in the field often point to fjords as ideal extensions of laboratory flumes (Hay et al., 1982). For instance, particle and flow dynamics in large open ocean prodelta environments are best studied in appropriate fjord settings because the sheer size of the open ocean examples does not permit accurate monitoring. Steep fjord walls and sills provide boundary conditions ideal for the modeling of bottom friction effects on circulation as well as for studies of the generation of large-scale turbulence from bottom roughness. The modeling of such fluid dynamic parameters provides insights into the laws governing sediment gravity flows, thus relating physical oceanographic models to sediment transport problems.

Fjord deposits have a good potential for providing a comparatively high-resolution sedimentary record that reflects both local terrestrial and marine processes. Stratigraphic interpretation of proxy climatological and paleoecological signals contained in well-dated and unbioturbated marine cores can provide insight into the impact of past and future climatic and environmental conditions. For example, the stabilizing influence of terrestrial vegetation—particularly grasses—is a major factor in moderating contemporary rates of fluvial erosion. Fjord sandurs offer actualistic models that offset these biotic effects that were absent until the upper Paleozoic (Syvitski and Farrow, 1983). Similarly, studies of modern tidal flats have been carried out most commonly in areas where a prolific infauna, particularly on the lower flats, tends to obliterate nearly all physically produced sedimentary structures. However, the combination of low-salinity estuarine waters and high sedimentation rates common to fjord deltas results in an impoverished macrofauna such that physical structures tend to remain intact. Fjord deltas may therefore provide actualistic models of sedimentary deltaic processes that were operative prior to the Cretaceous period. The effect of fjord topographic gradients, whether river thalweg or deltaic foreset slopes, can be compared with coastlines intersected by grabens (e.g., east Greenland in the Lower Jurassic, Surlyk, 1978; Millstone Grit deltas in the Carboniferous, Collinson, 1969). The modern-day "turbid association" of brachiopods, solitary corals, and sponges from Knight Inlet, British Columbia, has been compared with typical outer shelf–basinal shale facies of the Paleozoic (Farrow et al., 1983).

Fjords are also important environments for understanding land-sea interactions that occurred during the Quaternary glaciations. It would be expected that bio- and lithostratigraphic relationships could provide clues to circumscribing the extent and importance of pre-Quaternary glaciations. Such events, some of global extent, are of known importance during the early Proterozoic, Late Precambrian, Eocambrian, Varangian, and early and late Paleozoic (Hambrey and Harland, 1979).

The marine biology of fjords, especially of Scandinavian fjords, has been studied since the beginning of this century (e.g., Hjort and Gran, 1900; Petersen and Boysen-Jensen, 1911; Petersen, 1915; Soot-Ryen, 1924). Most of these early studies were concerned with estuarine rather than specifically fjordic communities, and some classic work—for example, Petersen's hypothesis concerning the effect of seagrass detritus exported from Danish fjords on production in the North Sea—referred to shallow fjord-estuaries of a type not generally treated in this book.

In recent years, however, biologists have recognized the opportunities afforded by deep fjords to discriminate among many of the controlling environmental factors and thus have been able to develop increasingly sophisticated ecosystem models (Brattegard, 1980). As deep estuaries, the fjord-basin benthos is separated from the euphotic and freshwater mixing zone. Since they are geologically young and physically stressed, fjords are also excellent locales for studying community structure and colonization succession (Pearson, 1980a). Benthic biologists can observe easily identifiable gradients that permit formulation of theories on faunal density and diversity as limited by nutrient availability, particle loading, sedimentation rates, water depth, and feeding modes (Farrow et al., 1983). Pelagic biologists can study the interaction of plankton and suspended sediment in a marine system dominated by the vertical flux of particles (e.g., Lewis and Syvitski, 1983).

Fjord types span a wide spectrum from very high to very low carbon input (see Chapters 6 and 7), and various modeling techniques have been applied to quantify the temporal and spatial distribution and flux of carbon (energy) within fjord ecosystems. Figure 1.9 (Pearson et al., in press) illustrates a generalized and simplified carbon flow schematic (using the energy circuit symbols of Odum, 1965). A specific example of this approach applied to a silled (but shallow) fjord has been given by Rosenberg et al. (1977). Beyer (1981) describes a comprehensive fjord hydrodynamic-ecological model (a multicompartment box model).

Geochemists are afforded the opportunity to investigate reaction zones on either side of the primary redox discontinuity within the sediments, and sometimes in the water column within or below the zone of photosynthetic ac-

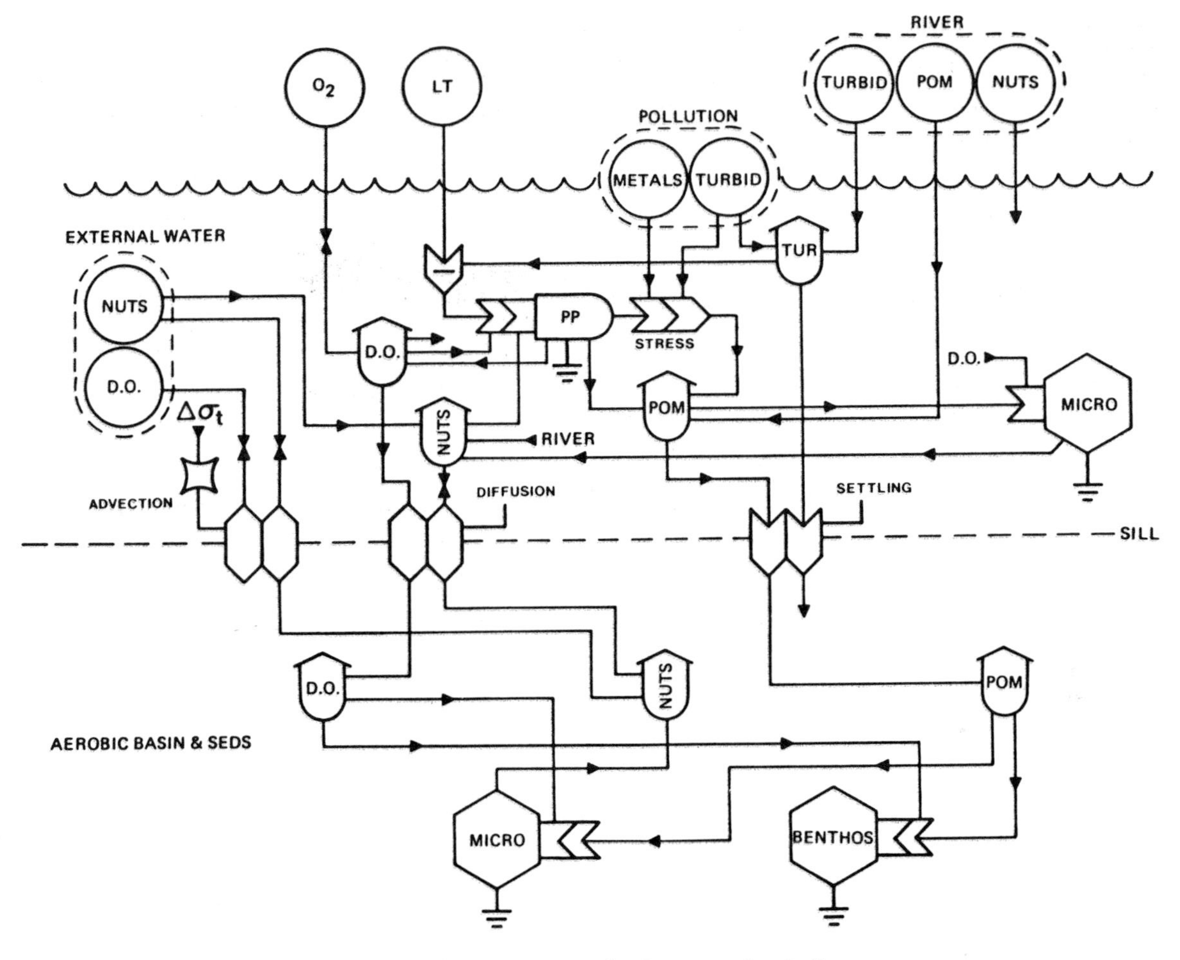

FIGURE 1.9. Generalized fjord energy circuit diagram.

tivity (Skei, 1983a,b). Migration of the redox zone, solubilization and precipitation of minor constituents and trace metals, cycling of carbon, nitrogen, trace metals, and other mobile species are all processes that may be studied in detail in anoxic fjord environments. Fjords with low organic carbon input, coupled with low sedimentation rates, provide proxy environments for deep sea hemipelagic sediments. Where the organic carbon input is much higher, anoxia within the benthos will eliminate the effects of bioturbation. Advective and diffusive forms of chemical transport across boundaries are thus more easily investigated. Also, fjords with permanent or persistent anoxic conditions are ideal for the study of equilibria facies and the kinetics of geochemical processes that were operative during early periods of the earth's history (Holland, 1984).

Geochemical investigations in fjord basins that are flushed at relatively long intervals may be facilitated because of the researcher's ability to model transport and reaction processes by means of simplified advection-diffusion or box models. Several such applications are outlined in Chapter 7. The fjord-type, layered near-surface circulation region should also provide ideal conditions for the application of tractable box models applied to the distribution of, for example, nutrient species.

Over the last few decades there have been numerous comprehensive, multiyear investigations within fjords (Table 1.3). Most commonly these have taken the form of multiple investigations within specific inlets, either predominantly single discipline (e.g., physical investigations in Knight Inlet, British Columbia; Freeland et al., 1980; Farmer and Freeland,

TABLE 1.3. Some multiyear/multidisciplinary fjord programs.

Region	Fjord	Latitude	Program	Reference	Notes
Greenland	Agfardlikavsâ	71.0	Multidiscipl.	See Section 8.2	Pollution studies
Alaska	Port Valdez	61.1	Interdiscipl.	See Section 8.4	Pre- & post-impact
	Boca de Quadra	55.3	Interdiscipl.	Burrell, 1982b	Pre-impact
Canada W	Kitimat system	53.9	Interdiscipl.	Macdonald, 1983	Post-impact
	Knight Inlet	50.8	Physical	Freeland et al., 1980 Farmer & Freeland, 1983	
	Rupert Inlet	50.5	Interdiscipl.	See Section 8.6	Post-impact
	Saanich Inlet	48.6	Chem/multi-discipl.	Menzel, 1977 Emerson, in press	Controlled ecosystem & discrete studies
Canada E	Baffin Island fjords	62–69	Geol.	Syvitski & Schafer, 1985	
	Saguenay Fjord	48.4	Multidiscipl.	See Section 8.7	Discrete studies
Norway	Balsfjorden	69.5	Biogeochem.	Eilertsen & Taasen, 1984	
	Hardangerfjord	60.3	Multidiscipl.	Braarud, 1961	
	Sørfjorden	60.3	Multidiscipl.	See Section 8.10	Pollution studies
	Korsfjorden	60.2	Biol.	Erga & Heimdal, 1984	
	Oslofjord (inner)	59.8	Multidiscipl.	See Chapters 4, 6, 8	
	Framvaren	58.2	Geochem.	Skei, 1983c	Anoxic water
	Rosfjorden	58.1	Biol.	Kattner et al., 1983	Controlled ecosystems
Scotland	Lock Eil	56.8	Interdiscipl.	See Section 8.12	Pollution studies
Sweden	Byfjorden	58.2	Multidiscipl.	See Section 8.13	Pollution studies
	Saltkällefjord	58.4	Multidiscipl.	See Section 8.12	Pollution studies

1983), or multidiscipline (e.g., the Hardangerfjord program; Braarud, 1961; Brattegard, 1966). Unfortunately, very often in the latter case the various disciplinary studies have not been closely coordinated, and this has tended to limit their potential value. Chemical and geochemical investigations have usually been an exception to this trend, however, because these generally require close integration of data from a number of ancillary fields. Information gained from work over many years in Saanich Inlet, British Columbia, is perhaps the most noteworthy example.

In recent years there has been an impetus towards genuinely integrated multidisciplinary (and frequently multinational) research programs in fjords, such as the program presently investigating the fjords of Baffin Island, Canada (Syvitski and Schafer, 1985; Schafer and Blakeney, 1984), and investigations in Framvaren, Norway, initiated to detail biogeochemical processes in permanently anoxic basins (Skei, 1983c). However, as is evident from Table 1.3, large-scale investigations in fjords are predominantly environmental impact and pollution oriented. Most usually, marine scientists are required to separate natural gradients and fluctuations from those created by man (primarily fishing, agriculture, forestry, urban, power generation, and industrial impacts). In a number of cases, events that the public perceives as pollution related, such as fish kills resulting from low-oxygen conditions, have been shown to be attributable to natural processes (e.g., Levings, 1980a). However, input of a wide range of anthropogenic waste products has severely impacted a number of fjords in many countries. Some specific examples are discussed in detail in Chapter 8.

Fjords are popularly considered to be environmentally sensitive because they are believed to have relatively slow flushing rates. However, this may be the case only if the pollutant impinges below the near-surface mixed zone. Soluble waste products introduced at the surface may be rapidly flushed from the inlet if the characteristic stratified flow is well developed.

The fate of pollutants with respect to transport, distribution, and influence on the biota has been the subject of numerous studies over the last few decades. Because fjords are sedimentary sinks, they tend to be sinks also for those pollutants having short estuarine residence times. High-resolution analysis of sediment cores has proved useful in tracing the history of industrial development around fjords, and in documenting long-term effects (e.g., Cato et al., 1980). There is an urgent need for a better understanding of the natural processes operating, and of the assimilation capacity and resilience of the estuaries.

Although many fjords in industrialized areas have been deleteriously exploited by man and polluted to various degrees, the overall prognosis is optimistic. As expressed by Bornhold (1983a):

> Fjords, those ragged incisions that penetrate a continent's edge as deeply as several hundred kilometers, are of immense importance to the handful of nations blessed with them.

Man's activities have been, and are still, intimately connected with the practical use of fjords: food, transportation, etc. When one activity impinges on another, practical solutions may be sought and implemented (Carstens and Rye, 1980). In recent years, cultivation of various fish and shellfish species (mariculture) has become an important industry in many fjord districts, establishing an economic base for people living in remote areas. Manipulation of ecosystems for the benefit of man in areas of natural high productivity, such as many fjords are, is of great benefit to a world requiring an ever increasing food supply.

There have been only a few previous texts devoted exclusively (Freeland et al., 1980) or substantially (e.g., Nihoul, 1978) to fjords, and these have predominantly emphasized physical circulation regimes. This present text is an attempt to present fully referenced state-of-the-art summaries of, primarily, fjord geomorphology, sedimentology, and biogeochemistry in the broadest sense, and in an interdisciplinary fashion. The interrelated processes are summarized in Figure 1.10. The subject matter contained in the following chapters is written in sufficient detail to exceed the immediate needs of fjord-orientated researchers and to include the earth science and oceanographic community in gen-

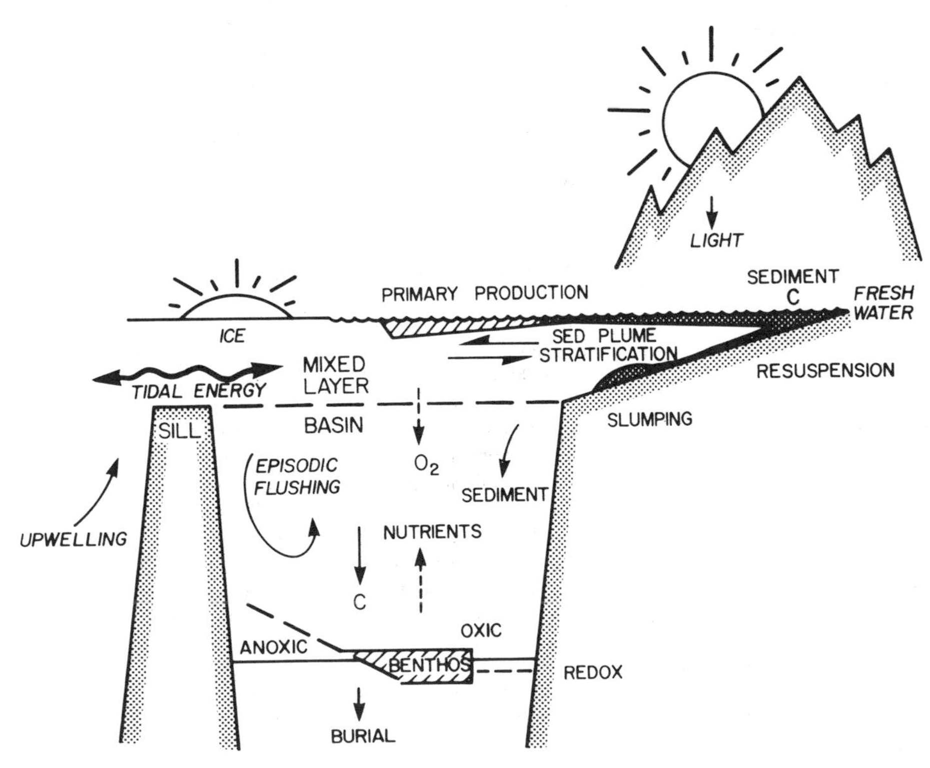

FIGURE 1.10. Interrelationship of physical, geological, biological and chemical processes in fjords.

TABLE 1.4. A few fjord comparisons.

	FJORDS	
(a) Rivers		
	FJORDS	World's 20 largest rivers[1]
Drainage basin area	< 0.1 (10^6 km^2)	0.9–6 (10^6 km^2)
Freshwater input	< 1 (10^2 km^3 yr^{-1})	2–63 (10^2 km^3 yr^{-1})
Sediment input	< 0.4 (10^8 t yr^{-1})	0.6–16 (10^8 t yr^{-1})
(b) Estuaries		
	FJORDS	Drowned-river estuary[2]
Physiography	Mountainous (rocky-high slopes)	Alluvial plain (soft sediment low slopes)
World location	Mid to high latitudes	All latitudes
Depth of water	Deep ($<$ 1500 m)	Shallow ($<$ 100 m)
Tidal resuspension	Low	High
Salinity	Relatively stable	Widely fluctuating
Stratification	High	Low
Suspended particulate matter	1–10^3 mg L^{-1}	10–10^4 mg L^{-1}
Turbidity maxima	Seasonal	Year-round
Sedimentation rates	1–10^3 mm yr^{-1}	1–10 mm yr^{-1}
Submarine volume	$< 10^{13}$ m^3	$< 10^{11}$ m^3
$\frac{\text{Annual freshwater input volume}}{\text{Submarine volume}}$	10^{-5}–10^1	10^{-1}–10^5
(c) Submarine canyons		
	FJORDS	Submarine canyon[3]
Width	1–30 km	2–30 km
Length	20–300 km	50–300 km
Water depth	0–1500 m	50–2000 m
Typical slopes (i) axis	0.1°–5°	0.1°–5°
(ii) side wall	10°–30°(+)	10°–30°(+)
Bottom water SPM	0.5–10 mg L^{-1}	0.1–10 mg L^{-1}
Sedimentation styles	Hemipelagic > gravity flows > slides	Gravity flows > slides > hemipelagic
Average sedimentation rates	10 mm yr^{-1}	0.1–1 mm yr^{-1}
Sedimentation events	Pulsed	Pulsed
(d) Lakes		
	FJORDS	Lakes[4]
Length: width	4 to 40	1.5 to 15
Fjord area	< 30 (10^3 km^2)	10 lakes > 30 (10^3 km^2)
Depth	$<$ 1500 m	$<$ 1610 m
Stratification	Salt-dominated	Thermal-dominated
Tides	Important	Negligible
Long-shore sediment transport	Not important	Important
Wave influence	Only locally important	Important
Inflows & underflows	Rare	Relatively common
Bottom water temperature	−1.7° to 15°C	3° to 30°C
Bottom water salinity	Marine (28 to 33‰)	Variable (0 to 300 ‰)
Dominant grain size control (exceptions)	Distance from source (Gravity flows)	Water depth (Gravity flows)
Average sedimentation rate	10 mm yr^{-1}	3 mm yr^{-1}
Authigenic mineral component	Very low	Low to high
Rhythmic bedding	Moderately common	Common

[1]After Milliman and Meade, 1983.
[2]After Farrow et al., 1983.
[3]After Heezen et al., 1964; Moore, 1969; Nelson et al., 1970; Drake and Gorsline, 1973; Shepard, 1973; Carter et al., 1982.
[4]After Picard and High, 1972; Lerman, 1978; Håkanson and Jansson, 1983.

eral. There are many marine basins that have a number of similar operative processes and products (Table 1.4). Some of the many present-day examples include: (1) basins of the California borderland where anoxic bottom waters exist, and where infilling by sediment is dominated by storm events leading to gravity flows (Savrda et al., 1984; Moore, 1969); (2) the Mediterranean Sea with its silled basin, two-layer circulation (Thomson, 1981), and dominance of gravity flow sedimentation processes (Stanley, 1972); (3) the Black Sea, another silled basin with stratified anoxic waters and well-documented biogeochemical reactions (Spencer et al., 1972); (4) submarine canyons where the major source of sediment is at the head, and ensuing gravity flows are constrained by sidewall boundaries (Allen et al., 1971; Nelson et al., 1970; Heezen et al., 1964); (5) glacial troughs that cut across the world's high-latitude continental shelves, and where Coriolis ponding of sediment and glacial histories have similarities with fjord sedimentation (Holtedahl, 1955); (6) island arc basins and trenches such as those found around the Sea of Japan (Aoki and Oinuma, 1981; Otsuka, 1976), or the Chile (Scholl et al., 1968) and Aleutian trenches (von Heune and Shor, 1969).

It is hoped that this text provides a comprehensive synthesis of current information on both fjords and fjordlike environments that will be useful to a broad disciplinary range of environmental scientists.

2

Environmental Setting

There are two major fjord regions of the world, a belt north of 43°N and a belt south of 42°S (Fig. 1.1). These fjord coasts are commensurate with areas that were previously or are presently glaciated; glacial erosion appears to have played a significant role in the mode of fjord formation. The environmental setting of fjords is closely interrelated with respect to geomorphology, climatological conditions, water circulation, and sediment sources. Some fjords are more typical than others, showing characteristic features that fit the definition of fjords (Fig. 2.1; see Chapter 1). Other high-latitude estuaries are more fjord-like, exhibiting only a few of the features associated with fjords, while the overall natural setting would suggest that they be classified as fjords.

This chapter sets out to discuss characteristics of fjord geomorphology, climate, oceanography, and sediment sources.The typical fjord areas of the world are described in terms of the natural progression in the glacial development of the fjord.

2.1 Geomorphology

Fjords and fjord valleys are genetically synonymous features, the only difference being that fjords are now submarine. Fjord lakes are a subset of fjords distinguished by the fact that they contain only fresh water. The geomorphology of a particular fjord may be related to local conditions of geology, climate, and glacial and postglacial history.

The classic description of fjord valleys is as elongated U-shaped valleys in glaciated mountainous terrain (Plate 2.1). The U-shaped cross-section is, however, more apparent than real. The actual bedrock cross-section profile is more of a parabola (Gilbert, 1985). The U-shape results from the sidewall talus cones and fluvial

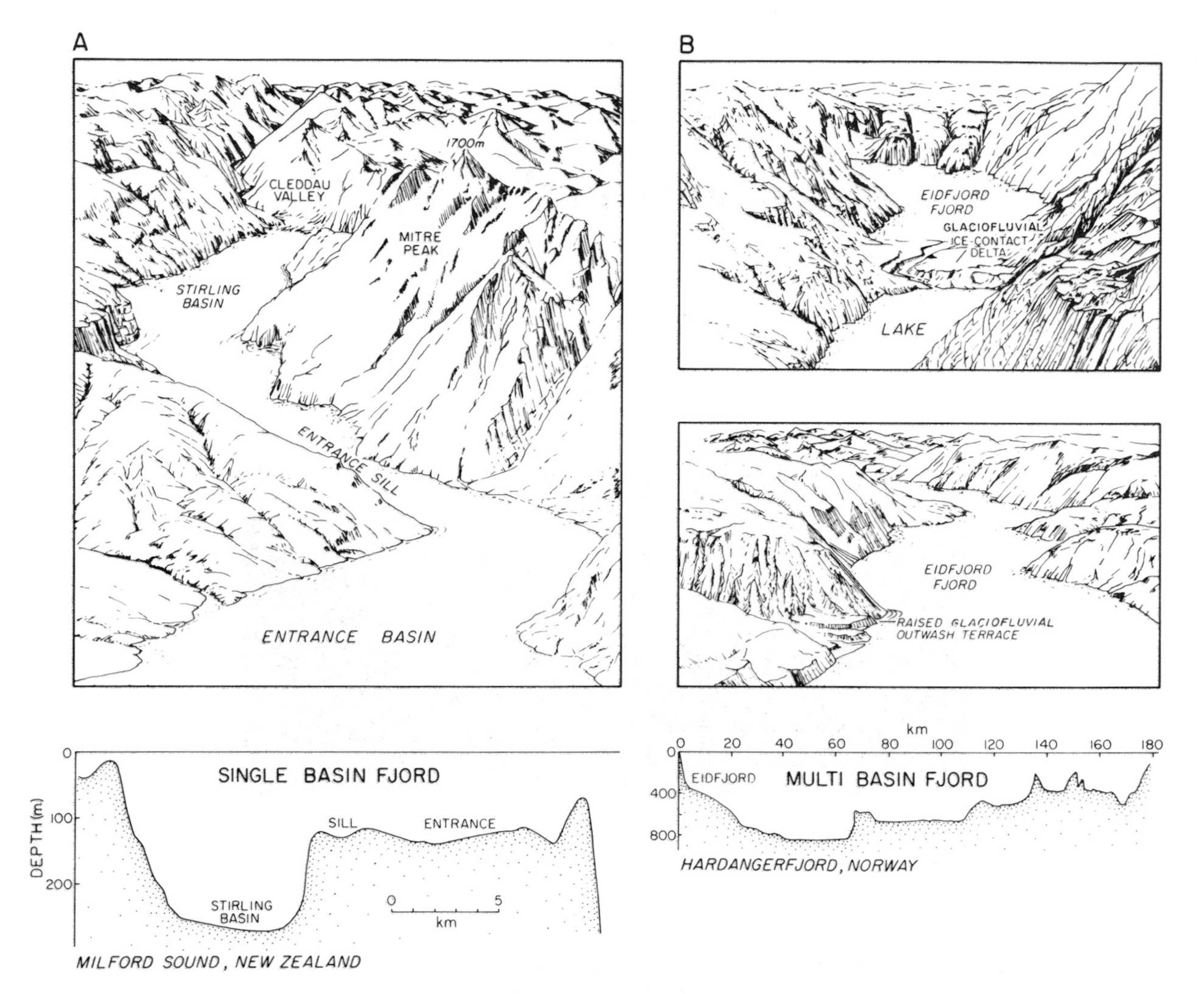

FIGURE 2.1. Schematic presentation and longitudinal bathymetry of two of the world's classical fjords: (A) the simplicity of the single basin Milford Sound, N.Z. (after Bruun et al., 1955); (B) the complexity of the multibasin Hardangerfjord, Norway (after Holtedahl, 1975, cf. Fig. 5.19).

sediment infill in the case of the fjord valley, and the intersection of the mountain slopes with the sea surface or marine sediment infill in the case of the fjord (Fig. 2.2). The longer the history of sediment infill, the more apparent the U-shape of the submarine cross-section will appear: compare the Hudson estuary profile, considering its +10,000-year history of infill, with the profile from Muir Inlet (Alaska), where sediment infilling has occurred within the last 25 years (Fig. 2.2). The subaerial profile will depend on how easily the local rock can be eroded, initially by glacier ice, later by normal mountain weathering processes. Erosion by ice also depends (among many parameters) on the thickness of the ice sheet, the slope, the plasticity of the ice, the basal thermal regime of the glacier, bedrock friction, and friability (Boulton, 1975). When cut in granite, the sides of some mountains may be steeper than 60° (Plate 2.2), or less than 30° when cut in schists (Glasby, 1978a). Fjord walls are often polished and striated. Pot holes are common to some Norwegian fjords (Plate 2.3), many found along the steep side walls (Holtedahl, 1967). Pot holes may indicate that subglacial fluvial erosion, by meltwater carrying rock material under high hydrostatic pressure, may have been important in the formation of fjords. Subglacial streams can be very powerful and are able to carry large amounts of debris (Mackiewicz et al., 1984).

Terraces are commonly found along the

PLATE 2.1. Characteristic U-shaped profile of McBeth Fiord, Baffin Island.

mountain slopes of fjords. They may result from: (1) older shoreline erosion or glaciodeltaic deposits formed when sea level was sometimes as much as 100 m above the present sea level (Fig. 2.2A; Donner, 1977); or (2) the formation of lateral moraines and kame terraces at the ice margin during a previous phase of glaciation. Moraine ridges may extend tens of kilometers along the fjords at altitudes up to many hundreds of meters above sea level (Fig. 2.2A; Funder, 1972). Occasionally evidence of multiphase glaciation is preserved, that is, the last phase was not as extensive or destructive as the one before. In such cases, more than one lateral moraine is preserved but at different elevations. Sometimes morainal ridges extend below sea level within the fjord and may even continue across the continental shelf (Fig. 2.3; Gilbert, 1982a; Løken and Hodgson, 1971).

Talus cones along the sides of fjords are often common (Plate 2.4). Their formation depends on a number of factors: (1) the number of freeze-thaw cycles, (2) bedrock friability, (3) mountain slope, and (4) the presence or absence of vegetation. Where the slope is too steep, or where vegetation shields the bed rock from ice-expansion damage, talus cones will not form. These cones are of postglacial age.

Cirques (corries) are common to intrafjord mountains that have been glaciated with alpine glaciers during the Holocene (since 10,000 years BP). They may be seen along fjords in northern Norway, Scotland, Alaska, Labrador, Baffin Island, mainland British Columbia, and Antarctica (e.g., Boyd, 1935; Mercer, 1956). There are several low-level cirques, a few still containing small glaciers and some with short valleys with submerged mouths that are incipient fjords. Cirques with floors below sea level are also known (Mercer, 1956).

Hanging valleys are also characteristic in fjord areas (Plate 2.5). Rivers may flow along these valleys and enter the main fjord as spectacular waterfalls (Plate 2.6). Hanging valleys may also occur as a subaquatic landform when a tributary fjord enters the main fjord at a shallower depth (Aarseth, 1980).

Below the water line the side walls continue at roughly the same angle (Fig. 2.2). Where the slope is greater than the angle of repose for submerged sediment (anywhere from 5 to 30° depending on the sediment type; see Chapter 5), the walls will have little sediment cover. Sediment accumulation increases normally to the center of the fjord basin. The steepness of the side walls has biological consequences. For instance, the lack of sediment cover on the steep slopes may result in the establishment of a different community compared with areas with gentle slopes and soft bottoms (Farrow et al., 1983; Chapter 6).

The submarine cross-sectional profile of a fjord depends primarily on the dominant sedimentary processes (Fig. 2.4). Where the sedi-

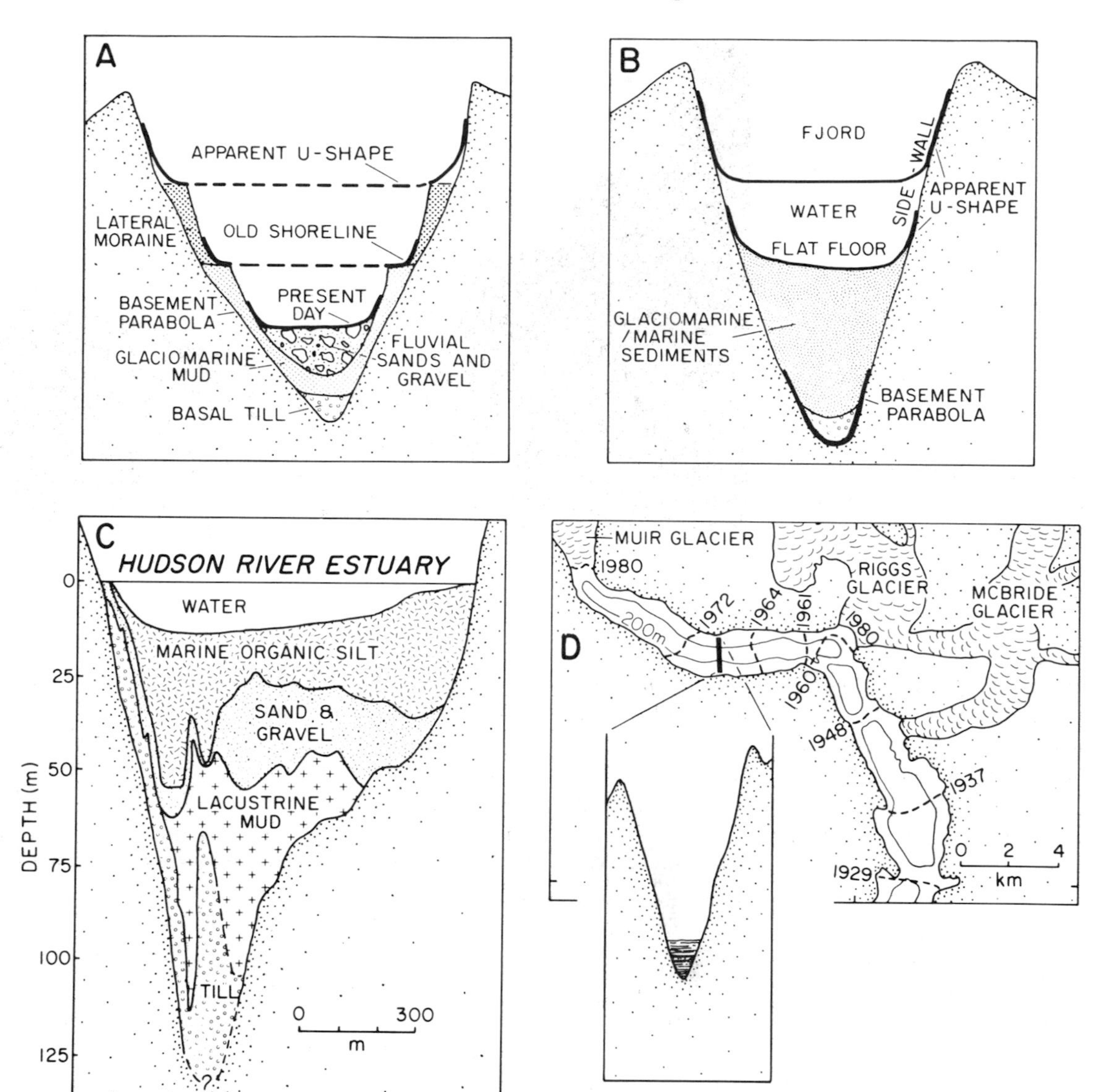

FIGURE 2.2. The bedrock parabola of fjord cross-sections and the apparent U-shaped profile (see text for details): (A) the fjord valley; (B) the fjord basin; (C) the Hudson River paleofjord infill (after Newman et al., 1969); (D) location and seismic profile across Muir Inlet (after Molnia, 1983).

mentation rate is high and bottom currents are weak, sediment infill may be deposited as a conformable layer to the bottom bathymetry (Fig. 2.4A). If the current energy was stronger but still decreased with water depth, basin fill could resemble a more onlapping structure (Fig. 2.4B). If the basin sedimentation was dominated by sediment gravity flows (e.g., turbidity currents, grain flows), sediments would appear ponded (Fig. 2.4C). Occasionally sediments are confined to one side of the basin (Fig. 2.4), and the maximum thickness does not follow the fjord thalweg (Fig. 2.5). This may be explained by the Coriolis effect, which tends to deflect down-fjord flows to the right in the northern hemisphere and to the left in the southern hemisphere (see details in Chapter 4). The best developed submarine U-shaped profile is observed where the fjord is wide. At narrow stretches, the transverse section approximates a V-shape (Fig. 2.4D). Here, the bottom currents are strong and little sediment can accumulate. Deformed

PLATE 2.2. Spectacular cliffs along McBeth Fiord, Baffin Island.

ponded sequences may result from sediment compaction (Bennett and Savin, 1963) and dewatering (Fig. 2.4E). Irregular seafloor surfaces are commonly the result of sidewall or down-fjord slides (Fig. 2.4F). It is also not unusual for a fjord to undergo a variety of sedimentation styles during its history of basin infill (von Huene, 1966; Bornhold, 1983b).

Most fjords have their main fluvial input at their head, with only minor contributions at widely spaced point sources from their side-entry drainage basins (e.g. Saguenay Fjord, Quebec; Fig. 2.6). There are notable exceptions, especially the archipelago fjords of Svalbard, Canada and Russia (e.g., Inugusuin Fjord, Baffin Island; Fig. 2.6). Fjord-valley rivers have large sediment-transport capacity. The coarser material is deposited within the valleys as sand plains (sandurs) and at the fjord margins, forming deltas and outwash fans (Molnia, 1983). These bayhead deltas may be partly supra-aquatic and partly subaquatic. Their morphology can be related to climate, paraglacial history, and fjord energetics (see Chapter 3 for details).

Sills are commonly found at the mouth or within a fjord, separating it into several basins, (Fig. 2.1). They may consist of morainal or other glaciomarine deposits mantling bedrock highs (Fig. 2.7; Brodie, 1964), and may appear as a series of islands or shoals, sometimes as a well defined ridge or a more lengthy threshold (Figs. 2.1, 2.8). Sills may result from glacial overdeepening of the fjord basin compared with the adjacent shelf (Fig. 1.3). Nuka Bay is a complex fjord system along the SW coast of Alaska. Based on present-day bathymetry, there are five basins (Fig. 2.8A). However, when an isopach map is constructed, the number of bedrock basins doubles (Fig. 2.8B): five basins have since filled and buried some of the original sills.

The strike (direction) of a fjord, its degree of bifurcation and curving are often determined by geologic structures such as fault zones, igneous dykes, sills and plutons, and changes in the composition of the rocks. The direction of the paleo-ice movement seems to have had little influence on the direction and configuration of the fjords: glaciers are plastic in their mechanical behavior (Dowdeswell and Andrews, 1985). The typical bifurcation of the fjord landscape may result from the uplift of the Tertiary peneplain. During the uplift, the peneplain was broken along its edges (Gregory, 1913).

Strandflats (Plate 2.7) are prominent erosion surfaces (Tiltze, 1973) common to a few fjord regions, particularly Antarctica and Norway (Nichols, 1960; Holtedahl, 1960). However, a strandflat is not observed along Baffin Island, Labrador, and the S.W. Alaskan fjord coasts. Strandflats occur near the mouths of fjords. They may be the result of marine abrasion and weathering in preglacial time, concomitant with

PLATE 2.3. Typical pot hole on the side walls of Hardangerfjord, Norway.

changes in sea level, and possible further erosion later by ice (Holtedahl, 1960; 1975).

2.2 Climate

Climatic variations have controlled the changes in sea level associated with glaciation and deglaciation (see Chapter 3) and therefore the origin and distribution of all types of estuaries (Schubel and Hirschberg, 1978). Climate remains of great significance for the present-day modification of fjord morphology and sediment input. Climate has a major effect on the rates of weathering of rocks, and hence on sediment yields. It also affects the vegetation development in the catchment area, which, to some extent,

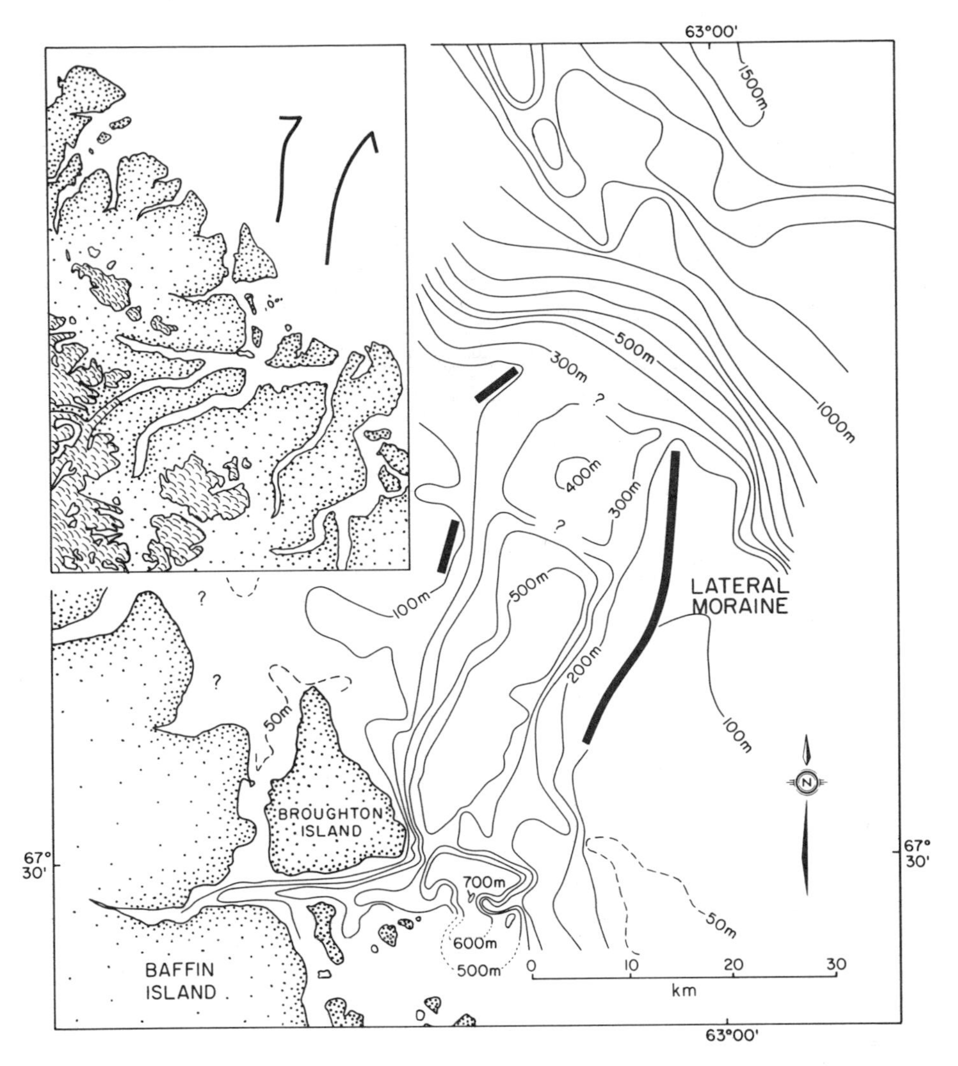

FIGURE 2.3. Bathymetry in meters of Broughton Trough (from Gilbert, 1982a). Note the present-day ice fields shown on the inset in relation to the shelf trough and the shelf lateral moraines.

PLATE 2.4. Coalescing talus cones along Coronation Fiord, Baffin Island.

PLATE 2.5. Typical hanging valleys along the walls of McBeth Fiord, Baffin Island.

PLATE 2.6. Waterfall into Knight Inlet, British Columbia.

will influence the availability of source sediments. Since climate governs the amount and type of precipitation, runoff, and water temperature, it also determines the style of estuarine circulation (Chapter 4).

The climate of fjords may vary from arctic desert to temperate maritime, and this has a considerable effect on environmental conditions. In areas where precipitation falls principally as snow, the spring melt and freshet are the major driving force within the estuary. Where precipitation is mostly as rain, the rivers are flood dominated (see Chapter 3.1). In a climate allowing the formation and retention of glaciers, the input of fresh water and sediment to fjords is different from areas where glaciers are absent. In the former, glacier melt is associated with sunny summer conditions and can transport high loads of suspended sediment.

Seasonal changes in climate directly affect the rhythms of sedimentation. Flood periods may give rise to sediment plumes or pulses, which in turn can cause discrete layers of sediment on the seafloor (Glasby, 1978b). In temperate areas, such pulses occur normally in the spring and autumn. In areas where freshwater discharge is primarily from glaciers, even the daily cycle may be important (e.g., Østrem et al., 1967; Hoskin and Burrell, 1972). In subarctic and arctic fjord areas, the spring snow freshet is important. The flow of several fjord-rivers are controlled for the generation of hydroelectric power, removing the

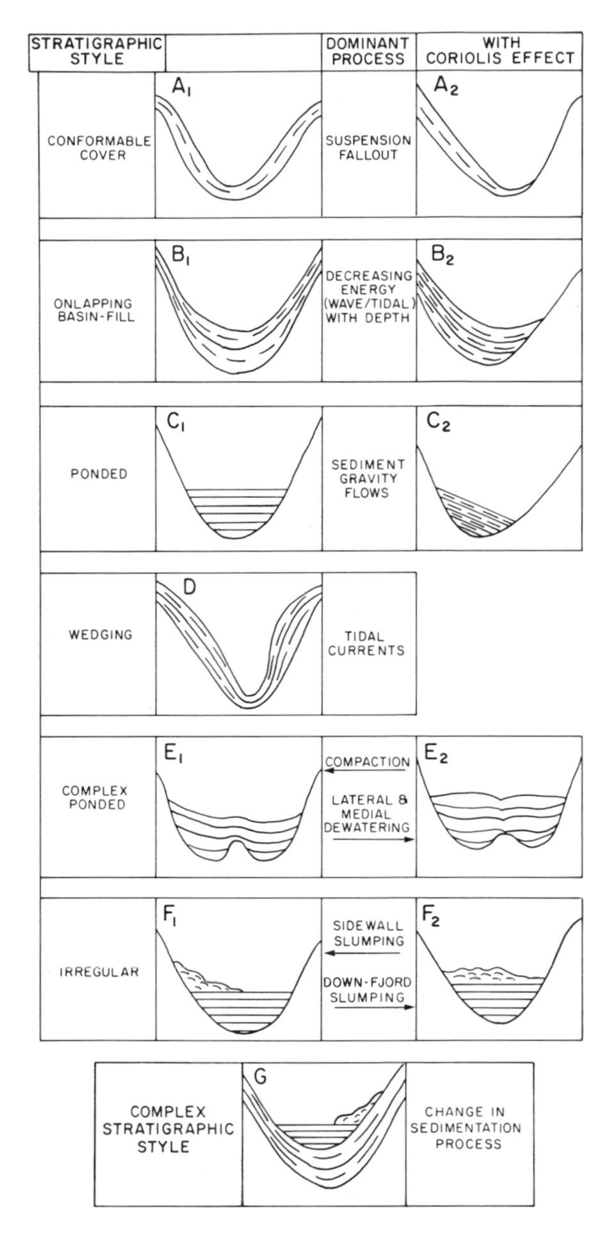

FIGURE 2.4. Cartoons of fjord cross-sections showing styles of basin infilling developed as a result of different dominant sedimentation processes and as affected by the Coriolis force (A2, B2, C2). Also shown are some postdepositional disturbances to basin layers, such as the result of compaction and dewatering.

typical flood peaks, and hence limiting the sediment load transported.

In many fjords a layer of ice may develop during the cold of winter. During this time, the sediment flux may drop 90 to 95% (Skei, 1983a). The ice cover also inhibits wind and wave action, retarding shallow water resuspension and supply of sediment to the deeper basins.

Fjords subject to strong winds may have a different sediment distribution pattern compared with fjords where the surface water is only influenced by the energy from tide and runoff. Wind induces waves and turbulence which, in addition to resuspension of bottom sediment, cause the breakdown of the stratification in the upper water and promotes sedimentation of particles that would otherwise remain on top of

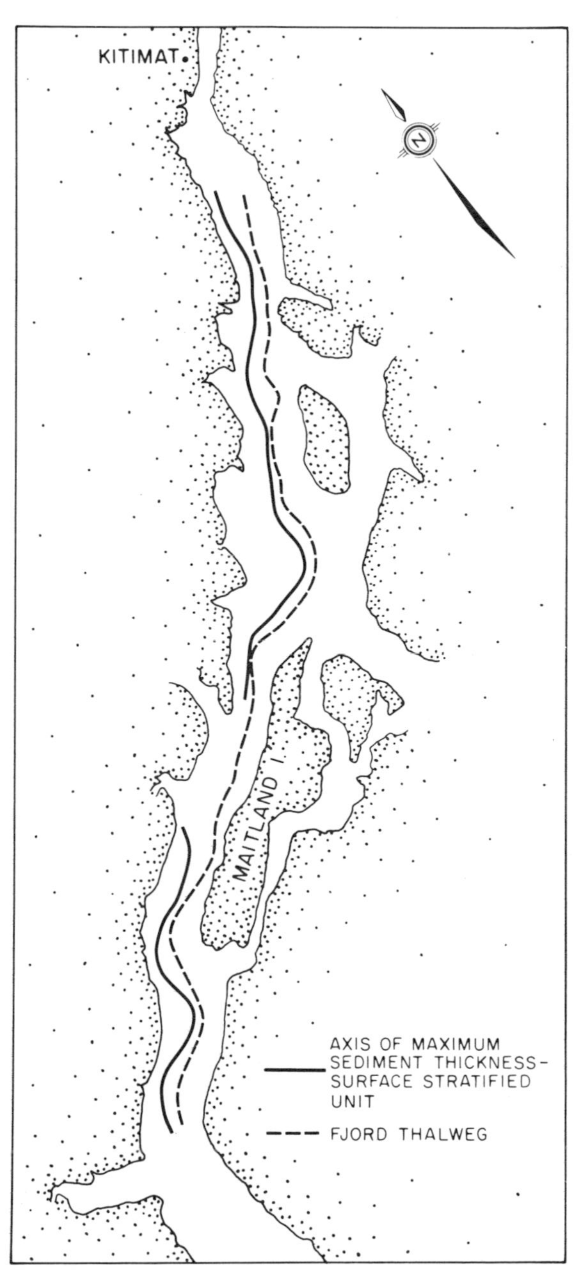

FIGURE 2.5. Maximum sediment thickness relative to fjord thalweg: Douglas Channel, B.C. (from Bornhold, 1983b).

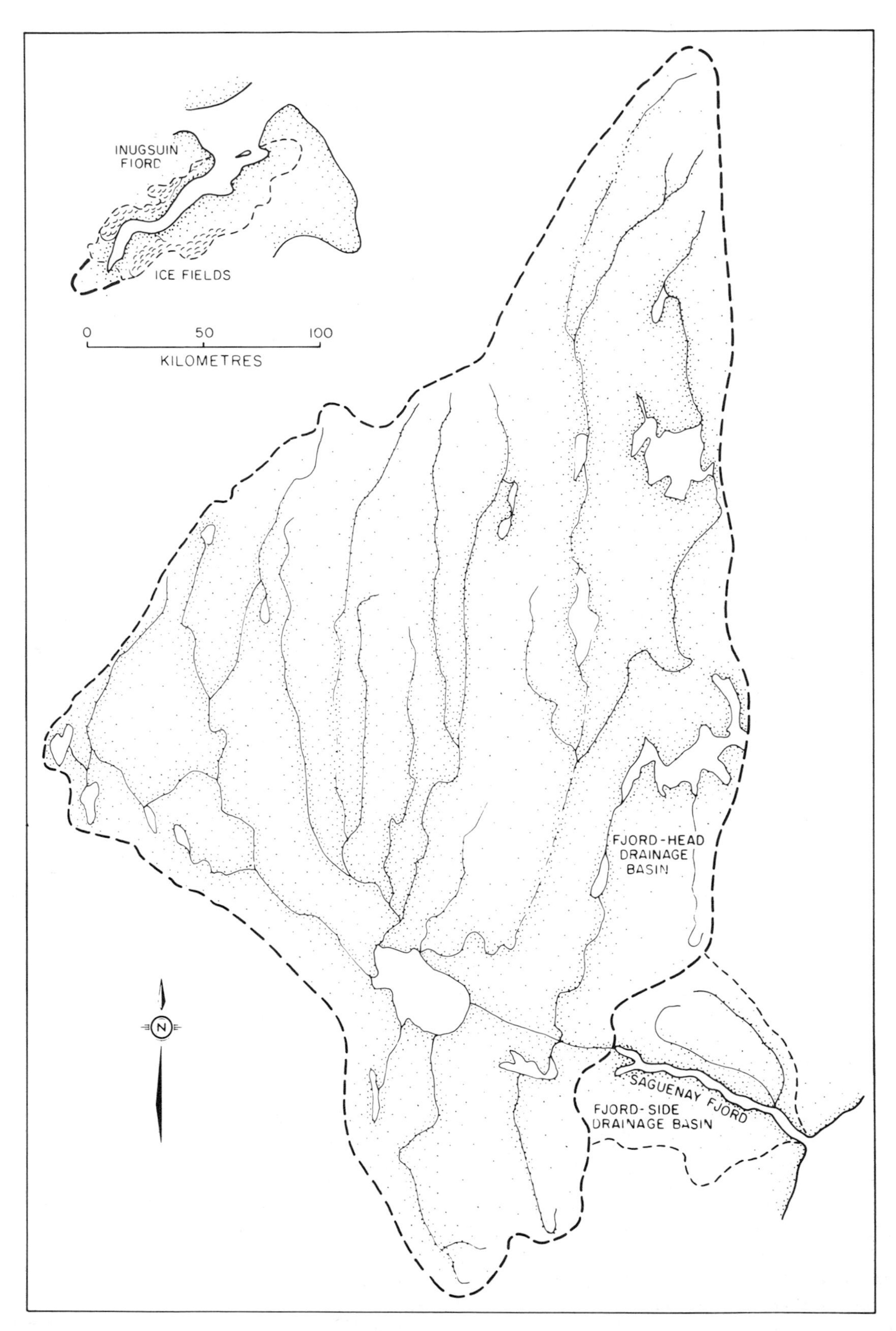

FIGURE 2.6. A comparison of drainage basins of two fjords of similar size but dissimilar hinterland area: Saguenay Fjord, Quebec, and Inugsuin Fiord, Baffin Island. The former is filtered by lakes and has no ice fields; the latter is meltwater dominated.

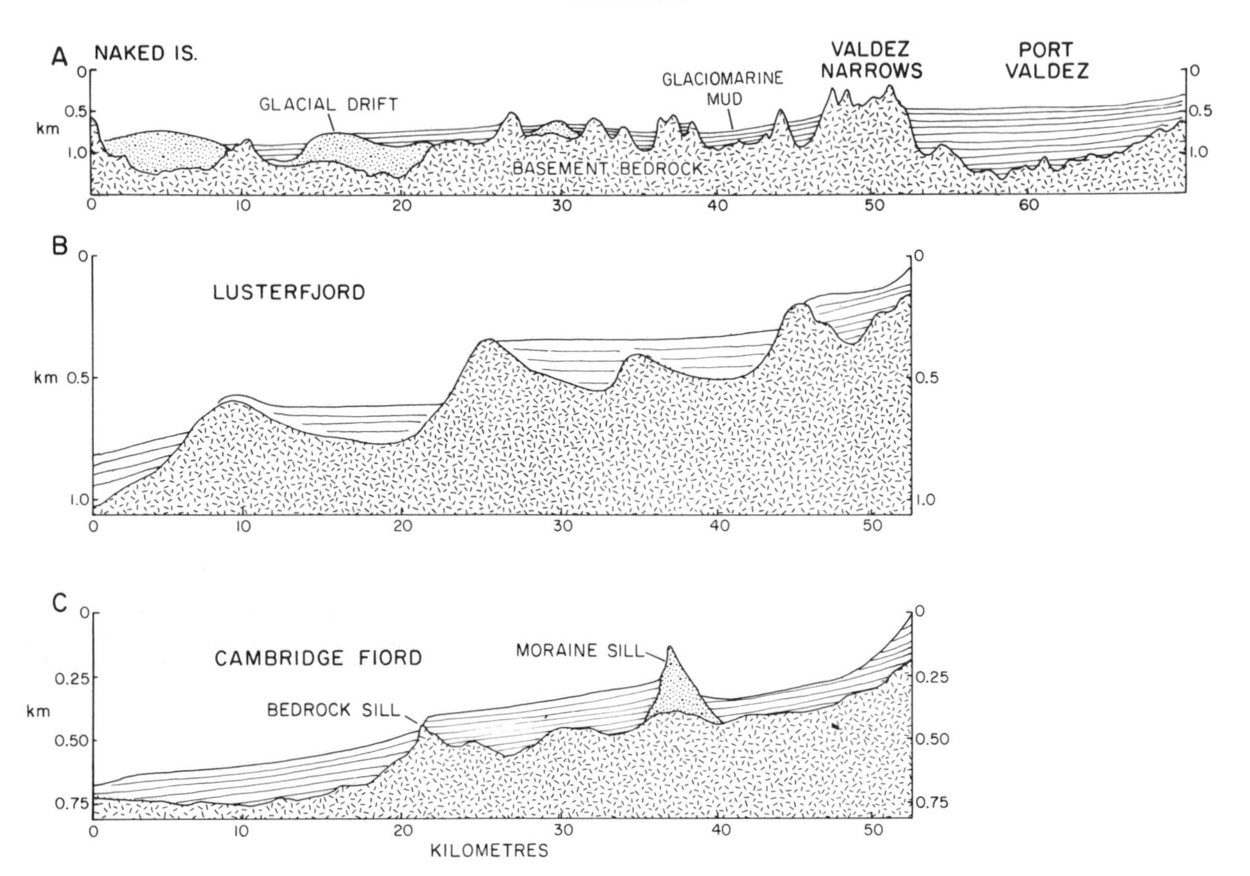

FIGURE 2.7. Interpreted down-fjord reflection seismic profiles: (A) PortValdez, Alaska (after von Huene et al., 1967); (B) Lusterfjord, Norway, a tributary fjord (after Aarseth, 1980); (C) Cambridge Fiord, Baffin Island. Note that the latter has less sediment in its inner basin as a result of a relatively recent glacier pull back.

the density boundary. Further, a strong wind blowing opposite the seaward flow of brackish water would retard the dispersal of sediment and increase the residence time of the surface water. Wind-induced waves may also trigger slumping of sediment off steep slopes (see Chapter 5).

Wind may also contribute to the subaerial transport of sediment in certain fjord areas, particularly in the arctic region where vegetation is limited. Strong winds blowing off sandurs or other alluvial outwash plains may carry substantial amounts of sediment into fjords (Chapter 3). In addition, atmospheric input of long-transport pollutants may contribute significantly to fjord sediments in areas of very low rates of sedimentation (Skei, 1983b). Fallout from airborne pollutants is particularly important during periods of calm, leading to a strong air stratification within the fjord. Air stratification develops as a result of the thermal differential (up to 30°C) between the sea surface and the tops of the fjord mountains (some exceeding 2000 m above sea level). Air stratification could cause the spread of industrial pollutants close to the sea surface (Plate 2.8).

Climate is one of the ruling factors with respect to weathering and soil development. In temperate fjord basins, the warmer temperatures and increased precipitation promote chemical weathering of bedrock and glaciofluvial deposits. Oxidation and leaching of minerals allow for relatively high dissolved-load discharge. Rapid changes in temperatures may cause extensive mechanical break-up of rock surfaces, increasing the potential sediment source. This is particularly important in the subarctic and arctic where there are high numbers of freeze-thaw cycles. Normally, sediment supply to fjords represents only slightly weathered material (O'Brien and Burrell, 1970), as most of the weathering has taken place within the last few tens to hundreds of thousand years (Boyer and Pheasant, 1974).

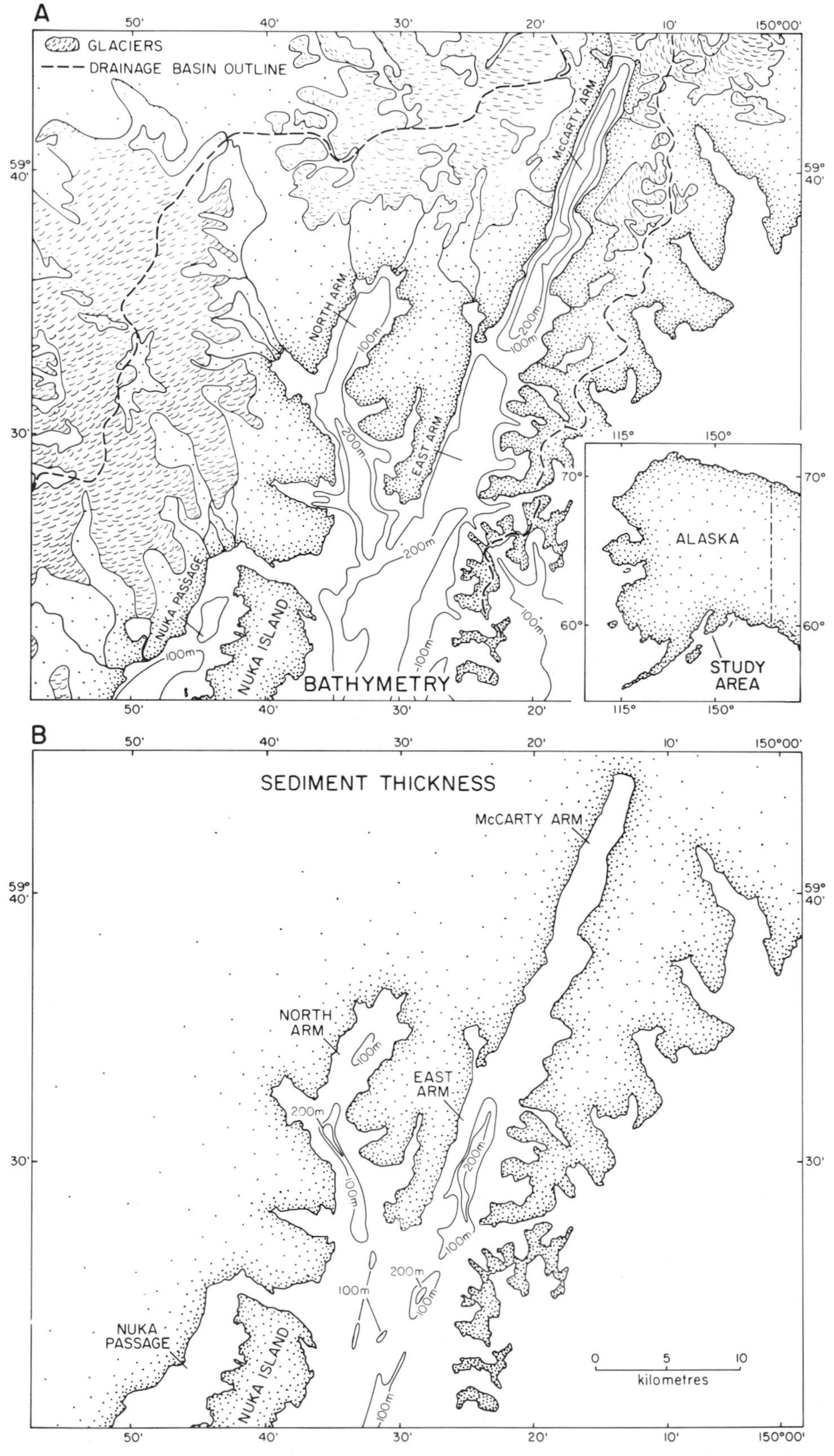

FIGURE 2.8. (A) Bathymetric map of Nuka Bay area, Alaska, showing present extent of glaciers and outline of the drainage basin (after von Huene, 1966). (B) Isopach map of stratified Holocene marine sediment in the bottom of Nuka Bay, Alaska based on a sediment acoustic velocity of 1500 m s^{-1} (after von Huene, 1966).

PLATE 2.7. Strandflat along the west coast of Norway.

The weathering products developed in preglacial times are no longer present in the catchment areas of fjords, having long since been removed to the continental shelves and slopes.

Soil development is a function of climate, topography, vegetation, source material, and time, and the soil cover is part of the zone of weathering. As fjord drainage basins are youthful in terms of soil development, they lack expansive areas of well-developed soil profiles. This is particularly true of polar and subpolar areas where the vegetation cover is thin or nonexistent and the soil profile consists predominantly of wind-blown sandy silt with low organic content.

PLATE 2.8. Lateral spread of industrial fumes as a result of air stratification in Powell Lake, British Columbia.

A developed soil tends to bind the underlying sediment, limiting its erosion. Hence, the sediment input to fjords without a proper vegetation and soil cover is much higher during heavy rainfalls. The content of organic matter in fjord sediments often reflects the soil development within the catchment area (Glasby, 1978b).

2.3 Oceanographic Characteristics

A fjord is a type of estuary: a semi-enclosed coastal body of water which has a free connection with the open sea and within which sea water is measurably diluted with fresh water derived from land drainage (Pritchard, 1967). Estuarine circulation systems have been classified principally by the degree of haline stratification within the column (for details see Bowden, 1967; Partheniades, 1972; Kjerfve, 1978; Dyer, 1979; Officer, 1983). The level of stratification is a simple balance between buoyancy forces set up by inflowing fresh water and the processes such as tides that work to mix the fresh water with the denser and saltier sea water. Hansen and Rattray (1966) devised a two-axis-estuarine classification scheme based on parameters representing the circulation and the stratification. The stratification parameter is denoted by the ratio of the tidally averaged salinity difference, ΔS, between surface water and bottom water, to the depth-and-tidally averaged salinity, S, at a given location. The circulation parameter was represented by the ratio of the tidally averaged (residual) surface velocity, $\overline{U}_s$, to the net river discharge velocity, U_o, cross-sectionally averaged. Four categories of estuaries are separated on a plot of these parameters (Fig. 2.9): (1) vertically homogeneous or well-mixed estuaries where mixing processes predominate; (2) partially mixed estuaries where the stabilizing influence of river flow is balanced by the destabilizing effect of tidal mixing (velocity shear); (3) two-layer flow with entrainment, where interfacial mixing processes result in an increased advective component of the landward salt flux; and (4) salt-wedge estuaries, where the river flow stratification is little influenced by the mixing between the freshwater and the marine-water layers. These typical estuaries are usually further complicated by: the Coriolis effect, which forces flow to the right in the northern hemisphere; the centrifugal accelerations along sinuous fjords; flow accelerations developed over bottom thresholds and through inlet constrictions; pressure gradients developed from meteorological conditions (changing wind structure or freshwater discharge); surface mixing from strong winds; energetics of breaking internal waves; and isohaline instabilities developed during the process of salt rejection ongoing with sea-ice development.

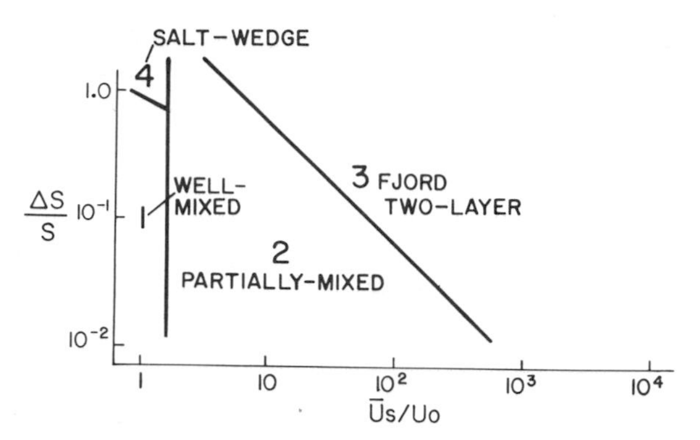

FIGURE 2.9. Stratification-circulation diagram for estuarine classification based on Hansen and Rattray (1966). Type 1: well mixed; Type 2: partially mixed; Type 3: two-layer (fjord); Type 4: salt wedge.

Two-layer flow with entrainment of marine water into the surface plume has become synonymous with fjord circulation: an outward flowing surface layer and an inward moving compensating current, compensating for the loss of salt entrained into the surface zone. There may, however, be a continuum of circulation styles for a given fjord at any one time, or variations according to the season. For example, the St. Lawrence Estuary changes seaward from a salt-wedge to a partially mixed, and finally toward the sea, into a fjord-type estuary (d'Anglejan and Smith, 1973). Other temperate fjords alternate between two-layer "fjord-style" circulation operative during the spring (snow-melt discharge), summer (ice-melt discharge), and fall (rain-storm discharge), and vertically homogeneous estuarine conditions of the winter (residual ground-water discharge). In the polar regions where runoff is limited to a few months, fjords lack estuarine circulation for a large portion of each year.

Still, it is the two-layer "fjord circulation" that is of environmental significance, causing maximum variations along sedimentological and biogeochemical gradients. The higher the discharge, the higher the transport of sediment into the fjords. Thus well-developed two-layer circulation is often associated with maximum rates of sedimentation (Syvitski and Murray, 1981) and in many cases biological productivity (Chapter 6). With the entrainment of marine water into the surface layer, salinity increases toward the mouth of fjords, assuming that the main freshwater discharge is situated at the head (Fig. 2.10). The force responsible for maintaining the flow of brackish water toward the sea originates from the pressure field associated with the seaward-sloping free surface (Gade, 1976).

The halocline is an important feature of the water structure in terms of transport and deposition of sediments. The depth of the halocline is dependent on the freshwater discharge and the morphology of the fjord. In narrow parts of the fjords, fresh water tends to pile up, creating a deep halocline. The opposite effect is observed where the fjord is wider. Normally, the halocline is shallower near the head because of minimal mixing between the rapid-flowing river water and the underlying sea water. As the mixing and entrainment increase seaward, the halocline gets deeper.

Deep fjords, in addition to the simple two-layer circulation, may have deeper circulation cells (Carstens, 1970) with alternating current directions (see Section 4.1.5). This complicates the dispersal of sediment (Syvitski and Macdonald, 1982).

Sills play an important role with respect to water structure, circulation, sediment transport, and biological life in fjords. Shallow sills hinder a free exchange of water with the open ocean (Fig. 2.10). In extreme cases, this may lead to stagnancy in the deeper parts of the water and depletion of oxygen (Richards, 1965). This may seriously affect all macroscopic life and cause a complete change in the chemistry of the water and sediment (Chapter 7). Replenishment of basin water with more oxygenated shelf water is governed by density differences, meteorological conditions, internal waves and fjord geomorphology (see Section 4.4). Deep water renewal may take place when the density of the water at sill depth exceeds the density of the basin water inside the sill. This may occur frequently, annually, or seldom. The sill depth and the density of the outside water are the critical factors. Some fjords have several sills and thus basins, and as the deep water overspills the first sill it gradually gets mixed with less saline water. Consequently, the innermost basins may not experience a deep water renewal. Deep water renewals do not generally exchange the entire volume of fjord water; replacements of 20 to 80% of the basin water are more common (Molvaer, 1980).

Even during "stagnant" periods, fairly intense vertical mixing of basin waters may take place (Gade, 1968). Tides are a key source of energy for this process, generating internal waves at the sill with their subsequent conversion to turbulence. In Indian Arm, B.C., eddy diffusion coefficients as large as 10^{-3} m^2 s^{-1} have been measured in the deep water (Gilmartin, 1962). More recent work in British Columbian fjords, shows Kz-values (based on 222Rn) in the range of 10^{-5} to 3×10^{-3} m^2 s^{-1} in the deep water (Smethie, 1981). The larger the eddy diffusion coefficient, the more intense the vertical mixing. In Norwegian fjords, these coefficients are normally lower. The lowest known vertical diffusivities in fjords are in the order of 10^{-6} m^2 s^{-1} (e.g., Framvaren, Norway; Gade, 1983).

The temperature structure in the shallow parts of fjord waters is not very different from that of other estuaries. A thermocline often corresponds with a halocline, creating a strong pycnocline in the near-surface water. The position of the thermocline may vary seasonally, with changes in the air temperature and the temperature of the river runoff. In deep fjords, basin water is distinctly different from other types of estuaries (Chapter 6), remaining of more constant temperature year-round. Often seasonal forcings result in the formation of a mid-depth temperature minimum that may last a number of months (Fig. 2.11). The minimum results from the winter cooling of the upper few tens of meters until the surface water is colder than the deep water (although less dense because of the lower salinity). Subsequent warming of the surface waters during the spring and summer results in the temporary mid-depth temperature minimum. Fjords that have more complex temperature profiles with a number of maxima and minima may be affected by submarine melt of glaciers and/or icebergs (e.g., Chilean fjords; Pickard, 1971).

The temperature of the deeper basin waters depends on the temperature of similar-depth coastal shelf waters. The polar regions have the coldest water temperatures (< -1.0°C), New

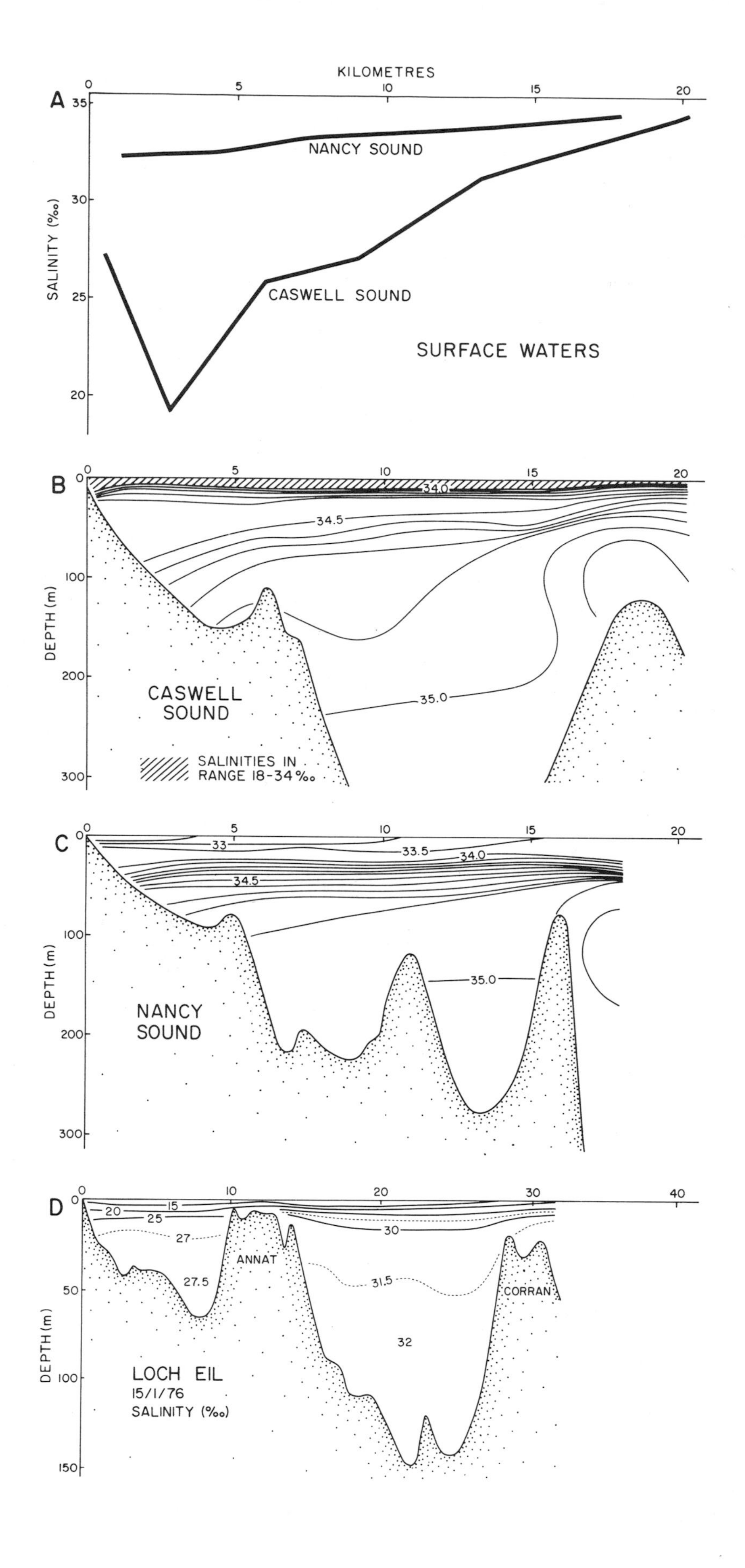

KILOMETRES
A
SALINITY (‰)
NANCY SOUND
CASWELL SOUND
SURFACE WATERS
B
DEPTH (m)
34.0
34.5
35.0
CASWELL SOUND
SALINITIES IN RANGE 18-34‰
C
33
33.5
34.0
34.5
35.0
NANCY SOUND
D
15
20
25
27
27.5
30
31.5
32
ANNAT
CORRAN
LOCH EIL
15/1/76
SALINITY (‰)

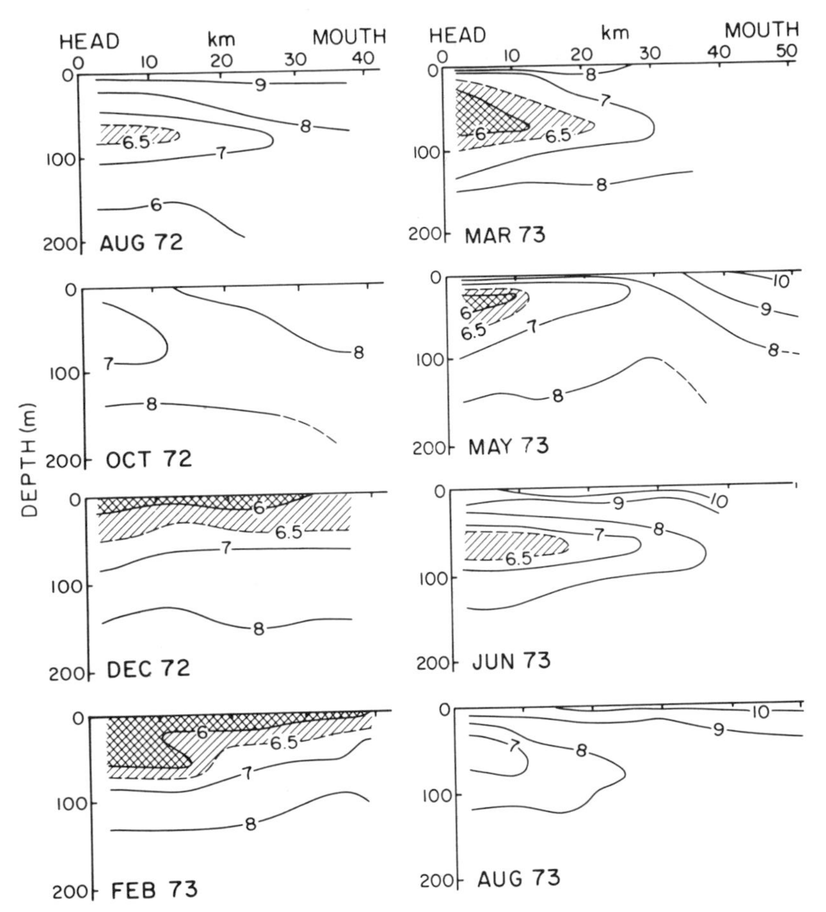

FIGURE 2.11. Synoptic time sequence of the longitudinal distribution of temperature showing the development and decay of the mid-depth temperature minimum for Bute Inlet, British Columbia (from Pickard and Stanton, 1980).

Zealand and Scotland have the warmest temperatures (> 10°C) (Table 2.1). For very deep fjords (> 500 m) the seasonal temperature variation in the deep basin waters seldom exceeds 1°C (Pickard, 1975); the shallow fjords of Scotland can have bottom-water temperature variations of 7°C (Milne, 1972). In fjords that have permanent anoxic bottom water, a temperature increase toward the seabed is often recognized (Skei, 1981b): a feature similarly observed in meromictic lakes (Williams et al., 1961). This may result from the escape of heat from the interior of the earth, creating a temperature gradient in the bottom water. In polar meromictic fjords, a mid-depth temperature maximum has been attributed to the trapping of solar radiation beneath a permanent ice cover (Hattersley-Smith et al., 1970).

The tidal amplitude in the different fjord regions of the world is also highly variable, ranging from almost zero in S Norway to 10 m in Cook Inlet, Alaska, or 9 m in Frobisher Bay, Baffin Island. This has a profound impact on sediment transport and deposition. In areas with large tidal prisms, sediments are moved back and forth, subsequently being deposited/eroded/redeposited. Of importance here are the residual transport pathways. The tidal energy resuspends sediments along the shores and deposits them in lower energy (usually deeper) regions.

The oxygen regime in fjords depends on the

◁
FIGURE 2.10. (A) Sea surface salinity distribution for two New Zealand fjords (from Stanton, 1978); (B) vertical salinity distribution for Caswell Sound, N.Z.; (C) vertical salinity distribution for Nancy Sound, N.Z.; and (D) salinity distribution in Loch Eil and Upper Loch Linnhe (from Edwards et al., 1980). Note the salinity distribution differences between these two basins as affected by the sill.

TABLE 2.1. Average temperatures at or near the 200-m depth of fjord regions.

Region	Temperature (°C)	Reference
Alaska	3 to 7	Pickard, 1967
British Columbia	6 to 9	Pickard, 1975
Chile	6 to 9	Pickard, 1971
New Zealand	10 to 12	Stanton & Pickard, 1981
Baffin Island	−1 to 0	Trites et al., 1983
Lock Eil, Scotland (155 m)	7 to 13	Milne, 1972
Labrador	−1.5 to −0.5	Nutt, 1963
Norway	6 to 8	NIVA data (unpublished)
Kongsfjord, Spitsbergen (Svalbard)	1.2	Elverhöi et al., 1983

periodicity of water exchange, organic matter supply, and rate of utilization (Chapter 7). In fjords with deep sills and frequent water exchange, oxygen saturation generally occurs above the pycnocline (sometimes is supersaturation resulting from primary production), and slight undersaturation occurs at depth, owing to organic matter breakdown and sluggish water movements. The oxygen content in the upper waters is also dependent on the extent of turbulence and mixing between river water and sea water. A mid-water oxygen minimum is sometimes observed (Fig. 2.12), representing the water mass where the main mineralization of organic matter takes place. Low oxygen at mid-water depths may also be caused by deep water renewal, which vertically displaces low-oxygen water upward. The residence time of the water is a critical factor with respect to oxygen consumption. In fjords with shallow sills and longer residence time, the basin water may become anoxic. This will allow the reduction of nitrate and sulfate, and the creation of poisonous hydrogen sulfide (Richards, 1965). The levels of sulfide accumulated in the bottom water depend on the residence time of the water and the rate of sulfate reduction. Wide ranges of sulfide levels have been measured in fjord basins (Strøm, 1936; Richards, 1965), with a maximum value of 8 mmol H_2S (Framvaren, Norway; Skei, 1983c). In intermittently anoxic fjords, the anoxic bottom water may be vertically displaced during water renewals (Fig. 2.13), and the sulfide-rich water mass may reside at an intermediate depth for a considerable time. In Frierfjord, Norway, hydrogen sulfide was observed at intermediate depth for several weeks after a major deep water renewal (Molvaer, 1980).

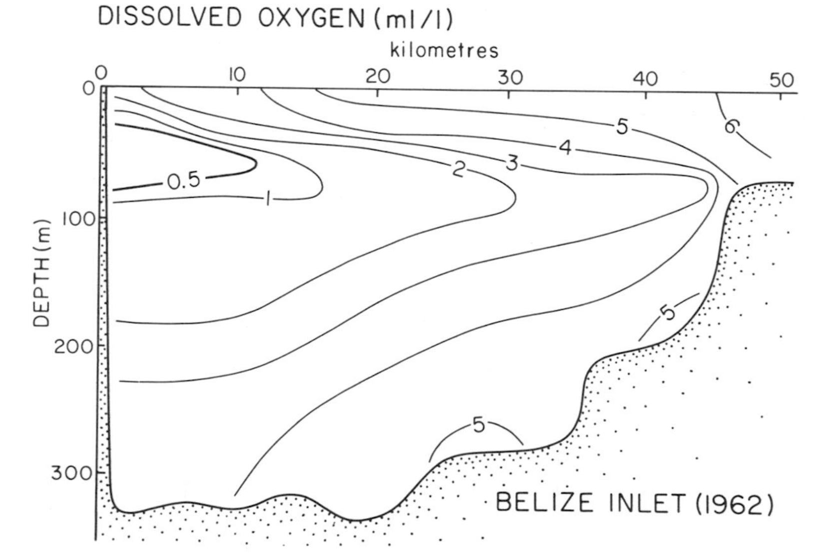

FIGURE 2.12. Longitudinal section for Belize Inlet, B.C., showing the mid-depth oxygen minimum (from Pickard and Stanton, 1980).

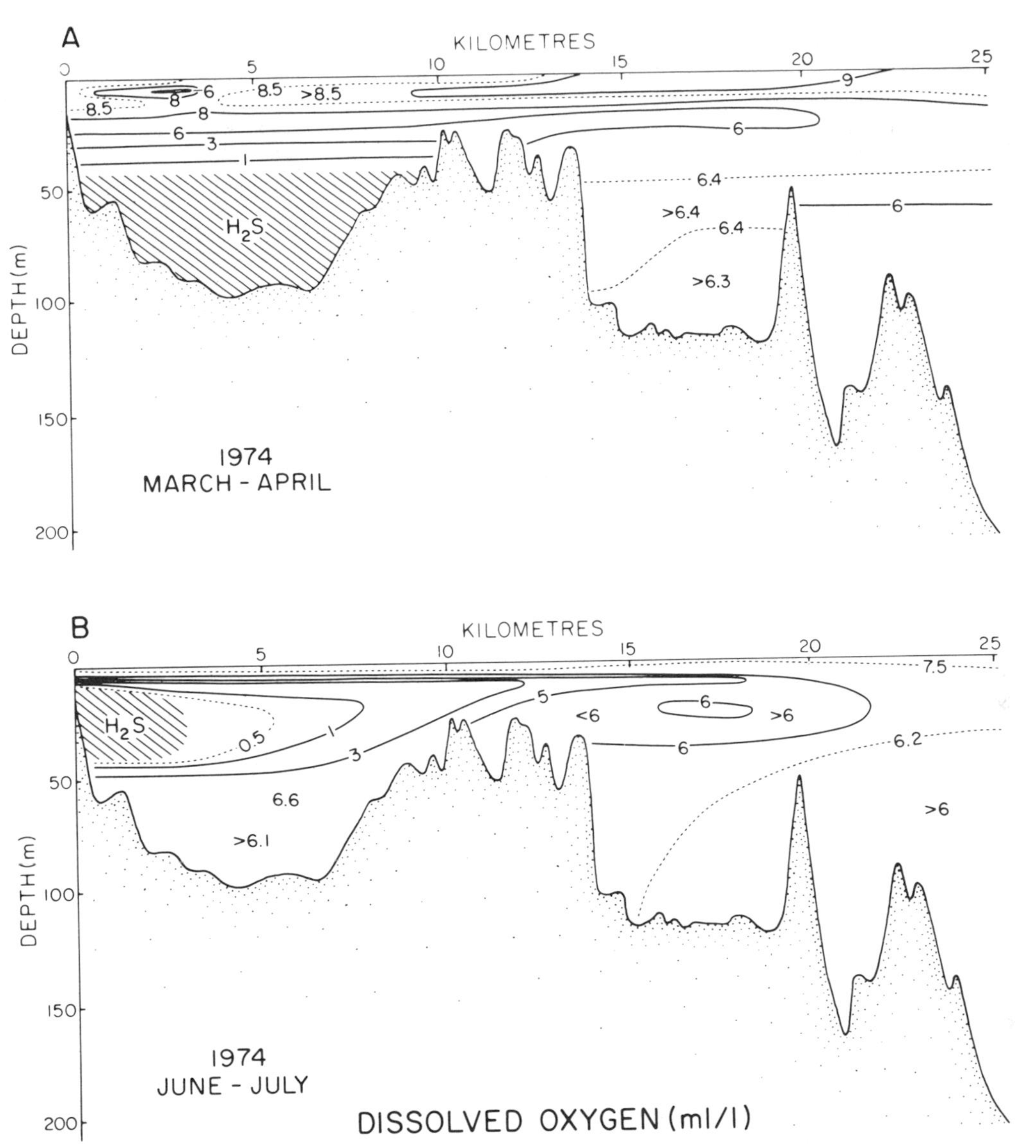

FIGURE 2.13. Longitudinal sections of oxygen content (ml/L) shortly before (A) and during (B) a deep water renewal in the Frierfjord, Norway (from Molvaer, 1980).

Occasionally, the vertically displaced anoxic water reaches the very surface of the fjord, creating a massive fish kill (Hellefjord, Norway; Brogersma-Sanders, 1957). Such episodes have a definite impact on the bottom sediments of fjords with excess input of organic matter.

2.4 Sediment Sources and Transport Mechanisms

Inputs of sediment to fjords in the temperate zone include river- and wind-transported terrestrial sources, anthropogenic sources, open ocean sources, and internal fjord sources (Fig. 2.14). Fjords that are ice dominated have additional sources and input mechanisms for sediment (see Fig. 4.30).

An oxic fjord contains sediment composed predominantly ($> 90\%$) of inorganic particles, of which fluvial sediment sources may account for most. Our knowledge of the quantitative transport of fluvial sediment into fjords is not very good, largely a result of the unpredictable seasonality of discharge (see Chapters 3 and 4). Annual sediment transport to individual fjords may reach 10^9 to 10^{10} kg from large glacial British Columbia rivers (Syvitski and Farrow, 1983) to less than 10^7 kg from the smaller glacial Baffin

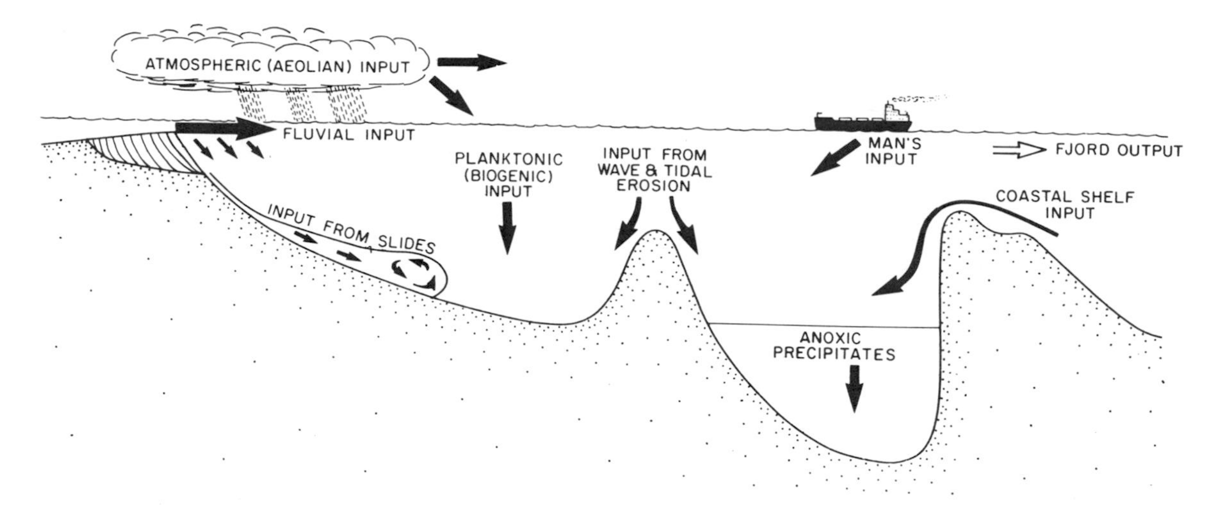

FIGURE 2.14. Primary sediment inputs to a nonglaciated fjord (cf. Fig. 4.30).

Island rivers (Syvitski et al., 1984b). Fjord rivers transport erosional products from weathering, reworked morainal/glaciofluvial/glaciomarine deposits and freshly produced glacial flour (Elverhöi et al., 1980). Additionally, fjord rivers supply terrestrial organic matter such as leaves, twigs, humic substances, etc. (Glasby, 1978b). The latter source is relatively more important in fjords situated at lower latitudes, where the vegetation cover of the catchment area is dense. The grain size of the fluvial sediment will vary according to the parent material, the extent of erosion, the inclination of the river, the energy of the river water, and the filtering effect of lakes. Hence, the sediment source material may range from clays to boulders.

There exists an important group of fjords that receive less than 25% of the sediment infill from rivers. Here the main sediment supply is derived from waves or tides reworking coastal deposits of older marine sediment or glacial till. Such fjords are typical along the east coast of Canada (Piper et al., 1983), where cliff retreat rates can exceed 1 m yr^{-1} near the fjord mouth, decreasing to 25 cm yr^{-1} in exposed inner-fjord areas.

Sediment within a fjord basin may have been emplaced by mass sediment transport processes that result from numerous failures around the basin. Fifty percent of the sediment column within the basins of Hardangerfjord is a result of slumps and turbidity currents (Holtedahl, 1965). Slope failures occur near the fjord-head delta, the side wall slopes, side-entry deltas, off of sills, and at junctions with tributary (hanging) valleys. Slide volumes may range from very small (10^3 m^3) to very large (10^9 m^3). The frequency of slope failures is controlled by the local rate of sediment accumulation and the frequency and force of the triggering mechanisms, and may range from annual events to rarer but catastrophic events (see Chapter 5).

Sediments derived from the continental shelf and transported into fjord basins are less abundant (Syvitski and Macdonald, 1982; Slatt and Gardiner, 1976) in comparison with other types of estuaries. The limiting factor for fjords is the effective barrier of the outer fjord sill. Additionally, the compensation current is not along the seafloor as in other shallow estuaries, and hence is not eroding and transporting sediment up-fjord. However, biological material such as plankton may be passively transported into fjords by the compensation current and may account for the deposition of allochthonous marine organic matter. Phytoplankton transported to areas of larger nutrient supply may cause plankton blooms and a substantial flux of organic matter to the sediments. According to Petersen (1978), Greenland fjords may act as a sink for organic matter, most of which originates from shelf waters.

The in-situ biomass production in fjords contributes significantly to the organic input to the sediments (Chapter 6). Many fjords are very productive areas, sustaining primary production from early spring to late autumn. The flux of organic carbon to the sediments has distinct seasonal variations (Burrell, 1983a). Similarly, seasonal fluctuations in C/N ratios of the deposited material reflect the influence of organic matter of marine origin during the productive season. Additionally, the primary production will also contribute siliceous and calcareous matter to the sediment during blooms of diatoms, radiolarians, foraminifera, and coccolithophorides. This has been demonstrated by Si/Al and Ca/Al anomalies of suspended particulate matter (Price and Skei, 1975).

Another sediment source of significance in certain stagnant fjords is from the precipitation of inorganic substances such as oxides, hydroxides, carbonates, and sulfides (Chapter 7). Although these substances are not quantitatively abundant in fjord sediments, they are important in a biogeochemical sense. These precipitates are normally controlled by changes in redox conditions and pH. Fjords are chemically as well as physically stratified, with steep vertical gradients. The redox boundary may be situated within the sediment, at the sediment-water interface, or in the free water mass within or out of reach of photosynthetic activity (Skei, 1983b). Such boundaries are sites of high chemical and bacteriological activity (Emerson et al., 1983; Jacobs et al., 1985). Iron and manganese may form oxidized precipitates above the redox boundary (Jacobs et al., 1985), while the same elements may form sulfides and carbonates, respectively, in the anoxic environment (Suess, 1979). These processes can therefore influence the chemical composition of the bottom sediments (for details see Chapter 7).

Atmospheric or aeolian input of solids to the fjord is not normally regarded as a significant contributor. However, if the total rate of sediment accumulation is very low, as a result of minimal fluvial sediment sources and low primary production, then the atmospheric contribution may be important. Measurements on the deposition rates of particulate matter from the atmosphere are scarce. Particle concentrations of 7 to 15 $\mu g\ m^{-3}$ in air have been measured at various sites in Norway (Semb, 1978). Atmospheric deposition rates in the North Sea have been assessed at 10% of the fluvial input to this region (McCave, 1973). In the polar desert regions of the Canadian archipelago, the contribution of wind-blown sand and silt to fjords is considered more important (Gilbert, 1983).

During the last century, manmade products have played a significant role as a source of sediment in estuaries and fjords (see Chapter 8). Discharge of organic substances from sewage plants, pulp and paper mills, and various solid wastes from the chemical and mining industries, has led to increased rates of deposition in some fjords—sometimes accompanied by a deleterious quality change (Pearson and Rosenberg, 1976; Nyholm et al., 1983; Skei et al., 1972). Population and industrial activities are traditionally concentrated along fjords, a result of near-perfect port conditions (deep water near the shoreline and limited fetch conditions). In some extreme cases, the fjord bottom is entirely covered by industrial waste and filled up to sill depth (Jøssingfjord, Norway). Dredging and dumping of dredged material in fjord basins is a modern way of sediment transport (Skei, 1983a). If erosion occurs at the dump site, the dredged material is transported elsewhere. The development of the coastal zone, including river channel stabilization programs, the loading of waste, tailings, or fill, and the development of port facilities on the steep coastal slopes, has led to numerous slope failures accompanied by property destruction and the loss of life (Bjerrum, 1971; see Chapter 5). Man's activities may also alter the sediment sources and the availability of these sources. For example, where dams have been built, sediments are trapped before they reach the fjord (Heling, 1977). The antithesis can be observed when land is cleared for cultivation or construction purposes, thus increasing the sediment yields.

2.5 Fjord History

Fjords are a dynamic coastal environment whose succession from formation to completion (sediment infill) can be related to the local glacial and paraglacial history of erosion and deposition, superimposed on a well-developed preglacial geologic setting. In prior discussions, we have noted that "true" fjords have at one time been subject to glaciation/deglaciation processes. Much of the sediment accumulation within these overdeepened coastal basins may be related to glacial/proglacial infilling during and after the last major ice advance (Aarseth et al., 1975; Gilbert, 1985) or may reflect several episodes of proglacial basin infilling. Basin deposits may include: (1) a basal (ice-contact) till complex (lodgment till, waterlain till, push and dump moraines); (2) proximal glaciomarine sediments dominated by gravity flow deposits alternating with hemipelagic layers; (3) distal glaciomarine sediments that tend to be fine grained and highly bioturbated; and (4) widely varying nonglacier-influenced sediments that depend on local supply and energy conditions. Not all fjords have yet reached the more mature nonglacial infill stage, others have been subject to a number of these complete cycles.

Some of the fjords of Greenland and Alaska, and many of those of Antarctica, are still at stage (1): completely or nearly ice filled. Many of the fjords of Greenland, Svalbard, Baffin Island, Ellesmere Island, and other high-latitude archipelago fjords, and some from Alaska and Chile, remain at stage (2): influenced by the presence of one or more tidewater (or floating) glaciers. A significant number of the world's fjords are presently at stage (3), where glaciers are subaerial but part of the drainage basin, or transitional to stage (4), where nearly all land-based glaciers are gone. Areas include much of the remaining high-latitude fjords, Alaskan and Chilean fjords, mainland British Columbia fjords, northern and western Norwegian fjords, and a few of the Iceland and northern Labrador fjords. Those areas that have been at stage (4) for some time include the remaining fjords of: Vancouver Island, Iceland, southern Norway, Scotland, New Zealand, Labrador, Newfoundland, Nova Scotia, Quebec, and Kamchutka.

The oceanographic characteristics of many fjords, whatever the stage, have remained unchanged and in steady-state equilibrium for hundreds and possibly even thousands of years. Other fjords, however, have undergone more recent changes—some are still in the process of change. These changes relate mostly to relative sea-level fluctuations controlled by glaciation/deglaciation processes. The following, then, are a few of the more typical scenarios associated with changes in a fjord's equilibrium condition.

The first scenarios concern a time when global sea level is falling as a result of a major ice-sheet advance. A coastal estuary, initially dominated by cold brackish water, is turned into a freshwater lake with the sea-level drop. With the retreat of the ice sheet, the sea level begins to rise locally and a deep salt-wedge estuary is produced. As the trend continues a deep-silled fjord results. An actualistic model is the Bedford Basin, Nova Scotia, that fluctuated from initial estuarine circulation (8000 years BP) to the present-day fjord condition (Fig. 2.15; Miller et al., 1982). A number of other examples on the east coast of North America began as proglacial lakes and have since made the transition to present-day estuarine or fjordlike conditions: (1) St. Margaret's Bay, N.S. (Piper and Keen, 1976); (2) Mahone Bay, N.S. (Barnes and Piper, 1978); and (3) Hudson River estuary, N.Y. (Fig. 2.2; Newman et al., 1969).

More common scenarios involve a fjord's change from marine to lake conditions, often a result of changes in sea-level fluctuations. However, instead of involving the eustatic sea-level component, the transition is associated with the isostatic component: the relative lowering of sea level resulting from the rebound of land in response to ice-sheet removal (see Chapter 3). As sea level drops, shallow sills will become exposed concomitant with a change from marine to brackish to freshwater conditions (Fig. 2.16A). Other mechanisms to cut a fjord off from the sea include: (1) cutoff by a prograding delta, for example, Pitt Lake, B.C. (Fig. 3.24; Clague et al., 1983); (2) cutoff by a prograding barrier spit, examples include the southern coast of Newfoundland and Lake McKerrow, N.Z. (Fig. 2.16B; Pickrill et al., 1981); (3) cutoff by a side-entry glacier, many Canadian arctic examples (Fig. 3.7; Hattersley-Smith and Serson, 1964; Gilbert et al., 1985); and (4) cutoff of a tributary fjord by a main trunk glacier, for example, Muir Inlet, Alaska (Fig. 4.26; Goldthwait et al., 1966).

If the input of fresh water is large, with associated high values of vertical eddy diffusivities (mixing energy), the trapped marine water will eventually become diluted and washed out of the basin. In other words, the replenishing salt water brought on by large tides and deep water renewals (if still possible) cannot keep up to the removal of salt during the period of closure. Lake Melville, Labrador, is in the initial stages of desalination. The Hamilton Inlet sill that partially encloses Lake Melville was 80 m deep 7000 years ago, 50 m deep 5000 years ago, and is 28 m deep at present. During that time, salinity of the fjord bottom has been reduced by 5‰ (Vilks and Mudie, 1983). The fjord bottom has remained oxygenated during that time—not so with most other fjords undergoing closure. Al-

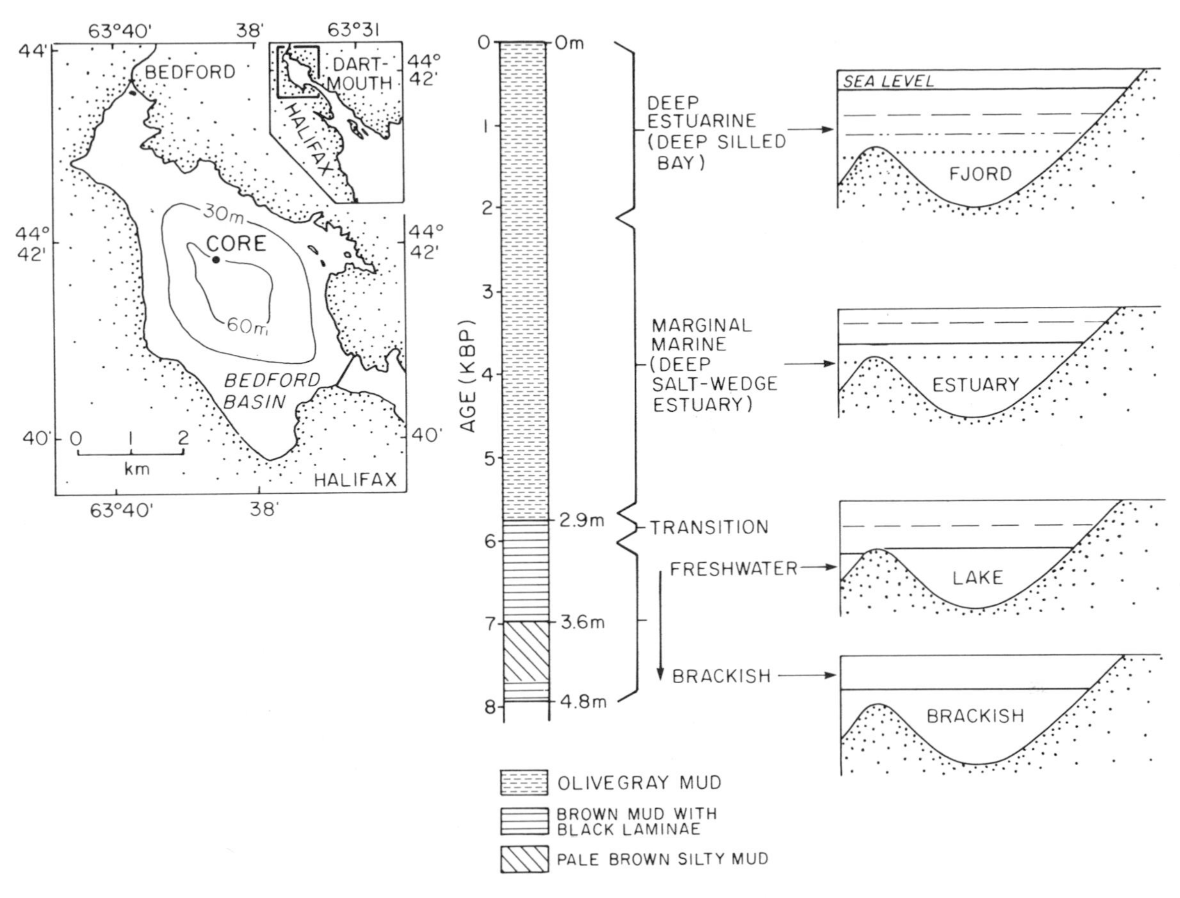

FIGURE 2.15. Schematic of a sediment core from the Bedford Basin, Nova Scotia (after Miller et al., 1982). High resolution biostratigraphic analysis has revealed a fjord influenced by a fluctuating sea level.

though Ogac Lake, Baffin Island, receives oxygenated sea water yearly, the bottom water becomes anoxic very quickly (McLaren, 1967). Lake McKerrow, N.Z., was closed off from the open sea by a Holocene barrier spit approximately 7700 BP (Pickrill et al., 1981). During that time, the lake has freshened, exhibiting its present maximum basin salinity of 10‰. The bottom waters are not, however, anoxic. Future probabilistic scenarios might see both Lake Melville and Lake McKerrow completing the conversion to fully freshwater oxygenated lakes. The conversion has already occurred for Pitt Lake, B.C., and many other Norwegian lakes and Scottish freshwater lochs.

Ogac Lake will probably go the route towards a meromictic lake where the hypolimnion is occupied with old anoxic sea water (e.g., Framvaren, Norway, Fig. 2.17). Such meromictic lakes are generally deep compared with the above freshwater lakes and lochs, and/or have small catchment areas and therefore limited freshwater supply. Both conditions combine to allow for low vertical eddy diffusivities and very limited mixing. The bottom saline waters may become trapped for several thousand years. Examples include: (1) Lake Nitinat, B.C. (Fig. 4.34B; Northcote et al., 1964); (2) Powell Lake, B.C. (Williams et al., 1961); and (3) a number of Norwegian lakes (Strøm, 1957, 1961; Bøyum, 1973; Bremmeng, 1974).

Sill depths may also be artificially changed by man. For example Framvaren sill was dredged in 1850, thus transforming a meromictic lake back into a fjord (Fig. 2.17; Skei, 1983c). Framvaren presently has two anoxic water masses, one of recent age overlying a much older layer. Another example of altering the hydraulic regime over a sill was the blasting of Ripple Rock through Seymour Narrows, B.C., in 1958 (the biggest nonnuclear peace-time detonation on record at that time; Thomson, 1981, p. 61). The blast increased the water depth of the sill, decreasing tidal velocities as a result. Similarly,

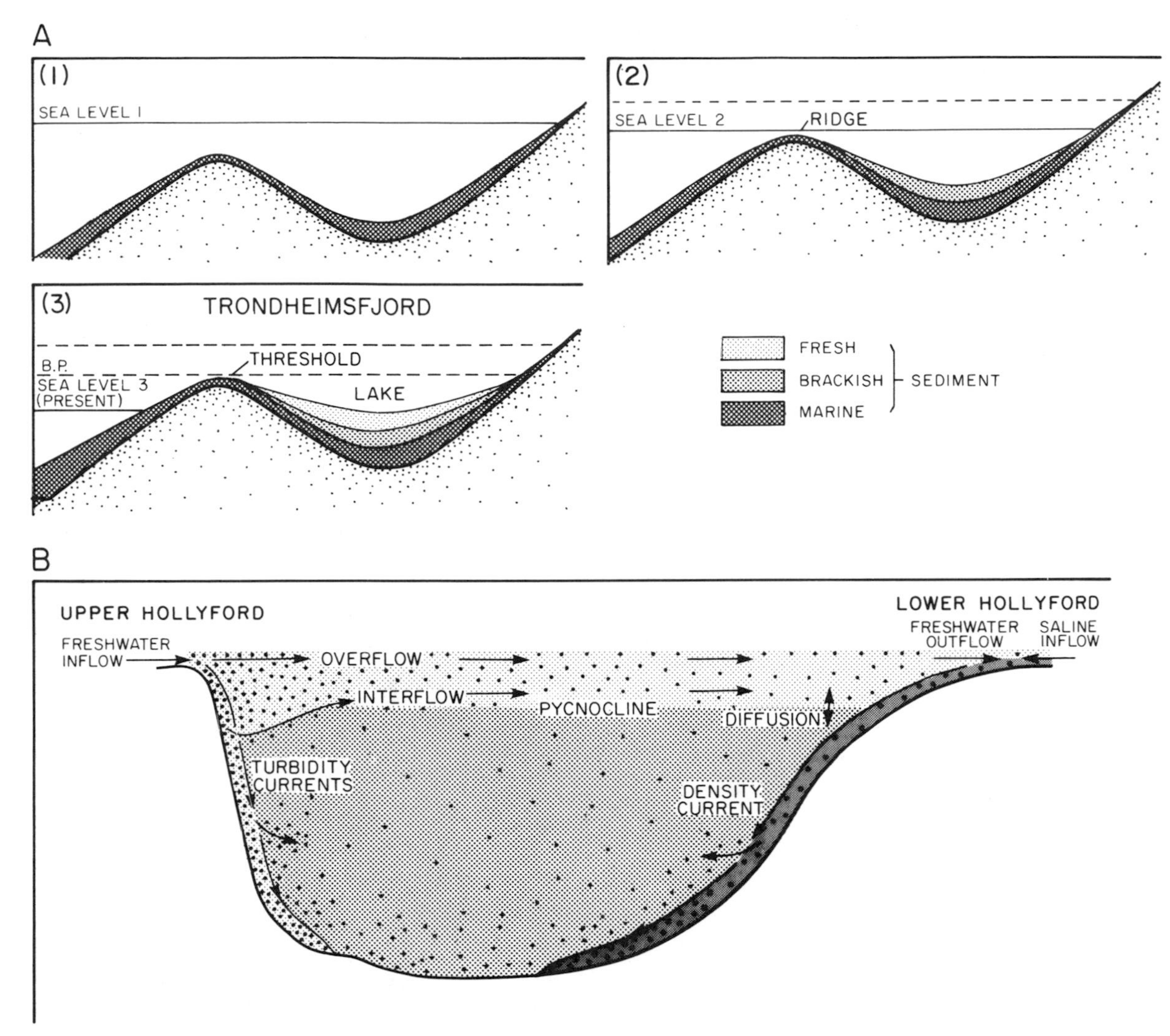

FIGURE 2.16. (A) The effect of falling sea level as a result of isostatic rebound on the development of sediment at the head of Trondheimsfjord, Norway (after Kjemperud, 1981). (B) Schematic representation of the circulation and sedimentation within Lake McKerrow, N.Z., a paleofjord (from Pickrill et al., 1981).

deglaciation of Glacier Bay, Alaska, some 250 years ago, has been associated with emergence of the sills and detectable increases in the tidal velocities (Matthews, 1981). Thus, through the action of man or through natural processes, some fjords are in a state of flux between one equilibrium condition and another. As we come to understand these flux transitions, we might predict new steady-state conditions with some certainty. In a most interesting paper, Reeburgh et al. (1976) note that Russell Fjord, Alaska, may be cut off from Disenchantment Bay by the Hubbard Glacier (Fig. 2.18). The following is their speculation on future conditions in Russell Fjord.

If the fjord becomes completely closed . . . rapid accumulation of fresh water runoff would be expected. The water level in the fjord would rise at a rate of about 28 m yr^{-1} until water flowed around or through the glacier or through channels in the terminal (sic) moraine at the southern end of the fjord . . . With accumulation of fresh water runoff and no renewal of bottom water, the fjord should stratify strongly and the bottom waters should become anoxic within less than two years. With strong salinity-maintained stratification, it should be possible for the surface waters of the fjord to cool strongly enough to freeze. The altered salinity and light conditions would be fatal to intertidal organisms and to those members of the plankton community unable to tolerate prolonged low salinities or low oxygen conditions . . . as long as the

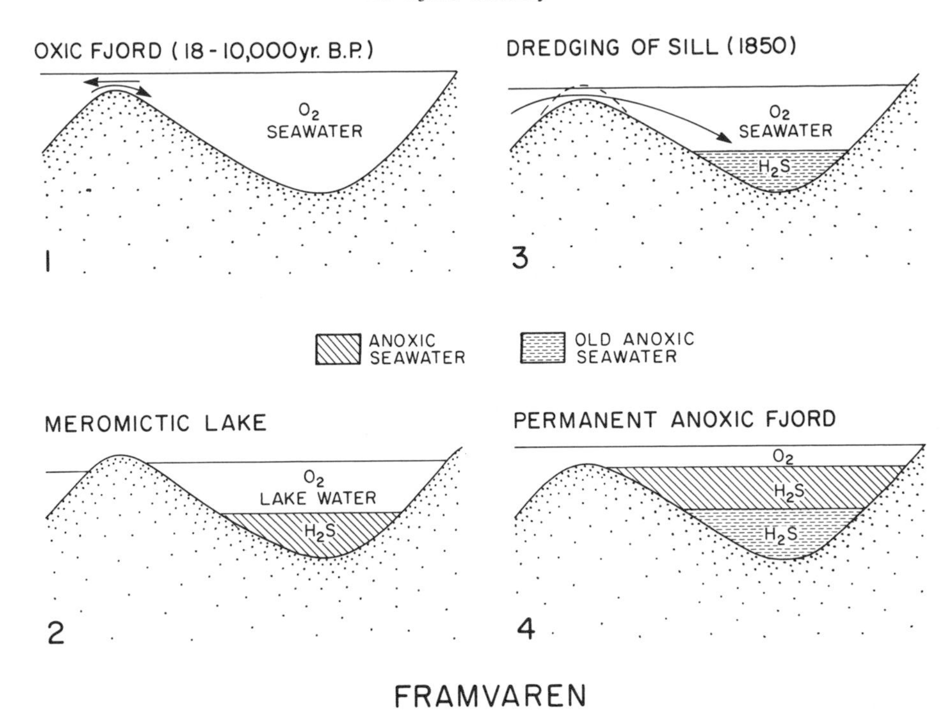

FIGURE 2.17. The changing water character of Framvaren, a fjord in southern Norway, as a result of natural sea-level fluctuations and man's interference (Skei, 1983c).

glacial dam remained intact, the fjord would evolve toward a (meromictic) lake . . . Under conditions of less complete closure . . . outflow currents would exceed 6 m s^{-1}, so tidal inputs might be overcome or greatly diminished (during the summer). The fjord might be expected to undergo a seasonal variation in water level coincident with runoff fluctuation. During winter, salt-water renewal could take place but whether the renewal rate would be high enough to prevent the deep water from becoming anoxic is unknown.

Sea-level fluctuations will also affect the depth of erosion that the seafloor is subjected to. Piper et al. (1983) identified a class of wave-dominated fjords on the Labrador coasts that are presently emerging (i.e., undergoing a marine regression). During this period of emergence, more and more of the seafloor has been brought into the domain of wave erosion (Fig. 2.19; Barrie and Piper, 1982). Three facies may develop during this type of sedimentation history, for example, Makkovik Bay, Labrador: (1) a lower basin-fill unit thought to be composed of coarse proglacial sediment; (2) a conformable cover unit deposited under high rates of hemipelagic sedimentation (ice proximal); and (3) an upper basin-fill unit as a result of increasing wave erosion concomitant with the land emergence.

In the beginning of this section, we listed four stages of glacial history that a fjord might be presently experiencing. There is a fifth and final stage, the death of a fjord. The typical scenario might consist of: (1) sediment infilling by delta progradation, hemipelagic rain, and sediment gravity flows, until the basin depth and sill depth are equal; (2) the development of salt-wedge or partially mixed estuarine hydraulics (depending on river and tidal conditions); (3) the dramatic increase in the supply of sediment to the once-starved continental shelf; (4) depending on local wave conditions, the formation of a barrier that may partially enclose the estuary and develop a salt-marsh lagoon; (5) with further fluvial deposition, complete infilling of the lagoon resulting in the formation of an open coastal delta; and (6) with the increased shelf sedimentation, bathymetric irregularities on the shelf would be smoothed out. A similar scenario was proposed for the Vancouver Island shelf/fjord complex by Carter (1973).

How close a particular fjord is to being infilled depends on its present rate of infill and its volumetric storage capacity. Since the last glaciation, a large range of infill thickness have resulted (Table 2.2). Considering the present

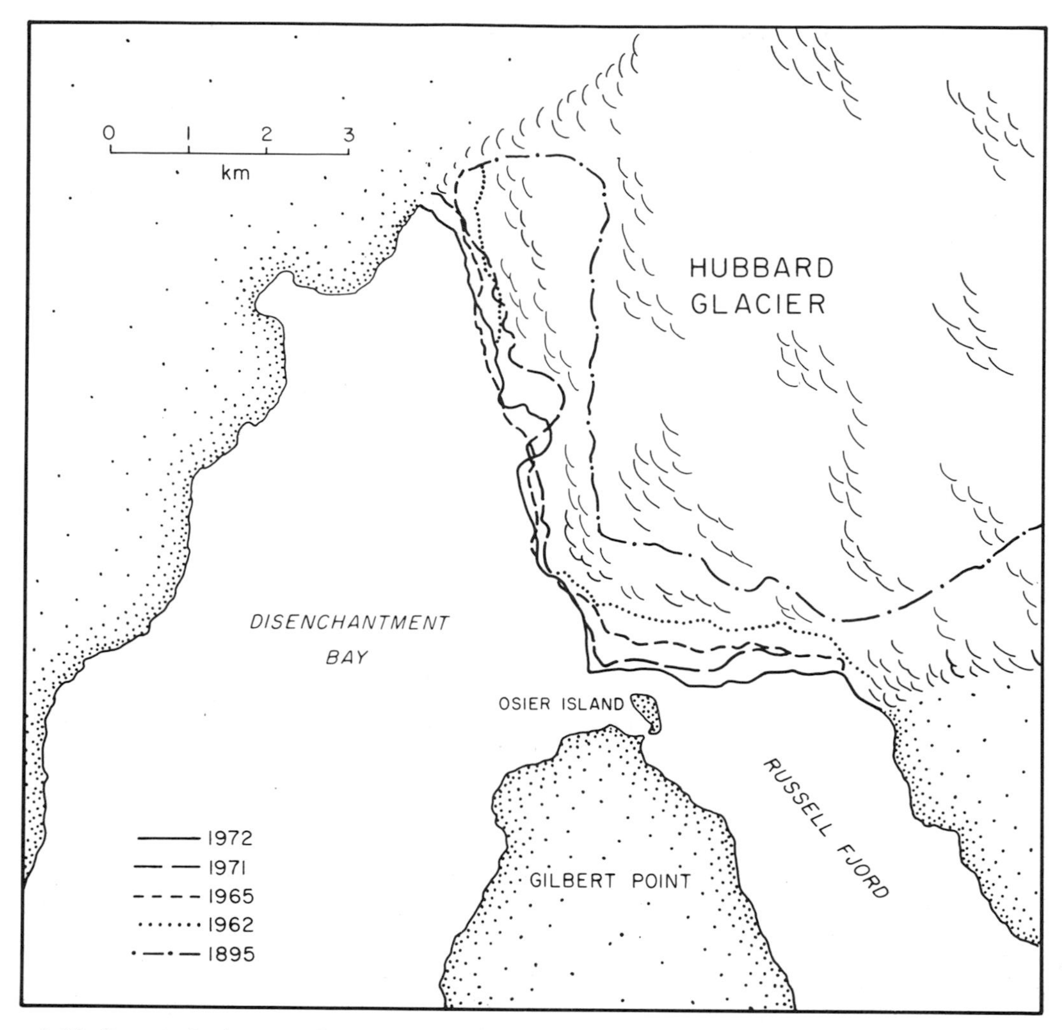

FIGURE 2.18. Stages of advance of the Hubbard Glacier from 1895 through 1972, that may eventually lead to cutoff of Russell Fjord (after Reeburgh et al., 1976).

accumulation rates within fjords (Table 4.2) and the basin depth to be infilled (Table 2.2), there will be few fjords left after the next million years, assuming no reglaciation occurs. Many of the New Zealand fjords have already infilled with fluvial sediments and are now fronted by a narrow outwash plain (Pickrill et al., 1981). The Mecatina coast of southern Quebec is fronted by a series of infilled fjords and overlapping deltas: only one, the Saguenay Fjord, remains to be infilled. The process of infilling may be accelerated by manmade discharges of solid wastes; for example, Jössingfjord, Norway, was filled up to sill depth (60 m) from an initial basin depth of 96 m over 25 years of titanium mining (see Section 2.4).

2.6 Characteristic Features of Fjord Coastlines

There are some interesting similarities and dissimilarities observed in the most characteristic fjord districts (Table 2.3). Given below is a brief but important scientific snapshot of these coasts. Details and exceptions as related to processes are described throughout the remainder of the book.

2.6.1 Greenland Fjords

The fjords of Greenland are geomorphologically similar to Norwegian fjords (Ahlmann, 1941), that is, fronted by a plateau—presumably a Late

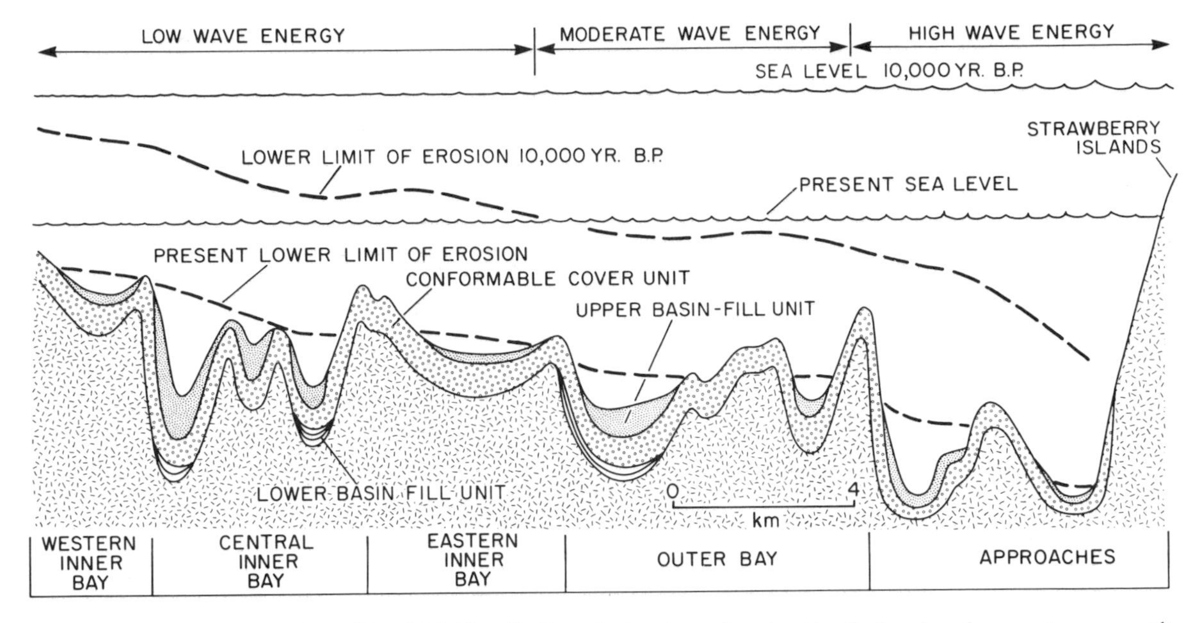

FIGURE 2.19. Longitudinal profile of Makkovik Bay, Labrador, showing the facies development as a result of sea-level lowering and increase in the effect of seafloor wave erosion (Barrie and Piper, 1982).

TABLE 2.2. Sediment infill within fjords.

System	Maximum thickness (m)	Basin depth (m)	Reference
Nuka Bay, Alaska	250	300	von Huene, 1966
Port Valdez, Alaska	800	280	von Huene et al., 1967
Glacier Bay fjords, Alaska	100	300	Powell, 1980
Knight Inlet, B.C.	600	540	Syvitski, unpubl.
Howe Sound, B.C.	750	250	Syvitski & Macdonald, 1982
Douglas Channel, B.C.	600	420	Bornhold, 1983
Mahone Bay, N.S.	80	50	Barnes and Piper, 1978
St. Margaret's Bay, N.S.	40	80	Piper and Keen, 1976
Makkovik Bay, Labrador	50	160	Barrie and Piper, 1979
Nain Bay, Labrador	112	75	Piper et al., 1975
Lake Melville, Labrador	400	250	Grant, 1975
Baffin Island fjords (10)	20–200	320–800	Gilbert and MacLean, 1983
Strathcona Sound, B.I.	90	300	Lewis et al., 1977
Søndre Strømfjord, Greenland	400		Larsen, 1977
Sermilik Fjord, Greenland	100		Heling, 1977
Lock Nevis, Scotland	30	40	Boulton et al., 1981
Maurangerfjord, Norway	70	280	Cone et al., 1963
Hardangerfjord, Norway	140	900	Cone et al., 1963
Oslofjord, Norway	130	340	Richards, 1976
Sognefjord, Norway	300	1300	Aarseth, 1980
Arnafjord, Norway	180	200	Aarseth, 1980
Lusterfjord, Norway	200	650	Aarseth, 1980
Kongsfjorden, Spitsbergen	100	100	Elverhöi et al., 1983
Van Mijenfjorden, Spitsbergen	20	100	Elverhöi et al., 1983
Cunningham Inlet, Somerset Is.	25	30	Hunter and Godfrey, 1975
Admiralty Bay, Antartica	160		Anderson, Personal Communication, 1985

TABLE 2.3. Typical characteristics of the world's major fjord coastlines.

Fjord District	Fjord stage[1]	Tidal range[2]	River discharge[3]	Climate	Sedimentation rate[4]	Man's influence[5]
Greenland	1,2	Low	Medium to high	Subarctic to arctic maritime	Medium to high	Low
Alaska	1,2,3,4	High	Low to high	Subarctic maritime	Medium to high	Low to moderate
British Columbia	3,4	High	Medium to high	Temperate maritime	Medium to high	Low to high
Canadian Maritime (N.S., Nfld., Quebec, Labrador)	4,5	Low to medium	Low to high	Subarctic to temperate maritime	Low	Low to moderate
Canadian Arctic Archipelago	1,2,3,4	Low to high	Low to medium	Arctic desert to maritime	Low to medium	Low
Norwegian Mainland	3,4	Low	Low to medium	Subarctic to temperate maritime	Low	Moderate to high
Svalbard	2,3	Low	Low	Arctic island	Medium	Low
New Zealand	4,5	Medium	Low to medium	Temperate maritime	Low to medium	Low
Chile	2,3,4	Low	Low to high	Temperate to subarctic maritime	Medium to high	Low
Scotland	4,5	Low to high	Low	Temperate maritime	Low	Moderate

[1]Stage 1: glacier filled; 2: retreating tidewater glaciers; 3: hinterland glaciers; 4: completely deglaciated; 5: fjords infilled.
[2]Low: < 2 m mean range; medium: 2–4 m; high: > 4 m mean range.
[3]Low: < 50 $m^3 s^{-1}$ mean annual discharge; medium: 50–200 $m^3 s^{-1}$; high: > 200 $m^3 s^{-1}$.
[4]Low: < 1 mm yr^{-1} averaged over the entire fjord basin; medium: 1–10 mm yr^{-1}; high: > 10 mm yr^{-1}.
[5]See Chapter 8.

Tertiary peneplain—and flanked by high sidewall mountain peaks. Cirques and strandflats are not observed. The fjords of Greenland are typically U-shaped, with very steep sides and flat bottoms, particularly in their deepest basins (Ahlmann, 1941).

The coast of Greenland can be divided into four quadrants (Fig. 2.20). The NE coast, from Victoria Fjord to the Scoresby Sund fjord complex (Fig. 2.21B), has approximately 50 major fjords, some of them the world's largest and deepest (Funder, 1972). The confining sidewall mountains have altitudes between 800 and 2500 m, giving a total relief from the fjord floor of over 3000 m. These fjords are cut in crystalline bed rock, and those with a N–S trend are controlled by the major fault zones. Fjords with prominent sills are not common (Funder, 1972). The largest fjords are (Fig. 2.21B): (1) Victoria Fjord–120 km long, 15 km wide; (2) Independence Fjord–190 km long, 18 km wide; (3) Danmark Fjord–180 km long, 15 km wide; (4) K. Franz Josephs Fjord–200 km long, 5 km wide, and (5) the Scoresby Sund fjord complex–300 km long, 50 km wide. The most northern fjords, such as Victoria Fjord, are "frigid" with a permanent sea-ice cover. Others, such as Nioghalvfjerdsfjorden, remain filled with glacier ice and await deglaciation. The climate varies along the NE coast from high arctic in the north, with mean annual temperatures below 0°C, rising to 7°C at Scoresby Sund (Lysgard, 1969). Consequently, postglacial frost action may be high. The climate is more maritime on the outer coast, becoming more continental inland with warmer and drier summers.

The SE coast, from Scoresby Sund to Kap Farvel (Fig. 2.20), has approximately 100 fjords. These are much shorter usually < 50 km long and < 3 km wide. The terrain is rugged, with Watkins Bjerge at 3100 m. Most of these fjords are partially under the Greenland ice cap. The climate is mostly maritime with annual precipitation of ≈ 0.6 m (locally up to 2 m), of which 60% is in the form of snowfall. The mean annual temperature is 10°C. This coastline is largely unpopulated and scientific information is, at best, poor.

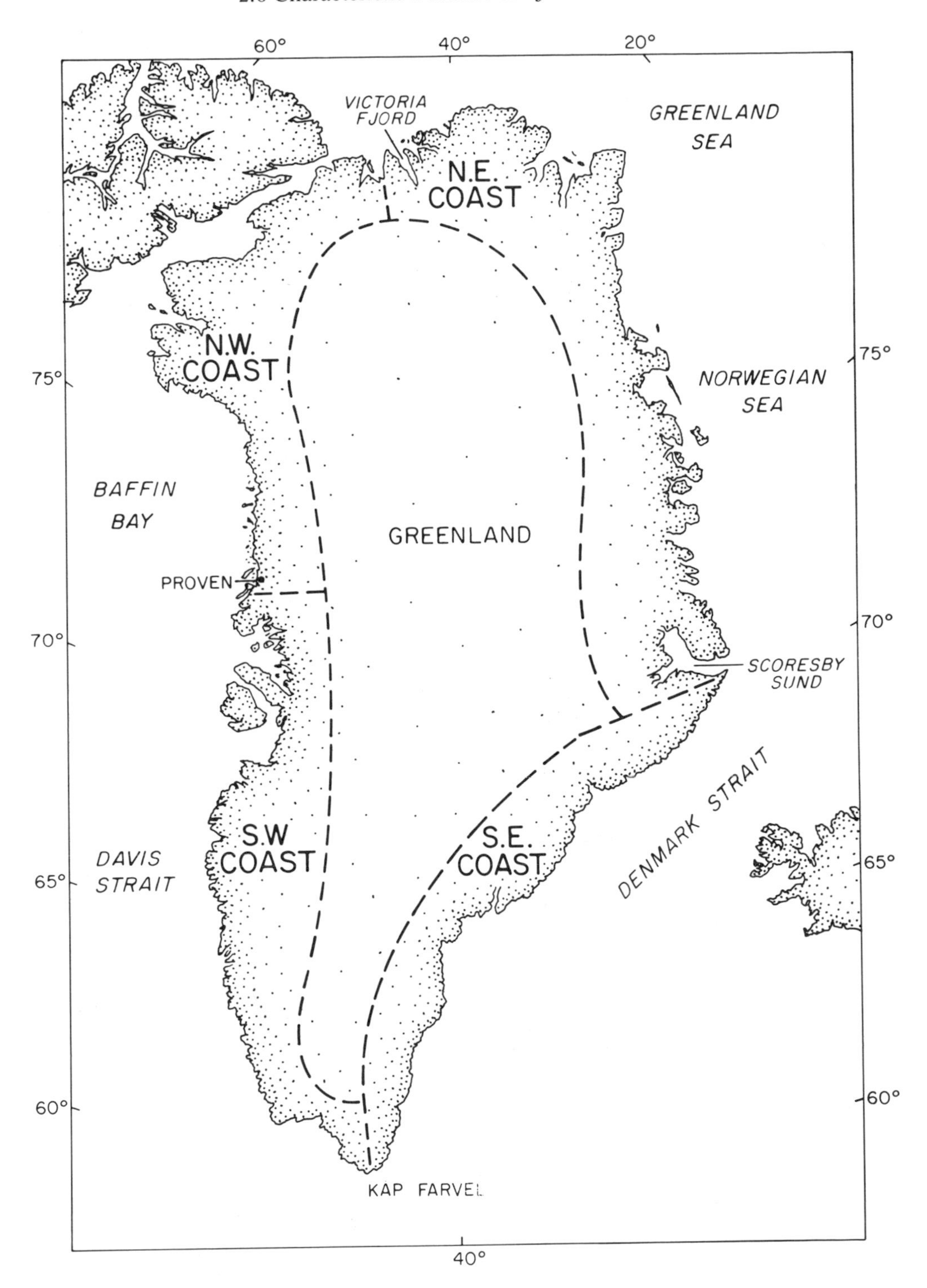

FIGURE 2.20. The coast of Greenland as divided into four quadrants.

The SW coast, from Kap Farvel to Proven (Fig. 2.20) is the best-developed fjord coast of Greenland, largely as a result of the most extensive pull back of the Greenland ice cap. Although most of 165 fjords along this coast remain part of the sediment/water transport pathway from the ice cap to the sea, a number of coastal fjords are completely glacier-free. The sidewall mountains rise to 2700 m. The coast is dominated by fjord complexes (Fig. 2.21A), such as

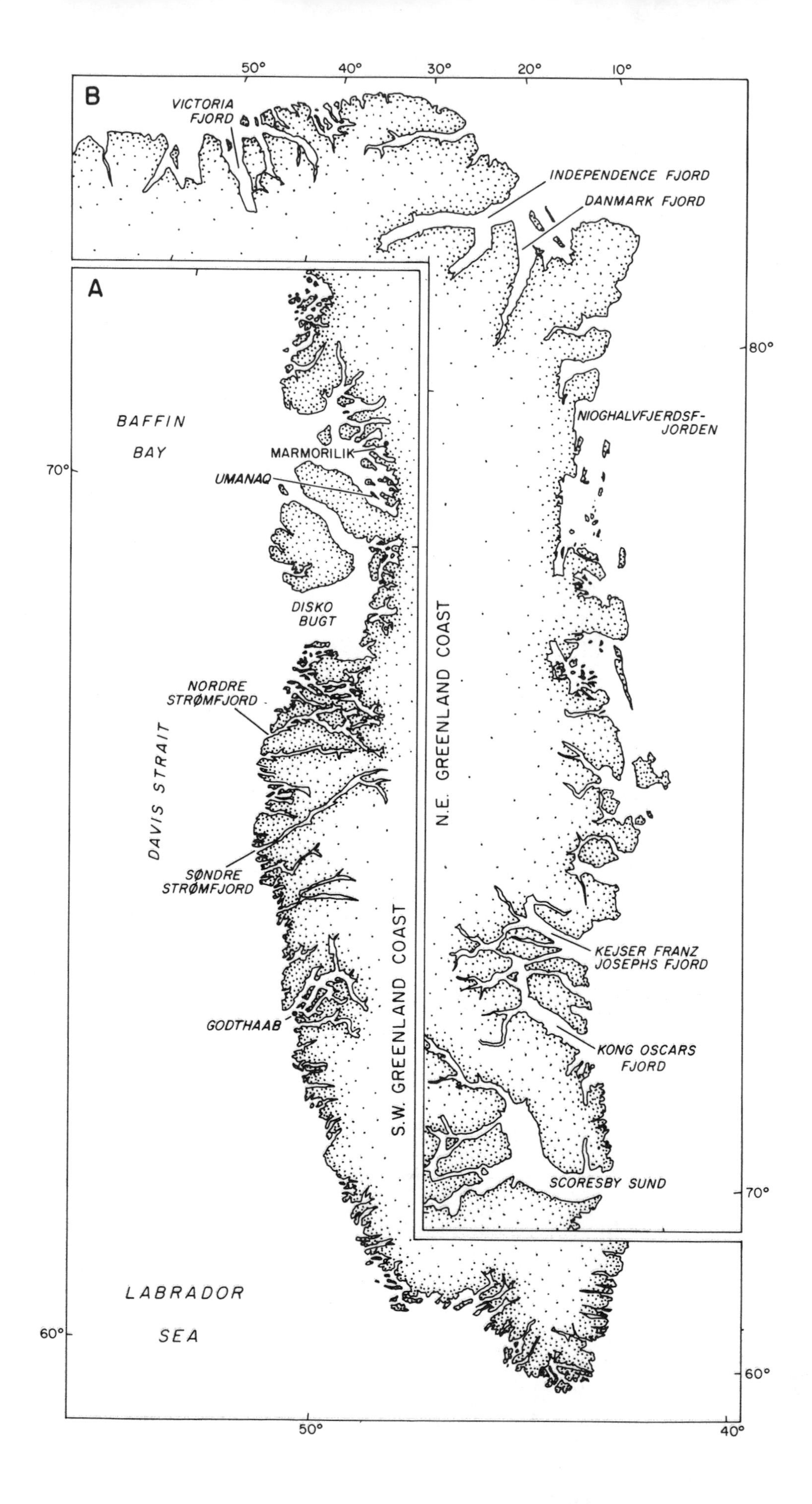

B
VICTORIA FJORD
INDEPENDENCE FJORD
DANMARK FJORD
NIOGHALVFJERDSF-JORDEN
A
BAFFIN BAY
MARMORILIK
UMANAQ
DISKO BUGT
NORDRE STRØMFJORD
DAVIS STRAIT
SØNDRE STRØMFJORD
GODTHAAB
S.W. GREENLAND COAST
N.E. GREENLAND COAST
KEJSER FRANZ JOSEPHS FJORD
KONG OSCARS FJORD
SCORESBY SUND
LABRADOR SEA
50°
40°
30°
20°
10°
80°
70°
60°

the Godthaab fjord complex, the Nordre Strømfjord complex, the Disko Bugt fjord complex, and the Umanaq fjord complex. Søndre Strømfjord is one of the straightest fjords in the world, being near straight for 220 km (Fig. 2.21A). Søndre Strømfjord is also typical of many SW-coast fjords in that it narrows toward the sea (from 10 km wide near the head to 3 km wide at its mouth). These SW-coast fjords often have prominent sills, many identifiable by the stranded icebergs at their mouths. Glaciers in these fjords are some of the most iceberg productive on earth. Investigations in Nordfjord, Mellourfjord, and Disko Bugt, all three outlets of large valley glaciers, have noted the presence of moraines and kame terraces along their mountain slopes (Donner, 1977). Kame terraces can be traced for tens of kilometers along the fjords. Frontal deposits, however, are rare. The steep slopes along the fjords have a cover of till and talus cones undergoing solifluction (Donner, op. cit.). The latter is a result of postglacial frost action. The climate of the SW coast is similar to the SE coast.

The NW coast, from Proven to Victoria Fjord (Fig. 2.20), has fewer fjords (25), largely a result of the ice cap extension to the coast, thus burying potential fjords under ice.

There have been limited scientific studies of the fjords of Greenland. Hydrographic surveys made prior to 1950 are summarized by Dunbar (1951). Typically, the fjord waters are very cold, with temperatures below 0°C at subsurface depths. During the summer, the surface salinities are low near the fjord head owing to the influx of meltwater. The presence of actively calving glaciers is an important parameter controlling the circulation of these fjords (Dunbar, 1951). Although the shallow sills of the SW-coast fjords inhibit free exchange of fjord waters with the ocean, waters with low levels of dissolved oxygen have not been observed.

Surface circulation during the summer is predominantly two-layer estuarine flow. The outflowing surface layer may be 3 to 4 m thick in Godthaab Fjord (Dunbar, op. cit.), a result of a large catchment area and considerable meltwater input. Variations in freshwater discharge are mainly a result of air temperature fluctuations (Möller, 1984), as the average meltwater contribution may reach 90%. The runoff is mainly between June and September. During winter, the fjords are mostly ice covered; wind-generated mixing and seiching are cut off (see Section 4.5.1). Circulation in the winter is mostly driven by the rejection of salt during the growth of ice (Möller, 1984).

Glacial flour, generated from the tidewater glaciers, is the principal source of sediment to Greenland fjords (Heling, 1974). Sediment transported by rivers is often trapped in lakes before reaching the fjords. Proximal glaciomarine sedimentation may, however, reach > 100 cm day^{-1} (Heling, 1974). Such high sedimentation rates depend on the activity of a glacier. For instance, in Bulisefjord an inactive glacier is associated with minimal sediment input into the fjord (Heling, 1977). Anthropogenic sediment input is only locally important. Over 450,000 tonnes of mine tailings are dumped annually into Agfardlikavsâ fjord and with considerable effect on the sedimentary environment (Asmund, 1980; Nyholm et al., 1983; Chapter 8).

◁ FIGURE 2.21. (A) Details of SW fjord coast of Greenland. (B) Details of the NE fjord coast of Greenland (cf. Fig. 2.20).

2.6.2 Alaskan Fjords

The fjord coastline of Alaska extends over 1500 km from the vicinity of Cook Inlet, merging into the contiguous fjord province of British Columbia (discussed below). However, only some 5° of latitude separate the northern and southern limits (Fig. 2.22), and there is little difference in the mean ambient climate (northern temperate or subarctic maritime), or in external oceanographic conditions impacting this coast. There are at least 200 major Alaskan fjords, few of which have been studied in any detail. Topography along the entire continental margin is very rugged: Mt. Logan rises to almost 6000 m. Fjords dissecting the coastal range are consequently steep sided and deep, but relatively shallow inlets also occur, especially on the islands. Most—but not all—Alaskan fjords consist of one or more topographic basins separated by sills that may be either bedrock or morainal deposits (Molnia, 1983).

Two primary geographic fjord areas are recognized: south central Alaska, which includes Cook Inlet and Price William Sound, and southeast Alaska, a narrow coastal strip (the pan-

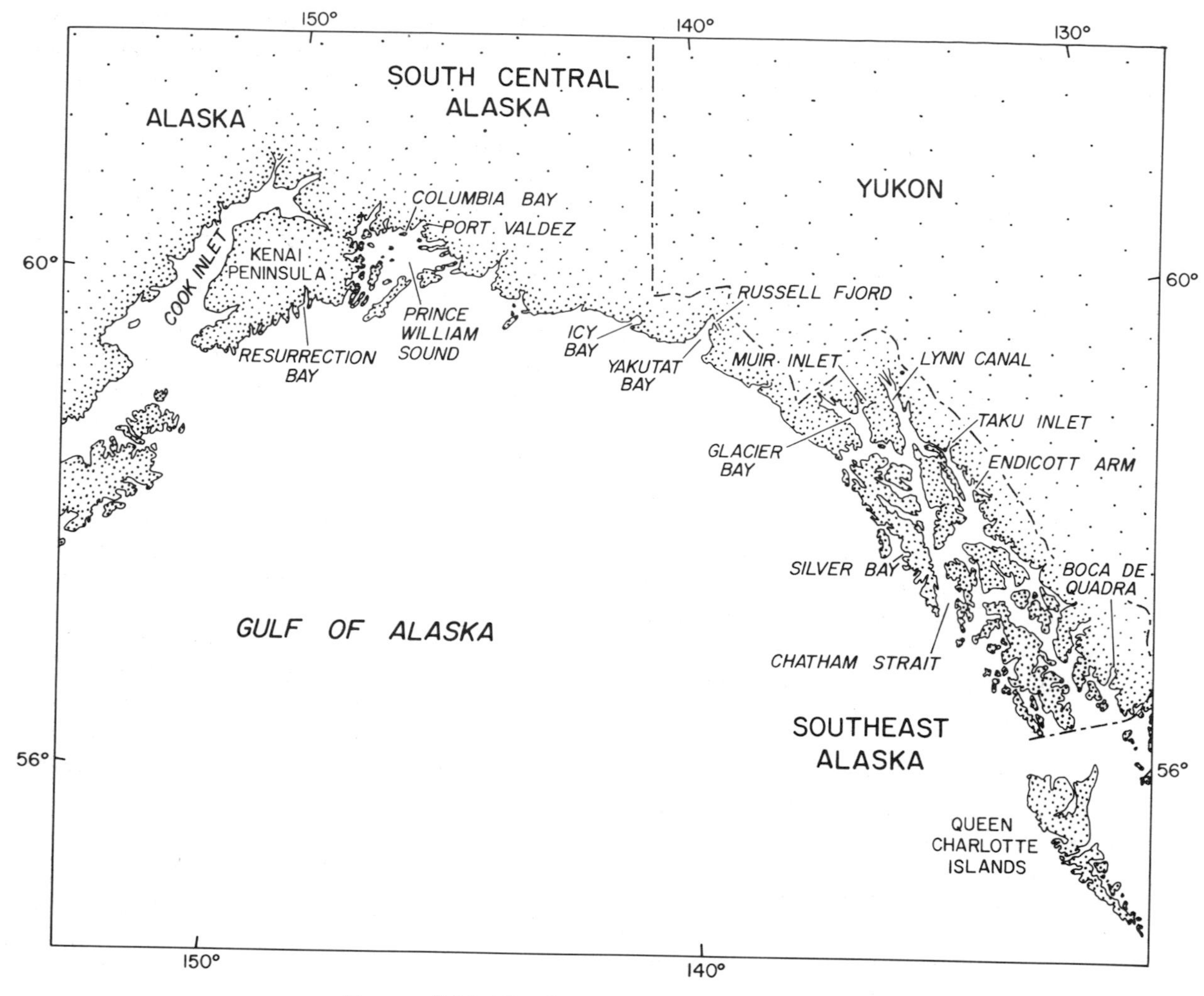

FIGURE 2.22. The fjord coast of Alaska, U.S.A.

handle) lying between part of continental British Columbia and the Gulf of Alaska. The more northern region includes a number of fjords opening directly onto the Gulf along the southern and eastern margins of the Kenai Peninsula. A portion of this region (some 2300 km^2) has been designated the Kenai Fjords National Park, and is preserved in perpetuity from development. Prince William Sound is a large and deep (in excess of 3500 km^2 and 800 m) fjord complex, separated from the shelf by narrow and relatively shallow passages, and containing approximately 25 tributary fjords. Most of these fjords are backed by permanent ice fields, and fed, directly or indirectly, by glacial runoff. Immense piedmont glaciers (Bering, Malaspina, and others) occupy much of the coast between Prince William Sound and the islands of southeast Alaska, and there are only a few fjordlike indentations, notably Icy, Yakutat, and Lituya bays.

Cook Inlet is the largest fjord-estuary in Alaska: 280 km long and covering more than 26,000 km^2. Approximately 75% of the freshwater influx into the inlet enters at the head, and two of the major rivers are seasonally highly turbid with glacial silt. Particulate sediment loads exceeding 2 g L^{-1} have been measured near the head in July (Sharma and Burrell, 1970). Apart from its size, Cook Inlet is not a typical Alaska fjord; it lacks sills and is relatively shallow, and—most notably—in the upper reaches, is a well-mixed-type estuary. Tides are amplified some 2.2 times up-inlet, yielding a mean tidal range at the head of around 10 m, the tenth highest in the world, and tidal currents well in excess of 2 m s^{-1} occur through the narrowest segment.

The mountainous coastal margin of southeast Alaska is bordered by a multitude of islands and coastal passages diastrophically formed, and subsequently shaped by glacial erosion and post-Pleistocene submergence (Sharma, 1979). The measured shoreline is in excess of 50,000 km, some 63% of the Alaskan total. Glacier Bay (National Park) is a spectacular fjord complex (765 km^2), partially ringed by permanent ice fields, and incorporating some ten glacial, subsidiary fjords. Glacial fjords also indent the northern mainland coastal range, but nonglacial fjords are the norm in the south (e.g., the Misty Fjords National Monument preserve).

Because of the mountainous terrain, mean annual precipitation along the entire fjord coast of Alaska and northern British Columbia is around 2.5 m yr^{-1} (Crean, 1967; Royer, 1979) and may locally be very much higher (Sugai and Burrell, 1984b). Much of this is deposited as snow and initially stored. Annual freshwater discharge in southeast Alaska is characteristically bimodal—as in northern British Columbia (Pickard and Stanton, 1980)—representing release of stored precipitation in the spring and maximum direct rainfall in autumn (Burrell, in press). The rugged terrain ensures that the residence time of precipitation in the restricted catchment areas is very short. Burrell (1983a) and Sugai and Burrell (1984b) have noted that, for two southern Alaskan fjords, approximately 90% of the annual precipitation falls in the space of a few weeks in the fall, and that the lag time prior to discharge into the marine fjord is on the order of days or less. In most fjords, precipitation stored as snow generates a secondary discharge peak in the spring. However, in the few cases where large river systems enter the fjord, maximum freshwater influx—from snow melt—is likely to occur in spring-summer. This pattern has been observed in a number of fjords in adjacent northern British Columbia (Crean, 1967; Macdonald et al., 1983).

In south central Alaska there is a preponderance of glacial and snow-field runoff, and a paucity of large rivers. Glacier meltwater discharge peaks in late summer-fall, coincident with maximum direct precipitation. Hence, on the Kenai Peninsula (e.g., Heggie and Burrell, 1977), and for Cook Inlet (Sharma, 1979), the freshwater discharge pattern is characteristically strongly unimodal. The single major river in this area, the Copper River, has maximum discharge in June-July; but its mean annual flow is only around 10% of the total regional flux estimated by Royer (1979, 1983) from precipitation-discharge box models. Royer's line-source hydrology models predict maximum freshwater runoff in October, coincident with maximum sea-level values along this coast. Royer (1979, 1983) has shown that freshwater input is the dominant influence on the circulation of the upper mixed layer of the Gulf of Alaska in this region.

Because of the relatively low mean freshwater discharge and the large regional tidal range, near-surface transport in the few Alaskan fjords studied in detail to date, including Cook Inlet, appears to be tidally dominated (e.g., Muench and Nebert, 1973), and freshwater input is characteristically less than 5% of the tidal prism (Burrell, in press). Tidal mixing at sill constrictions is likely to be an important mechanism promoting turbulent mixing within the body of the fjord, and hence helping to prevent anoxic conditions from developing in the basins. Intense down-fjord katabatic winds ("Taku" winds in southeast Alaska) frequently occur in the winter (e.g., Reeburgh et al., 1976). These may transport surface waters out of the fjord and generate shallow upwelling at the head.

The Aleutian low-pressure system, dominant through the winter, is the major geostrophic force driving circulation within the near-surface waters of the Gulf of Alaska (Xiong and Royer, 1984). Prevailing cyclonic winter winds generate onshore Ekman transport along the Alaskan fjord coast. Burrell (in press) has described influx of "intermediate zone" (above sill) water into a southeast Alaska fjord in fall and early winter. Subsill fjord basin water is replaced when higher density external water is available for transportation in over the sill. During the summer, relaxation of the intense winter downwelling condition along the Gulf of Alaska coast permits run-up of denser water onto the shelf and replacement of resident waters in fjord basins bounded by relatively deep sills (e.g., Resurrection Bay, south central Alaska: Heggie and Burrell, 1981). Conversely, shelf surface water has a minimum density in the summer because of freshwater dilution and increased insolation, and shallow-silled fjords turn over in the winter. The density of the source water available to each fjord basin is therefore largely a function of the barrier sill height (Muench and Heggie, 1978): the region of minimum seasonal density variation is centered at around 150 m. This regular seasonal pattern of shelf water circulation (multiyear patterns of Gulf circulation are poorly known) ensures that the subsill waters of Alaskan fjords are generally flushed at least annually. (Only once has complete overturn not been recorded; but few Alaskan fjords have been monitored for more than a year or two). Anoxic bottom water has not been observed to form in any Alaskan fjord, except in Skan Bay (Reeburgh, 1980).

Glacial sedimentation is of primary importance in the majority of Alaskan fjords. Fjords directly influenced by active glaciers may be subdivided into "tidewater glacier" and "turbid outwash" types (Burrell and Matthews, 1974). In the latter variety, the glacier terminates on land and morainal sediment is seasonally discharged into the stratified marine environment as a turbid surface plume (e.g., Burrell, 1972). Sedimentation rates in the range 1 to 2.5 m yr^{-1} have been estimated (Hoskin and Burrell, 1972; Hoskin et al., 1976) for one tributary Glacier Bay fjord (Queen Inlet) where the glacier front is within 2 km of tidewater. Very high rates of deposition—up to 9 m yr^{-1}—have been measured (Powell, 1983; Mackiewicz et al., 1984) in front of the retreating, tidewater Muir Glacier. In the latter case, subglacial fluvial supply is presumed to be the primary transporting agent. In Queen Inlet, axial submarine channels are evident (Hoskin and Burrell, 1972), which may result from underflows, or from slumping off the steep slope at the head. Mean, long term sedimentation rates in the range 0.2 to 2.0 m yr^{-1} are estimated from bathymetric measurements (Molnia, 1983; von Heune, 1966) for coastal embayments along the northeast Gulf coast fed by both terrestrially grounded and tidewater glaciers.

Sedimentation rates in nonglacial fjords are significantly less than in glacial fjords; this is largely a reflection of the limited catchment areas. Mean sedimentation rates are less than 0.5 cm yr^{-1} in the basins of two adjacent nonglacial fjords located near the southern border of Alaska (Sugai and Burrell, 1984b). Sediment focusing in the deep basin regions from hanging-valley tributaries and episodic slumping from the heads of the fjords have been identified. In this seismically active region, slumping may occasionally be triggered by earthquakes, as identified in Nuka Bay by von Heune (1966).

Alaskan fjords are geologically youthful features, but all the fjord stages listed in Table 2.3 are represented. For example, Columbia Glacier (Prince William Sound) is presently in the process of retreating back from a sill to create a new fjord. Most fjord glaciers in Alaska are currently in retreat, and ice has cleared from Yakutat and Glacier Bays in historical times (Muir Glacier, for example, is receding at a mean rate of 0.35 km yr^{-1}; Powell, 1983). The fjords of the Kenai Peninsula, within Prince William Sound, and along the Yakutat coast generally appear to be in advanced states of infill (von Heune et al., 1967; Molnia, 1983). However, at presently measured rates of deposition, the nonglacial fjords of the southern Alaska panhandle region will maintain their present identity for a long time.

2.6.3 Canadian Fjords

Canada has the longest coastline in the world, a result of the large number of fjords and islands along its coast. For instance, the fjord coastline of the Queen Elizabeth Islands is longer than that of Norway; the fjord coasts of Newfoundland, British Columbia, and Baffin Island are each even longer. Canada has more fjords than all other fjord countries combined, the next largest being Greenland (Denmark). A result of this geographic immensity is that it contains the full range of fjord types, from temperate drainage basins free of snow and ice to frigid, permanent ice-covered fjords, from tide- or wave-dominated fjords to river-dominated fjords, from permanently anoxic fjords to continuously oxygenated systems, and from low to high sedimentation rate fjords.

The west coast, British Columbia fjords (Fig. 2.23), can be divided into two districts: the 75 mainland fjords and the 75 island fjords indenting the coasts of Vancouver Island, Pitt Island, Princess Royal Island, and the Queen Charlotte Islands. In general, precipitation falls over a 160- to 300-day period and amounts to 2 to 3 m yr^{-1}.

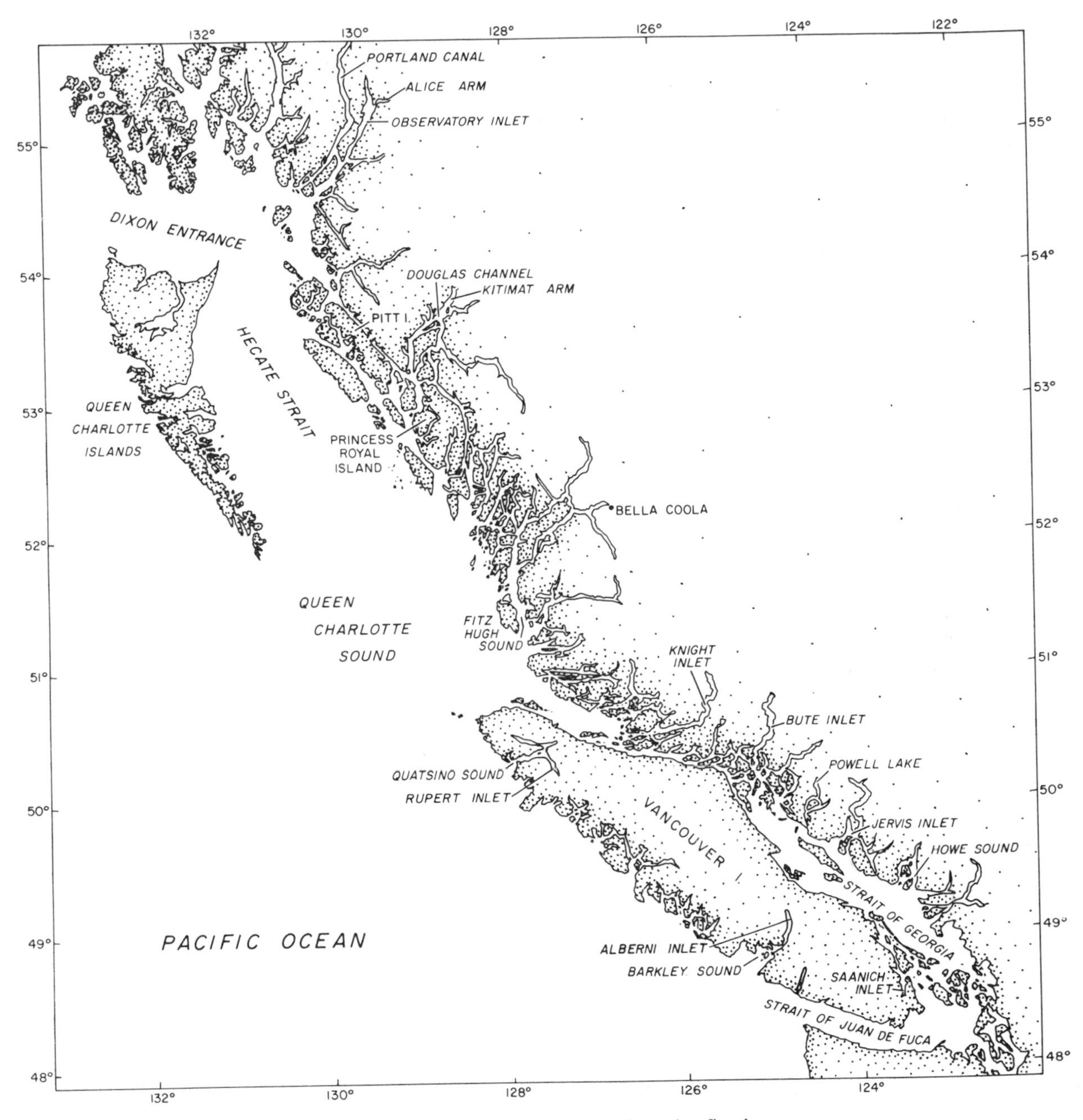

FIGURE 2.23. The British Columbia, Canada, fjord coast.

Except in the high-altitude ice and snow fields, there is no permafrost. Thus baseline (groundwater) discharge is high compared with arctic fjords.

The island fjords have a temperate maritime climate. Their inflowing rivers receive little or no snowfall. With the absence of snow or ice fields, precipitation storage is limited to lakes and freshwater input is directly related to storm events (Fig. 3.1B). Similar to the above Alaskan fjords, two-layer circulation is best developed during the rainy season (late fall-winter-spring) when fluvial discharge is highest. During the drier summer season, circulation within these island fjords is largely driven by the wind (Farmer and Osborne, 1976). Many of the island fjords occur as tributaries to a larger complex, and most fjords are usually short (< 40 km) and narrow (< 2 km). They have limited detrital sediment input (small drainage basins, no ice fields), relatively high terrestrial organic input (lush rainforest vegetation) and high marine organic production. Many of the industrially impacted fjords such as Alberni, Rupert, Cowichan, and Neroutsos Inlet have been studied in detail. Some, such as Saanich Inlet and Lake Nitinat, have been used as natural laboratories to study biogeochemical reactions under anoxic conditions (e.g., Emerson et al., 1979). Most of the island fjords, however—especially those of the Queen Charlotte Islands—remain to be investigated.

The mainland British Columbia fjords are larger than their island counterparts, receive large snowfalls, and, with some exceptions such as Sechelt Inlet, have medium to large ice fields in their drainage basins. Their generalized characteristics are: (1) high input rates of detrital-sediment, much of which is fluvially transported glacial flour; (2) high discharge (Fig. 3.1D) during the drier summer months (a result of snow and ice melt), although a minor discharge mode occurs during the fall rainy season (before the advent of snow); (3) "fjord-estuarine" circulation is developed best during the summer months, but breaks down during the winter (a result of low fluvial discharge and energetic mixing of the surface waters by strong winds); (4) strong winds and high tides control much of the circulation in the outer fjord, even during the summer; and (5) the basin waters seldom go anoxic. Exception to the latter feature include Howe Sound (intermittently anoxic) and Powell Lake (an anoxic meromictic lake). Most fjords are narrow, < 3 km, but longer than their island counterparts. Six are more than 100 km in length. The fjord topography consists of lower altitude forested mountains at the fjord mouth (200 to 500 m), to higher confining mountains along the fjord length (1000–2000 m), to hinterland mountains that may reach 4000 m (Mt. Waddington). The fjord-head drainage basin(s) account for most of the freshwater and sediment delivery (70–95% of a fjord's total), with minimal side-entry sediment delivery. Many of the mainland fjords are not deep (average maximum depth is 385 m), although Jervis Inlet has a maximum depth of 730 m. This may be a result of the fjords having thick sediment infill sequences (400–800 m sediment thickness for the larger fjords; Table 2.2). Knight and Bute Inlets may compete for having the world's largest input of fluvial sediment to a fjord ($< 10^{10}$ kg yr^{-1}; Syvitski and Farrow, 1983). Subaqueous slope failures are common and are a consequence of these high sedimentation rates and moderately intense seismic activity (see Chapter 5). The larger mainland fjords have received intense scientific study (possibly second only to some Norwegian fjords). For general background information on B.C. fjord oceanography, readers should refer to Pickard and Stanton (1980) and Thomson (1981). Like some of their island fjord counterparts, many mainland fjords are presently affected by anthropogenic waste disposal, pulp and paper and mining being the largest industries on the coast.

The fjord district of the east coast of Canada can be divided into four zones: the approximately 100 fjords of Newfoundland, the 35 Nova Scotian fjords, the 5 Quebec fjords, and the approximately 60 Labrador fjords. In general, there is little known on most of these fjords. The Newfoundland fjords (Fig. 2.24) have highly variable morphology, although most are less than 30 km long and < 2 km wide. Their basin depths range from 50 to 500 m. Sill depths are also variable from subaerial in the case of Holyrood Pond to absent in many other cases, such as, Halls Bay (Slatt, 1975). The hinterland topography may range from 200 to 800 m. Drainage basins are small (usually less than 1000 km^2) and fluvial sediment input is minimal. Precipitation may range upward of 2 m yr^{-1} and can occur over 200 days of the year. The highlands of the NW (i.e., Long Range Mountains) receive 60% of their precipitation as snow; freshet discharges and floods are locally important. At the other extreme, the Avalon Peninsula receives little snowfall. There are no permanent snow or ice fields in Newfoundland. Circulation within the inlets is strongly influenced by shelf storms; 10-m waves may locally come in contact with the shoreline during winter storms. With negligible modern (Holocene) sediment input, much of the sediment distribution within these fjords reflects wave reworking of glacial (Pleistocene) sediments (Slatt, 1974). Organic matter, locally as high as 30%, is an important component to any Holocene infill. Where sills are absent or buried, shelf sediment may be an important component of Holocene sediment. Locally, icebergs from Greenland and ice floes from arctic Canada may ground or concentrate in these fjords.

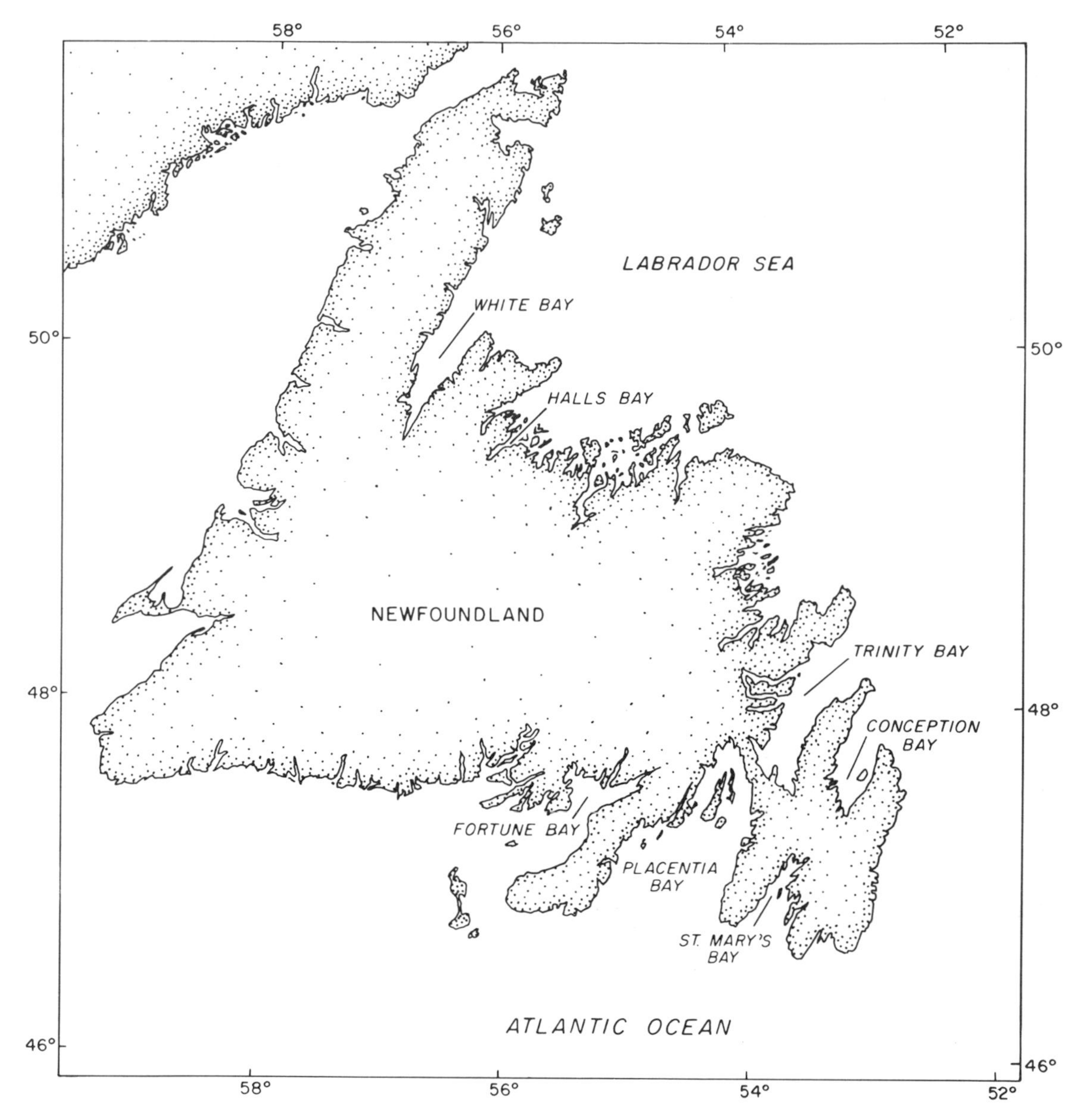

FIGURE 2.24. The Newfoundland, Canada, fjord coast.

Nova Scotian inlets are best described as fiards, that is, similar to fjords in glacial origin except that they occur along lower altitude coastlines. Nova Scotian fiards seldom exceed 100 m in basin depth and many have sills. Local topography is < 300 m. Like Newfoundland fjords, they are wave dominated with fluvial sediment contribution accounting for < 25% of their infill (Piper et al., 1983).

There are few remaining Quebec fjords (the Saguenay River, Baie de Gaspé, Lac Sale; Gros Mecatina River, Baie de Jacques-Cartier), as a result of high isostatic rebound (120–200 m) and associated rapid fjord infill. The northern Gulf of St. Lawrence coast is one of the longest infilled paleofjord coasts (the other being in New Zealand). The rapid infill is a result of large drainage basins (10^4 to 10^5 km^2) that have changed little since the time of deglaciation. The Saguenay Fjord (Fig. 2.6) is a classical fjord in every sense: deep basin (275 m), shallow sill (20 m), long (120 km) versus wide (3 km), major fjord-head fluvial input (1600 m^3 s^{-1}), and two-layer estuarine circulation (Côté and La Croix, 1978; Sundby and Loring, 1978). Modern sedimentation rates decrease exponentially with distance from the Saguenay river mouth, from values in excess of 10 cm yr^{-1} to less than 0.1 cm yr^{-1} (Smith and Schafer, 1985). The St. Lawrence Estuary-Laurentian channel complex has been considered by many to be a fjord. If it is, it is the world's largest (from head to sill its length is 1200 km, average width of 50 km, maximum depth of 550 m, with one of the world's largest rivers draining into its head). The size of the St. Lawrence Estuary, however, leads to circulation complexities, such as cross-fjord currents, that are normally not important in fjords (El-Sabh, 1979).

The Labrador coast is dissected by numerous fjords and bays, often with many islands lying offshore (Fig. 2.25). The fjords of Labrador have diverse morphological features, some with virtually no sills and some with shallow sills or restricted channels (Nutt, 1963). Most of these fjords have small drainage basins; others have extensive drainage from the interior (the Churchill River that debouches into Lake Melville and Hamilton Inlet drains an area of 92,500 km^2). To the south, the fjords are usually shallow (< 200 m) and confining mountains are less than 700 m. To the north, the Torngat fjords are deeper, > 200 m, and the 1600 m high mountains may contain small alpine glaciers. At St. Lewis Inlet, in the south, the drainage basin contains no permafrost, but the fjord is still considered subarctic as sea ice forms in the winter. Farther north, the fjords are within the zone of semi-continuous permafrost, and still farther north the Torngat fjords are within the zone of continuous permafrost and north of the treeline. As a result, the moderating effect of ground-water storage and flow is only important in the south. The southern fjords receive more precipitation (1.2 m yr^{-1}) much of which falls as snow, compared with the northern fjords (0.7 m yr^{-1}), much of which falls as rain. The prevailing northwesterly airflow from the continental interior in winter can cause air temperatures as low as $-30°C$; the coastal waters are frozen over for five to seven months of the year, often with a meter or more of ice being formed (Nutt, 1963). With the arrival of spring (late May–early June), fresh water is added to the fjords from freshet runoff and the melting of sea ice; two-layer estuarine circulation is well developed for a 30- to 60-day period. Basin waters are very cold (see Table 2.1) and may reach $-1.76°C$ in the case of Hebron Fiord (Nutt, 1963). Basin salinities vary from 28‰ for Lake Melville to 33‰ for Hebron Fiord. The tidal range along the coast is < 2 m. Only Tessiarsuk, a coastal meromictic lake in northern Labrador is known to have anoxic waters (McLaren, 1967). Most of the Labrador fjords remain pristine with low rates of sedimentation, partly derived from wave erosion (Piper et al., 1983). One exception is Lake Melville that receives an annual fluvial input of 2500 $m^3\ s^{-1}$ (Fig. 3.1C). The inflowing Churchill River, although presently regulated for hydroelectric power, remains the largest contributor of sediment infill in Lake Melville (Vilks and Mudie, 1983).

Over 50% of Canada's fjords are located in its arctic archipelago, being located between 62°N and 84°N (Fig. 2.26). Hinterland topography ranges from 400 m for the western arctic islands to over 2000 m for the eastern islands of Baffin and Ellesmere. In general, these archipelago fjords have a smaller length to width ratio than most other of the world's fjords (Lake and Walker, 1975). Their mean maximum depth is 400 m, although some may exceed 1000 m deep. The largest system is the Nansen Sound–Greely Fiord system (450 km long, 25 km wide, maximum depth of 1052 m). As a consequence of isostatic rebound of land, the fjords have experienced over 200 m of relative sea-level fall in the south central islands, to less than 60 m along the outer margins of the archipelago. Some sills have therefore become subaerially exposed (e.g., Tromso and Ekalugad fjords), resulting in the development of anoxic fjords (e.g., Ogac Lake). A number of fjords have no sills (Syvitski and Schafer, 1985).

At the northern extreme of Ellesmere Island, the fjords are "frigid," being ice covered year round and enclosed from the sea by the Ward Hunt Ice Shelf (Fig. 4.22). These frigid fjords have very low tides (< 0.5 m range), experience no wind-generated mixing, and as a result of the ice shelf have developed a thick (up to 45 m) layer of fresh water above the marine water. Tidal range increases to the southeast and reaches a maximum of 12 m in Frobisher Bay (McLaren, 1967). Likewise, precipitation increases from desert conditions (< 0.2 m yr^{-1}) in the high arctic to more maritime conditions (> 0.6 m yr^{-1}) in the southeast. With the southward increase in mean annual temperature, most

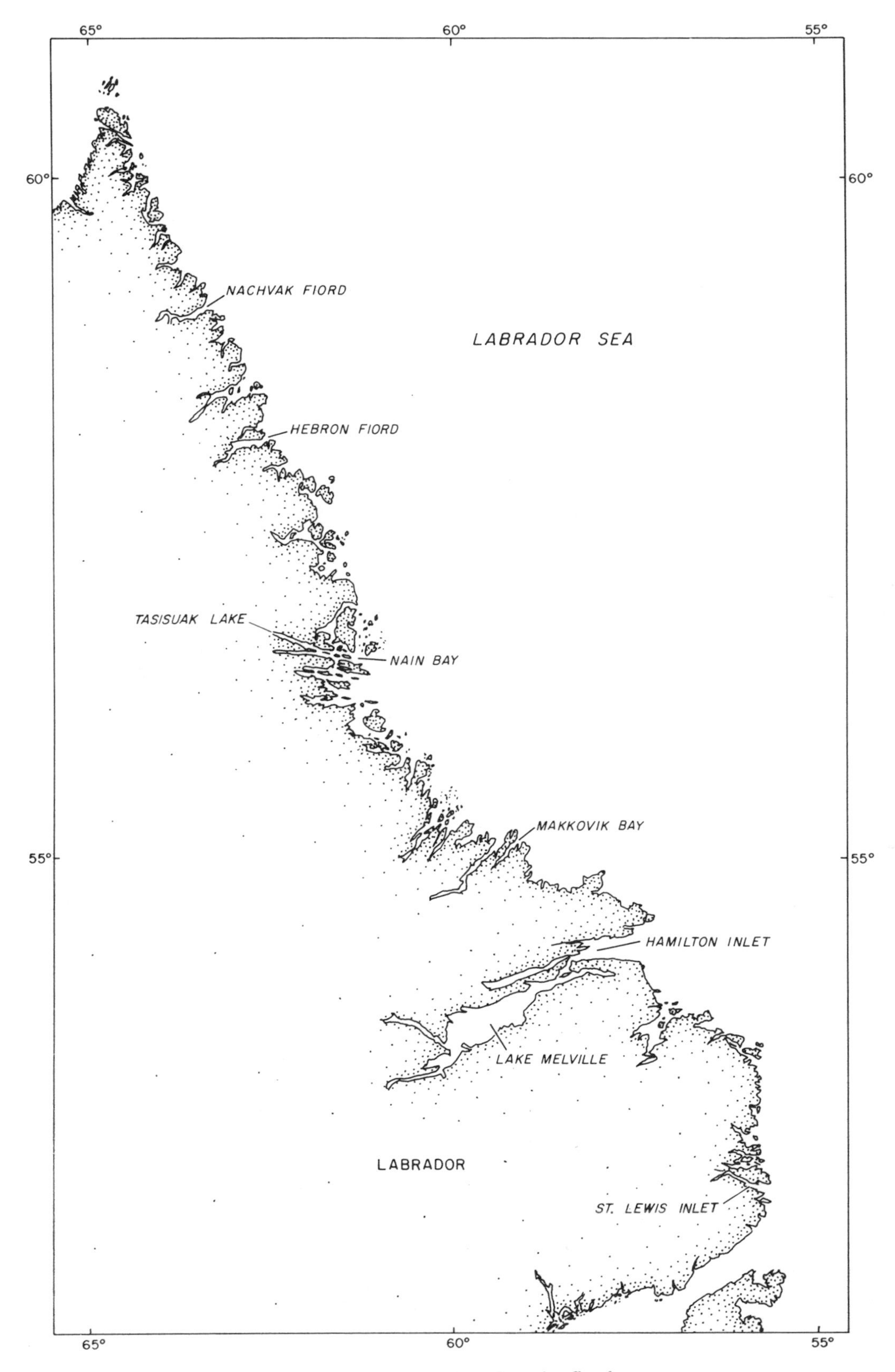

FIGURE 2.25. The Labrador, Canada, fjord coast.

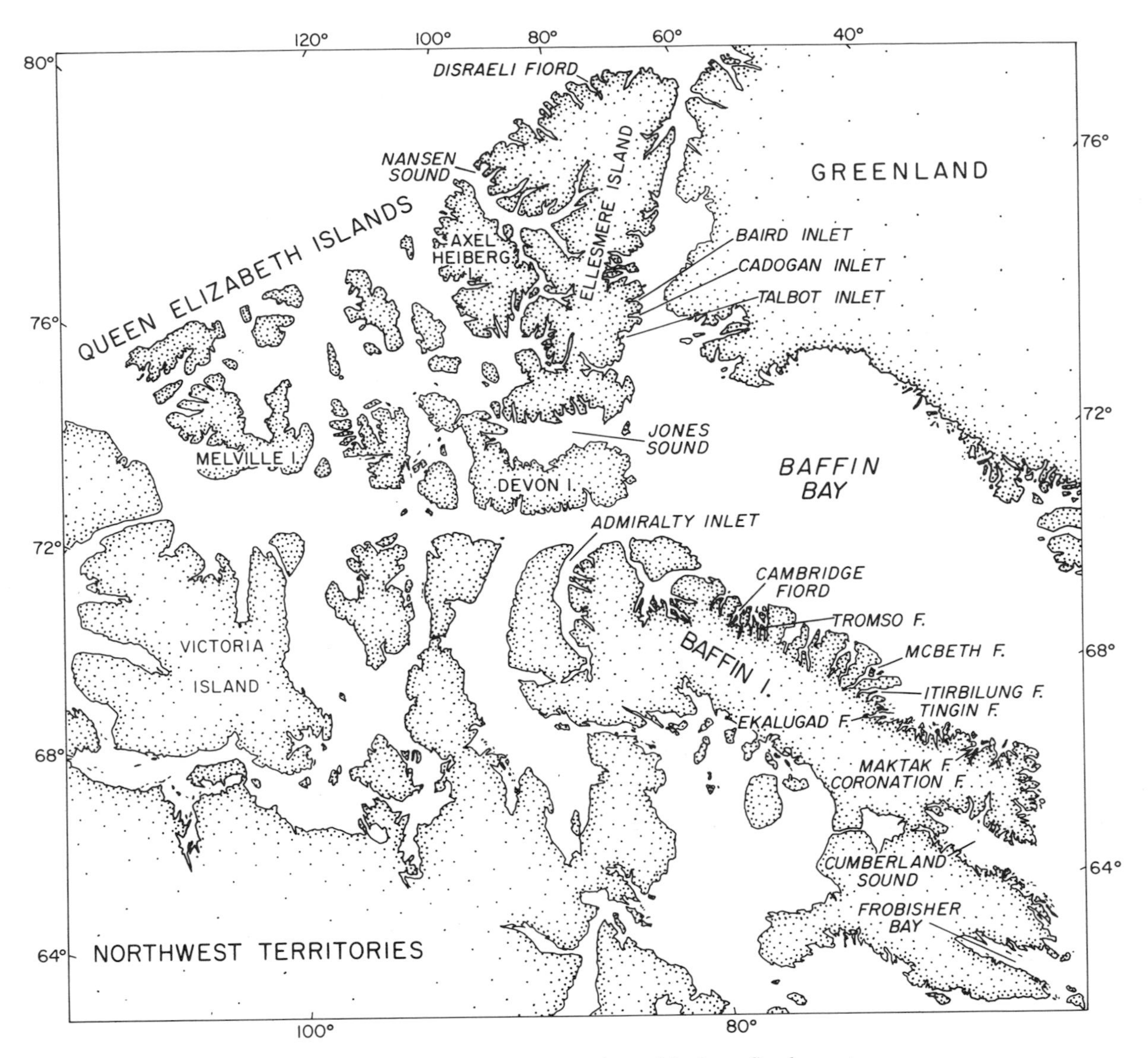

FIGURE 2.26. The Canadian arctic archipelago fjord coasts.

arctic fjords do undergo seasonal freeze and thaw of sea ice: from a typical 15-day open season for high arctic fjords to more than 80 days for the Frobisher Bay fjords. However, the influx of shelf pack ice may extend the presence of sea ice year-round for some fjords. The eastern islands of Ellesmere, Devon, and Baffin have fjords with tidewater glaciers; some are still ice filled (e.g., Codogan, Talbot, and Baird Inlets, Ellesmere Island). Some of the western islands, such as Axel Heiberg Island, have large hinterland ice fields that influence the fjords through their turbid meltwater streams. Other western island fjords, such as those found on Bathurst, Victoria, and Prince of Wales islands, contain no hinterland glaciers.

Sediment thickness in these archipelago fjords rarely exceeds 300 m (Table 2.2), much of which is related to glacial events before the advent of the Holocene (10,000 yr BP). Sediment dynamics are best known for the Baffin Island fjords (see Syvitski and Blakeney, 1983; Syvitski, 1984a). These fjords are dominated by subaqueous slope failures and sediment gravity flow deposits, and, like the British Columbia and Alaskan fjords, are within a high-magnitude earthquake zone. The high-arctic fjords of Ellesmere and Axel Heiberg do not experience the same earthquake intensity, and it is expected that less energetic hemipelagic sedimentation takes place. As a result of the cold climate, small drainage basins, and permafrost conditions, the discharge curves are spikey (Fig. 3.1A). Tundra vegetation does little to limit the erosion of sediments within the hinterland, especially in the more alpine terrain of the eastern islands. Sedimentation occurs predominantly during the short discharge season. Estuarine circulation lasts for less than 25% of an annual cycle.

Population in the Canadian Arctic is sparse, and anthropogenic influences are so far minimal. The biggest concerns at present are those associated with the development of oil and gas reserves under polar conditions.

2.6.4 Norwegian Fjords

The Norwegian coastline is the most fjord dissected of any country in the world (Gregory, 1913), and appropriately the term fjord is of Norwegian origin. The type examples were originally described for this region, and Norwegian fjords have been more extensively studied. There are some 200 principal fjords along the mainland coast (Fig. 2.27; from 58°N to 71°N) and another 35 along the Svalbard islands (Fig. 2.28A; from 76.5°N to 80.5°N). The total fjord coastline is 21,000 km long. Although the country is within the same latitudinal range as Greenland or the Canadian archipelago, the Gulf Stream conveys warm water from the Gulf of Mexico along the entire length of the Norwegian coast. As a result, the climate is not unlike the fjord coast of western North America. Environmental concerns are more important with respect to Norwegian fjords than for any other fjord country, as most of the Norwegian population is situated at the coast.

Along the southern coast, the fjords are short, relatively shallow, and with shallow sills, sometimes causing stagnant bottom water, intermittently or permanently (Strøm, 1936). Minimal tides (< 0.5-m amplitude) also contribute to basin-water stagnancy. Along the southeast Skagerrak coast, the embayments are small and irregular.

The west coast fjords are long, narrow, steep sided, deep, and often have relatively deep sills (Holtedahl, 1967). Sognefjord (200 km long and 1300 m deep) and Hardangerfjord (180 km long and 900 m deep) are among the world's largest and deepest fjords. Sognefjord and its tributary valleys form a dendritic pattern. It has several basins with sediment thickness up to 300 m (Aarseth, 1980). The basins are extremely flat with gradients $< 1 \text{ m km}^{-1}$. Distribution of glaciofluvial terraces along the fjord reflects several glacial intervals. Most of the basins are within a short distance of the terminus of paleomeltwater channels (Aarseth, 1980). The direction of Hardangerfjord coincides with the bedrock (Caledonian) structure (NE–SW). In the outer fjord complex, tributary fjords form hanging valleys to the main fjord (Holtedahl, 1965). Several rocky sills separate the fjord into basins. According to Holtedahl (1965), Hardangerfjord is overdeepened by 700 m through glacial erosion. The fjord has characteristically steep side walls (up to 40°–50°) and a flat bottom, with a sediment cover of 140 m in the main fjord (Cone et al., 1963) and up to 240 m in one of its tributaries (Klosterfjorden). The landmass surrounding Hardangerfjord is 1600 m high in the central, becoming lower near the fjord mouth and forming part of the Norwegian strandflat (see Plate 2.7).

In the north, the mainland fjords are somewhat shorter, wider, and often without pronounced sills. Larger tidal amplitudes provide efficient flushing of the fjord basins, and stagnant bottom water is rarely recorded. Surrounding topography may reach 2000 m. The Spitsbergen (Svalbard) fjords are still wider (most > 5 km wide) and somewhat longer (five exceed 100 km). Most are shallower than 300 m.

Generally, the narrowest, most steep-sided fjords occur in areas with hard, jointed rock while the softer bedrock terrain disposes for broader and less steep basins (Holtedahl, 1960). Additionally, the extent of glacial erosion as a consequence of the thickness of the ice cover and basal shear stress, and the distance from the center of the ice cap, further influences the fjord geomorphology. The fjord-valley direction was formed in the Tertiary and influenced by the oblique Tertiary uplift.

Norwegian mainland fjords only locally receive meltwater from glaciers, hence fluvial discharge is mainly governed by snow melting in early spring and summer and an autumn peak attributable to high rainfall. Annual precipitation is variable along the coast ($< 1 \text{ m yr}^{-1}$ to > 2 m

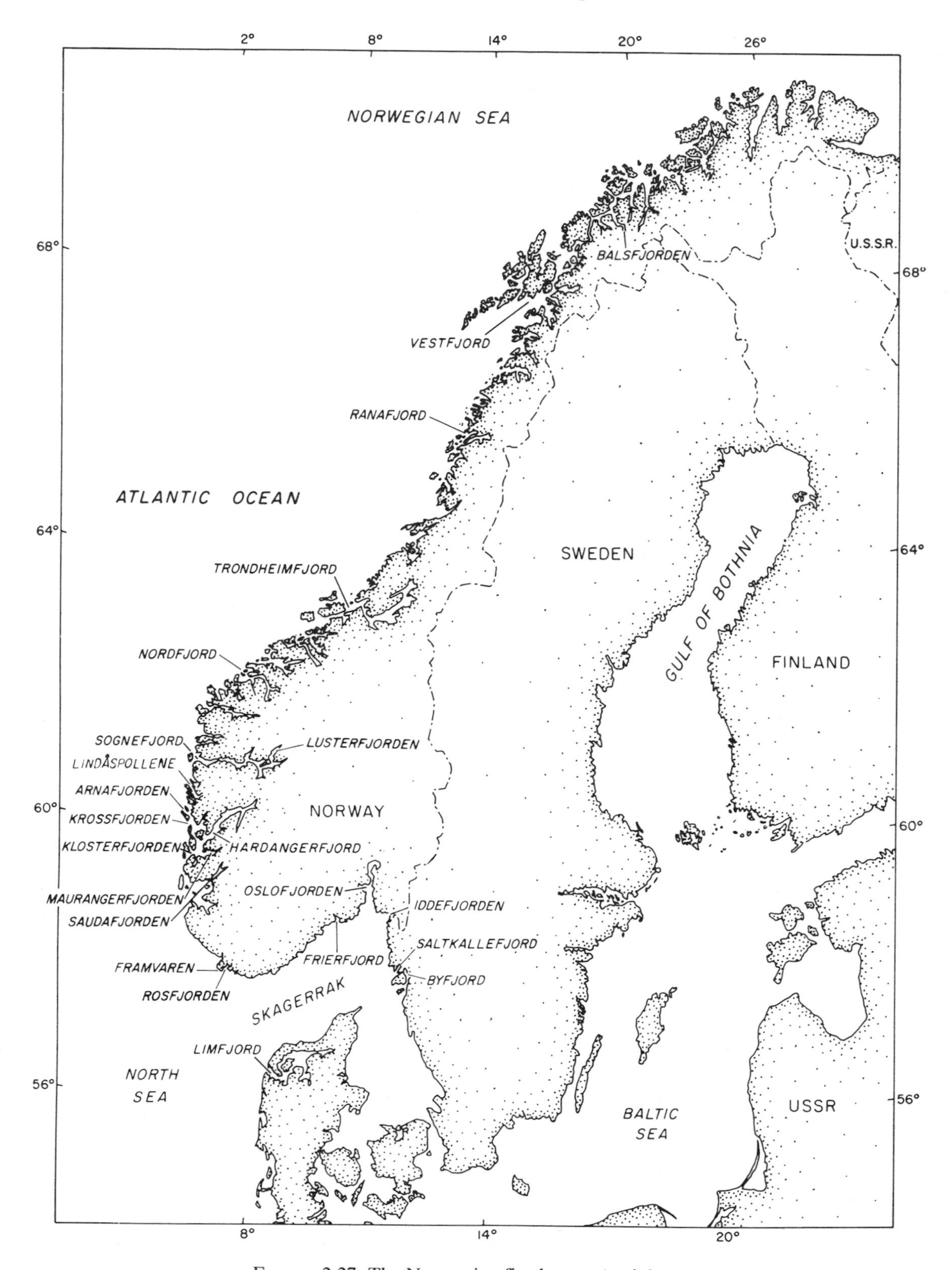

FIGURE 2.27. The Norwegian fjord coast (mainland).

yr^{-1}), being greatest along the SW coast. The fjord-head drainage basins are usually small $< 2000\ km^2$, and rivers are filtered by numerous lakes. As a result, most Norwegian fjords have low sedimentation rates (1 to 7 mm yr^{-1}: Skei, unpublished data), the exception being those that drain ice fields. Gaupnefjord, a branch of Sognefjord, receives meltwater from the Jostedal Glacier and sediment transport is high compared with other Norwegian fjords, although its peak discharge of 110 $m^3\ s^{-1}$ is more typical (Bogen, 1983). Sedimentation rates in the glacier-dominated fjords of Spitsbergen are considerably higher than mainland fjords (Table 4.2; Elverhöi et al., 1983), even though Spitsbergen has near-desert conditions. The fjords of Spitsbergen are influenced by both sea ice and iceberg rafting of sediment, and some of the best examples of surging tidewater glaciers are found in this area.

The regulation of river flow for hydroelectric power greatly influences the more normal seasonal effects of freshwater and sediment discharge into many Norwegian fjords (Tollan, 1972). Under normal conditions discharge occurs in pulses. With regulation, discharge is increased in the normally low winter period and decreased during the summer. Consequently, a much weaker, yet more constant, fjord circulation is set up.

The circulation regimes of several fjords in Norway have been studied, of which Hardangerfjord, Nordfjord, and Oslofjord are well known (Saelen, 1967, 1976; Gade, 1970). The upper water circulation is conditioned by the freshwater supply and best developed during spring-summer and periods of rain. During winter, the estuarine circulation is practically nonexistent in many fjords (Saelen, 1976). Entrainment and upward turbulent diffusion of salt lead to increased salinity of the surface layer seaward. The tide is predominantly a standing wave and contributes little to the salt balance in Norwegian fjords (Saelen, 1976). Deep water circulation is to a large extent governed by the conditions outside the fjord sill. Coastal shelf water lies as a wedge on top of the Atlantic water. Oscillations of this wedge may cause inshore upwelling of the dense Atlantic water, which may occasionally spill over the sill into the fjord basins. The frequency of these deep water renewals is variable. The onshore-offshore wind may provide the final impetus (Gade, 1970). In Nordfjord, basin renewals occur at 9-year intervals (Saelen, 1967; Fig. 4.19B). Basin salinities from Norwegian fjords are close to 35‰, hence somewhat higher than for Canadian or Alaskan fjords (Saelen, 1976). Exceptions are those fjords with shallow sills. Framvaren, with a sill depth of 2 m and a basin depth of 180 m, shows a bottom salinity of only 23‰ (Skei, 1983c).

2.6.5 New Zealand Fjords

The southwestern coastline of New Zealand's South Island is incised by approximately 30 fjords (Fig. 2.28). The fjords range between 15 and 45 km long and are always < 2 km wide (Skerman, 1964). Confining topography ranges between 1000 and 2000 m, and basin depths range between 250 and 420 m. The fjord walls are steep, being glaciated through successive periods. Glaciers are presently restricted to the hinterland of the northernmost fjords, and, in particular, Milford Sound. Shallow sills at the fjord mouth restrict the free circulation of water (Glasby, 1978b).

Some of the fjords are deepest towards their mouths (e.g., Nancy and Caswell Sounds), others are deepest towards their heads (e.g., Milford Sound, Fig. 2.1A). Drowned glacial valleys extend across the narrow continental shelf out from Milford and Dusky Sounds (Brodie, 1964). Several basins appear overdeepened at the junction of tributary fjords. Deepening was accompanied by widening so that each fjord is widest at the major basin (Irwin, 1978). Even glacially eroded Lake Wakatipu, an inland lake, is overdeepened 70 m below sea level.

The New Zealand fjords experience extremely high rainfalls, with annual values of 6 m and extremes of 9 m (Garner, 1964). The snowline is approximately 1000 m, and thus rainfall, combined with the melting of winter snow, results in maximum freshwater inflow in the spring and summer period; the monthly averages do not, however, fluctuate much over the annual cycle (Pickard and Stanton, 1980). The drainage basins are small (most less than 200 km^2), and lush vegetation is often water saturated (Poole, 1951). Runoff is normally rapid, and there is a close correspondence between peak rainfalls and peak river discharge (Stanton and Pickard, 1981).

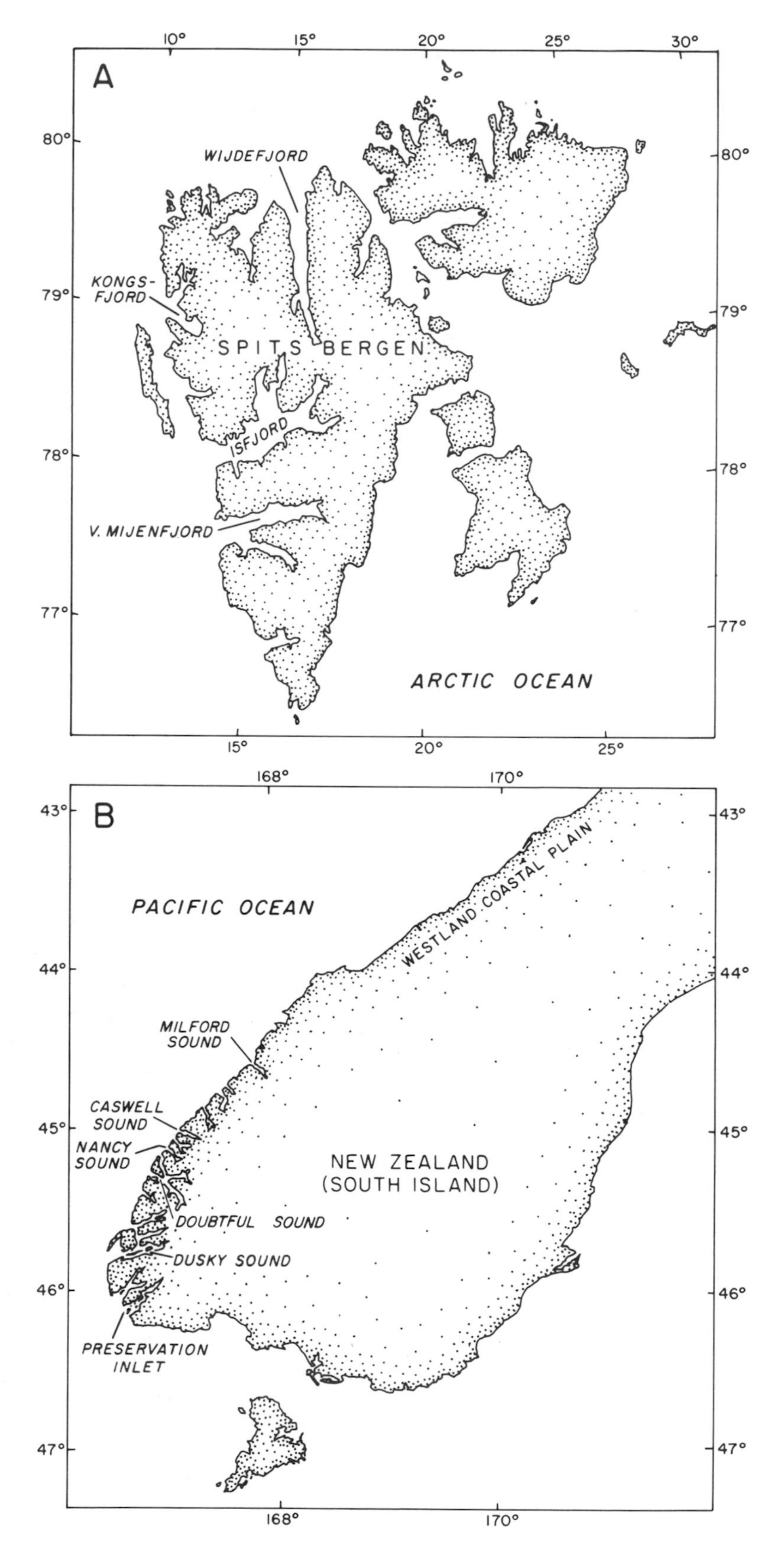

FIGURE 2.28. (A) The fjord coast of the Svalbard islands (Norway). (B) The New Zealand fjord coast.

The surface salinities reflect estuarine conditions with limited freshwater discharge, that is, salinities increase toward the fjord mouth but remain relatively high near the fjord head. The up-fjord compensating current is relatively weak. The tidal range for the fjords is between 1 and 2.5 m. Tidal currents can be high over the sills and may drive deep water renewal events (Stanton, 1984). Although complete anoxic conditions have yet to be observed, they are considered possible for Milford Sound.

North of Milford Sound there are several fjords in various stages of sediment infilling; still further north, fjords have been completely infilled and converted to the Westland coastal plain (Adams, 1980; Pickrill et al., 1981). Lake McKerrow is an example of a tidal meromictic lake similar to those Canadian examples described by McLaren (1967), except that the former does not go completely anoxic. The lake is presently being infilled by turbid water associated with: (1) freshwater overflows and interflows; (2) tide-induced saline underflows; and (3) earthquake-induced turbidity currents (Fig. 2.16B; Pickrill et al., 1981).

As a result of the warmer oceanic climate, the sedimentation rhythm of New Zealand fjords and fjord-lakes differs from those more affected by continental climate. In the latter case, cold winters and hot summers may result in seasonal varves (e.g., Saguenay Fiord: Schafer et al., 1983). New Zealand fjords are subject to more erratic sediment layering, a result of unpredictable storms (Pantin, 1964; Pickrill and Irwin, 1983).

The fjords of New Zealand comprise the Fiordland National Park and have a pristine character. Only in Deep Cove, part of the Doubtful Sound complex, is the effect of man noticeable: water diverted from Lake Manapouri through a power station is nearly an order of magnitude higher volume of water than normal for the local drainage basin (Stanton and Pickard, 1981).

2.6.6 Chilean Fjords

The fjord region of Chile extends from about 41°S to 55°S and is also known as the Patagonia fjord coast (Fig. 2.29). There are more than 200 fjords, of which the mainland coast has 70 fjords over 15 km in length, and the offshore island archipelago has at least another 50. These fjords are normally < 100 km long and < 6 km wide. They have basin depths frequently greater than 150 m and may reach 1050 m deep (Canal Baker; Pickard, 1971). Sills may or may not be present and are of variable depth. The confining topography is between 500 and 1500 m, and hinterland mountains may reach 4100 m. North of 46.5°S, the fjord mountains are covered in trees; the treeline altitude descends towards the south to < 500 m. Major ice fields are present between 46°S and 52°S, although small ice fields are nearly ubiquitous in the high altitudes. The tidal range for most of the coast is < 2 m; increasing to 6 m near the northern coast of Puerto Montt.

The climate ranges from temperate maritime in the north to subarctic maritime in the south. There are few sunny days along this coast. Here, rainfall may range from 2.7 to 5.1 m yr^{-1} and may be much higher up-fjord toward the Andes (Pickard, 1971). Rainfall reaches a maximum during the winter (May–August). The fluvial discharge is expected to be greatest in the spring and summer as the snow progressively melts in the Andes, some places with contributions from glaciers. The rivers are very turbid, especially where they drain major ice fields. Although sedimentological information is lacking, the flat floors and turbid rivers suggest high rates of sedimentation—particularly for the mainland fjords with large drainage basins (usually > 1000 km^2). Pickard (1971) noted that Chilean fjords contained abundant glacial silt in the surface waters, especially those with large tidewater glaciers.

Although the drainage basins are relatively small, the high precipitation and concomitant runoff allow two-layer estuarine circulation to develop. The presence of tidewater glaciers and icebergs is associated with complicated vertical temperature distributions (Pickard, 1971, 1973). On average, the icebergs are small "bergy bits." This is a result of the highly dissected tidewater ice fronts that form under the near continuous and sometimes warm rain (J. Stravers, personal communication, 1984). Sediment rafting by these small icebergs is locally important. Since the sills are generally deep, and no stagnant basins have yet been reported, none may exist (Pickard, 1971).

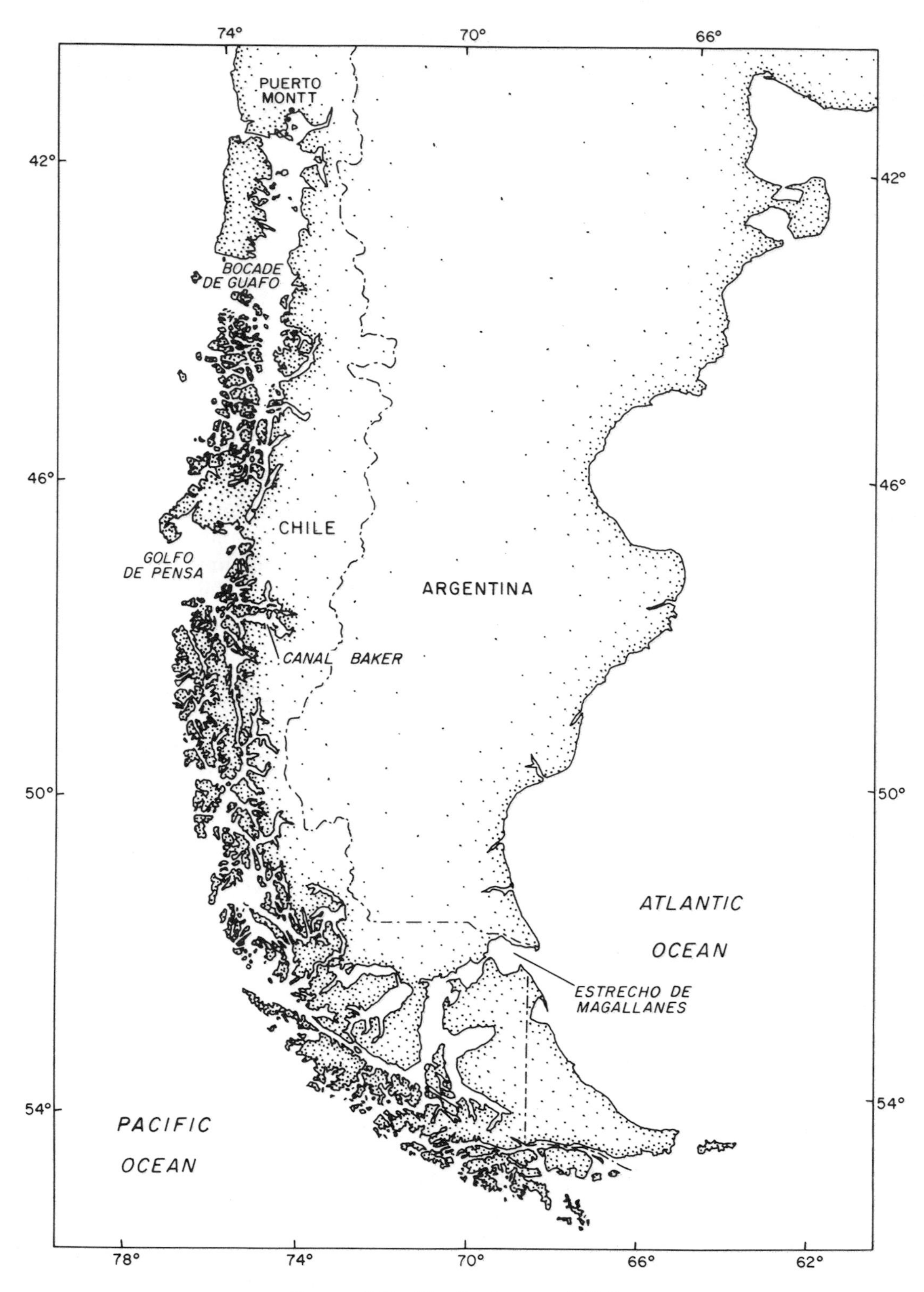

FIGURE 2.29. The fjord coast of Chile.

2.6.7 Scottish Fjords

The west coast of Scotland, including the outer Hebrides, is indented with some 50 fjords (known locally as "lochs"). There may have been more on the west and east coasts, but some have since infilled with sediment and have converted to partially mixed estuaries (e.g., Firth of Tay; Cullingford, 1979). As a group, the Scottish fjords are small, mostly < 20 km long and < 2 km wide (Fig. 2.30), and are also relatively shallow (< 150 m). Most contain a shallow sill at the mouth; some contain an additional internal sill (e.g., Lochs Teacuis, Creran, Etive, Feochan, Leven, Linnhe, and Tarbert); and still others have no sill (e.g., Lochs Laxford, Fyne, Chadh-fi, Eynort, and Torridon). The hinterland mountains range between 400 m to 1343 m (Ben Nevis) in altitude. There are no permanent snow or ice fields.

The climate can be described as island maritime. Annual precipitation may range from 1.5 m to 3.5 m, of which 60% falls during the winter months. Snow is usually confined to the upper mountain slopes. The pattern of fluvial discharge is similar to Vancouver Island and New Zealand and can be related to peaks in rainfall. Discharge lags can occur locally where drainage basins contain lakes or when precipitation falls as snow. Most drainage basins are small (< 400 km^2), especially the island lochs. Some drainage basins are influenced by flow enhancement as a result of industrial needs (e.g., Loch Leven receives enchanced flow from an aluminum plant) or from hydroelectric development (e.g., Loch Etive). In the latter case, the larger freshwater flow allowed a thin ice covering to form at the head of the loch during the 1969 winter: the brackish layer was 5 m deep (Milne, 1972). Although most fluvial discharge occurs at the fjord head, exceptions occur: Loch Sween has its main freshwater inflow at its middle.

Circulation in Scottish lochs depends on sill depth, tidal range, freshwater inflow, and wind events (Milne, 1972). The range in spring tides along the coast is from 4.5 m around the inner Skye fjords (Lochs Torridons and Carron) to less than a meter around the Jura fjords. Tidal currents may be large: Loch Eynort has currents of 3 m s^{-1} over its sill (Milne, 1972). Stagnant conditions of basin waters seldom last longer than a few months and occur most frequently in summer. Although basin waters never go completely anoxic (only Loch Etive and Loch Kanaird have values less than 50% of oxygen saturation: Craig, 1954; Milne, 1972), many lochs are floored with anoxic surface sediments (e.g., Lochs Uskavagh, Keiravagh, Linnhe Arm, Moidart, and Craignish).

The temperature of basin bottom waters fluctuates more in Scottish lochs than most other fjords because of their small basin size, maximum depth, and water volume. In general, water temperature follows the pattern of ambient air temperature but with a short lag: the smaller the tidal range, the greater the temperature range (Milne, 1972).

Although some Scottish fjords were initially affected by man, anthropogenic disposal has become increasingly regulated and this impact has ceased or been ameliorated. Lochs Linnhe, Eil, Leven, and Etive remain affected. Disposal needs in these and other lochs must compete with a growing mariculture industry (Milne, 1972).

2.6.8 Other Fjord Provinces

The remaining fjord districts of the world include those of Iceland, the U.S.S.R., Kerguelen Islands, South Georgia, Antarctica, those of northern Europe not previously discussed (Ireland, Denmark, Sweden), and Washington and Maine (U.S.A.).

The coast of Iceland is indented with approximately 80 fjords (Fig. 2.31A). Most are between 50 to 100 km long and 3 to 5 km wide. Two of the largest are Isafjördur (300 km long, 50 km wide) and Eyjafjördur (240 km long, 22 km wide). Most of the fjords are less than 200 m deep. Hinterland mountains range from 500 to 1000 m. The climate can be described as subarctic island maritime; variability around the island is large. Average January temperatures range from -10 to 0°C; average July temperatures range from 5 to 15°C. These relatively high temperatures for this latitude (63.5°N to 66.6°N) result from influence of the warm Gulf Stream waters (10°C). Formation of sea ice is rather rare. The annual precipitation is also variable, ranging from 0.4 to 4 m yr^{-1} and highest along the warmer southern coast. The spring tidal range is mostly low (< 1 m) except along the western coast (> 4.3 m); wave action is strong on all coasts, but especially along the southern and western coasts. Eight fjords are still fed by

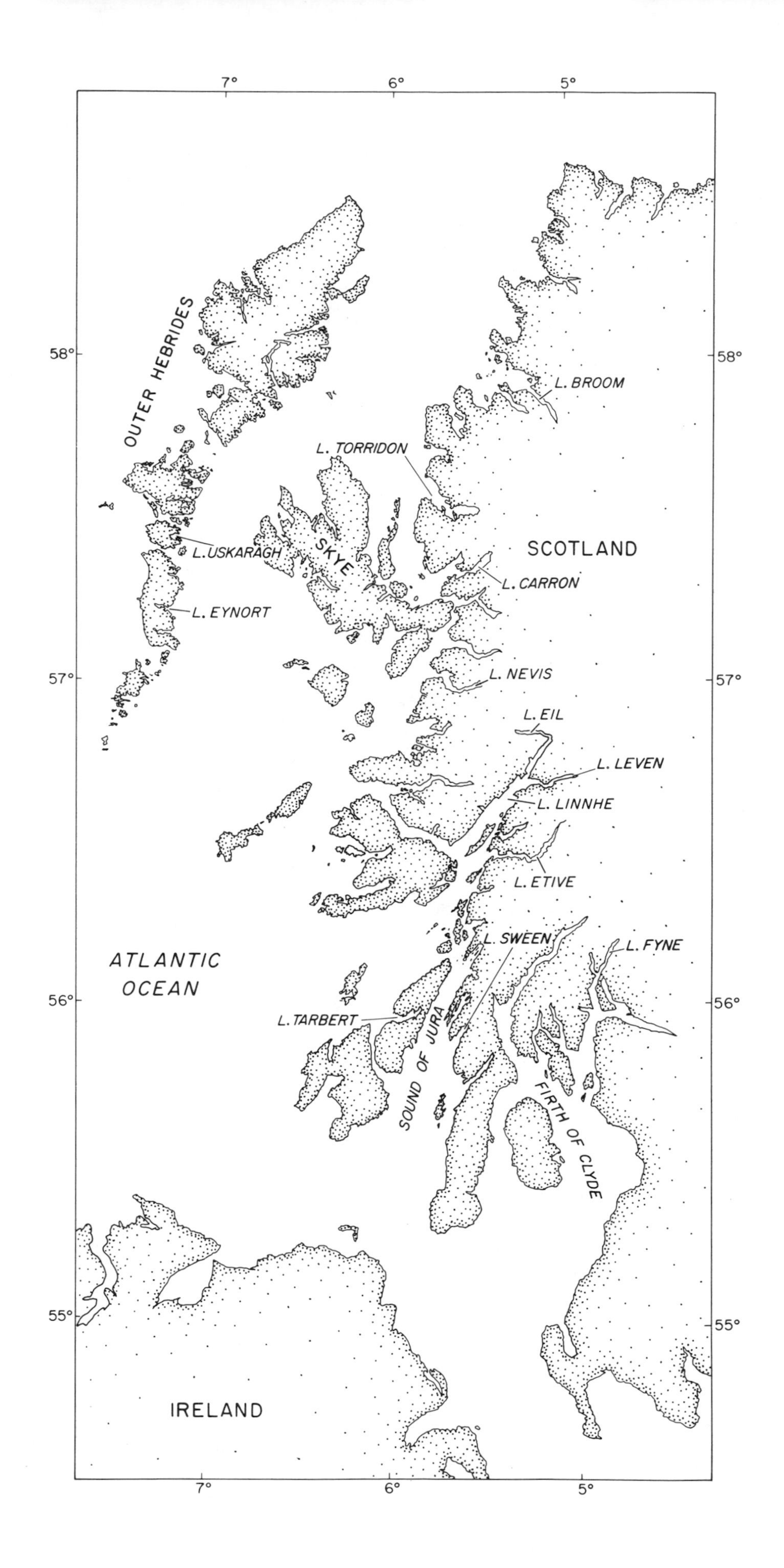
7°
6°
5°
58°
57°
56°
55°
OUTER HEBRIDES
L. BROOM
L. TORRIDON
SCOTLAND
L. USKARAGH
SKYE
L. CARRON
L. EYNORT
L. NEVIS
L. EIL
L. LEVEN
L. LINNHE
L. ETIVE
L. SWEEN
L. FYNE
ATLANTIC
OCEAN
L. TARBERT
SOUND OF JURA
FIRTH OF CLYDE
IRELAND

ice caps (Reykjafjördur, Furufjördur, Kaldalon, Borgarfjördur, Skagafjördur, Skjalfandi, Axarfjördur, and Heradsfloi), and many have characteristics of rapid sediment infill. The latter five also contain large drainage basins (> 10^4 km^2). Many Icelandic fjords have classical sandur deltas at their heads. There are few lakes to filter sediment transport, and delta fronts are coarse grained, some as a result of catastrophic jökulhlaup events (see Chapter 3). As a result of these high fluvial loads, many fjords along the southern coast are completely infilled with sediment.

Along the Soviet coastline there are nearly 40 fjords; most are part of the arctic archipelago of Novaya Zemlya (20 fjords) and Severnaya Zemlya (4), and the Kamchatka and Barents Sea coasts (10). Their characteristics are described in a rather obscure book by Kaplin (1962). The Russian term for fjord is F'yord. The archipelago fjords are < 25 km long and between 3 and 10 km wide. Many of these fjords either drain or are partly filled with glaciers (Fig. 2.31B). The climate is that of an arctic desert, and these fjords may be similar to archipelago fjords found within the Canadian high arctic. Little is known of the Kamchatka and Barents Sea fjords (Kudusov, 1973). They are usually shallow, some being meromictic lakes (Propp et al., 1975).

The Kerguelen Islands, in the south Indian Ocean, have approximately 10 fjords with water depths between 100 and 200 m. Most are fed by the Glacier Cook Le Dôme. The island of South Georgia, in the south Atlantic Ocean, has 6 fjords with prominent sills at their mouths. They too are fed mostly from glacier melt during the summer, and by seasonal drainage runoff.

There are a number of fjords along the Antarctica coast. Most are deep with steep sides and flat floors; sills may or may not be present. Many are still ice filled with tidewater glaciers; others are fenced in from the sea by thick ice shelves (Nichols, 1960). The insignificant quantity of alluvium, the small size of the meltwater streams in the summer months, and the absence of deltas suggest that fluvial transport of sediment is of little importance. Estuarine circulation is weak and is primarily driven by meltwater from the submerged portion of tidewater glaciers and floating glaciers. The Antarctica fjords may have very low sedimentation rates (0.004 mm yr^{-1}: Barrett, 1975; or lower: Drewry, 1976). The relative contribution from iceberg rafting is apparently much higher in Antarctica than in the Arctic (Elverhöi and Roaldset, 1983). However, during operation "Deep Freeze 85," seismic profiling revealed relatively thick infill of sediment (160 m) for two King George Island fjords (John Anderson, personal communication, 1985). In the long glacial history of Antarctica, at least one fjord has been infilled (Taylor Valley: Pewe, 1960; Powell, 1981b).

There are a number of shallow water fjords in northern Europe outside of Scotland including: Northern Ireland (e.g., Stratford Lough), Shetland Island (e.g., Sullom Voe), the Faroes, Ireland, Denmark, and Sweden. These have been referred to as fiards by some and have similarities with Nova Scotian inlets (Piper et al., 1983). Almost all are < 50 m deep with even shallower sills. As such they are prone to anoxia. Since they are in the final stages of sediment infill, most have lagoon-like features (Pearson, 1980a). Lough Ine is typical of many on the SW coast of Ireland (Kitching et al., 1976): it has a maximum depth of 48 m and intermittently goes anoxic in its basin waters (Rosenberg, 1980). Three fjords have been studied extensively in Denmark: (1) Isefjord (Rasmussen, 1973); Flensburg Fjord (Fig. 4.39; Exon, 1972); and Limfjord (Larsen, 1980). Limfjord is the largest, being 170 km long with a maximum depth of 28 m: basin waters go anoxic in the summer (Rosenberg, 1980). The fjords of Sweden are even smaller and include Byfjord, Kosterfjord, Gullmarsfjord, Kungsbakafjord, and Iddefjord on the border with Norway. Byfjord is typical with dimensions of 4 km by 1.5 km, basin depth of 50 m, and sill depth of 11 m. Byfjord receives low freshwater discharge and experiences very small tides: basin stratification is maintained by outside influences (Dyrssen, 1980; Svensson, 1980). Many of the Swedish fjords experience basin-water anoxia as a direct result of pollution (e.g., Saltkällefjord and Iddefjord; Rosenberg, 1980). Much of the Swedish and German research has concentrated on the fjordlike basins of the Baltic (e.g., Baltic Sea, Gulf of Bothnia, and Kiel Bay). Prominent sills do not allow for continuous deep water exchange, and these basins go anoxic regularly (Rosenberg, 1980).

There are approximately 20 fjord basins in

FIGURE 2.30. The fjord coast of Scotland, U.K.

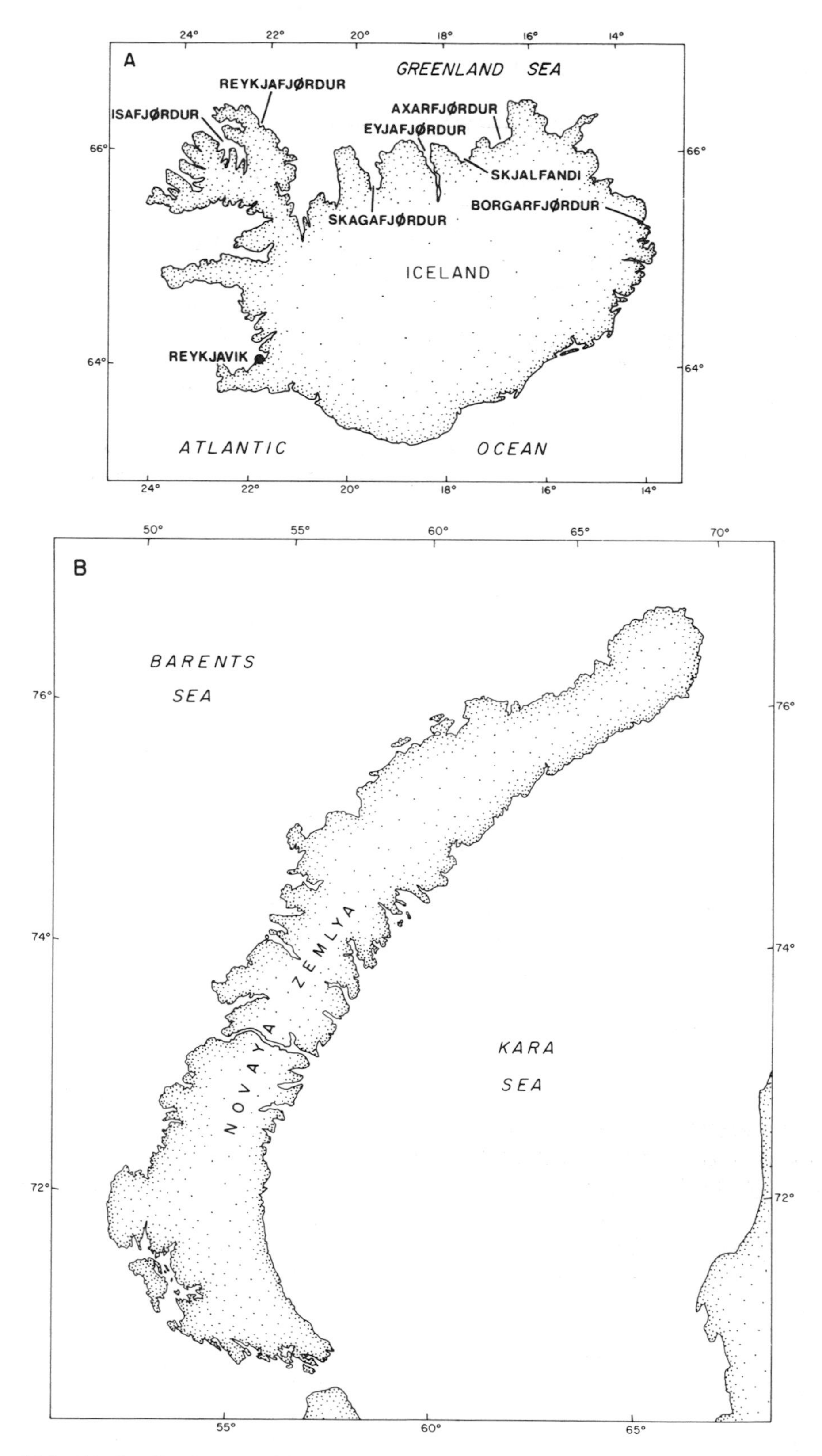

FIGURE 2.31. (A) The fjord coast of Iceland. (B) The fjord coast of Novaya Zemlya, U.S.S.R.

Washington and another 30 fjordlike basins in Maine (U.S.A.). The Washington fjords, on the west coast of North America, are similar to many of the fjords situated along the coast of Vancouver Island (see preceeding discussion above). Together the fjords are part of one large complex (130 km long) whose main basin has a depth < 300 m and sills between 40 and 70 m (major basins included Admiralty Inlet, Puget Sound, Hood Canal, and Dabob Bay: Fig. 4.5D). Sixty percent of the water is supplied near the fjord mouth rather than the fjord head (for oceanographic details see Cannon and Laird, 1980, and Strickland, 1983). The east coast fjords of Maine are similar to those of Nova Scotia: they are relatively shallow, with hilly rather than mountainous terrain. Many have a limited supply of fresh water. Basin sediments are mostly anoxic (Folger et al., 1972).

2.7 Summary

With this chapter we have taken a descriptive approach to the general environmental setting of fjords, including some particulars on the world's fjord districts. Most of these particulars are highly variable (Table 2.3), making generalizations on the nature of fjords difficult. Fjord similarities include: (1) having been a glaciated valley at some stage in their development; (2) based on point (1), some of their basin infill is composed of glacial or glaciomarine sediment; (3) they may contain one or more of the following features: a basin deeper than the adjacent shelf, sills, cirques, hanging valleys, U-shaped transverse profile, pot holes, terraces, talus cones, and strandflats; (4) sedimentation rates are mostly related to the fjord's glacial stage (see Table 2.3 and Section 2.5) and the river discharge; and (5) the style of estuarine circulation is predominantly but not exclusively two-layer flow, especially during the main discharge season. We have also discussed the wide range of fjord end-members, depending on the glacial stage of the fjord, the state of sediment infill, and local geography, including climate and drainage basin characteristics. The quantitative approach of the following chapters will continue to re-emphasize the important interactions between geomorphology, climate, water structure, and sediment sources.

Part 2 Processes and Products

3

The Fluvial-Deltaic Environment

Many of the processes and products in fjord systems are closely related to the movement of water and sediment down fjord valleys. For instance, the rate of sediment accumulation is directly related to river dynamics in many fjords. Fjord circulation and the transport of natural and anthropogenic sediment are highly dependent on the hydrologic cycle. Quaternary sequences exposed in fjord valleys also provide information on the paraglacial framework of fjord infilling. (By paraglacial, we refer to the effects of glacial-proglacial sedimentation, dynamic sea-level fluctuations, and historically significant climate shifts.) Within this chapter, we explore the variations found in fjord-valley runoff, the principles of sediment erosion and transport with specific fjord-river examples, and the many influences that complicate the paraglacial growth of fjord sandurs and deltas.

Symbol Notation

(All other symbols as defined in the text)

A	cross-sectional area of a river channel
B	channel width
C_s	suspended sediment concentration
d	grain diameter
E	evapotranspiration loss
e_b	bed-load efficiency
f	the limiting angle of repose
g	acceleration due to gravity
g_s	bed-load rate
P	precipitation
Q	discharge
Q_s	suspended sediment discharge
qs	bed-load transport
R'_h	channel hydraulic radius
S	channel slope
U	flow velocity (defined)

u_* shear velocity
V volume (defined)
x horizontal distance
Y topographic elevation
ΔS change in water storage capacity
ρ fluid density
ρ_s grain density
τ_o shear stress
$(\tau_o)cr$ critical shear stress
ϕ intensity of shear on the particles
ψ intensity of bed-load transport

Runoff

The balance of water in a drainage basin is the simple balance of inputs and outputs with a slight modification for storage, that is,

$$P = Q + E + \Delta S \tag{3.1}$$

where P is the total precipitation input; Q is the total stream discharge; E is the total evapotranspiration loss; and ΔS is the change in storage capacity (a complex function of soil moisture storage, ground-water storage and outflow at depth, changes in depression storage such as lakes, and changes in snowpack and ice-field storage). Values of the terms P, Q, E, and ΔS are highly variable between fjord-valley drainage basins and can vary dramatically with time in a given basin. Such fluctuations are typical of mountainous hinterlands, especially where ΔS may depend on the timing of ice jams, log jams, sudden drainage (jökulhlaups), or the ablation and accumulation rates of glaciers.

Discharge is usually measured directly by stream gauging techniques rather than by solving equation (3.1). Stream-water level is recorded and converted to cross-sectional area, A, with the assumption that the channel perimeter is fixed. Stream velocity, U, is calculated from the integration of velocity-depth profiles taken across the channel cross-section. Discharge is then given as:

$$Q = A \cdot \overline{U} \tag{3.2}$$

Errors in Q are related largely to the conversion of water level to cross-sectional area, especially during winter months where river ice changes the perimeter profile in an indeterminable manner. The channel cross-section may also change during and after high discharge events, with bed erosion or deposition, respectively. Gauging is usually done in highly turbulent and straight reaches of the river where the velocity is more uniformly distributed over the channel cross-section.

Once discharge has been calculated, it can be plotted against time to produce a streamflow hydrograph. The shape of the hydrographs provides a means of comparing the runoff dynamics between drainage basins. Figure 3.1 gives four highly divergent examples of stream hydrographs that were obtained from fjord-valley drainage basins. The Weir River is an example of an arctic, nonglacial, nival regime (Fig. 3.1A; Ambler, 1974). The river flows down steep slopes (4.6%) from a small drainage basin (30 km^2) into d'Iberville Fiord, Ellesmere Island, Canada. The annual precipitation is low (140 mm), of which 50% is stored through the winter as snow. The river is not regulated and has no lakes to moderate its flow. The hydrograph is simple: a large spring discharge from snow melt (known as a freshet) followed by lower summer flows punctuated by periodic rain-storm floods that are induced orographically. The steep slopes, absence of lakes, sparse vegetation, and pervasive permafrost make storage nearly nonexistent during the summer months. Lag between rainfall and river-mouth discharge maxima is of the order of tens of minutes; this is significantly shorter than lags of hours or days characteristic of larger and lower latitude basins. Permafrost at shallow depths inhibits deep infiltration of water and encourages immediate runoff once the active layer becomes saturated (Church, 1974). In the autumn, some ground-water seepage may persist through the riverbed gravels after the surface has frozen. When the advancing seasonal frost reaches the permafrost table, thereby cutting off the drainage, hydrostatic pressure may cause water to break through onto the surface at a weak point, to form an icing.

The Somass River is an example of a maritime, nonglacial, pluvial regime (Fig. 3.1B; Water Survey of Canada, 1982). The Somass flows into Alberni Inlet and is one of the largest rivers on Vancouver Island, British Columbia, with a drainage basin area of 120 km^2. The river discharge is moderated by large lakes and thus the response time between peak rainfall and peak discharge is of the order of one or two days.

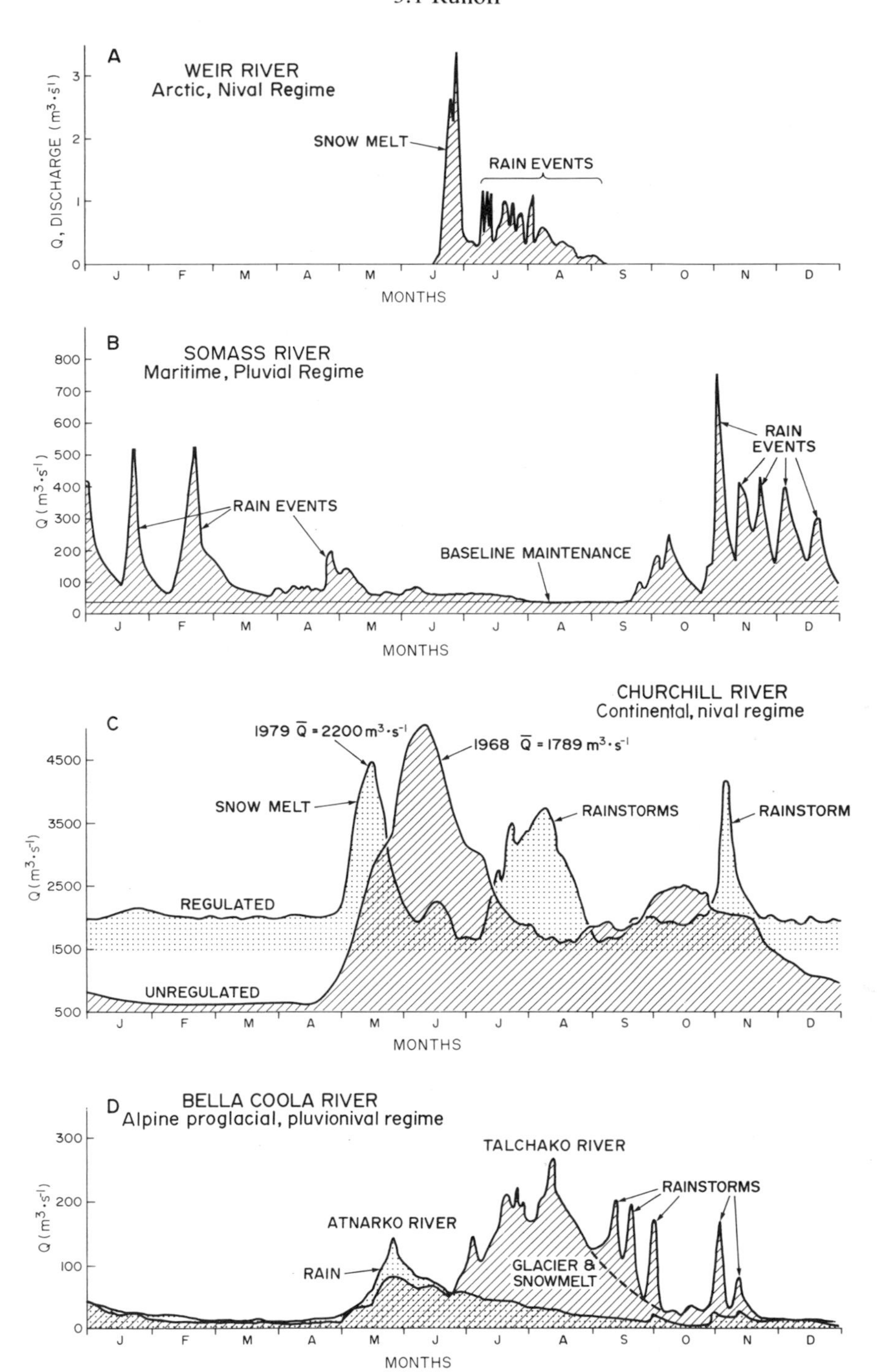

FIGURE 3.1. Streamflow hydrographs from four radically different fjord-valley rivers (data from Water Survey of Canada). (A) Arctic nival regime: Weir River, Ellesmere Island. (B) Maritime pluvial regime partly regulated to ensure minimal flow: Somass River, British Columbia. (C) Continental nival regime before and after regulation: Churchill River, Labrador. (D) Alpine proglacial, pluvionival regime with a complex streamflow hydrograph from two tributaries that have distinct climate and water storage effects: Bella Coola River, British Columbia.

The hydrograph indicates the absence of a spring freshet because significant winter snow storage does not normally occur. When it occurs, the storage lasts only days to weeks before the snow accumulation melts. Hydrograph peaks are therefore directly related to precipitation events, and the lowest discharge occurs during the summer months. This type of fjord-valley hydrograph appears to be limited to basins from southern Vancouver Island, Scottish lochs, Nova Scotia inlets (Delure, 1983), and many of the New Zealand fjords (Stanton and Pickard, 1981). The Somass River is also an example of river regulation for the purpose of maintaining a baseline flow. Prior to 1956, the Somass River discharge varied considerably, and at times of low discharge the oxygen levels in the estuary decreased (Morris and Leaney, 1980). River regulation was undertaken to maintain an adequate supply of dissolved oxygen necessary to stabilize the pulpmill effluent.

The Churchill River provides an example of a continental nival regime (Fig. 3.1C; Water Survey of Canada 1969, 1980). Such fjord-valley basins can be very large. The Churchill River has a drainage area of 92,500 km^2 with a mean annual discharge exceeding 2000 m^3 s^{-1}. It empties into Lake Melville and Hamilton Inlet on the coast of Labrador. The discharge hydrograph indicates stable winter conditions (snow storage) and a large spring freshet followed by a wet autumn. With a discharge three orders of magnitude greater than the Weir River, the Churchill River flows year-round as a result of large ground-water and lake storage capacity. Also, the hydrograph is considerably smoothed by the river's slow response time; individual rain events are seldom registered unless they are very large (details of Norwegian counterparts are given in Tollan, 1972). Such rivers are ideal for large-scale hydroelectric production. In 1972, the Churchill River was regulated; the largest effect of its increased storage capacity was to increase the winter discharge by a factor of 4. Summer discharge decreased correspondingly, except for the excess flow released during sudden snow melt or prolonged rain storms (Fig. 3.1C). Further examples of the effects of fjord-basin regulation can be found in Asvall (1976).

The last hydrograph example is that of the Bella Coola, a river that drains into the head of North Bentinck Arm, British Columbia (Fig. 3.1D; Water Survey of Canada, 1981). The Bella Coola is an example of an alpine, pluvionival, proglacial regime. The Bella Coola has in fact two tributary drainage basins, the Atnarko River with a drainage area of 2430 km^2, mean slope of 0.75%, and mean annual discharge of 30 m^3 s^{-1}; and the Talchako River with a drainage area of 1300 km^2, mean slope of 1.5%, and mean annual discharge of 60 m^3 s^{-1}. The Atnarko drains a relatively dry interior plateau, and the river peaks in May–June owing to heavy rains. The Talchako peaks in July–August, first from snow melt, followed by glacier melt, and last by severe autumn rain storms. A common hydrological phenomenon is the devastating autumn flash flood, which starts typically with an early frost followed by heavy snowfall, rapid thaw, and warm rain. The result is rapid runoff of surface water that is unable to permeate the still-frozen ground. A rare (200-yr) flood might discharge 30 times more than the mean annual flood discharge. If, as is usually the case during flood conditions, trees on the stream bank are uprooted and deposited as log jams further downstream, the valley floor can be flooded with several meters of water. Rivers with similar flood characteristics are dyked in their lower reaches to protect roads and towns (Andru and Allen, 1979). Similar flood conditions on the steeper and smaller rivers that flow into the margins of a fjord are known to build sediment dams (rather than log jams) in areas of reduced slope. When the dam eventually breaks, impressive water and mud surges (debris flows) occur (Russell, 1972; Eisbacher and Clague, 1981).

Proglacial regimes such as the Talchako River exhibit a discharge that continues to rise until late summer, as progressively higher zones on the glacier melt and become effective contributing portions of the watershed. Discharge from glacier melt is, however, highly variable between drainage basins, depending on the ablation characteristics of the individual ice field. The Decade River, flowing into Inugsuin Fiord, Baffin Island, drains a basin that is 68% glacier covered, yet precipitation appears to control the discharge hydrograph (Østrem et al., 1967). At the other extreme, the Jostedal River, draining into Gaupnefjord, Norway, has only 27% of its watershed covered by glaciers. Here the runoff responds more directly to the glacier melt with a distinct diurnal periodicity (Relling and Nordseth, 1979).

Proglacial rivers are prone to sudden releases of water from ponds or lakes that were held back temporarily behind ice or snow dams (jökulhlaup). When the dam is breached, the peak discharge is enormous, up to 50,000 m^3s^{-1} in the 1934 Grimsvotn jökulhlaup, Iceland (Nye, 1976). Clague and Mathews (1973) examined a number of such events and found that peak discharge of each jökulhlaup is approximately proportional to the two-thirds power of the total volume of water released during the flood, that is,

$$Q_{max} = 75 \cdot V_{max}^{0.67} \ (r^2 = 0.96) \quad (3.3)$$

Ice thresholds at lake outlets can also dam a considerable volume of water that can be quickly released after the first substantial melt. Østrem et al. (1967) found that lakes in the Inugsuin Fiord drainage system were 29 to 46 cm higher than normal in the spring owing to ice dams.

In summary, runoff response in fjord rivers is highly variable depending on local climatic conditions, storage of winter snow, presence of glaciers, and the modulating effect of lakes. Fjord rivers are also highly susceptible to near-instantaneous floods that can be more than an order of magnitude larger than normal peak flows.

3.2 Sediment Transport

Given an infinite supply of transportable sediment, particle transport is known to increase with a stream's discharge. The sediment may be transported either in continuous suspension, known as the wash load, or through interaction with the stream bed, known as the bed-material load. Whereas the concentration of wash load is related largely to source areas, bed-material load is related closely to the stream's hydraulic and dynamic character. The bed-material load may be further divided into: (1) bed load, coarser particles transported by tractive processes (rolling, skipping, or sliding along the channel bed); and (2) suspended load, finer particles held temporarily in suspension by fluid turbulence.

3.2.1 Bed Load

In bed-load transport no upward impulses are imparted to the particles other than those attributable to successive contact between each particle and the bed (Bagnold, 1973). Unfortunately, both the predictive theoretical-empirical formulae and field measurements are known to provide only first-order approximations of actual bed-load transport. Nevertheless, such approximations are still useful in understanding the sediment yield in fjord valleys and the subsequent growth of sandurs and deltas. One reasonable, albeit time-averaged, method for the calculation of bed-load transport rates comes from the sequential mapping of delta advances (Adams, 1980; Østrem et al., 1970; Sundborg, 1956).

There are three classes of theoretical-empirical bed-load formulae: (1) The DuBoys-type equations based on the mean tractive force or shear stress; (2) the Schoklitsch-type equations based on a slope/discharge relationship; and (3) the Einstein-type equations based upon statistical considerations of the lift forces, a result of stream line convergence.

DuBoys (1879) first suggested that bed-load transport, qs, is related to the tractive force,

$$qs = fct\ (\tau_0) \quad (3.4)$$

Later, Shields (1936) proposed the concept of threshold level for the initiation of grain movement and derived a dimensionally homogeneous transport function based on the critical shear stress for initiation of motion. Kalinske (1947) developed this concept by including the notion of turbulence in the flow above the bed. A dimensionless form of this bed-load equation is given by

$$\frac{qs}{u_* d} = fct\left[\frac{(\tau_o)cr}{\tau_o}\right] \quad (3.5)$$

where u_* is the shear velocity, a measure of the intensity of turbulent fluctuations, and d is the grain diameter. Church (1978) notes that this approach is highly dependent on the state of the bed. In an "overloose" condition, the particles are resting in a dilated state because of the presence of a large upward flow of water within the sediment that exerts a dispersive stress; in an "underloose" condition, particles are closely packed or imbricated. Such possible variation in bed conditions is consistent with an order-of-magnitude range in bed-load predictions.

Schoklitsch (1950) considered that bed-load motion would be initiated at a certain critical velocity, which in turn would be related to discharge (equation 3.2). Bagnold (1966) more specifically involved the hydraulics of the bed (roughness terms and the angle of grain repose) but still considered stream power (directly related to stream discharge) the important term, that is,

$$qs = \left(\frac{\rho_s}{\rho_s - \rho}\right) P_A e_b \, [g \tan(f)]^{-1} \qquad (3.6)$$

Where $P_A = \rho g Q S B^{-1}$ and is the mean available power to a column of fluid over a unit bed area. Q is discharge, S is channel slope, and B is channel width.

The third approach to bed-load transport was developed by Einstein (1942, 1950) and is based on an equilibrium model of the exchange of bed particles between the bed layer and the bed. This approach relates the hydrodynamic lift forces acting on the particle to its weight and thus provides a relationship between ψ, the intensity of bed-load transport and ϕ, the intensity of shear on the particles,

$$\phi = \text{fct}\,(\psi) \qquad (3.7)$$

with

$$\phi = \frac{g_s}{\gamma_s}\left[\left(\frac{\rho}{\rho_s - \rho}\right)\frac{1}{gd^3}\right]^{1/2} \qquad (3.8)$$

and

$$\psi = \left(\frac{\rho_s - \rho}{\rho}\right)\frac{d}{SR'_h} \qquad (3.9)$$

where g_s is the bed-load rate, γ_s is the specific weight, and R'_h is the channel hydraulic radius. For low rates of sediment transport the relation was given as:

$$\phi = (e^{-0.391\psi})/0.465$$

and similarly the Meyer-Peter and Muller (1948) equation has been given by Chein (1954) as:

$$\phi = \left(\frac{4}{\psi} - 0.188\right)^{3/2}$$

Bed-load deposition is rapid once the velocity of a stream falls below a corresponding threshold value for deposition of a particular grain diameter. Thus changes in the river gradient (thalweg), and the length of river over which these changes occur, control the level of sorting that sediment undergoes. Fjord-valley rivers tend to have their thalwegs described by simple exponential equations of the form:

$$Y = c_2 e^{-c_1 x} \qquad (3.10)$$

where x is horizontal distance, Y refers to the elevation, and c_1 and c_2 are form constants. It is no surprise then, that there is a close relation between thalweg form and the associated exponential decrease in mean grain size and its standard deviation down-river (e.g., Church, 1972). Persistent deviations in the graded river profile (Foster and Heiberg, 1971) may result from disequilibrating influences of new down-valley sediment sources or sinks. Temporary deviations may arise because bed load moves much less rapidly than water. Since many discharge events in fjord-rivers are short lived (Fig. 3.3C), much of the bed load must move stepwise down-valley in trains. The distance between steps would depend on the duration of a given event and coarseness of the bed-load particles. The bed-load train would not be remobilized until there is a new discharge event of equal or greater magnitude, especially in areas of decreasing valley slope. At rest, these bed-load trains act as armor to the river bed, out of equilibrium with normal mean seasonal flows. The next period of movement may be days, or, in many cases, years away.

Abrasion is also important in decreasing grain size downstream (Adams, 1980). Ninety percent of the abrasion occurs during the period of bed armoring, transforming boulder trains into suspended load. In the New Zealand rivers, half of the total bed load is lost to the suspended and wash load by abrasion during the course of fluvial transport (Adams, 1980).

Bed load, then, is essentially a capacity load controlled by the stream discharge, hydraulic slope, bottom roughness, bed compaction, and grain properties (of which grain diameter is most significant). Bed-load transport can range from less than 5% of the total load for lowland fjord-valley rivers to 95% for proglacial mountain streams (Østrem et al., 1970; Church, 1972; Ziegler, 1973; Adams, 1980; Syvitski and Farrow, 1983; Bogen, 1983). The highest percentage of bed-load transport has been found in arctic proglacial fjord-sandurs (Church, 1972). There, bed-load transport was two orders of magnitude larger than suspended load transport, apparently related to the flood-dominated discharge events (Syvitski and Schafer, 1985) and small basin areas.

3.2.2 Suspended Load

The nature of suspended bed–material load and its measurement is such that it cannot be differentiated satisfactorily from wash load, the theoretical difference being related to the permanency of a particle remaining in suspension. In many theoretical discussions, wash load is chosen arbitrarily as a small percentage of the total load (e.g., Syvitski and Murray, 1977) or as load below a certain grain size (e.g., Sundborg, 1956). Still others use the term suspended load inclusive of wash load (e.g., Nordseth, 1976). We therefore re-emphasize that suspended bed–material load is a capacity load dependent on stream discharge: as Q increases, so does the quantity and coarseness of the suspended load population. Wash load, however, is a non-capacity load and highly dependent on source areas and supply conditions. For instance, many rivers need flows three orders of magnitude greater than those of a glacial stream to reach similar suspended sediment concentrations (compare findings of Østrem et al., 1970 with Farrow et al., 1983). Thus, unlike bed load, suspended sediment discharge, Q_s, (including both suspended and wash load) cannot be theoretically predicted from water discharge. However, the relative precision and ease of field measurement have allowed detailed data bases to be obtained. These have provided a rather precise understanding of the interrelationships between Q and Q_s for specific drainage basins. The general relationship between suspended sediment concentration, C_s, (or its discharge load, Q_s) and stream discharge can be represented by a rating plot (Fig. 3.2A). For statistical reasons it is more meaningful to consider C_s rather than Q_s in rating plots, because Q_s is by definition a product of stream discharge (Q) and sediment concentration (C_s), and the relationship between Q_s and Q would therefore involve the common variable Q.

Concentration of suspended particles is usually log-linear with discharge best fitted by the power function:

$$C_s = aQ^b \tag{3.11}$$

The coefficient, a, indicates the turbidness of a stream at times of low discharge. The exponent, b, characterizes the sensitivity of a particular catchment area in terms of sediment yield to changes in runoff. The higher the exponent, the greater the proportion of the annual sediment yield removed from the basin during extreme high discharge conditions. Proglacial streams, for example, may have 60 to 70% of the annual sediment yield occurring during one day (Nordseth, 1976; Østrem et al., 1967). In general, the exponent, b, is usually less than 2 for fjord-valley rivers: it is highest for glacial streams, lower for lowland streams draining silt and clay deposits (a function of the erodability of the sediment), and lowest for high mountain streams because of restricted access to fine-grained material (Nordseth, 1976).

Rating curves may change with the season (Fig. 3.2). Four separate Q/C_s relationships were found for the Lierelva in Norway, accounting for the early snow-melt season, the late snow-melt season, rising summer glacial discharge conditions, and constant or falling discharge in late summer and fall (Fig. 3.2A). Such changes in rating relationships with the season usually indicate new sources or changes in the sediment supply. For instance, nival rivers having a marked spring freshet have the greatest sediment yield in the spring (Fig. 3.2B). The initial sediment supply comes with the erosion of the recently deposited winter fines (cf. Hjulström, 1935). The sediment supply decreases drastically once these fines have been washed out of the system. This results in a summer suspended load more closely in tune with the degree of bed erosion. The rating pattern is reversed for proglacial streams with suspended concentrations increasing proportionally as the contribution of glacial meltwater increases in the late summer (Gaddis, 1974). This pattern will occur even though the spring runoff takes place over a disproportionately larger area. Many fjord-valley rivers are further complicated by a hysteresis effect between concentration and streamflow. Sediment concentrations are greater on the rising river stage for a given level of discharge than on the falling stage, and the rating plot for a series of discrete discharge events will assume the form of clockwise loops rather than a straight line (Fig. 3.2D; Nordseth, 1976;

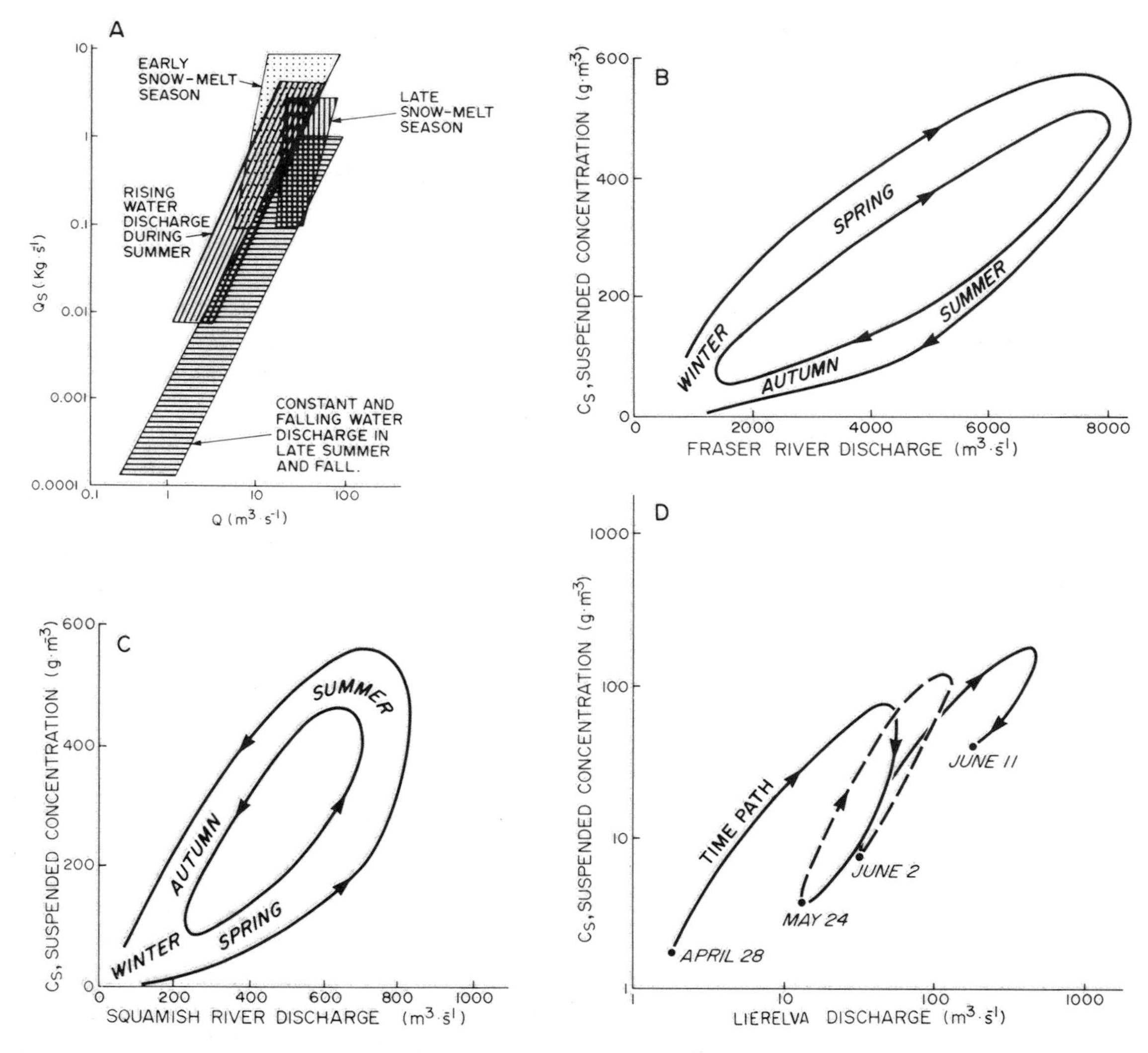

FIGURE 3.2. Time dependence of fluvial rating curves that compare suspended load discharge, Q_s, or suspended sediment concen tration, C_s, with water discharge, Q. (A) and (D) after Nordseth (1976) data on the Lierelva, in Norway. (B) after Milliman (1980) data on the Fraser River, British Columbia. (C) from unpublished Water Survey of Canada data on the Squamish River, British Columbia.

Christian and Thompson, 1978). This feature is best explained in terms of a reduction of the erosive effect of the rainfall on the falling limb and the increased volume of subsurface runoff contributing to the recession flow.

In the above discussion on rating curves we have made the implicit assumption that peaks in suspended concentration and water discharge are coincident. This is not necessarily or usually the case; there are many documented examples of the peak suspended concentration lagging behind the flood peak, or, conversely, preceding the flood peak. On a gross level, using monthly averages of C and Q, the Squamish River in 1974 had peak suspended concentration in August, yet peak water discharge occurred in July (Fig. 3.3A). In 1975 the peaks were coincident. Detailed analysis of the daily values reveals the complexity inherent in many fjord-valley rivers. The suspended concentration peak can be found coincident (e.g., day 146), lagging (e.g., day 155), preceding (e.g., day 165) and non-existent (e.g., day 203) to the stream discharge peak (Fig. 3.3B). Even the variation in suspended concentration in one day is remarkable, with changes between one and two orders of magnitude being possible (Fig. 3.3C).

The coarseness of suspended particles de-

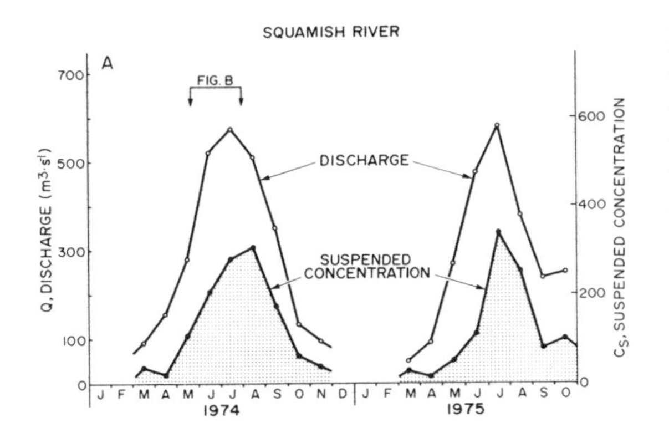

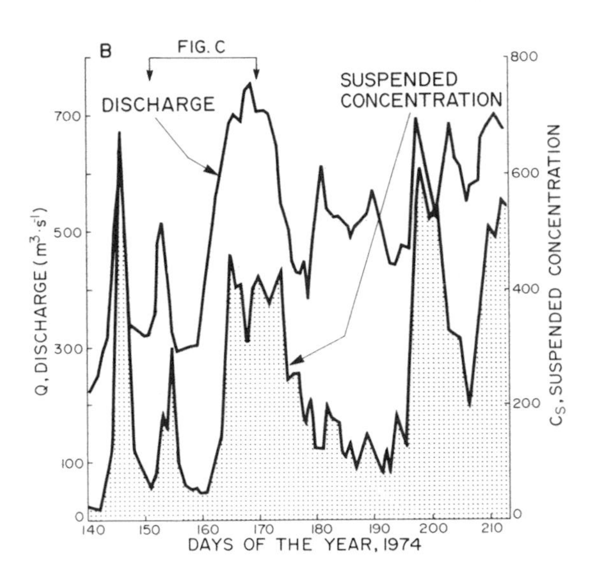

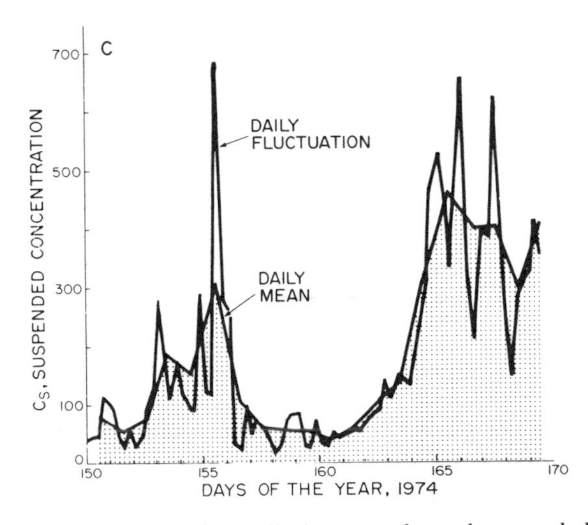

FIGURE 3.3. Timing of discharge peaks and suspended concentrations peaks for the Squamish River, British Columbia (unpublished data, Water Survey of Canada). (A) Monthly averages; (B) Daily averages; and (C) Daily fluctuations in C_s.

pends on whether C_s reflects hydrodynamic equilibrium conditions or not. At hydrodynamic equilibrium, the wash-load component of Q_s is more likely to be minor, thus allowing the fluctuations of the suspended bed–material load to dominate. As such, concomitant increases in suspended concentration and stream discharge will result in log-linear increases in the suspended sand fraction (Fig. 3.4A). If, however, flood conditions bring the rising water in contact with a potential wash-load source, then the relationship inverts and the sand component becomes diluted by the new wash-load mud (e.g., Adams, 1980; Fig. 3.4B).

3.3 Paraglacial Sedimentation

Denudation is that process which, if continued long enough and in the absence of further mountain-building processes, would reduce all surface inequalities of the globe to a uniform base level. Catchment basins having a "normal" denudation rate would transport and dispose of the same amount of material that is made available by weathering processes in the local basin. In areas of paraglacial sedimentation, vast quantities of glacial, proglacial, and/or exposed marine sediment are available and may account for abnormally high rates of fluvial erosion and transport, leading to high rates of denudation (Church and Ryder, 1972). Fjord valleys are particularly susceptible to redistribution of sediment because of their high relief. The position of modern fjord-valley rivers on the paraglacial sediment yield curve (Fig. 3.5) depends on three contributing factors: (1) their glacial history, (2) their history of sea-level fluctuations, and (3) local paleoclimatic conditions. For instance, fjord rivers that have no glaciers in their drainage hinterland, and which drain into fjords that have had little sea-level change over the past few thousand years, may already be near steady-state equilibrium. Examples include the Vancouver Island fjord rivers (Mathews et al., 1970). At the other extreme are fjord systems that recently began a new paraglacial cycle at the onset of the Little Ice Age (circa 1700 AD), or those high arctic fjords that have just begun to be deglaciated from the last major Pleistocene advance (Price, 1965; Boyd, 1948; Misar, 1968).

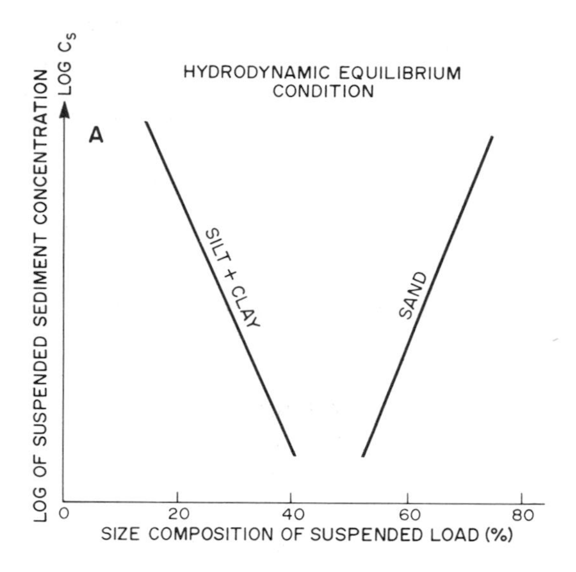

FIGURE 3.4. The variation in the coarseness of suspended particles with increasing suspended sediment concentration for (A) riverbed equilibrium condition; and (B) nonequilibrium condition where flood waters come into contact with and begin to erode mud deposits.

3.3.1. Glacial/Proglacial Sediment

Our general view of fjord-valley glacial sedimentation processes will mainly follow the ideas of Boulton and Eyles (1979). A valley glacier carries debris at its base that is derived from the subglacial bed; at a higher level and on its surface, debris is derived from flanking mountain walls. The resultant deposits comprise subglacial till, englacial eskers, supraglacial moraines and kames, and proglacial outwash in front of dump moraines. Figure 3.6 illustrates how they are related to the parent ice mass in a glacial valley. Subglacial, englacial, and to some extent extraglacial deposits have a relatively high proportion of silt and fine sand (glacial flour), at least compared with supraglacial debris or till. Till is defined as an aggregate whose components are brought together by the direct agency of glacier ice and which does not undergo subsequent disaggregation and redeposition.

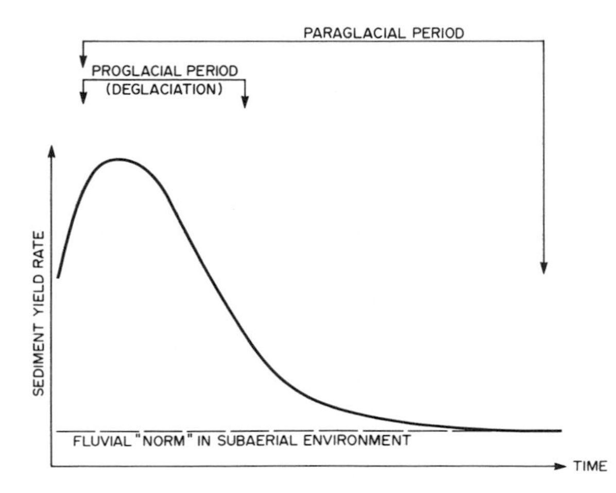

FIGURE 3.5. Schematic to indicate the abnormally high sediment yield for fjord-valley rivers during the paraglacial and especially during the proglacial period (after Church and Ryder, 1972).

Using this framework, two main valley-glacier sedimentary facies can be identified, depending on whether the glacier is advancing or retreating. Facies 1 results from deposition along lateral and latero-frontal margins during stationary or advance phases. The main deposits are dump moraines that accumulate mostly as scree from the steep glacier front. Mud flows and water-washed sediment are often intimately associated with this scree, often infilling the large voids within these open-textured bouldery sediments (Fig. 3.6).

Facies 2 comprises deposits laid down along the frontal margin during retreat. If the supraglacial till cover is thin, the material is slumped off during retreat as a relatively thin and sporadic veneer over the progressively exposed subglacial surface (lodgment till, bed rock, or outwash). The thickness of the veneer is proportional to the rate of retreat. If the supraglacial till cover is thick enough to slow the melting rate of the underlying ice, till is let down very

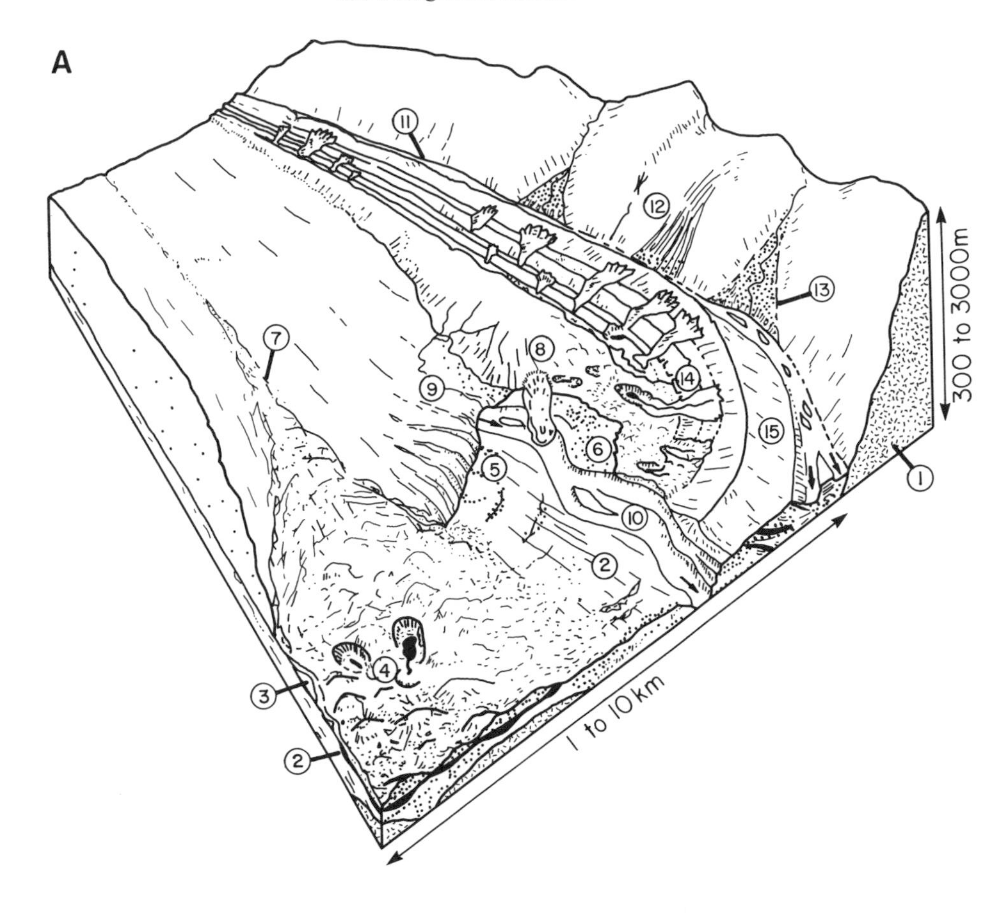

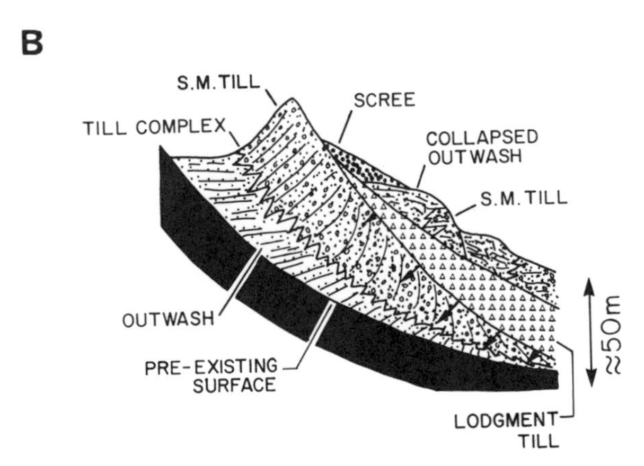

FIGURE 3.6. Landforms and sediments associated with the retreat of a fjord-valley glacier (after Boulton and Eyles, 1979): (1) bed rock; (2) lodgment till with fluted and drumlinized surface; (3) supraglacial morainic till; (4) ice-cored and kettled supraglacial morainic till; (5) crevasse fills; (6) bouldery veneer of supraglacial morainic till; (7) supraglacial medial moraine; (8) supraglacial lateral moraine; (9) fines washed from supraglacial till into crevasses and moulins; (10) proximal meltwater streams; (11) kame terrace; (12) truncated scree; (13) truncated fan; (14) gullied lateral terrace; and (15) latero-terminal dump moraine. The lower diagram gives the distribution of sedimentary types typically associated with a major lateral dump moraine formed during a single phase of glacier advance and retreat. The arrows show a zone of glaciotectonic deformation.

slowly onto the subglacial surface. The result is hummocky stagnation topography (i.e., thaw lakes and kettles). If the outwash contains ice blocks (i.e., an ice-contact outwash), melting of the buried ice results in a pitted kame plain or outwash surface.

The availability of large quantities of readily transported sediment and the rapid buildup and decay of flood discharges has a strong influence on the character of glaciofluvial sediments. The derived sediment is often redeposited as planar, matrix-supported, outwash beds with a particle-size distribution that closely resembles the parent till. In such a case, all particles are transported and deposited simultaneously. The outwash beds are poorly stratified, with individual units topped with a gravel armor and scour bowls. Within a few kilometers of the ice front, normal fluvial hydrodynamics take over concomitant with a decrease in grain size and increase in sorting down-valley of the glacier (Church and Gilbert, 1975).

Facies 1, plastered high along the valley walls, has a relatively high preservation potential in the short term and is frequently found grading upward into normal mountain-slope talus cones. These morainic terraces may slump from the valley side to produce a conical pile of poorly sorted material several meters above the flood plain (Hattersley-Smith, 1969). Facies 2 can be buried by fluvial deposits and preserved; however, if the valley floor becomes braided and incised across its width it has little chance of survival (Price, 1980). Depositional features such as eskers can survive erosion so long as the meltwater streams in the proglacial area follow restricted courses. Glacier re-advance may or may not obliterate previous depositional units, a factor that depends largely on the basal thermal regime of the glacier (Eyles et al., 1983). Temperate and wet-based valley glaciers are capable of reworking previous glacial deposits, especially where the valley gradient is steep. If the glacier's basal thermal regime is cold and dry, the glacier must advance through internal plastic deformation rather than by erosive basal sliding. Alexandria Fiord, Ellesmere Island, is an example where a cold-base glacier advanced and retreated during a 400-yr period, preserving the underlying vegetation (Bergsma et al., 1983). Where valley gradients are not steep, even wet-base glaciers can override glacial deposits in subsequent surges. The external weight of the glacier could cause structural deformation of the deposits (Boulton et al., 1976).

A glaciolacustrine facies is not uncommon in fjord valleys. The lakes are usually found in bedrock depressions formed during the glacial advance and exposed during retreat. Latero-frontal dump and push moraines, where extensive, can also form dams for valley lakes. Lake depths can vary from a few tens to several hundred meters. During their initial infill with water, such lakes tend to be in contact with the glacier tongue (Østrem, 1975; Gustavson, 1975). Since water may enter the lake at a number of possible depths, depending on the position of any ice tunnel, discharge dynamics and flow structure within the lake can be highly complex and variable. The freshwater discharge plume from an ice tunnel could spread out at its entry depth, or rise or sink (Fig. 3.7), depending on the density characteristics of the lake (ρ_l) and the density of the input source (ρ_g). The water density in each case is determined by properties such as salinity and temperature and by suspended particulate matter concentrations. If the inflow is denser than the deepest lake water ($\rho_g > \rho_{l\ max}$), then it flows along the lake bottom as an underflow. If the inflow density is greater than the surface waters yet less dense than the deeper waters ($\rho_{l\ max} > \rho_g > \rho_{l\ min}$), it will sink or rise until reaching an equilibrium depth and then traverse the lake as an interflow. If $\rho_g < \rho_{l\ min}$, then the inflow will rise to the surface (if entering from depth) and traverse the lake as an overflow.

Whatever the mechanism, the lake quickly becomes turbid during the melt season, and sedimentation processes cover the lake floor with varves. Varves are deposits in which a coarse-grained layer related to the summer discharge maximum alternates with a thinner and finer-grained layer related to the late-summer or autumn low-discharge period (Church and Gilbert, 1975; Østrem, 1975; Pickrill and Irwin, 1983). The salt content of glacial lakes is usually negligible (less than 1‰), and enhanced settling of suspended matter through the process of flocculation does not normally occur. However, many glacial lakes become populated quickly by plankton, and ingestion of the suspended particles by zooplankton will cause increased settlement of the fines through their packaging and

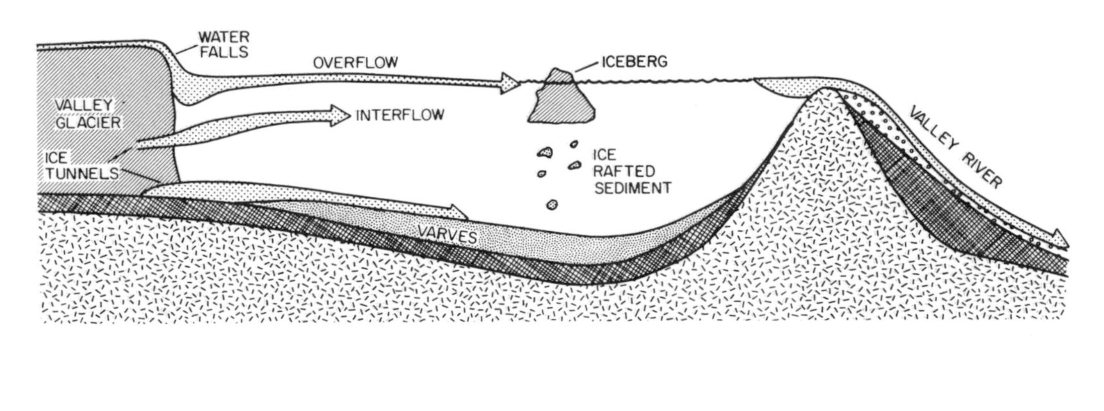

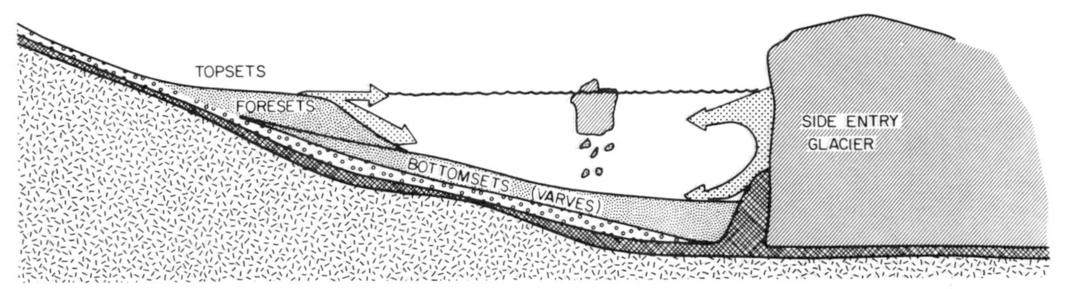

FIGURE 3.7. Schematic of processes and products associated with glaciola-custrine facies common to certain sections of the fjord valley. The upper figure represents a bedrock silled lake in contact with a retreating valley glacier. The lower figure represents an ice-dammed lake formed from a side-entry glacier (see Fig. 3.8 for further details and stipple explanation).

egestion in the form of fecal pellets (Smith and Syvitski, 1982). Thus although the varve cycle is annual, both coarse and fine sedimentation occur mainly during the runoff season. Underflows are subjected to gravitational acceleration according to the difference in density between the flow and the surrounding medium (Gilbert, 1975). In such cases, sediment may be deposited in the form of ripple drift cross-lamination between varve couplets (Gustavson, 1975).

Many valley glaciers calve into water bodies (Hattersley-Smith, 1969). Thus the sediments of true glacial lakes are apt to contain ice-rafted particles of all grain sizes, which are dropped sporadically onto the varved lake floor. Icebergs in lakes tend to ground quickly and ice-rafted piles or dumps are common (Harris, 1976).

Many modern arctic fjords receive hanging side-entry glacier tributaries that expanded during the Little Ice Age from the interfjord ice-field plateau (Fig. 3.8). They may extend out onto the valley floor forming large lateral or push moraines comprising bouldery ridges 50 to 100 m high (Boyd, 1948). These moraines may provide little transportable sediment but can completely disrupt normal up-valley river dynamics. Lakes may form behind the terminal bulb of such side-entry glaciers, becoming temporary sites of glaciolacustrine sedimentation (Fig. 3.7B). The up-valley end of the lake becomes infilled with a Gilbert-style delta (Gilbert, 1890) trapping all bed load. Retreat of the hanging-glacier may be accompanied by lake drainage, leaving behind easily erodable lacustrine muds. Ice blocks produced from calving terrestrial glaciers may occasionally be transported seaward by flood discharges of meltstreams (Slatt and Hoskin, 1968) and may temporarily or even permanently divert the river (Boyd, 1948). Side-entry glaciers may also discharge massive amounts of sediment into the main valley, building up large alluvial fans (Fig. 3.8; Syvitski et al., 1983a). The fans partly block the normal river course, thus altering its hydrodynamic equilibrium and flooding the river with newly produced proglacial sediment (Boyd, 1948). Similar results are produced by nonglacial processes such as rock or snow avalanches that can occur daily along the main fjord valleys in arctic environments (Church et al., 1979).

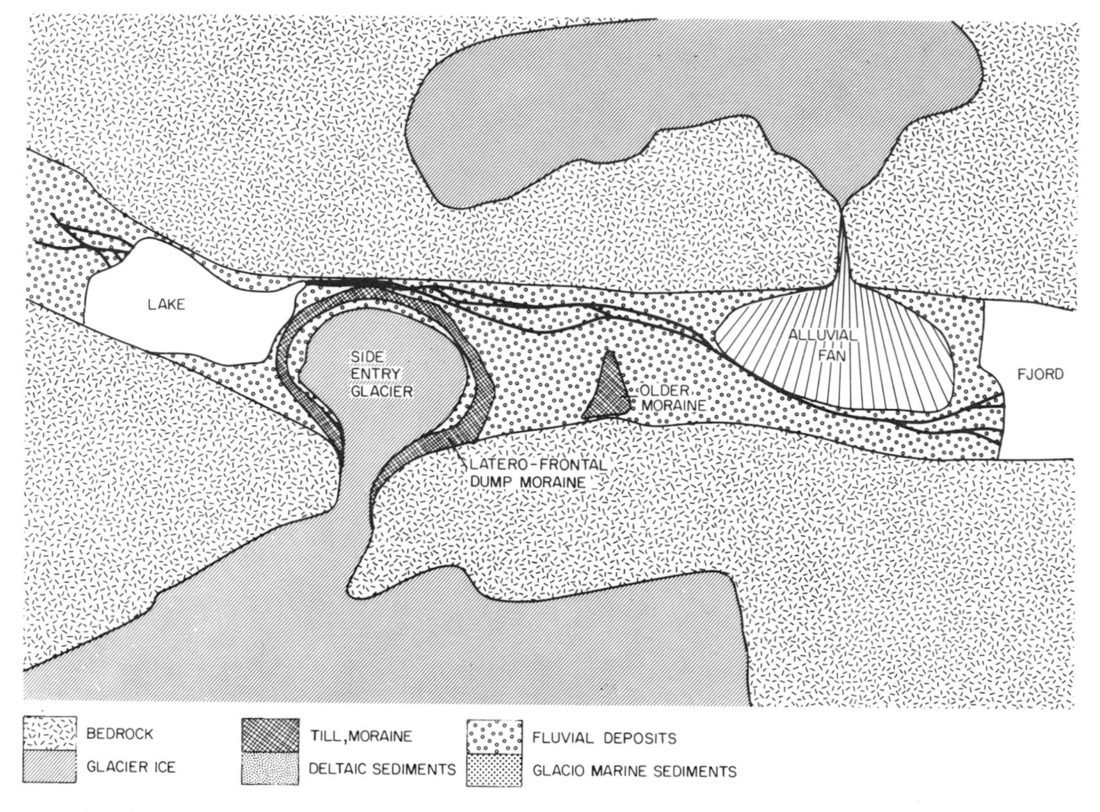

FIGURE 3.8. Disuptive effects of hanging side-entry glacier tributaries on normal fjord-valley sedimentation include formation of valley lakes, alteration of the river course, and addition of sediment to the valley.

3.3.2. Relative Sea-Level Fluctuations

The dominant external process affecting the relative height of mean sea level and the solid earth is glacial growth and decay. This is at least an order of magnitude greater than any other process for time scales of 10^2 to 10^4 years (Table 3.1). Changes in the earth's climate during the Quaternary allowed periodic growth, then melting, of large ice caps on parts of the earth's surface. In particular, all fjord regions were subjected to glaciation. Waxing and waning of the ice caps caused shifts of the earth's surface load and induced a mechanical response in the solid interior. For instance, ice caps larger than 300 km in diameter can depress the earth's crust

TABLE 3.1. Ice movement, accretion, denudation, and diastrophism in fjord valleys during periods within the past 10,000 years BP.

Parameter	Range	Reference
1. Glacier movement	10^3–10^7 mm yr^{-1}	Cooper, 1937; Price, 1965; Gilbert, 1980a,b
2. Deltaic advance	10^3–10^4 mm yr^{-1}	Church and Ryder, 1972; Bell, 1975; Adams, 1980
3. Deltaic accretion	10^2–10^3 mm yr^{-1}	Church & Ryder, 1972; Bell, 1975, Syvitski & Farrow, 1983; Bartsch-Winkler et al., 1983
4. Denudation		
(a) Glacial	10^1–10^2 mm yr^{-1}	Anderson, 1978
(b) Paraglacial	10^0–10^1 mm yr^{-1}	Church and Ryder, 1972; Nordseth, 1976
(c) Normal	10^{-3}–10 mm yr^{-1}	Nordseth, 1976; Adams, 1980
5. Diastrophism		
(a) Normal	10^0–10^1 mm yr^{-1}	Kaula, 1980; Riddihough, 1983; Adams, 1980; Adams, 1981
(b) Tectonic	10^3 mm (instantaneous)	Bartsch-Winkler et al. 1983

isostatically up to a depth one-third of the ice cap thickness (J.T. Andrews, personal communication, 1983). During the growth of ice caps, water is removed from the world's ocean through evaporation. Sea level falls by an equivalent water volume, and this process is known as eustasy.

Before the early 1970s most researchers employed a simple relationship between isostatic adjustment, eustatic sea-level change, and the resulting relative sea-level change, that is,

$$RSL(x,y,t) = s(x,y,t) - R(x,y,t) \quad (3.12)$$

where RSL(x,y,t) is the relative sea-level change at latitude x, longitude y, and time t; s(x,y,t) is eustatic sea-level at (x,y,t); and R(x,y,t) is the vertical position of the earth's surface at (x,y,t). R is the summation of both isostatic and diastrophic influences, the latter being mostly ignored in tectonically stable areas. In effect, these models implicitly considered the earth as an infinite plate and no consideration was given to the need to conserve mass. We now know that glacial isostatic deformation and eustatic response are more complex (i.e., Peltier, 1976; Peltier and Andrews, 1976; Farrell and Clark, 1976). It is usually impossible to sort out the individual s and R components based on field data (e.g., Quinlan and Beaumont, 1981). The relative heights of the sea surface and the solid earth are now believed affected by three components (Kaula, 1980): (1) laterally transferred surface loads; (2) changes in the solid surface as a result of interior motions in response to surface load shifts; and (3) change in the geoid height owing to both of the aforestated mass shifts. Figure 3.9 summarizes these factors. Note that "eustasy" is redefined by Mörner (1980) as the summation of ocean-water volume changes, ocean-basin volume changes (i.e., the changes in the gravitational attraction of ocean water to new land mass, such as ice caps, and the isostatic response to changes in the crustal water loading), and ocean-level distribution changes (gravitational and rotational changes in the ocean level and thus geoid shape). The geoid shape (geodetic sea level) is an equipotential surface determined by the attraction and rotation potentials and can be affected by changes in the hydrosphere, crust, mantle, and core. Farrell and Clark (1976) showed that the explicit expression for sea-level change, RSL, at time t

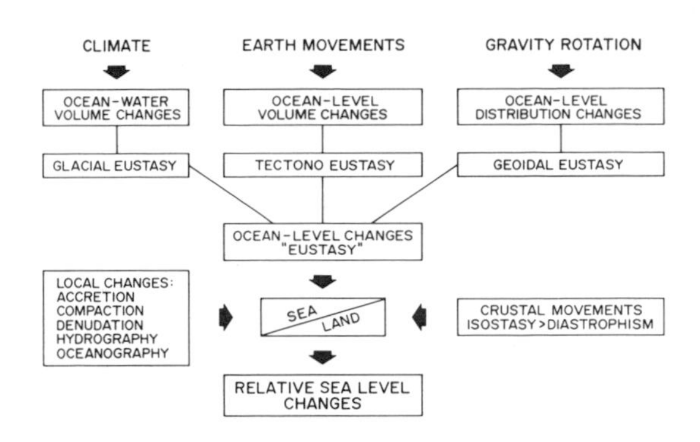

FIGURE 3.9. The factors affecting ocean-level changes and crustal movement that combine with local conditions to determine the position of the land-sea boundary (after Kaula, 1980).

and position r is:

$$\begin{aligned} RSL(t,r) &= \iint_{ocean} G^E(r - r')\rho_w RSL(t,r')\, d\Omega' \\ &+ \iint_{ice} G^E(r - r')\rho_I I(t,r') d\Omega' \\ &+ \int^{t}_{18{,}000\ BP} d\tau \iint_{ocean\,+\,ice} G^V(t - \tau, r - r')\,[\rho_w s(\tau,r') \\ &+ \rho_I I(\tau,r')]\, d\Omega' - K_e(t) - K_c(t) \end{aligned} \quad (3.13)$$

where G^E is the immediate response elastic Green function (see Peltier, 1974; Peltier and Andrews, 1976); G^V is the time-dependent viscous Green function (ibid.), I is the ice thickness; $d\Omega'$ is an element of area; K_e corrects for the oceanwide average rise in sea level; K_c ensures that mass is conserved; and $r - r'$ notation indicates the angular distance between position vector r and r′. The first term is the immediate sea-level change caused by changes in water load. The second term accounts for the immediate elastic sea-level response that is due to the change in ice load. The third term represents the slow deformation of the earth from changes in both ice and water loads from 18,000 years BP, the approximate time of the last main ice-sheet advance.

The above model has been used to divide the globe into six zones of differing relative sea-level response (Clark et al., 1978; Clark, 1980). Because of the recent glacial history of modern fjords, zones 1 (glaciated areas) and 2 (collapsing forebulge submergence) and the transition between them are of prime importance. Zone 1 includes glaciated areas that emerge continuously, to about three times the eustatic sea-level rise (isostatic response to ice cap melting). Zone 2 can be found just out from the ice cap margins, seaward of the peripheral bulge formed by sub-lithospheric material extruded from beneath accreting and isostatically subsiding ice sheets (Fig. 3.10; Quinlan and Beaumont, 1981). As the ice melts, the resulting isostatic equilibrium causes the material to flow back from the bulge toward the rebounding ice center. This migration involves viscous mass redistribution in the earth's interior and continues to take place after the comparatively rapid deglaciation. This slow rebound affects a transition zone between zones 1 and 2 with distinct sea-level curves (Fig. 3.10) because the bulge retreats more slowly than the ice margin.

Many fjords exhibit zone 1 type of uplift curves. Some are remarkable in their close approximation to exponential curves, especially those fjords found along Greenland and Baffin Island coasts (Kelly, 1973), that is:

$$a_t = a_o e^{-kt} \tag{3.14}$$

where a_t is the amount of uplift remaining to be achieved at time t after uplift began, and a_o and k are decay constants. Simple one-term exponential curves do not always model the observed field data. Walcott (1970) suggests a two-term curve of the form:

$$a_t = a_1 e^{-t/1000} + a_2 e^{-t/50000}$$

where the exponential terms have relaxation times of 1000 and 50,000 years, and a_1 and a_2

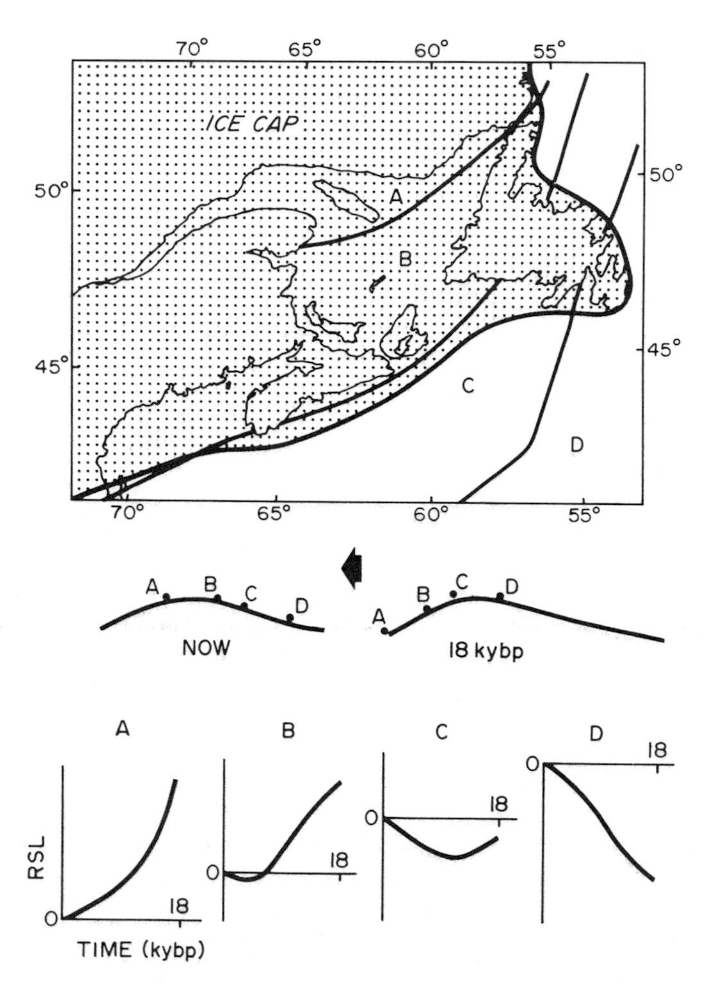

FIGURE 3.10. Types of relative sea-level (RSL) response associated with fjord terrain as indicated from modeling global glaciation specifically to proposed conditions for the east coast of Canada (after Quinlan and Beaumont, 1981). Indicated are (A) zone 1 RSL response for glaciated areas; (D) zone 2 RSL response for the collapsing forebulge submergence; and (B) and (C) are RSL response for the transition between zones 1 and 2.

are amplitude constants (a_1 = 150 m and a_2 = 450 m fit Walcott's gravity data).

Complications to the global model given by equation (3.13) depend on local variations in ice extent, and timing and rates of glacier advance and retreat (Andrews, 1975). As an example, England (1983) discussed two contrasting models to depict changes in relative sea level of arctic fjords (Fig. 3.11A). Model 1 depicts glacier ice extending initially to the mouth of a fjord, followed by slow retreat to the fjord head. The marine limits, that is, the highest level recorded by the late glacial sea, become progressively younger toward the head (Price, 1965; Dyke, 1979; Miller, 1980). Model 2 depicts a stable ice mass near the fjord head. Seaward of the ice front, the peripheral depression is occupied by the sea. Along this coast, the marine limit may

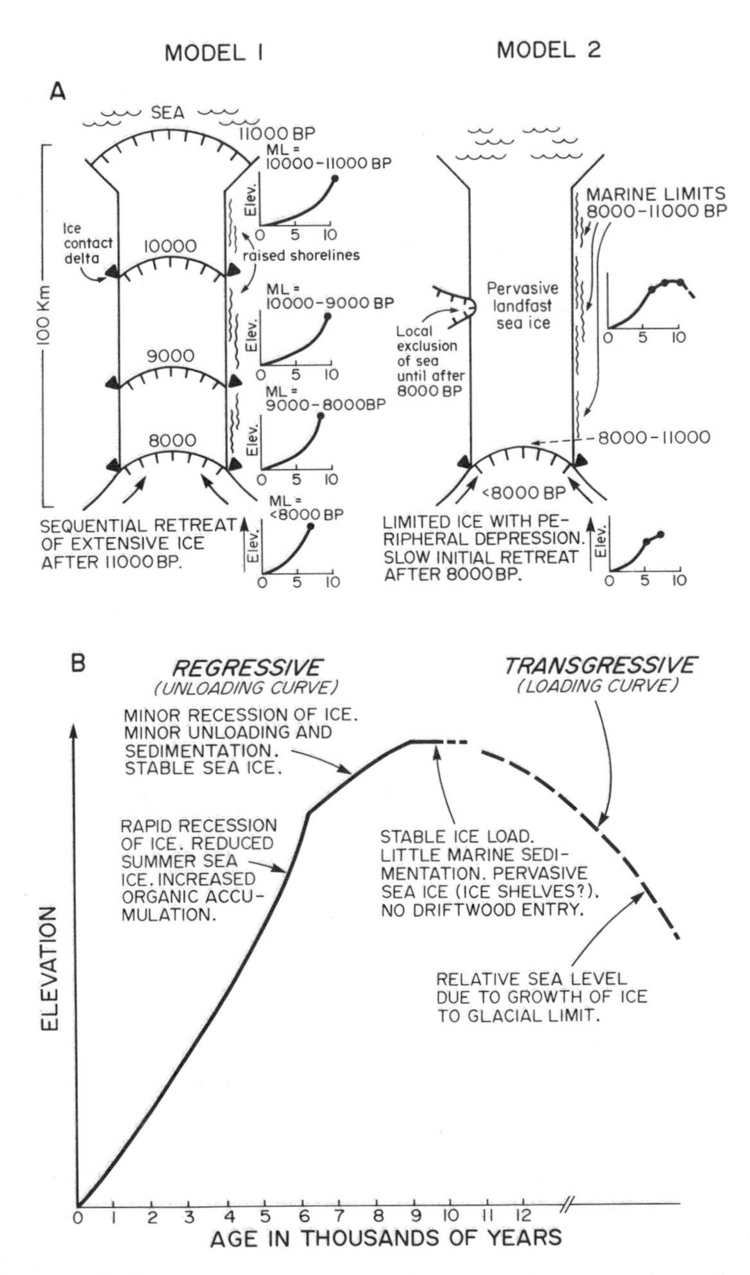

FIGURE 3.11. (A) Two contrasting models to depict changes in the RSL of arctic fjords (after England, 1983). (B) Details of the RSL curve from model 2 and with proposed transgressive and regressive phases to the curve.

have been established long before the initial emergence or synchronous with emergence. Both transgressive and regressive marine deposits may be preserved. Model 2 is applicable in some Ellesmere and Baffin island fjords (England, 1983). Relative sea-level curves from the peripheral depression may contain all or part of the following segments (see Fig. 3.11B): the glacial loading curve depicting a marine transgression of a full glacial sea; isostatic equilibrium depicting stable relative sea level during a period of no change in the glacial loading; slow emergence owing to minor retreat of ice; and/or rapid glacial unloading causing increased glaciomarine and fluviomarine sedimentation.

Land areas relatively close to each other (tens of kilometers) may be emerging or submerging at different rates. In other words, shoreline tilting is a ubiquitous feature of fjord coasts (Andrews et al., 1970; Mörner, 1971; Hodgson, 1973; Blake, 1975). Variations in gradient are caused by three isostatic effects (Mörner, 1971). (1) Glacio-isostatic effect is caused by variations in the distribution of ice at any given time. If the ice load were greater in the inner parts of the ice sheet, the isostatic effect would be stronger there than in the outer parts. (2) Another glacio-isostatic effect is caused by changes in the ice load over time, but at a given location. Ice advances tend to retard upheaval, thus resulting in a smaller gradient of tilting. (3) A hydro-isostatic effect is caused by a eustatic transgression loading the ocean basin and resulting in peripheral subsidence and upheaval farther inland. A fourth factor (4) caused by geoid deformation as a result of ice-ocean attraction could account for a further 5% of the shoreline tilting (Fjeldskaar and Kaneström, 1980).

The discussion above has assumed that diastrophic uplift or subsidence has a comparatively minor influence on sea-level change through the Holocene because of quantitatively more important glacially induced crustal changes. For most fjord areas of the world this is true, and only in recent times have diastrophic influences become important now that the crustal response of glaciation is near completion (Riddihough, 1983). New Zealand fjords are unique in that Holocene uplift has been dominated by the diastrophic influences of the Alpine Fault (90%) with only a small glacio-isostatic imprint (Mathews, 1967; Adams, 1981). In that environment, the rate of uplift is highly dependent on the proximity of a particular fjord to the fault line (range of < 0.5 to 20 mm yr^{-1}).

The effects of sea-level change on fjord-valley sedimentation are many and varied. Given below is an example of the interaction of sea level and fjord-valley sedimentation.

As the ice caps form, geodetic sea level falls because of eustatic processes. However, relative sea level rises as the land is depressed to an even further extent under the newly formed ice load. The result is a marine transgression along the fjord coast. With the river valley occupied by a valley glacier, sea level may adjust its position against the tide-water glacier (Fig. 3.12A). The erosive nature of valley glaciers may destroy any record of this initial (glacial) marine transgression (Fig. 3.12B). Because it is the most dynamic force operating during that time (Table 3.1), glacier movement would determine the land-sea boundary. Since glacier retreat can be rapid (10^6 to 10^7 mm yr^{-1}) and because crustal response to glacier retreat is slow (10^2 mm yr^{-1}), the sea transgresses up the fjord valley as ice retreats (Fig. 3.12C), for distances as great as 60 to 100 km up-valley (Armstrong, 1966; Schafer et al., 1983). Crustal rebound, initially greater than eustatic-related sea-level rise (10 mm yr^{-1}), slows and proglacial sediment is rapidly deposited (10^3 mm yr^{-1}). Thus the marine transgression will slow, then halt. At the marine maximum (marine limit), sediment is deposited either conformably or disconformably on the glacial/proglacial sediments as shallow water marine deposits. As uplift continues, denudation (i.e., sediment erosion and transport) is increased. The newly exposed marine deposits become incised, leaving terraces plastered along the valley walls. In areas of shoreline tilt, river gradients can be increased further, and such fluvial rejuvenation brings increased denudation, delta building, and basin filling (Fig. 3.12D).

River incision into uplifted fluvial and marine beds exposes unstable terraces. Hillslope collapses are common events in fjord valleys and depend on the stability of the terrace slope and the magnitude of the triggering event (Fig. 3.13A). The effects of a given slope failure depend partly on its size and partly on its timing relative to previous events. Three types of slope failure (Foster and Heiberg, 1971) are: (1) shal-

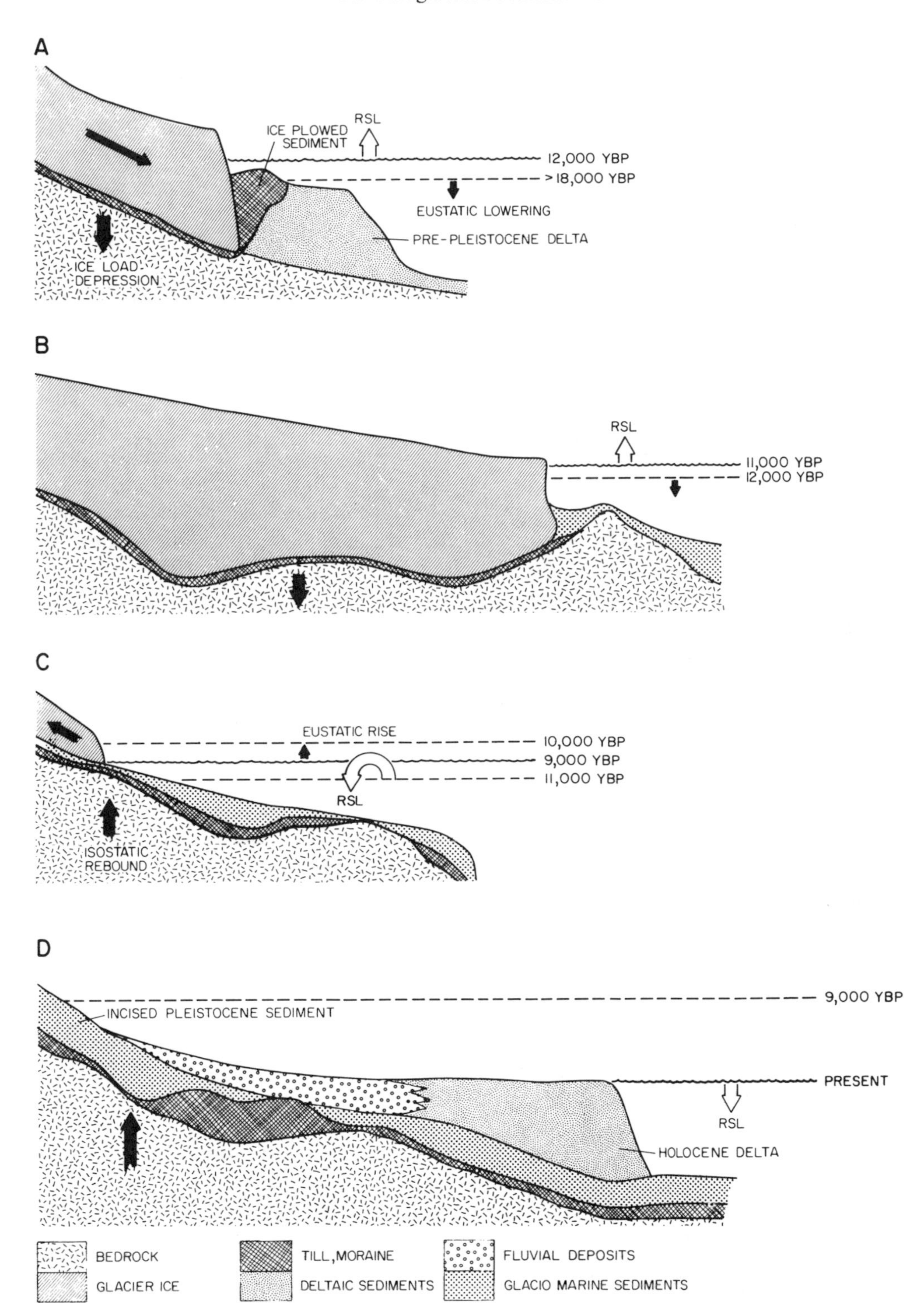

FIGURE 3.12. One of many possible scenarios indicative of the interaction of sea-level fluctuations and fjord-valley sedimentation.

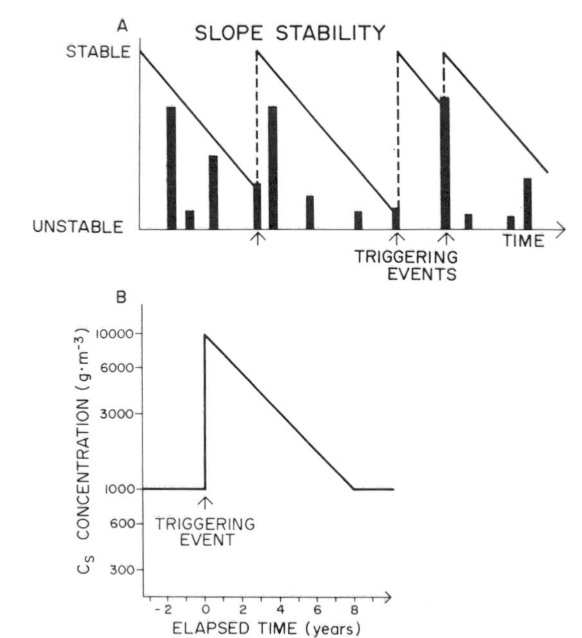

FIGURE 3.13. (A) Dependence of slope stability on the magnitude and timing of the triggering event (after Adams, 1980). (B) The response of suspended sediment concentration in a river after a hillslope collapse of sediment onto the river course (after Adams, 1980).

low sheet slides or solifluction failures; (2) small rotational failures occurring in the weathered crust or in the masses resedimented from a quick clay slide; and (3) large quick clay slides caused by the failure of a quick clay lens. Additionally, the melting of ground ice in the high arctic may form thermokarst retrogressive thaw/collapse failures. Quick clay deposits, formed by the adjustment of marine sediments through consolidation and leaching while exposed to the atmosphere and fresh water (Torrance, 1983), are particularly susceptible to slumping. The original deposits have a stiff desiccated and weathered crust underlain by lenses of quick and sensitive clay. Most landslips are only a few meters thick, although massive quick clay slides may involve the placement of large masses (10^9 to 10^{10} kg) of sediment into the valley rivers for subsequent transport to downstream areas over the next few years (Schafer et al., 1983; Adams, 1980; Jörestad, 1968). The two main causes of landslips are high intensity rain storms and earthquakes. A particular earthquake may result in thousands of slips, each dumping material into the river: the suspended load increases initially by an order of magnitude and then gradually returns to normal (Fig. 3.13B, Adams, 1980; Ovenshine and Kachadoorian, 1976).

In summary, Holocene sea-level changes in fjord areas are a result of complex land-sea-ice interactions that have a major impact on river hydraulics and the supply of sediment to the river.

3.3.3 Climatic Fluctuations

By controlling the governing parameters in equation (3.1), those of precipitation, evapotranspiration, and water storage, climate influences the characteristics of river discharge and thus the pattern of river hydraulics. In turn, river hydraulics control the rate of sediment erosion, transport, and deposition. Climate also controls the type and density of land cover by vegetation, an important factor in stabilizing paraglacial sediment. In addition, the growth or retreat of valley glaciers is dependent on climate-induced firn limit changes.

Climate can be divided into: (1) a time-averaged global signal, (2) the variable local or regional signal, and (3) microclimate relating to the individual elements of the landscape. Global weather patterns are thought by Fisher and Koerner (1980) to be triggered by some complex combination of: (1) variations in solar constant related to changes in the earth's orbit or the sun's activity; (2) atmospheric turbidity or particulate load changes arising from changing volcanic activity and/or continental dust erosion; (3) changes in atmospheric chemistry, that is, CO_2, NO, water vapor, and ozone levels; and (4) resonance in the atmosphere/ocean system. The local signal can enhance or mask the effect of global changes in climate so as to advance or delay the inevitable change in local climate. The local signal is dependent on latitude, altitude, nearness to the sea or ice, topography, ocean currents, prevailing winds, and the permanence of local high- and low-pressure areas.

Since the last Pleistocene glaciation, the earth has undergone a warm climatic optimum: the Hypsithermal period is a geologic-climate unit and is regionally time transgressive (13,000–8000 yr BP for Alaska and 7000–3500 yr BP for Baffin Island). It was a time of warm, dry summers with valley glaciers ablating rapidly. Major outwash deposits formed in front of the retreating ice (Church, 1978; Church and Ryder, 1972;

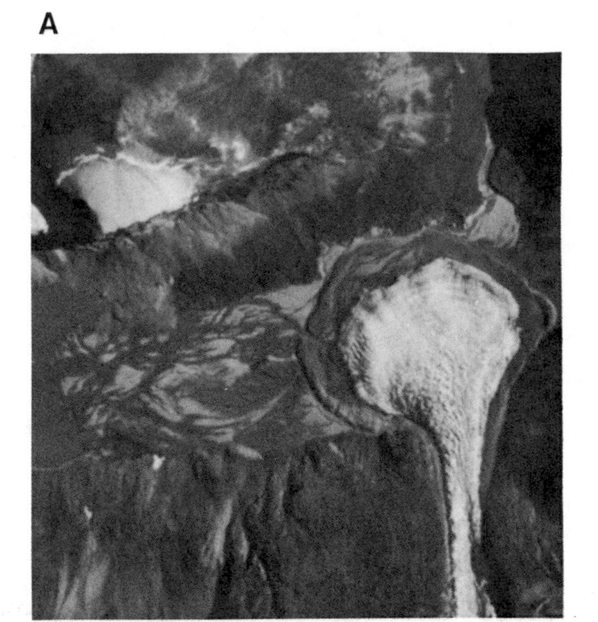

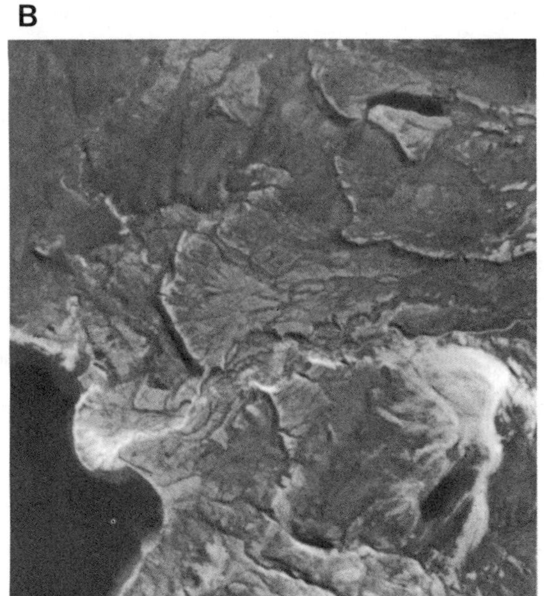

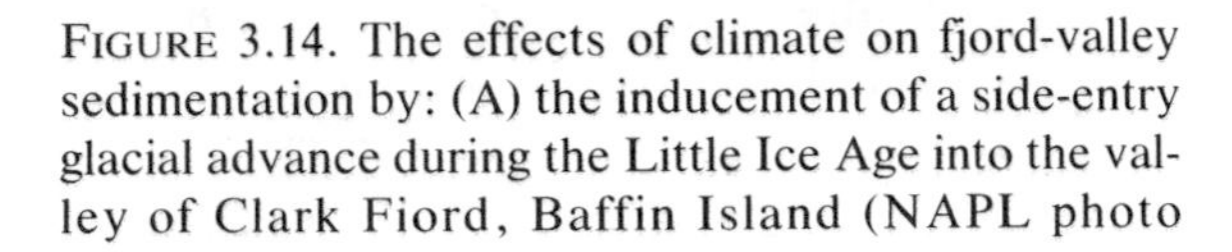

FIGURE 3.14. The effects of climate on fjord-valley sedimentation by: (A) the inducement of a side-entry glacial advance during the Little Ice Age into the valley of Clark Fiord, Baffin Island (NAPL photo A17044-18)—the result has been substantial delta progradation; (B) discharge fluctuations during the Holocene resulting in pulsed delta progradation (Tay Sound, Baffin Island; NAPL photo A16264-60).

Goldthwait et al., 1966). Church (1978) studied the grain size analysis of these deposits and employed tractive force theory (equation 3.5) to calculate that discharges attributable to glacial melt during the Hypsithermal; those discharges were about an order of magnitude larger than presently observed.

The Hypsithermal period was followed by the Neoglacial, a period characterized by a general cooling trend with minor oscillations that ended about 300 to 100 yr BP. The Neoglacial period began as early as 4800 yr BP. It is marked by the alternation of cool moist climes (and glacial advances; Fig. 3.14A) with warmer dry periods (growth of outwash deposits; Fig. 3.14B). The most recent ice advance, known as the Little Ice Age, occurred between 300 and 100 yr BP (circa 1700 AD; Powell, 1980) and appears to be globally synchronous. In the Canadian Arctic, these outwash deposits are followed by fluvial incision and peat development. Church and Ryder (1972) calculated that half of the annual sediment yield during the Neoglacial was derived from erosion of the isostatically uplifted Hypsithermal deposits.

In summary, a period of high rates of glacial-marine sedimentation is followed by a time of abnormally high rates of fluvial erosion, transport, and deposition (Fig. 3.14B). Glacial and proglacial processes leave behind vast quantities of easily eroded and transported sediment along the length of a fjord valley, during both glacier advance and retreat. Glacier movement is also accompanied by changes in the relative sea level. Marine transgressions allow the deposition of fine-grained sediments onto the drowned river valley; when later subaerially exposed, these clays become geotechnically unstable. In addition, uplift and gradient tilting is accompanied by fluvial rejuvenation that produces increased denudation, delta building, and basin infilling. The major factor controlling paraglacial sedimentation is climate, for climate directly controls the growth and recession rates of glaciers, river discharge intensities, and the character of the vegetation cover.

3.4 Fjord-Head Deltas

Previous sections reviewed the principles governing the movement of water and sediment down fjord valleys and show how the paraglacial history of a river can determine its present state of equilibrium. This section outlines the factors influencing the growth of marine deltas at the head of fjords with emphasis on the morphology of topset deposits. The sedimentology of prodelta and basin deposits is discussed in Chapters 4 and 5.

Topset deposits, in the classic Gilbert delta (Gilbert, 1890), are formed as a result of the radical change in hydraulic conditions near the river mouth; bed load is deposited as stream competence decreases sharply. Bed-load deposition can occur supratidally, intertidally, and subtidally during periods of maximum discharge. Factors that strongly influence the style and rate of development of topset deposits include: (1) the strength and periodicity of the fluvial discharge; (2) the river thalweg slope (gravity potential energy); (3) climate (periglacial vs. temperate conditions); (4) relative sea-level history; (5) sediment supply; (6) wave energy and direction; and (7) tidal energy. Because fjord-head deltas are confined by steep-sided mountains, they have virtually no littoral transport and are thus nearly always in a state of construction or progradation.

Nearly all fjord deltas have unique morphologies, each reflecting its own response to the above seven factors and the paraglacial history of the incoming river(s). This fact was recently emphasized by Syvitski and Farrow (1983) in their discussion of two British Columbia fjord-head deltas, the Klinaklini and the Homathko, which possess similar climate, sea-level history, tidal, wave, and discharge characteristics, yet have different styles of progradation (Fig. 3.15A,B). The variation is related to the gradient of the incoming rivers and the direction of waves onto the deltas. The Klinaklini River with its steeper thalweg results in an increased bed-load discharge with subsequent destabilization and migration of its distributary and submarine channels. The Homathko delta has its dominant fetch at an angle to a stable single channel discharge resulting in expansive wave-built linguoid bars.

If, however, there had to be a straightforward division of fjord-head deltas, a good separation would result from the use of climate. The wet and temperate fjords have many deltaic features common to their open ocean counterparts. Arctic deltas (sandurs) are strongly influenced by their lack of stabilizing vegetation, by glaciers, and by unique periglacial landforms.

3.4.1 High-Latitude Fjord-Head Deltas and Sandurs

Sandurs, commonly found in arctic fjord valleys, are not exclusive to high-latitude fjords, but they share many of the same features as arctic fjord deltas. Common features include strong winds, incomplete vegetation cover, intermittent discharge pattern, and high competence resulting in large bed load transport during short-lived events. By definition, sandur refers to an alluvial outwash plain formed by rivers carrying meltwater away from glaciers. Sandurs are areas of rapid aggradation, which are crossed by braided streams that continually shift their pattern and course as local erosion and deposition occur (Church, 1972).

Periglacial landforms develop from the combination of intense frost action, permafrost, nivation, strong winds, incomplete vegetation cover, and intermittent discharge pattern. The result is a unique morphogenetic landscape that includes sandurs. The weathering activity of numerous freeze-thaw cycles dominates the process of both detritus production along the fjord walls and attrition of material on the sandurs (Church et al., 1979; Church, 1972). The materials produced by hydrofracturing are mostly sand size or larger, and chemical weathering is not sufficiently active to contribute much clay size mineral material. On fluvially inactive isostatically raised terraces, frost-heaved boulders are concentrated on the ground surface by the expansion and contraction of pore water in the sediment. Ice-wedge polygons can develop on both the active and inactive portion of the delta surface (Syvitski et al., 1983a). They form from cracks that are a result of stresses generated by a drop in ground temperature from summer to winter. The cracks become water saturated during the summer and freeze during the winter. An ice lens develops in the permafrost zone, which does not thaw during the summer but rather undergoes some expansion causing a surface depression. As the process continues year after year, the ice wedge increases in size as does its surface expression (Fig. 3.16A). It has been proposed that sand-filled polygons (rather than ice-filled) may result, in part, from aeolian sand transport that infills the fractured activation surface (Pewe, 1959).

Under certain hydrologic conditions, ice intrusions can arch the overlying ground on san-

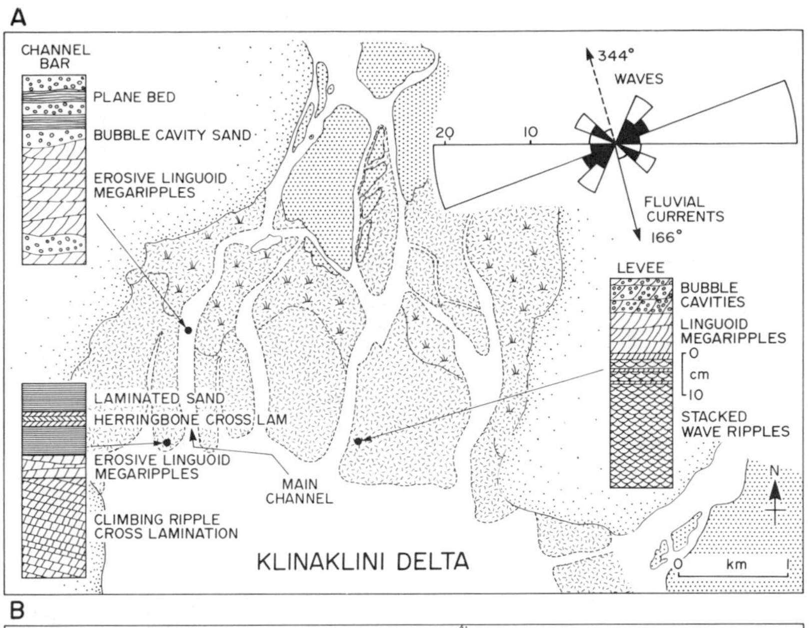

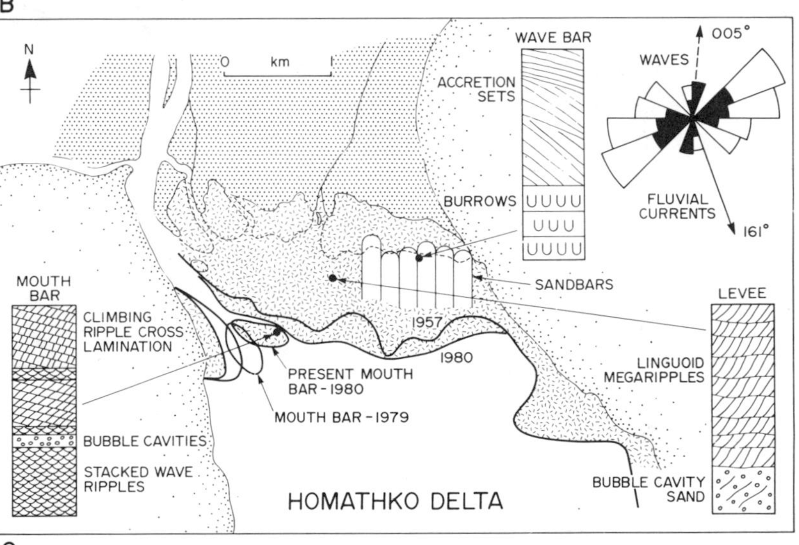

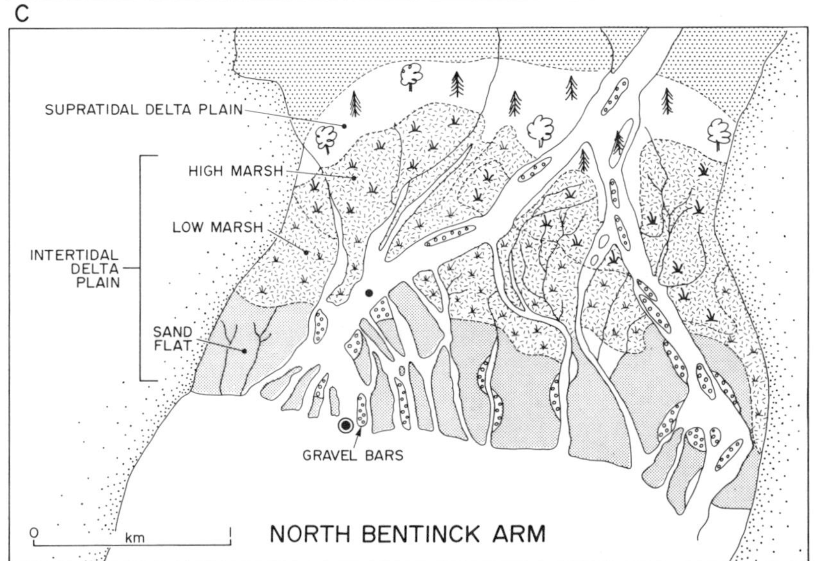

FIGURE 3.15. Structures and facies of three temperate fjord-head deltas found in British Columbia. The Klinaklini and Homathko (A and B) have nearly identical factors influencing their style and rate of progradation. The differences in morphologies relate to variations in river thalweg and direction of waves onto the delta (after Syvitski and Farrow, 1983). (C) Typical subdivision of facies and position of high-tide (solid dot) and low-tide (circled dot) mouth bars (after Kostaschuk and McCann, 1983).

FIGURE 3.16. (A) Ice-wedge polygons on raised terraces of the Keel River delta, Baffin Island. (B) Overbank clays temporarily protecting the sandy fluvial sediment beneath from wind erosion on the McBeth delta, Baffin Island. (C) Aeolian sand accumulation on the lee side of protruding bedrock headland on the McBeth delta, Baffin Island. (D) Raging dust storm overhead, while in the foreground sand drifts on wind-blasted surface (Itirbilung Delta, Baffin Island).

durs into fractured domes called pingos. They have been identified in the fjord valleys of Spitzbergen (Piper and Porritt, 1965) and Greenland (Muller, 1959; Boyd, 1948; Allen et al., 1976). They can rise 10 to 30 m above the stream bed with ellipsoid diameters of up to 400 m and 200 m, respectively. Sandur pingos are "open system" forms that are generated by ground water of meteoric origin flowing under considerable hydraulic head through or under the permafrost layer. At some weak point in the permafrost layer, ice pressure builds up sufficiently to arch the overlying permafrost, producing the pingo's surface expression. The ice core may be less than 1 m below the pingo surface and can be exposed in places. Springs and lakes may form at pingo crests if the ground water can break through the arched permafrost. The greatest significance of pingos to delta development is as sources of sediment as the fluvial channels erode their sides.

Low precipitation, freeze-drying of exposed sediment, sparse vegetation cover, and strong winds combine to make aeolian transport of sediment an important modifier on sandur deltas. Fjords by the nature of their shape make excellent wind tunnels. Katabatic and regional winds reach maximum velocities of up to 250 km h^{-1} in the Canadian and Greenland arctic fjords, making arctic fjords one of the more windy areas of the world. Such wind speeds are capable of moving grains of coarse sand and fine gravel. Since grain size decreases down-valley as a result of fluvial sorting and abrasion, the distal (seaward) end of sandurs shows the greatest impact of aeolian action. Erosion forms include large deflation bowls and lags (Fig. 3.16B), wind-blasted boulders (Gilbert, 1978), and polished frozen fluvial bed forms (Syvitski et al., 1983a). Depositional forms include ripples, small dunes, large extensive sand banks, and lee deposits that form behind everything from bedrock outcrops (Fig. 3.16C) to tundra grasses and fluvial channels (Syvitski et al., 1983a, 1984a). For instance, a 2.1-km^2 area of sand dunes covers an aeolian bank that rises 100 m above the Maktak Fiord sandur surface (Gilbert, 1983). Silt and clay size particles (loess) are blown into suspension and carried up many hundreds of meters (Fig. 3.16D); they are deposited many kilometers downwind, either directly into the fjord waters, or onto sea ice that may cover the fjord. The movement of sand is more local. Through the process of saltation, the bed load can extend several meters above ground level. During summer, saltating bed load will be trapped in the marine waters immediately seaward of the sandur. During the period of winter sea-ice cover, the bed-material load may move farther out into the fjord environment on the ice platform. Gilbert (1983) observed that the main season of aeolian action is winter because: (1) during the summer discharge season much of the sandur surface is wet; (2) during the winter the surface is dry, and erosion is unrestricted; and (3) sediment frozen before drying may be dried through sublimation during the drier winter season. Alternating layers of sand and snow, which accumulate annually on banks along the Maktak sandur, attest to the importance of winter transport.

Fluvial transport of bed load dominates the development of sandurs (Church, 1972), and flood events dominate the discharge pattern owing to the very high proportion of surface runoff. Between 25 and 75% of the total sediment transport may occur during the 4 or 5 peak flow days (Church, 1972). Because of the discontinuous nature of sediment movement, there is considerable storage of material along the channelways; sandurs are aggradational features by the very nature of their inherent sediment transport processes, unless sea level is rapidly falling. The sandur surface can be divided into three zones: (1) a proximal zone with a few well-defined streams with deep and narrow channels; (2) an intermediate zone that possesses the classical attributes of wide and shallow sandur channels; and (3) a distal zone where the channels may merge into a single sheet of flowing water that runs out over the delta front and on into the fjord. The distal zone is the major zone of modern sedimentation.

The following is a general model of sandur fluvial sedimentation processes (Church, 1972). During a flood event, the sediment load may become too great for normal channel processes to adjust to them. Extraordinary changes result; local aggradation, channel division, and braiding may occur, so that flow is diverted to new areas of the sandur. Thus the surface consists of amalgamated flood deposits: river bars and channel fill, sandur levees, and sheet deposits. The channel fill and bar deposits form on the falling river stage as bed load falls out with the rapidly dropping stream competence. Sandur levees are not well defined and appear as laterally graded bars along the river. If a major portion of the river is diverted behind a levee, the entire river channel may change course. Sheet deposits form when the distal portion of the entire active sandur surface is wetted and overflow processes dominate sedimentation (Fig.3.17D). Grain size decreases and sorting increases toward the sea, yet there is a lack of pattern in the fines. Sediment sorting is achieved on the river bed by selective erosion rather than by selective deposition. Sediment size changes little downstream on the stream bed; homogeneous conditions pertain throughout the channel length during normal flow. Sediment sizes below those found in the channel bed can be transported right through the system at normal flow levels. At normal flows the river is purely a transporting feature rather than a feature of aggradation.

Although sandurs (in the classical sense) have braided channels, there is a continuum of channel types to be found in arctic fjord deltas. The continuum may be explained using the alluvial channel classification scheme of Rust (1978). The scheme makes use of a "braiding parameter," which is defined as the number of braids per mean meander wavelength. Braids are defined by the mid-line of the channels surrounding each braid bar. Channels can then be divided into low and high sinuosity at a braiding parameter of 1.5 and into single and multichannels at 1.0. There are then four channel types. The two most common are: (1) meandering (single channel, high sinuosity); and (2) braided (multichannel and low sinuosity). The less common types are: (3) straight (single channel, low sinuosity), and (4) anastomosing (multichannel and high sinuosity). As the thalweg slope of a river decreases, so does the mean bed-shear stress; bedload transport decreases, allowing channel deepening and meandering. Braided channels can then become anastomosing (Fig. 3.17E). The same effect can be initiated with slow decrease in discharge (over years) or where channel reworking cannot keep pace with local uplift (Fig. 3.17C). Where the drainage basin is large and the discharge is more modulated, there is a likelihood of better sediment sorting and decreased bed load transport in the distal reach of the river. If the fjord river system has a large lake to modulate its summer discharge, the hydrograph shows less severe flood events, longer season flows, and at least initially, a bed load trap in the lake. The result is a single channel, straight to meandering river, with overbank sedimentation representing the dominant process of fjord-valley aggradation (Fig. 3.17A, B).

The distal end of periglacial deltas, whichever the river type, is mostly a continuation of the river form into the sea. This is true especially of fjords that have a low tidal range. Deposition at the sandur delta front, although localized to the area around the river mouth, often extends relatively uniformly across the fjord width as a result of frequent channel switching (Fig. 3.17D, E, F). Exceptions occur when the river enters at an angle to the main fjord axis (Fig. 3.17C) or through a tributary fjord valley (Fig. 3.17B). Shorefast ice can sometimes form an icefoot that protects the delta from ice gouging by sea ice (Petersen, 1977). In addition, sea ice can limit sediment reworking by wind-induced wave action on the delta front for a large part of the year. With the stabilizing influence of river incision during the Holocene, however, some sandurs have had portions of their delta front reworked into beach ridges. These storm ridges are preserved on the isostatically rebounding sandur surface. The result is a series of raised beaches sometimes numbering in the dozens, the oldest reaching an elevation some 50 to 150 m above sea level.

Grain size decreases rapidly over a short distance (Fig. 3.18A). In addition, there is an abrupt decline in sorting across the delta face owing to the large proportion of material that arrives at the delta front as bed load and immediately slides down the delta front (Fig. 3.18B). Ice rafting of delta gravel contributes to this poor sorting offshore (Knight, 1971; Aitken and Gilbert, 1981). Mineral sorting is most profound on the distal portion of sandurs with an extremely marked fallout of heavy minerals (Fig. 3.18C; Knight, 1971), including a concentration of gold (Reimnitz et al., 1970).

3.4.2 Temperate Fjord-Head Deltas

Temperate fjords, being both warm and wet, support a dense vegetation cover in their river valleys, usually a mixture of conifers and deciduous trees. The vegetation growth is partly successful in stabilizing river banks as is driftwood in stabilizing the delta surface. The result is fewer channels that are both deep and narrow compared with the arctic sandur situation. Valley vegetation and a wet climate are also effective in eliminating aeolian transport. Temperate fjord rivers have more complex and longer duration discharge events compared with the above arctic counterparts, primarily a function of increased precipitation storage (see Section 3.1). These rivers are also closer to an equilibrium profile; river thalweg, channel width, and hydraulic radius have exponential forms with dis-

FIGURE 3.17. Variations in the style of delta progradation for high-latitude fjord-head deltas and sandurs (as seen from vertical photography). (A) Single channel and low sinuosity (straight) delta front, Clyde Fiord, Baffin Island (NAPL A17015-104). (B) Single channel but increased sinuosity delta front from tributary valley in Walker Arm, Baffin Island (NAPL A17047-50). (C) Delta growth through meander incisement into raised Hypsithermal glaciomarine terraces, Sam Ford Fiord, Baffin Island (NAPL A17046-49). (D) Multichannel and low sinuosity (braided) delta front, Itirbilung Fiord, Baffin Island (NAPL A17011-16). (E) A combination of braided and anastomosing channels over the North Pangnirtung sandur front (Baffin Island; NAPL A24200-45). (F) Classic braided channels on a sandur front, Maktak Fiord, Baffin Island (NAPL A24200-34). Note large aeolian sand bank on northern edge of valley.

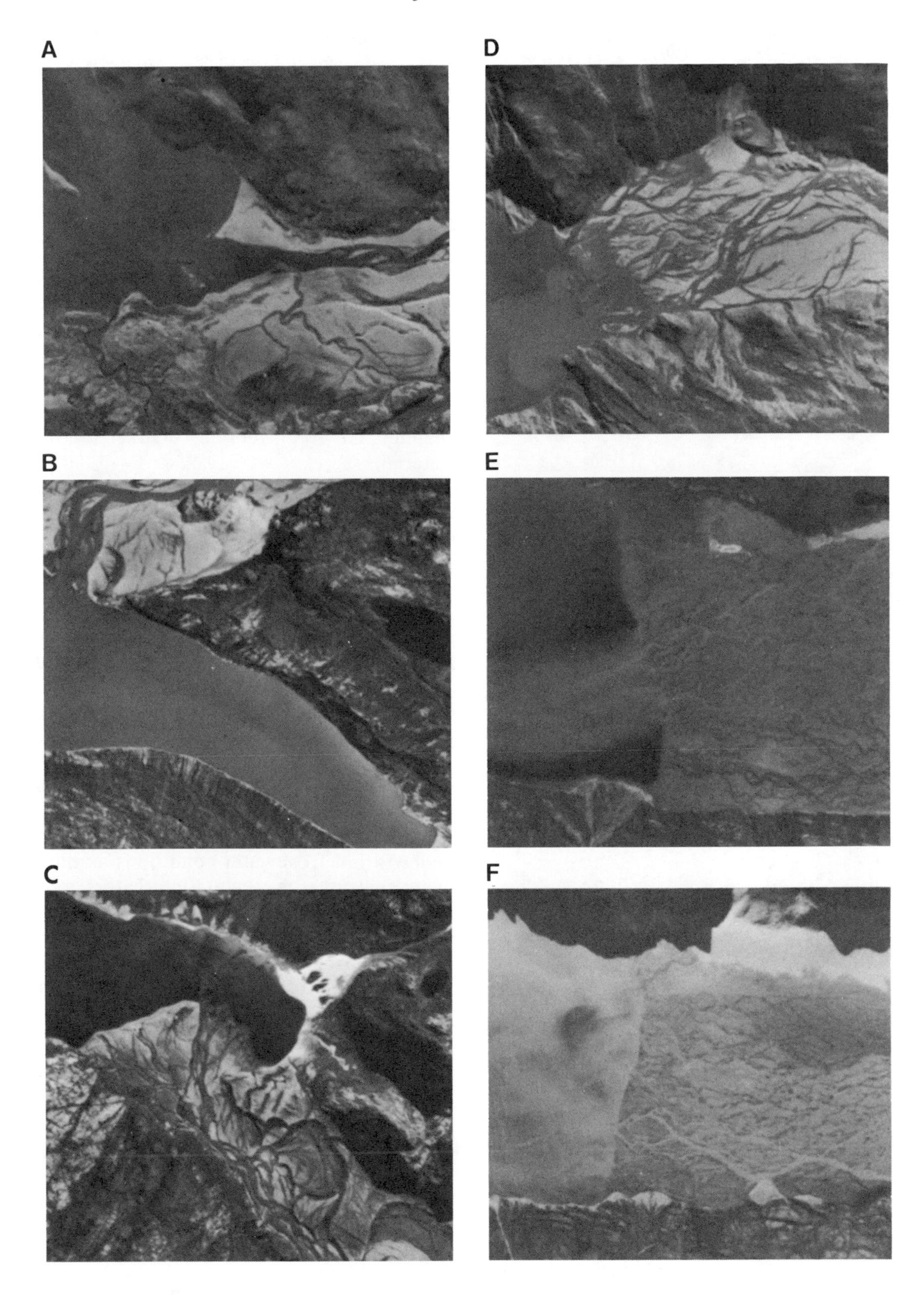
A
B
C
D
E
F

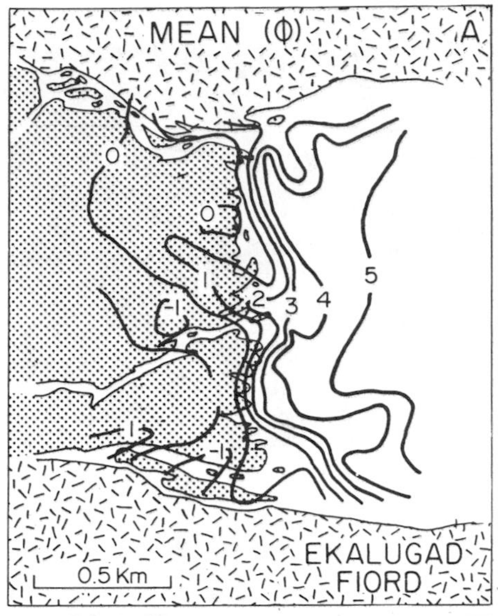

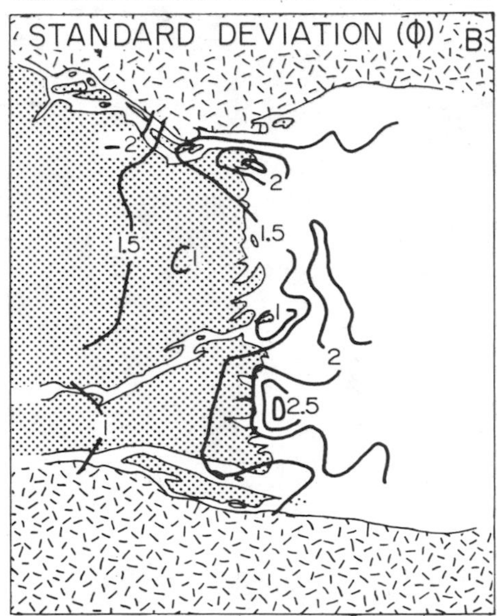

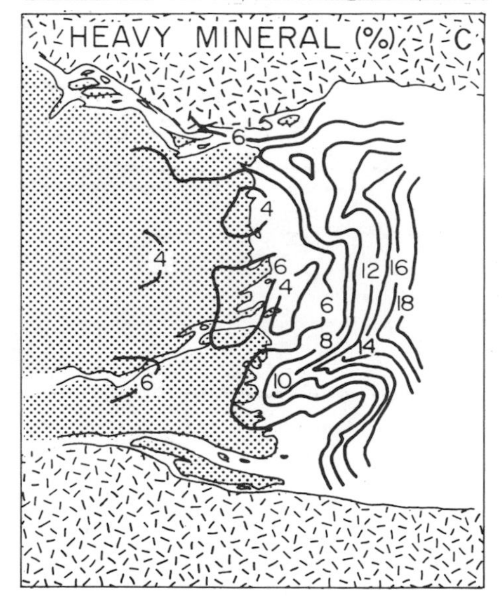

FIGURE 3.18. Grain size characteristics and heavy mineral content of topset and foreset deposits on Ekalugad delta, Baffin Island (after Knight, 1971). The mean and standard deviation measures are contoured in standard phi units $[-\log_2(\text{mm})]$.

tance down the fjord valley (equation 3.11; Syvitski and Farrow, 1983). The channels widen and their depths decrease toward the sea, resulting in a rapid falloff of bed load transport toward the delta. The climate of temperate valleys is more suitable to chemical weathering and large amounts of clay-sized material are produced. In addition to hydrofracturing as a means of detritus production, the temperate valleys have greater sediment production from fluvial bed load abrasion, a fact that is related mostly to the increase in annual duration of discharge and thus bed load transport. Flood events result in overbank processes involving relatively fine-grained sediments (levee development, crevasse splay, and flood-plain deposits), while river channels have considerably coarser textured material.

Temperate fjord-head deltas are zones of high sedimentation, and result in rapid progradation and accretion. The delta plain usually covers the entire fjord width and the intertidal length is a simple function of tidal range and river thalweg slope. The delta plain can be divided into supratidal and intertidal components (Fig. 3.15C; Kostaschuk and McCann, 1983). Supratidal deposits include freshwater (riverine) silt and sand deposits over a forested plain, developed during periods of high discharge. Bell (1975) divided the intertidal zone into three divisions. The *upper* tidal flats (or high marsh on Fig. 3.15C) mark the transition of marsh to shrub to forest. Sediment is deposited during the flood-tide stage under turbid freshwater conditions as the effluent spills over the delta. Bioturbators are absent and horizontal (silty) laminations are preserved. The *intermediate* zone (or low marsh on Fig. 3.15C) contains sedge grasses that trap considerable quantities of fine silts and clays during periods of flooding. This zone becomes inundated with saline water during periods of low river runoff and can support minor bivalve and worm burrowers. The *lower* flats consist of mouth bar and sand flats that are reworked by tidal and wave forces (Fig. 3.15C). Bioturbation is noticeably absent in this zone as a result of rapid sedimentation. The intertidal zone may have one or more river channels and tidal channels that intersect the plain.

Where the tidal energy exceeds the fluvial energy, the river channel broadens or flares out seaward (e.g., Kitimat Delta; Bell and Kallman, 1976). However, when the channel is dominated by fluvial discharge, the channel form undergoes little change across the delta plain until near the delta edge where mouth bars develop. At times of high tide, distributary bars may form farther up the channel as the sea water intrudes as a salt wedge along the river bed. The liftoff point at the head of the salt wedge is a place of rapid bed load deposition and a broad radial distributary mouth bar forms. Over the bar there is a seaward transition from higher energy to lower energy bed forms with a concomitant decrease in grain size. This reflects the deceleration of the effluent over the distributary mouth bar. In the Bella Coola (Fig. 3.15C), Kostaschuk and McCann (1983) found that the deceleration effect gave rise to gravel armor in the proximal portions of the bar. This in turn, gave way to: (1) flow-parallel lobes of gravel advancing over current rippled sand; (2) straight to slightly sinuous crested current ripples of medium sand, having small-scale planar tabular cross-beds; and (3) long and straight crested, rounded symmetric to asymmetrical wave-generated ripples combined with current-produced lunate and linguoid ripples (Fig. 3.15B). The low tide outlet (Fig. 3.15C) has one or more mouth bars that extend across the channel mouth: the bars slope gently landward and steeply seaward. The bars form on the leading edge of the delta and become subaerially exposed only during extremely low tides. The proximal part of the bar is composed of imbricated gravel grading distally into straight crested ripples of medium sand. Mouth bars are ephemeral features (Syvitski and Farrow, 1983), and their positions may change from year to year (Fig. 3.19).

Levees of active distributary channels can form a series of overlapping accretionary terraces formed initially as starved ripples. The terraces are best developed where the tides are semidiurnal and when the river is at bankfull stage: the effective width and depth of the channel changes over the flats in accordance with the tidal stage. Preserved sedimentary structures then are a series of fluvially deposited sand overlying stacked wave ripples formed on the tidal flat prior to levee development (Figs. 3.15A, 3.20A). The fluvial sand is preserved as linguoid current ripple bedding and as bubble cavity sand (indicative of rapid deposition). On stable levee deposits farther upstream, linguoid current ripple or megaripple bedding persists throughout with no preserved tidal-flat bedding.

Tidal channels migrate across the tidal flats in meander form; shallow dipping strata of accretionary point bars are commonly preserved. Well-consolidated muds may show prominent cracks (some more than 1 m deep) lined with iron hydroxide. They represent tensional failure on the inner bank of creek meanders (Syvitski and Farrow, 1983). The more fluid muds show a range of biogenic structures including green algal mats with gas heave domes and surface tracks and skid marks of birds.

Where the fetch is angled to the river discharge, wave-induced accretionary bars can form (Kostaschuk and McCann, 1983; Syvitski and Farrow, 1983). Coarse sand and gravel will move landward by wave action as large asymmetrical ripples. Internally the bars show the low-energy tidal-flat regime at the base, overlain by small current ripples produced by ebb drainage at the front of the advancing bar, and capped by upward flattening and upward fining co-sets with frequent internal truncations (Figs. 3.15B, 3.20B).

Although the previous discussion on the morphology and structures within temperate fjord-head deltas is based on a wide selection of examples, there are some noteworthy exceptions, for example, Turnagain Arm, Alaska, and Pitt River delta, British Columbia. Turnagain Arm at the head of Cook Inlet, Alaska, is 75 km long and up to 26 km wide. Of that area, 60% of the arm is exposed at low tide, which is unusual even considering its 11-m tidal range. The delta apparently has grown, not from the rivers at the fjord head, but from an outside and seaward sediment source (Fig. 3.21A; Ovenshine et al., 1976). Tidal energy is dominant, with surface tidal currents of 2.5 m s^{-1} and a 1.5 m high tidal bore that travels with speeds up to 16 km h^{-1}. The tidal bore marks the change in tides with the onset of a turbulent solitary wave. The tidal flats have few major tidal channels (Fig. 3.21B), and tidal bars (Fig. 3.21C) are the dominant morphological features of the flats. Energy levels correlate with bar topography: crests of bars are finer grained than the lower bar sediments. The flats have three categories of sedimentary structures: (1) straight crested to slightly sinuous ripples; (2) lunate and linguoid ripples; and (3)

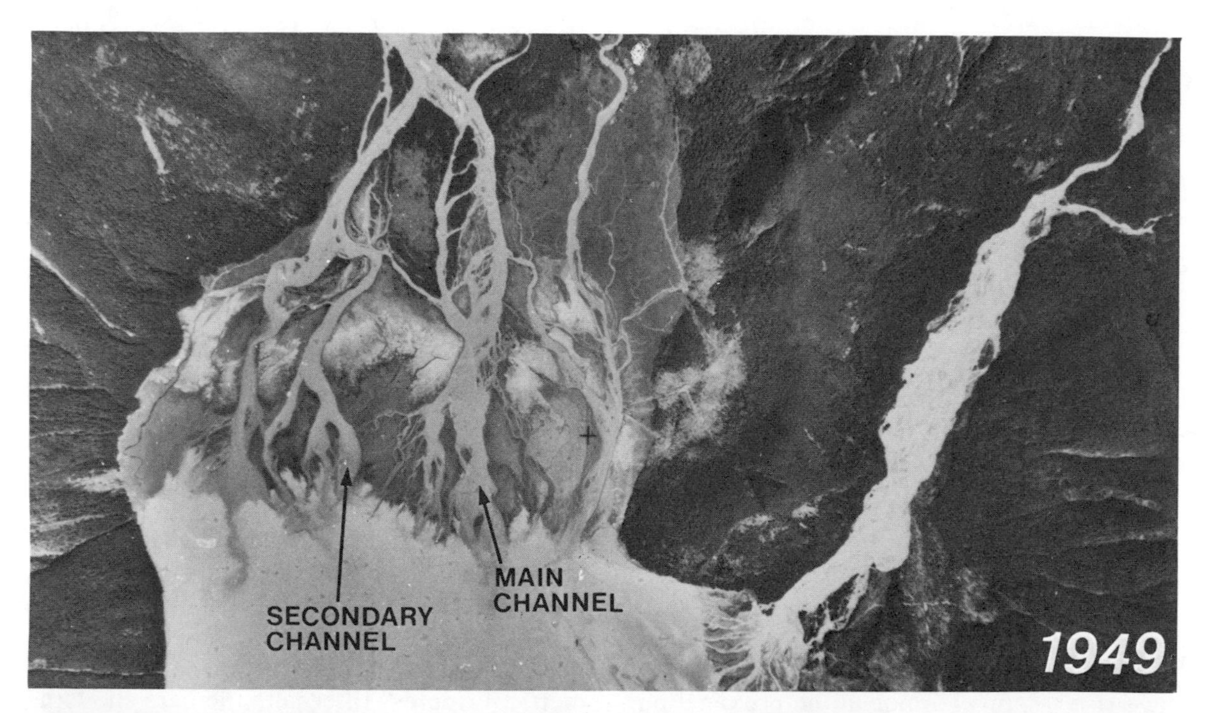

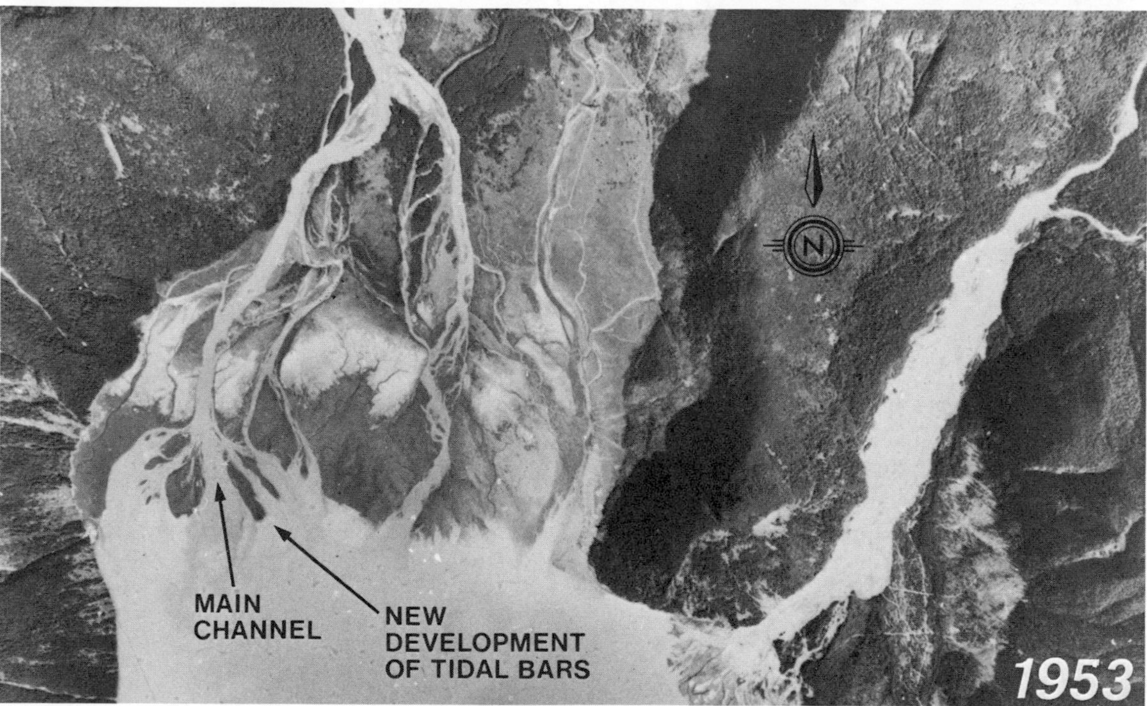

FIGURE 3.19. Vertical air photos of the Klinaklini delta, British Columbia (see Fig. 3.15) taken four years apart (1949: NAPL A12131-4; 1953: NAPL A13795-4). Dominant changes include a shift in the main channel and formation of a new set of mouth bars (after Syvitski and Farrow, 1983).

ladder ripples of wind-wave origin. A detailed examination of box cores for internal sedimentary structures revealed four types: (1) planar parallel lamination and ripple-generated small-scale cross-lamination; (2) small scour surfaces from the migration of starved ripples; (3) bubble cavity sand; and (4) waterlogged plant debris and rip-up clasts. Together these features were organized into four facies by Ovenshine et al. (1976; Fig. 3.21C). The lower flats (zone A, Fig.

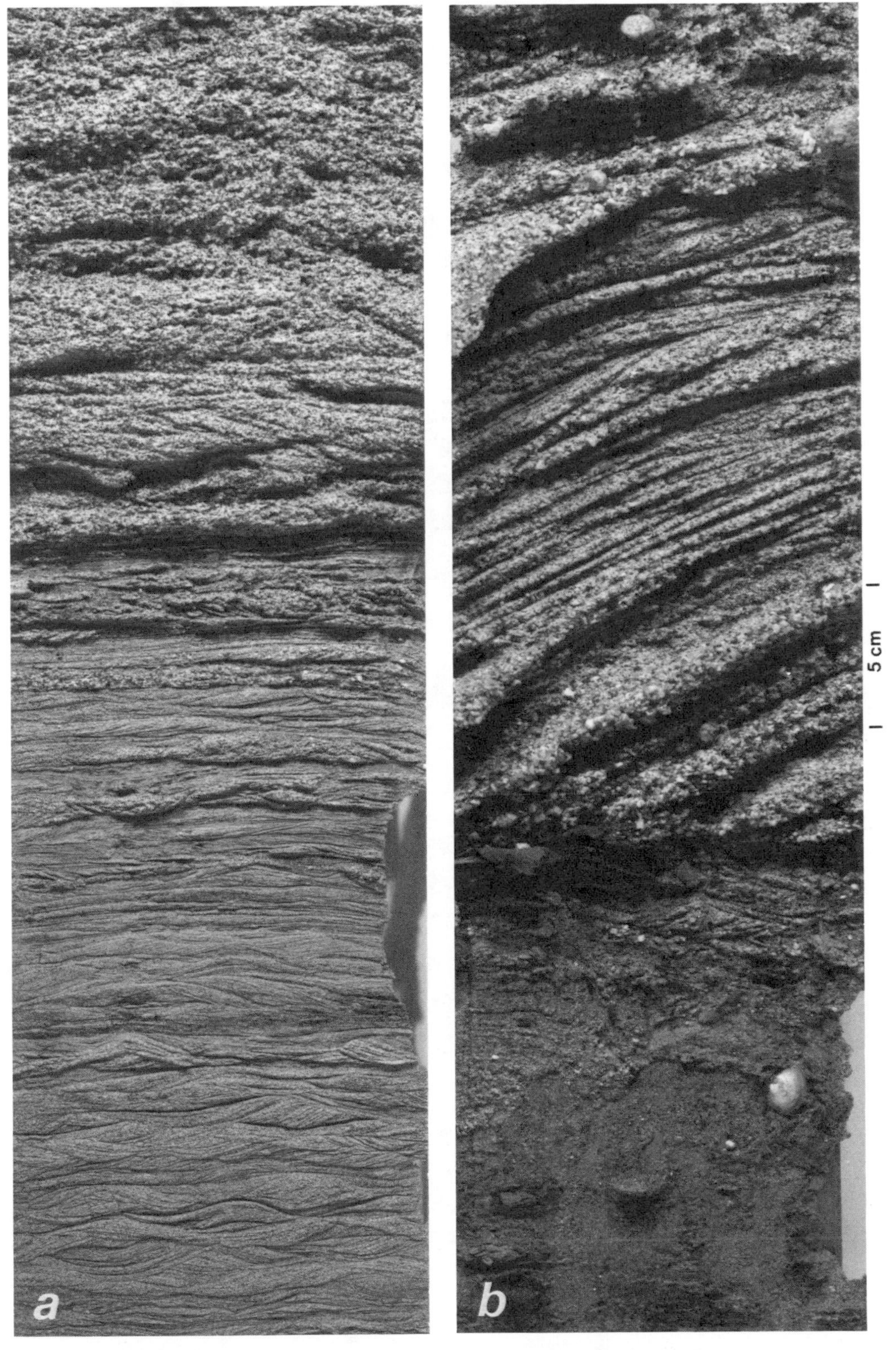

FIGURE 3.20. Epoxy resin peels of box-core samples showing contrast in style of upward coarsening of sequences. (A) Lower unit is of stacked wave ripples, sometimes with preserved tidal-flat form, followed by successively thicker units of fluvially deposited sand (see Fig. 3.15A, Klinaklini delta). (B) Lower unit is of upper tidal-flat burrowed sands with *Macoma* in life position sharply overlain by upwardly convex accretion cosets of wave-built bar (see Fig. 3.15B, Homathko delta) (after Syvitski and Farrow, 1983).

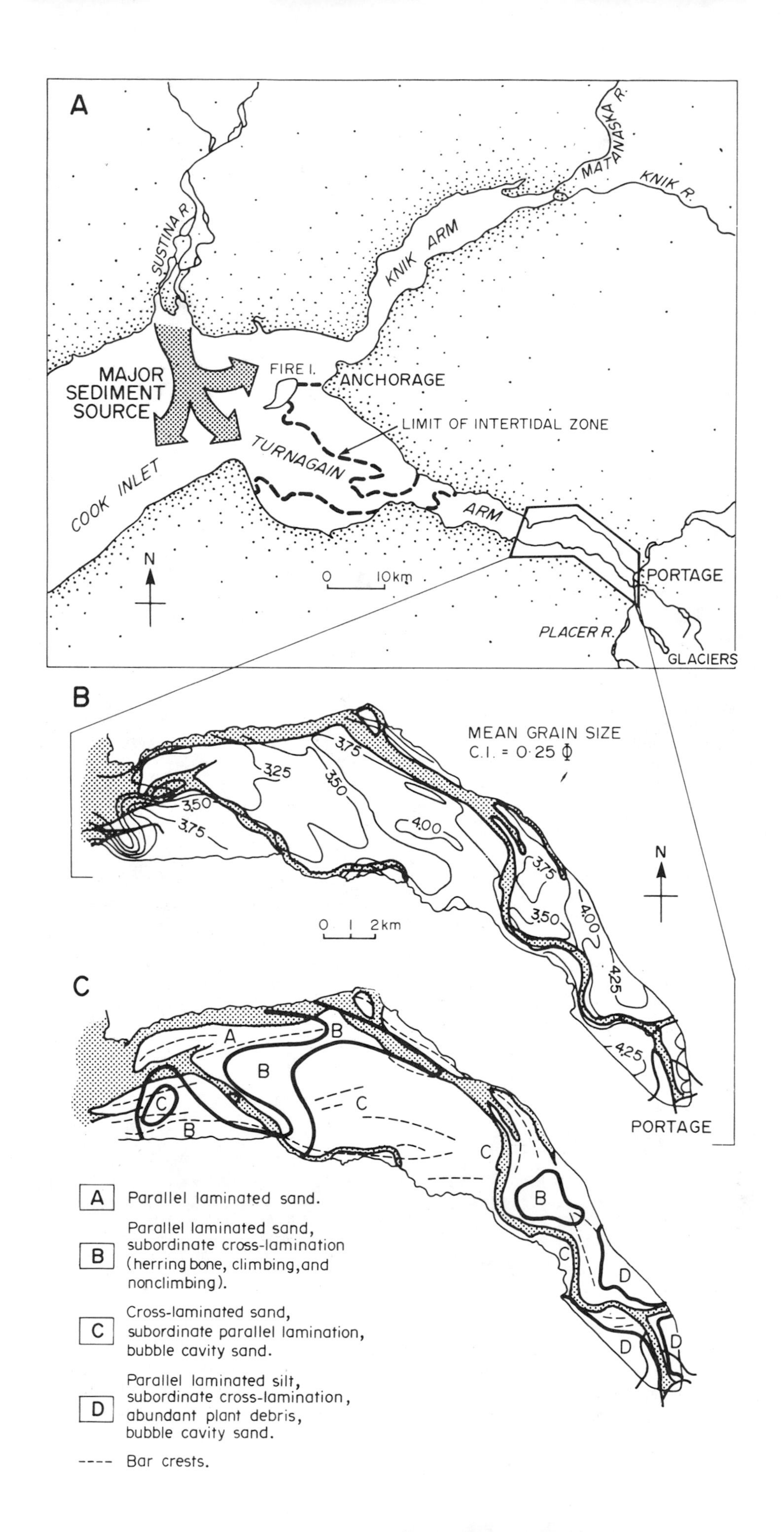
A
SUSTINA R.
MATANASKA R.
KNIK R.
KNIK ARM
MAJOR
SEDIMENT
SOURCE
FIRE I.
ANCHORAGE
LIMIT OF INTERTIDAL ZONE
TURNAGAIN
ARM
COOK INLET
N
0 10km
PORTAGE
PLACER R.
GLACIERS
B
MEAN GRAIN SIZE
C.I. = 0·25 Φ
3.25
3.50
3.75
4.00
4.25
N
0 1 2km
PORTAGE
C
A
B
C
D
A Parallel laminated sand.
B Parallel laminated sand, subordinate cross-lamination (herring bone, climbing, and nonclimbing).
C Cross-laminated sand, subordinate parallel lamination, bubble cavity sand.
D Parallel laminated silt, subordinate cross-lamination, abundant plant debris, bubble cavity sand.
---- Bar crests.

3.21C) are an area of rapid deposition producing parallel-laminated sand. Next is a transition zone B of parallel-laminated sand with subordinate cross-lamination (herringbone, climbing, and nonclimbing). Zone C, the intermediate flats, have primarily cross-laminated sand with subordinate parallel lamination and bubble cavity sand. The high flats (zone D) consist of parallel-laminated silt, subordinate cross-lamination, plant debris, and bubble cavity sand.

Turnagain Arm is also of interest in that four deep well cores have been drilled there: the deepest, 300 m, covers an earth history of 14,000 years BP (Bartsch-Winkler et al., 1983). Figures 3.22 and 3.23 give the lithology, depositional environment, and sedimentation rates over the last 8000 years BP. The core indicates cyclic episodes of a prograding shoreline with subaerial, intertidal, and subtidal deposition. The cyclicity is a function of: (1) sea-level changes whose net component, although modified by glacial re-advances, is mostly that of emergence; (2) consolidation of sediments from shaking during earthquakes; and (3) tectonic adjustment of the earth's surface after earthquakes and/or volcanism (Bartsch-Winkler et al., 1983). The dominant feature is a remarkable reduction of sedimentation rate by an order of magnitude since 5740 years BP, that is, from an area of sediment deposition to an area of sediment bypassing. The most reasonable explanation is a seaward shift in the high-tide water line.

Turnagain Arm is a classic example of the effects of a large earthquake (Richter magnitude 8.5 in 1964), which caused the tidal flats to subside by 2.4 m. As a result, some 20×10^6 m^3 of silt were spread over 18 km^2 of land previously above tidewater. The area is now one of quicksand because of the uncompacted nature and predominance of silt; the sediment is readily liquefiable when saturated with water (Ovenshine and Bartsch-Winkler, 1978). The cycle of earthquake, deposition, and land rejuvenation took only 10 years.

Pitt River delta represents a rare form of a tidal delta situated at the mouth of a relict fjord that is now a freshwater lake. Pitt Lake lies 30 km inland, adjacent to the Fraser River valley in southern British Columbia. The Fraser River has built an expansive delta since the late Pleistocene. Growth of this delta has sealed off a number of fjords during the Holocene, relict fjords that have long since been flushed of all saline water (Fig. 3.24; Ashley, 1978; Clague et al., 1983). Nevertheless, the present-day effect of the salt wedge extending up the Fraser River during flood tide is to retard flow of the Fraser, backing up its water into the Fraser-Pitt confluence. The consequence is to cause Pitt River to reverse its flow and rise by 2 m each tidal cycle (Fig. 3.24; Ashley, 1979). Although the net discharge is seaward, Pitt River is a flood-dominated system in terms of basal shear stress, stream power, and net sediment transport. The present delta surface contains a single river-tidal channel with a right-angle bend (Fig. 3.24B). A number of minor ebb-flowing drainage channels cover the delta surface, eventually leading back to the main distributary channel. The main channel is bordered by well-developed levees, occasionally broken by crevasse splay channels that represent the main source of sediment to the delta surface. The topset surface is flat, containing horizontally stratified and highly cohesive beds of silt and sand (Ashley, 1979). The cohesion comes from floral growth (macrophytes) and mucus-binding diatoms. In the main channel, point bars are accreting on the "upstream" side owing to flood-water influence (Ashley, 1980). The mean grain size of bed material decreases toward the lake, away from the Fraser influence.

Sediment characteristics of temperate fjord deltas are similar to their arctic sandur counterparts in that they exhibit a wide range in grain size and are petrologically very immature with a high lithic (rock fragment) component. Grains are generally of low sphericity and roundness with little indication of weathering: they are typical of first-cycle glacially eroded particles. Exceptions include local deposits of reworking, such as wave-generated accretion bars (Fig. 3.15), or the severely reworked tidal deposits of Turnagain Arm, Alaska. Such reworked sediment is composed of abraded particles having increased grain roundness. Mineralogy tends to be size related. Coarser grains are rich in quartz, feldspar, and lithic fragments; micas are more concentrated in the mud fraction.

◁

FIGURE 3.21. Intertidal zone of Turnagain Arm, Alaska, showing (B) contoured patterns of mean grain size (in phi units), and (C) four dominant facies based on sedimentary structures (after Ovenshine et al., 1976). See text for details on facies.

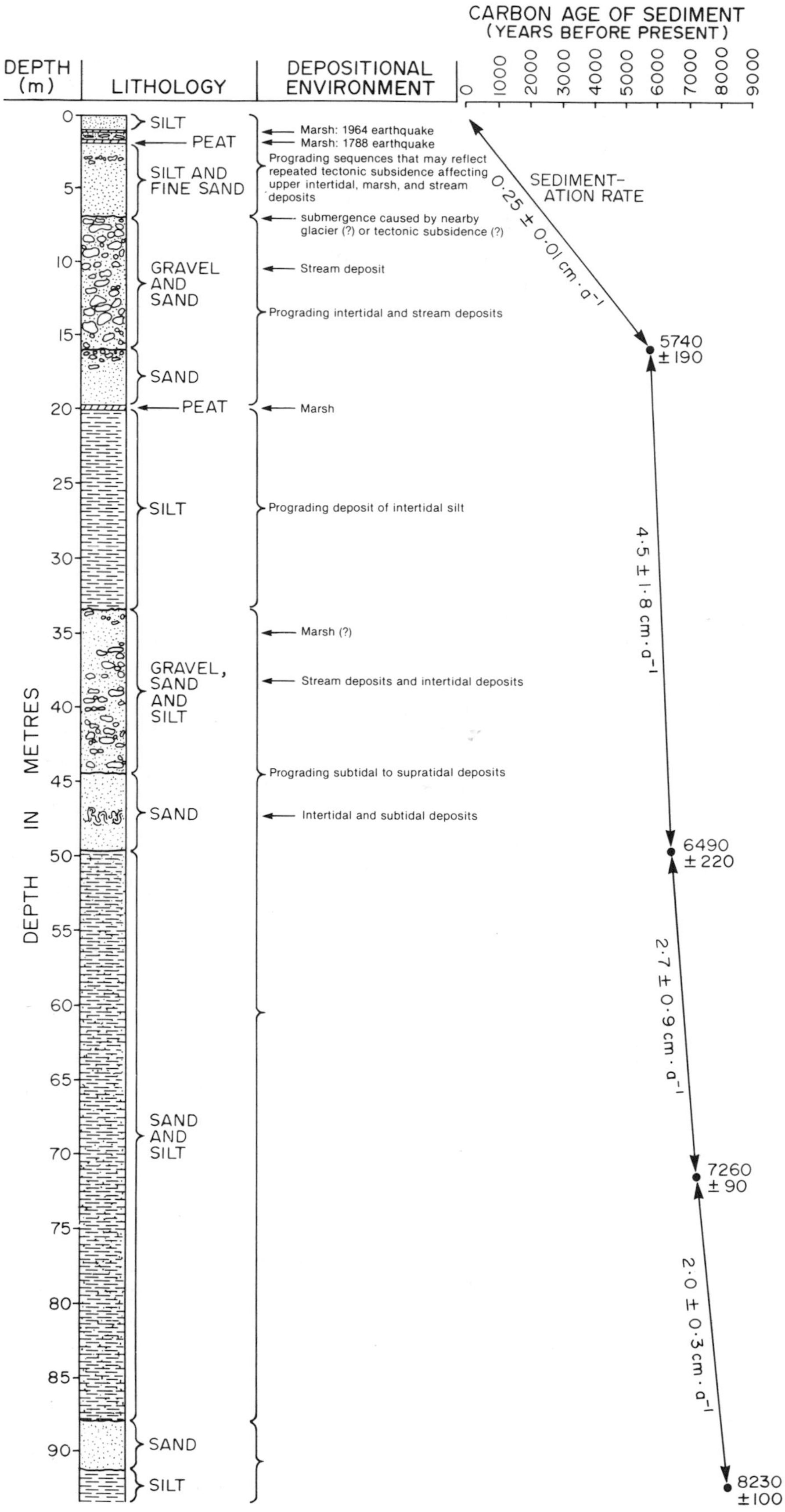

CARBON AGE OF SEDIMENT
(YEARS BEFORE PRESENT)

0 1000 2000 3000 4000 5000 6000 7000 8000 9000

DEPTH (m) | LITHOLOGY | DEPOSITIONAL ENVIRONMENT

SILT
PEAT
SILT AND FINE SAND
GRAVEL AND SAND
SAND
PEAT
SILT
GRAVEL, SAND AND SILT
SAND
SAND AND SILT
SAND
SILT

Marsh: 1964 earthquake
Marsh: 1788 earthquake
Prograding sequences that may reflect repeated tectonic subsidence affecting upper intertidal, marsh, and stream deposits
submergence caused by nearby glacier (?) or tectonic subsidence (?)
Stream deposit
Prograding intertidal and stream deposits
Marsh
Prograding deposit of intertidal silt
Marsh (?)
Stream deposits and intertidal deposits
Prograding subtidal to supratidal deposits
Intertidal and subtidal deposits

SEDIMENTATION RATE
0.25 ± 0.01 cm·a^{-1}
5740 ± 190
4.5 ± 1.8 cm·a^{-1}
6490 ± 220
2.7 ± 0.9 cm·a^{-1}
7260 ± 90
2.0 ± 0.3 cm·a^{-1}
8230 ± 100

DEPTH IN METRES
0 5 10 15 20 25 30 35 40 45 50 55 60 65 70 75 80 85 90

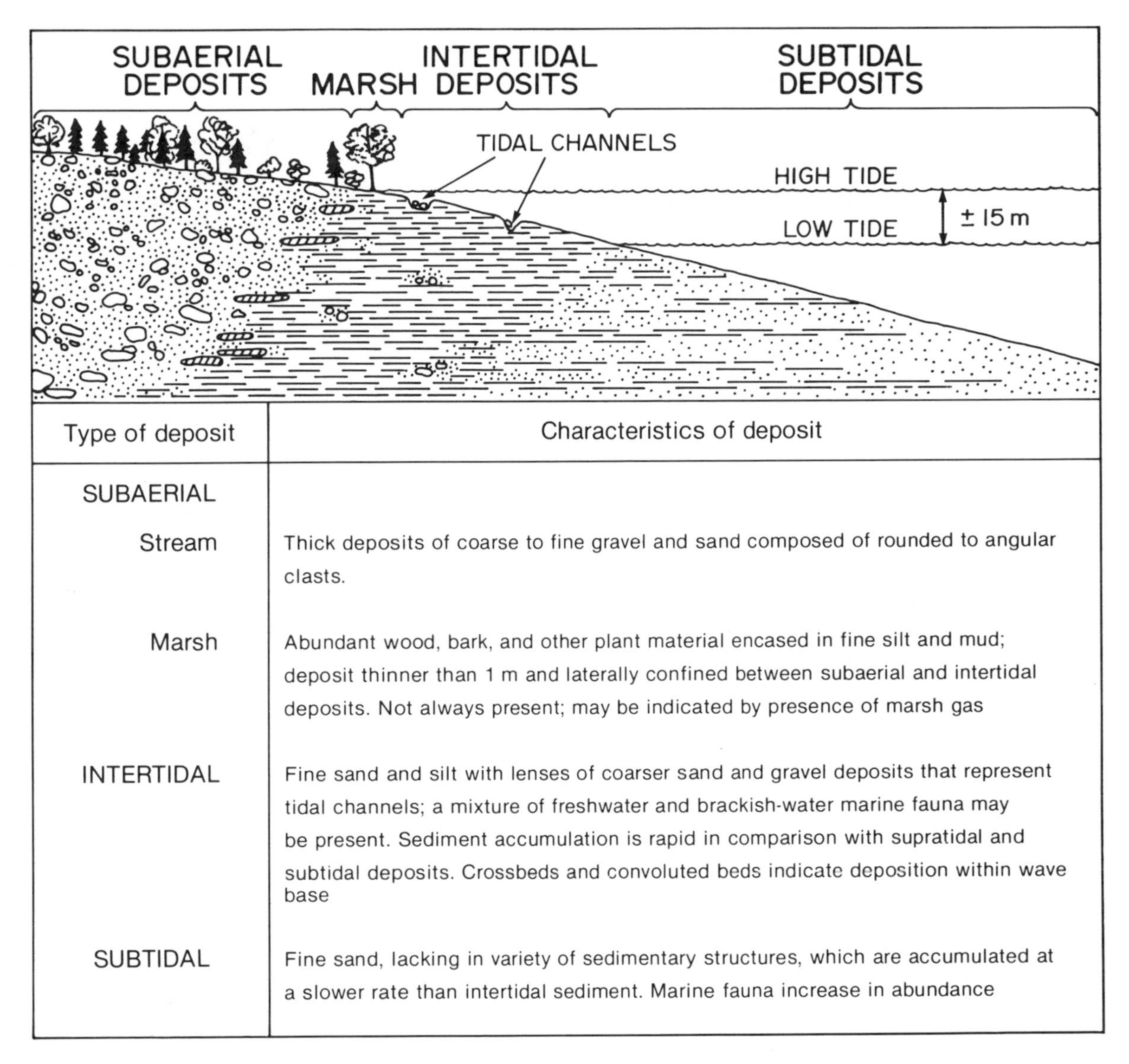

Type of deposit	Characteristics of deposit
SUBAERIAL	
Stream	Thick deposits of coarse to fine gravel and sand composed of rounded to angular clasts.
Marsh	Abundant wood, bark, and other plant material encased in fine silt and mud; deposit thinner than 1 m and laterally confined between subaerial and intertidal deposits. Not always present; may be indicated by presence of marsh gas
INTERTIDAL	Fine sand and silt with lenses of coarser sand and gravel deposits that represent tidal channels; a mixture of freshwater and brackish-water marine fauna may be present. Sediment accumulation is rapid in comparison with supratidal and subtidal deposits. Crossbeds and convoluted beds indicate deposition within wave base
SUBTIDAL	Fine sand, lacking in variety of sedimentary structures, which are accumulated at a slower rate than intertidal sediment. Marine fauna increase in abundance

FIGURE 3.23. Cross-sectional schematic of the subaerial and intertidal depositional environments of Turnagain Arm (after Bartsch-Winkler et al., 1983).

Three types of hydraulic-equivalent size frequency distributions are preserved commonly within the deltaic sediment (Fig. 3.25; Syvitski and Farrow, 1983). Type I, with unimodal and lognormal distributions, can be found in samples collected from all subenvironments (river, tide, or wave dominated) and is not restricted by grain size or sorting. The raison d'être of lognormal size frequency distributions is thought to be based on a multiplicative sorting phenomenon found in natural processes (Middleton, 1970). Type II, coarse-end truncated and unimodal distributions, are found in environments rich in moderately well-sorted 88- to 175-μm particles. Such deposits can be found on levees of tidal and fluvial channels and on tidal flats. As the water level drops with ebb tide, the current strength decreases and the last particles deposited are these easily moved grains (Syvitski and Farrow, 1983). These deposits result from single events that have not been subjected to the mul-

◁ FIGURE 3.22. Long core drilled through the tidal-flat environment of Tunagain Arm, Alaska (cf. Fig. 3.21) described in terms of lithology, depositional environment, and sedimentation rate (after Bartsch-Winkler et al., 1983).

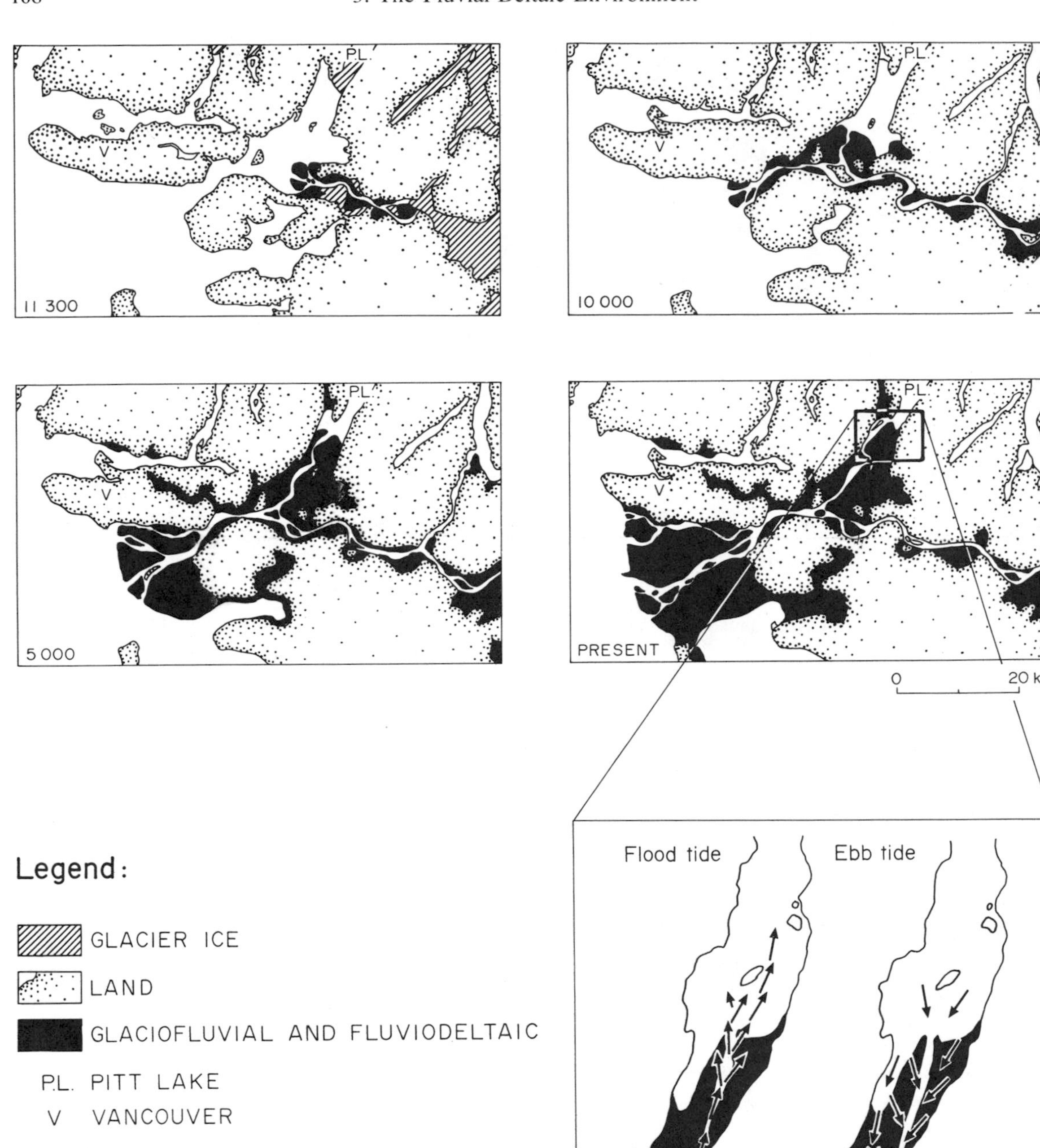

FIGURE 3.24. Growth of the Fraser River delta, since 11,300 years BP, has sealed off a number of fjords including Pitt Lake (after Clague et al., 1983). The present Pitt delta is tidally influenced through the effect of the salt wedge extending up the Fraser River during flood tide (after Ashley, 1979).

tiple sorting envisaged in the formation of type I deposits. Type III distributions result from the combination of the first two types to give bimodal and multimodal combinations. Such samples are from environments where the thickness of one depositional event is smaller than the sampling method. Such environments may include: starved sand ripples migrating over tidal flats (lenticular bedding); high-tide suspension fallout of mud onto current-rippled sand occupying flood-plain terraces (flaser bedding); and horizontal laminations where the mean grain

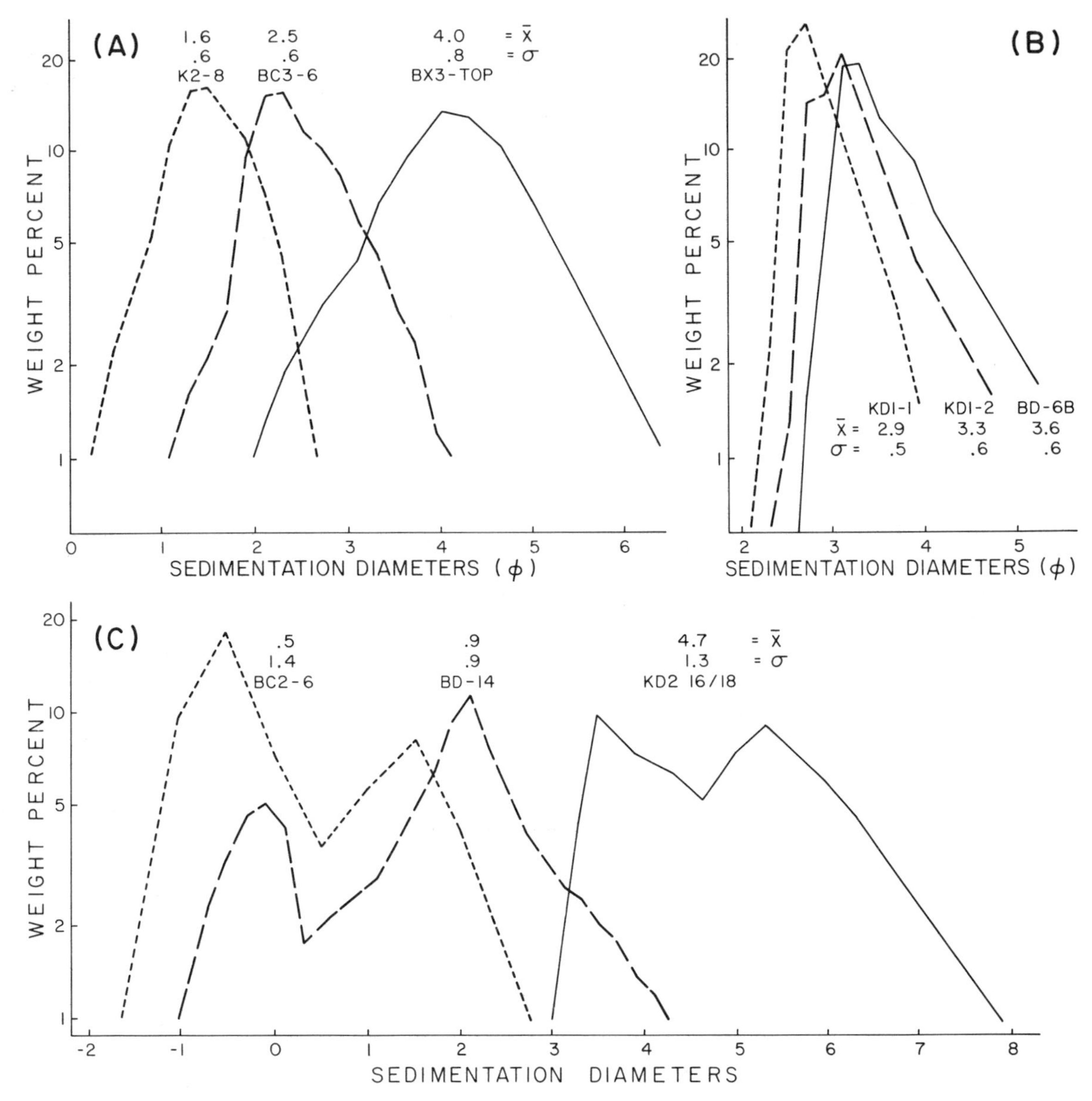

FIGURE 3.25. Three types of hydraulic-equivalent size frequency distributions common to fjord-head deltas (after Syvitski and Farrow, 1983). (A) Unimodal and lognormal distributions (common to all turbulent subenvironments). (B) Coarse-end truncated and unimodal distribution (common to levees and tidal flats). (C) Multimodal distributions, that is, combinations of A and B types (sampling induced). Mean, $\bar{x}$, and standard deviation, σ, are given in units of phi diameters.

size of a lamina may significantly differ from those surrounding it (tidal bedding).

3.5 Summary

The discharge of water down fjord-valley rivers depends on local climatic and storage conditions. A variety of distinctive hydrographic regimes are found in fjord-valley rivers, including: (1) arctic-nival; (2) continental-nival; (3) maritime-pluvial; (4) alpine-pluvionival; (5) alpine-proglacial. Only rivers with a continental regime, with their larger drainage basins, have substantial year-round flow. Other fjord rivers that are not regulated have discharges that are event dominated (melting snow freshet, rain storms, ice melt). Such rivers are susceptible to near-instantaneous floods.

Bed load is a capacity load controlled by the stream power, river hydraulics, and sediment properties. The delivery of bed load to the marine environment is high for many fjord rivers compared with other fluvial environments and, for proglacial streams, may reach 95% of the total sediment transport. Fjord-valley lakes, where present, effectively filter out the bed load. Finer particles can be transported either temporarily or permanently in suspension and in quantities dependent on the river discharge and sediment sources. Proglacial streams and streams that flow over isostatically exposed silt and clay deposits are the most turbid, especially during the warm periods of summer. Nonglacial rivers are more turbid during the spring freshet from the melting of snow. The relationship between the suspended sediment concentration and river discharge is complex: peak levels of both may or may not be coincident.

Fluvial delivery of sediment may or may not be in steady-state equilibrium with local sediment production by weathering and will depend on three paraglacial factors: (1) glacial history; (2) relative sea-level fluctuations; and (3) paleoclimatic history. Glacial and proglacial processes leave vast quantities of easily transported sediment within a fjord valley. Much of this sediment is continually re-exposed as the land emerges from under the ice load. Exposed terraces may contain glacial sedimentary sequences, transgressive marine muds, lacustrine or lagoonal deposits, and fluviodeltaic deposits. Failures along these terraces may contribute extraordinary amounts of sediment to the river.

Factors that influence the style and rate of development of fjord-head deltas include: (1) the strength and periodicity of the fluvial discharge; (2) the slope of the river; (3) climate; (4) relative sea-level fluctuations; (5) sediment supply; (6) wave energy and direction; and (7) tidal energy. Arctic deltas and sandurs are strongly influenced by their lack of stabilizing vegetation, presence of glaciers, and periglacial conditions. Of particular importance is the transport of sediment by wind and sea ice and the rapid surface runoff as a result of permafrost conditions. Temperate fjord-head deltas, being both warm and wet, support a dense vegetation cover; river banks are partly stabilized and river channels are both deep and narrow compared with the arctic sandur.

4

Circulation and Sediment Dynamics

Fjords, by their nature of being overdeepened basins, contain large volumes of saline water, which if protected by a sill, may lie almost stagnant (Fig. 4.1A). Exchange between the outer shelf waters and these protected deeper fjord waters tends to be intermittent. Estuarine circulation is common to many fjords, although its presence is seasonally regulated. Confined to a thin interval of the upper water column, the freshwater plume provides an important influence on the dispersal of dissolved and particulate matter. When the influx of fresh water is low or absent, sediment distribution may be strongly modified by wave or tidal mixing. Beneath the zone of estuarine circulation, there may exist larger, less-dynamic circulation cells that can be poorly coupled to the surface layer. The dynamics of polar fjords are further influenced by the effects of sea ice, icebergs, and tidewater glaciers. Chapter 4, then, sets out to discuss fjord circulation in relation to the dispersal, deposition, and resuspension of sediments.

Symbol Notation

(All other symbols as defined in the text)

A	surface area (defined)
A_p	light absorbance by particles
a	net accumulation rate
a_c	centrifugal acceleration
a'	amplitude (defined)
B	fjord width
B'	width of fjord narrows
B_p	light scatterance of particles
B	sediment accumulation rate from non-hypopycnal sources
B_o	width of river plume
C	concentration of suspended particulate matter
C_d	drag coefficient
C_p	specific heat of water
d	standard derivative
$\overline{d}$	mean grain size
E	net erosion rate
F	shape factor (defined)

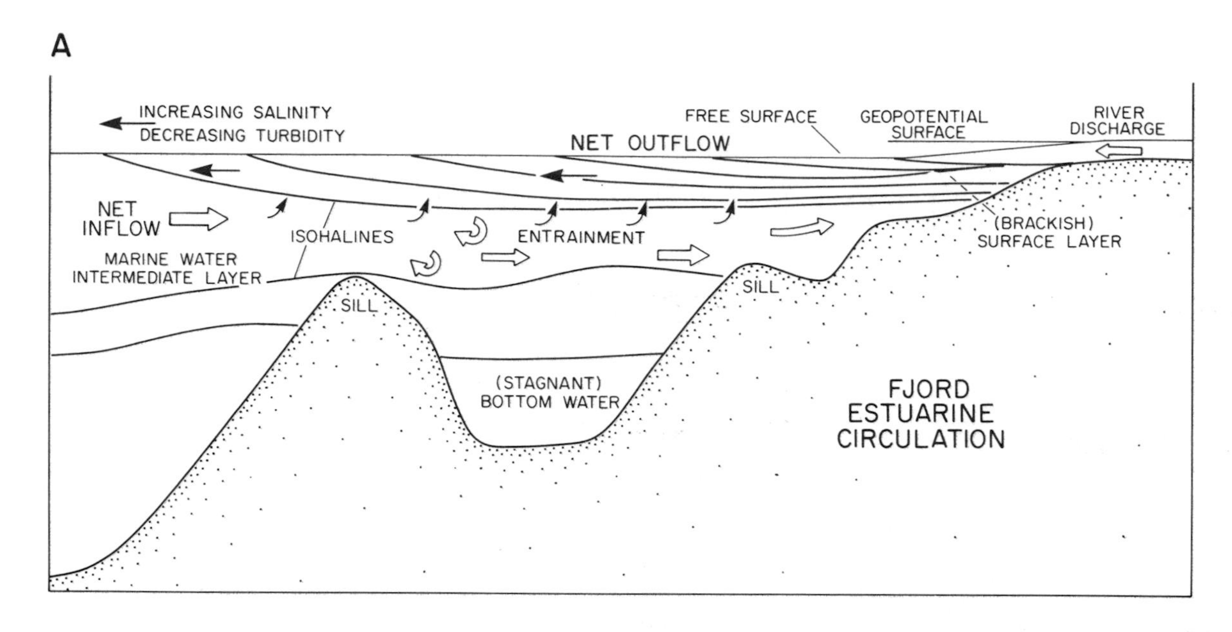

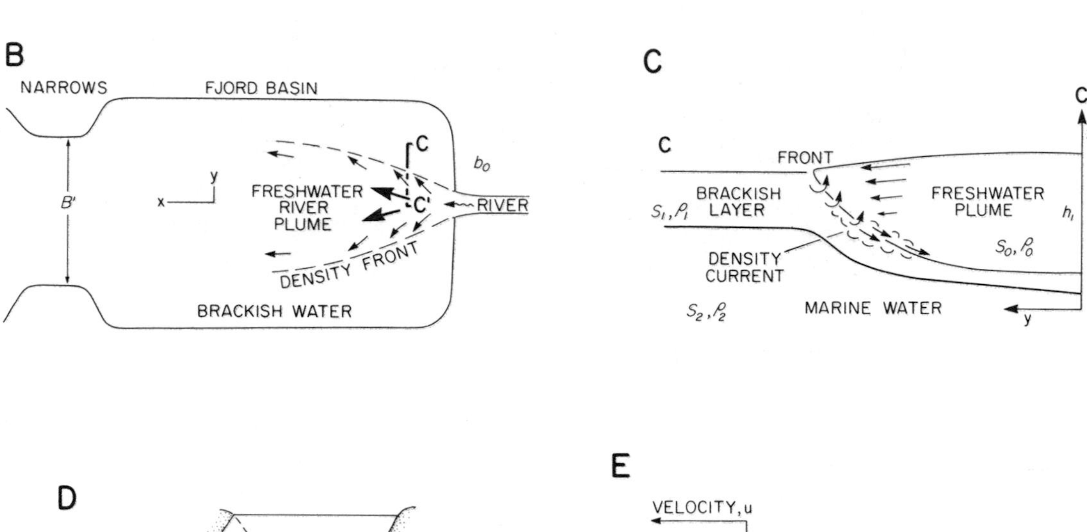

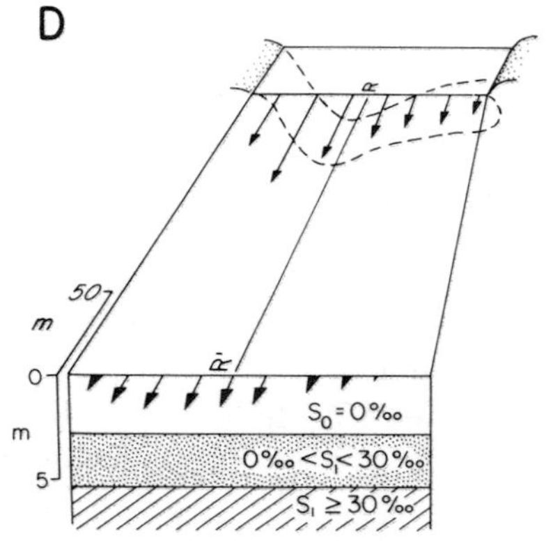

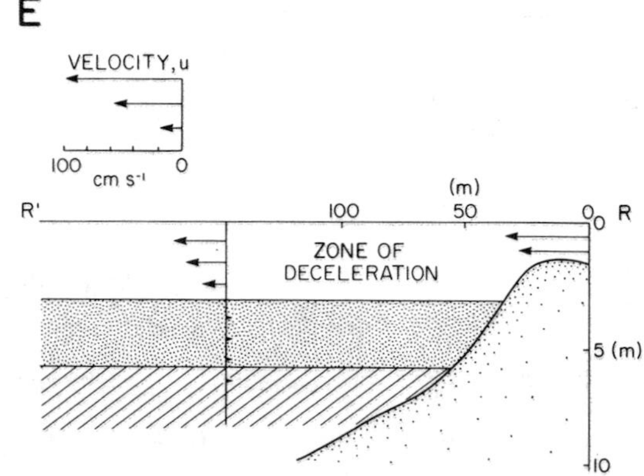

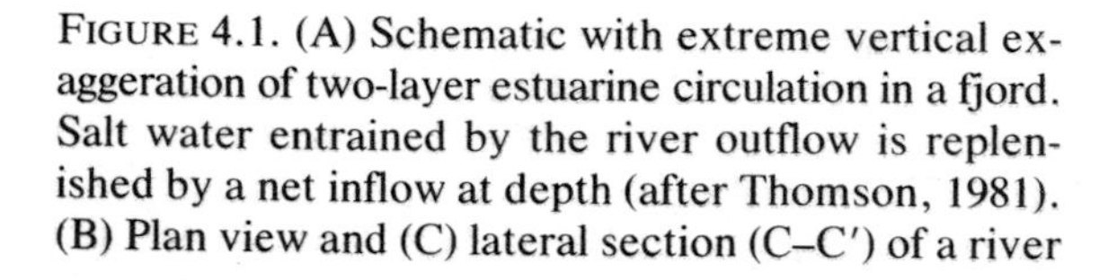

FIGURE 4.1. (A) Schematic with extreme vertical exaggeration of two-layer estuarine circulation in a fjord. Salt water entrained by the river outflow is replenished by a net inflow at depth (after Thomson, 1981). (B) Plan view and (C) lateral section (C–C′) of a river plume in a fjord basin, and with pertinent text nomenclature (after McClimans, 1979). (D) Aerial view and (E) cross section (R–R′) of current vectors of the Bella Coola River surface plume (North Bentinck Arm, B.C.: after Kostaschuk and McCann, 1983).

Fr	Froude number
Fr_d	densimetric Froude number
g	acceleration due to gravity
g'	reduced gravitational acceleration
H	reference thickness (defined)
H_f	surface heat flux
h_o	thickness of the surface layer
h_1	thickness of the brackish layer
H_f	thickness of the integrated freshwater cap
I_o	incident light intensity
I_z	resultant light intensity
K	mixing efficiency parameter (defined)
K_z	vertical eddy diffusivity
P	vertical flux of fecal pellets
Q	discharge (defined)
R	radius of curvature
Rf	Richardson number
r	particle radius
r'	Rossby radius
S	salinity
u	long axis velocity
u_o	velocity of the river discharge
$\overline{u}$	mean horizontal velocity
V_f	volume of fresh water in a fjord
W	wind velocity
W_e	vertical entrainment velocity
$\overline{w}$	mean vertical velocity
w_e	single particle settling velocity
w_o	in-situ settling velocity of marine particles
x,y,z	co-ordinate along the fjord, across the fjord, and water depth, respectively
x_b	distance from river mouth to the point that divides the upper and lower prodelta
Y_f	modified Froude number
Z	vertical flux of suspended particles
α	total attenuance coefficient
α_t	coefficient of thermal expansion
β_f	buoyancy flux
Δ	difference
∂	partial derivative
ρ	reference density (subscripts a for air, f for fluid density, s for particles, o or river water, 1 for brackish water, 2 for marine water
η	dynamic viscosity
Ω	angular velocity of the earth
Φ	Bernoulli constant
ϕ	degrees of latitude
Θ	angle (defined)
σ	root mean square of turbulence velocity
λ	wavelength (defined)
τ_b	basal shear stress
τ_f	residence time for fresh water
$1/\tau$	Secchi depth

4.1 Fjord Estuarine Circulation

Fjords are a type of estuary and, as such, the circulation of the surface waters depends strongly on the level of stratification that the influx of fresh water can maintain. The level of stratification is a balance between the buoyancy flux, set up by the discharge of fresh water, and those processes such as tidal mixing that work to homogenize the water masses. Stratification in a fjord is very seasonal, and the style of circulation may alternate between two-layer estuarine to partially mixed, to well mixed (see Section 2.3). However, the influx of fluvially transported sediment is highest during those periods of high river discharge when stratification is best developed.

The freshwater plume flowing within a fjord is commonly divided into two zones (McClimans, 1978b, Syvitski et al., 1985): a near zone (upper prodelta) in which the energy of the river discharge controls the spreading and mixing of the surface plume with its surroundings; and a far zone (lower prodelta) where external agents control transport and mixing. Such external agents include tidal currents, wind, shoreline morphology, and the earth's rotation.

Mixing between fjord water masses often occurs across sharp, well-defined pycnoclines. Farmer and Freeland (1983) note five main sources of kinetic (mixing) energy in fjords: (1) wind stress, (2) tidal interaction with topography, (3) double diffusion instabilities, (4) surface cooling and sea ice formation in winter, and 5) kinetic energy associated with fronts. Thus, this section sets out to review the formation and maintenance of two-layer estuarine circulation, including details associated with near and far zone dynamics of the river plume and factors affecting its thickness and velocity. This will be followed by a discussion of multilayer circulation, an estuarine circulation style unique to fjords and a consequence of energy imbalances to the "ideal" two-layer regime.

4.1.1 Two-Layer Circulation

Discharge of fresh water initially creates a hydraulic head near the river mouth and the effluent effectively flows downhill toward the sea. The gradient is calculated from the level or geopotential surface and the free (actual) surface and is typically of the order of 1 cm per 10 km (Farmer and Freeland, 1983). As the surface water flows seaward, it undergoes acceleration and entrains marine water into its outflow (Fig. 4.1). Random eddy motion from the shear between the surface outflowing layer and the underlying marine waters will involve both the downward movement of brackish water parcels into the deeper water, as well as the movement of saline water parcels upward into fresher water. Although restoring forces on the parcels are nearly the same in either case, mixing and erosion are more likely to occur in the surface layer, since turbulence is more intense there. Surface layer turbulence arises initially from river flow instabilities and later by interlayer friction-induced turbulence, breaking of internal waves along the boundary between the two layers, and wind-induced surface turbulence. Entrainment of saline water is the process of one-way transport of fluid from a less turbulent to a more turbulent region. The effects of entrainment and acceleration balance to maintain a relatively uniform thickness of the surface layer along the fjord (McAlister et al., 1959). As saline water is entrained into the outward-flowing surface layer, new sea water must enter the fjord at depth, thus satisfying the conservation equations of continuity:

$$\frac{\partial}{\partial x}(B\bar{u}) + \frac{\partial}{\partial z}(B\bar{w}) = 0 \quad (4.1)$$

and momentum:

$$\frac{1}{B}\frac{d}{dx}(\bar{u}^2 B h_o) = \frac{d}{dx}\left(\frac{1}{2} g h_o^2 \Delta\rho\right) - \tau_b + \bar{u}\, W_e \quad (4.2)$$

The return or compensating current, otherwise known as gravitational circulation, is in turn driven by a reverse internal pressure gradient arising from the generally sloping density field (Gade, 1976). It is generally assumed that the internal (baroclinic) pressure balances that of the sloping free surface (barotropic). Represented in terms of the buoyancy flux, β_f,

$$\beta_2 = \beta_1 = \beta_f \quad (4.3)$$

where β_2 is the buoyancy of the compensating current, β_1 is the buoyancy of the surface outflow and

$$\beta_f = \int_o^B \int_o^H g[\rho(z) - \bar{\rho}]\, u(z)\cdot n \cdot dz \cdot dy$$

(Farmer & Freeland, 1983)

where $\bar{\rho}$ is the reference sea-water density, n is the unit outward normal. During high runoff

$$\beta_f \approx g\,(\rho_o - \bar{\rho})Q \quad (4.4)$$

An indicator of stratification within a water column is the Richardson number, defined as the ratio between the stabilizing effect of density stratification and the destabilizing effect of velocity shear. The level of stratification within a fjord can be approximated by an "estuarine" Richardson number, R_{fE} where:

$$R_{fE} = (\beta_f\, b_o^{-1})\,(\rho u_t^3)^{-1} \quad (4.5)$$

The first term is the freshwater buoyancy as defined in equation (4.4) averaged over the width of the river mouth, b_o; the second term represents the mixing power of the tides where u_t is the mean tidal velocity. Most fjords have R_{fE} values between 0.1 and 5.

The mean freshwater residence for freshwater, τ_f, can be given as:

$$\tau_f = V_f/Q \quad (4.6)$$

where V_f is the volume of fresh water in a fjord, with:

$$V_f = A\left(\frac{S_2 - S_1}{S_2}\right)h_f$$

and

$$h_f = H - \frac{1}{S_{(H)}}\int_o^H S(z)dz$$

where h_f is defined as the thickness of fresh water which would occur if the measured water column were separated into a surface layer of fresh water and a lower layer whose salinity is equal to that found at the greatest depth of measurement H; and S(z), S_z, S_1 are the salinities at depth z, near the seafloor, and at surface, respectively. The mean residence time decreases as Q increases and is usually less in stratified fjords compared with well-mixed fjords. The above equations are useful in the development of geochemical box models and, in particular, the rate of a pollutant flushing through a fjord (see Chapter 8).

4.1.2 Upper Prodelta Dynamics

Most fjord-valley rivers have relatively steep thalwegs (Chapter 3). Thus these rivers tend to flow turbulently into the fjord and with a densimetric Froude number, Fr_d, greater than unity (McClimans, 1978b), that is

$$Fr_d = u_o/(g'H)^{1/2} > 1 \quad (4.7)$$

The first term is the inertial velocity of the river discharge; the second term represents the gravitational potential in the baroclinic field where H is the depth at the river mouth; g' is the reduced gravitational acceleration $g(1 - \rho_0/\rho_2)$. Thus near the river mouth, the surface layer of the fjord is well mixed. In most fjords, a brackish layer surrounds this well-mixed surface layer. In extreme cases, where the fjord has a very narrow connection with the sea, the freshwater layer may be so thick that the mixing energy of the surface flow has little influence on the deeper saline water. In that case, the salinity of the brackish layer approaches zero, and the surface layer becomes a neutral jet in a homogeneous layer, with the river mouth effectively at the sill (McClimans, 1978b). Examples include Ekalugad Fiord, Baffin Island (Knight, 1971), and Dramsfjorden, Norway (Beyer, 1976).

Once outside the confines of the river bed, the river plume spreads laterally to some dimension B′, a width determined by down-fjord narrows (Fig. 4.1B). During its lateral spread, the surface plume passes through a zone of deceleration (Fig. 4.1D, E; Kostaschuk and McCann, 1983). As an example, the drop in velocity away from the river mouth in Gaupnefjord was empirically modeled by:

$$u_x = 2.82\, u_o(x/b_o)^{-1.18}$$
(Relling and Nordseth, 1979)

where u_x is the surface layer velocity measured some seaward distance, x, from the river mouth; and b_o is the width of the river mouth.

Deceleration is also a function of the significant lateral mixing that takes place between the recently discharged water and the surrounding brackish layer (Fig. 4.1C; McClimans, 1979). A feature along this surface plume front is parcels of river water that peel off into the more quiescent brackish water, eventually to be laterally entrained back into the plume.

The upper prodelta zone can thus be defined as that zone near the river mouth where the surface layer is: (1) characteristically homogeneous and dominated by river flow instabilities (note Fig. 4.2A); (2) very fresh ($S < 3‰$); and (3) subject to transverse gradients. The extent of this near zone is seasonally dependent. For Knight and Bute inlets, British Columbia, the zone may extend some 10 km during the period of maximum discharge (August); during the cold of winter the near zone can disappear (Syvitski et al., 1985).

4.1.3 Lower Prodelta Dynamics

Fjord-estuarine circulation may also be influenced by the effects of the Coriolis force, centrifugal acceleration, topographically induced vorticity shedding, wind, and tides. In the outer fjord these may all play a significant role. The surface plume may migrate from shore to shore and vary greatly in character. The surface layer is no longer well mixed, rather is distinctly stratified with salinity increasing seaward and downward (Fig. 4.2A).

The Coriolis effect is the geostrophic result of the earth's rotation in curving the path of a freely moving parcel of fluid. When the width of the fjord exceeds the Rossby radius of deformation, a significant deflection of the water may be expected (Huppert, 1980). The Rossby radius, r', is a function of latitude, ϕ, and current velocity, u, as given by:

$$r' = u/(2\Omega \sin\phi) \quad (4.8)$$

where Ω is the angular velocity of the earth. The nomogram provided in Figure 4.3A (Gilbert, 1983) shows the variable effect of Coriolis force on fjords from different parts of the earth. Currents in even moderately small fjords, 3 or 4 km wide, will be influenced by the Coriolis effect when flowing at speeds of up to 0.3 to 0.4 m s^{-1}.

In sinuous fjords, where the radius of curvature, R, is comparable to or less than the internal Rossby radius, lateral variations in the

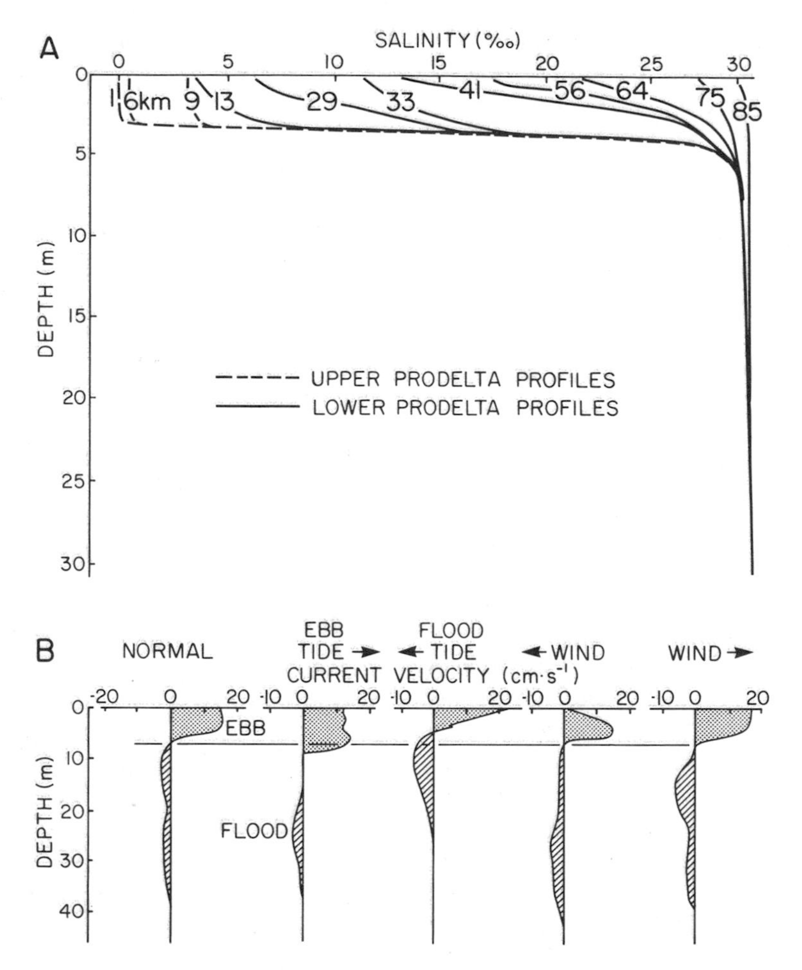

FIGURE 4.2. (A) Selected salinity profiles through the upper water column collected down-fjord during the summer freshet in Bute Inlet, B.C. (after Syvitski et al., 1985). Note the changes in profile gradient between those in the upper and lower prodelta. (B) Idealized current velocity profiles found in Bute Inlet and as affected by ebb and flood tide, and up- and down-fjord winds.

flow field can be influenced by the centrifugal acceleration, a_c, (Johns, 1962) as given by:

$$a_c = u^2/R \tag{4.9}$$

Howe Sound, British Columbia, is an excellent example of a fjord where centrifugal acceleration affects the circulation of the surface layer (Figs. 4.3B; 4.4D). In addition, the sinuous nature of the Howe Sound has allowed the strong katabatic winds and tidal oscillations to set up two semipermanent gyres of brackish water within the surface layer (Buckley, 1977). Wind or tidal interactions on topography can also induce vortices, shed from irregular shoreline (Yoshida, 1980). The vortices incorporate fresh water into the brackish layer (Fig. 4.4E). Shear between opposing river plumes can also result in a three dimensional current structure (Fig. 4.3C, McClimans, 1978a; Fig. 4.4C).

Similarly, a development of up-inlet winds can result in opposing cores of brackish water (Fig. 4.3B, Buckley, 1977; Fig. 4.4A). Up-inlet winds can also impede or reverse the surface outflow (Fig. 4.2B). The direction of the surface layer, in the lower prodelta of some fjords, is best related to wind direction except in cases of high runoff (Farmer and Osborne, 1976; Buckley and Pond, 1976).

4.1.4 Surface Layer Thickness and Velocity

The thickness of the surface plume can depend on discharge dynamics, wind mixing, and fjord

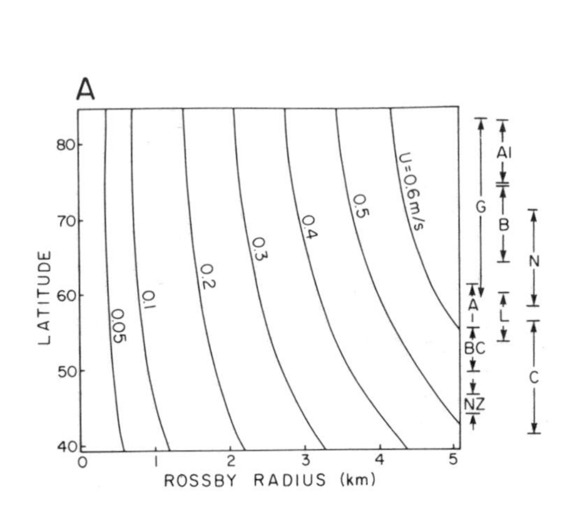

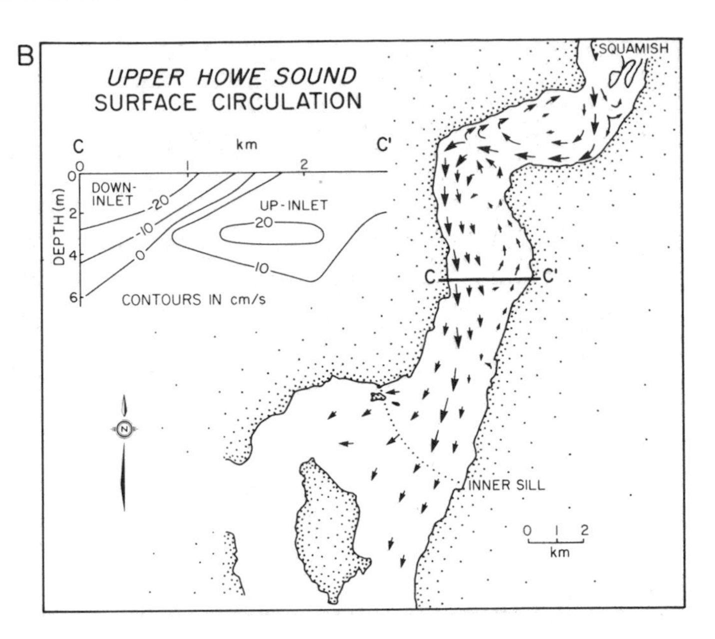

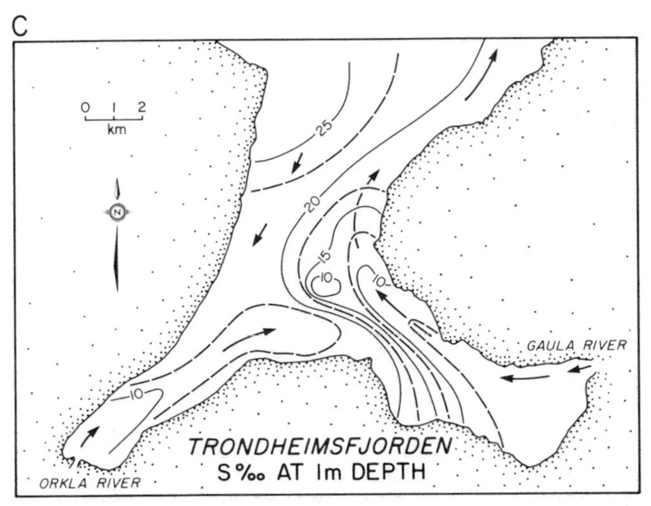

FIGURE 4.3. (A) The Rossby radius plotted as a function of latitude and current velocity. Also shown is the range in latitude of fjords in Greenland (G), the Canadian high arctic (AI), Baffin Island (B), Labrador (L), Norway (N), Alaska (A), British Columbia (BC), Chile (C), and New Zealand (NZ) (after Gilbert, 1983). (B) Plan view and lateral section (C–C′) of the surface current pattern for upper Howe Sound, B.C., deduced from surface drogue studies (after Buckley, 1977). (C) Surface plumes as found in Trondheimsfjorden, Norway, and as deduced from salinity at 1-m depth (after McClimans, 1978a, b).

morphology. At the river mouth, surface layer thickness, h_0, can be approximated by:

$$h_0 = \left(\frac{Q^2}{g'_o b_o^2}\right)^{1/3} \qquad (4.10)\ \text{(McClimans, 1979)}$$

where $g'_o = g(\rho_2 - \rho_0/\rho_2)$ with symbols defined in Fig. 4.1C. The thickness of the brackish layer, h_1, that surrounds the plume depends on discharge, Q, width of the narrows, B′, and salinities S_1 and S_2 as defined on Fig. 4.1. h_1 can be approximated by:

$$h_1 = 3/2\left[\left(\frac{S_2}{S_2 - S_1}\right)^2 \frac{Q^2}{g'_1(B')^2}\right]^{1/3} \qquad (4.11)\ \text{(McClimans, 1979)}$$

where $g'_1 = g(\rho_2 - \rho_1)/\rho_2$. Thus for large freshwater discharge events the surface layer thick-

FIGURE 4.4. (A) Areal photograph of tide-influenced surface vortices as shed from shoreline topography (Toba Inlet, B.C.). (B) Dilution of surface plume of main Homathko plume by side-entry river source (Bute Inlet, B.C.). (C) and (D) LANDSAT satellite composite (bands 5, 6, 7) of Knight Inlet and Howe Sound, B.C., surface plumes, respectively. Also shown in (D) is the Pitt River delta (cf. Fig. 3.24). (E) Topographically induced vorticity shedding of surface plume Bute Inlet, B.C. (F) Diffuse surface sediment bands flowing seaward from the sandur tidal flats of Pangnirtung Fiord, Baffin Island.

D

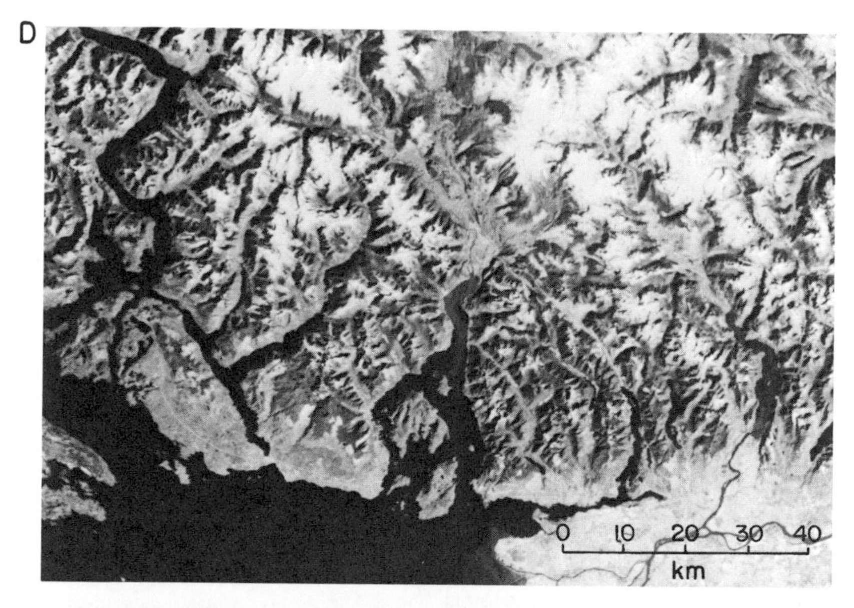

E

F

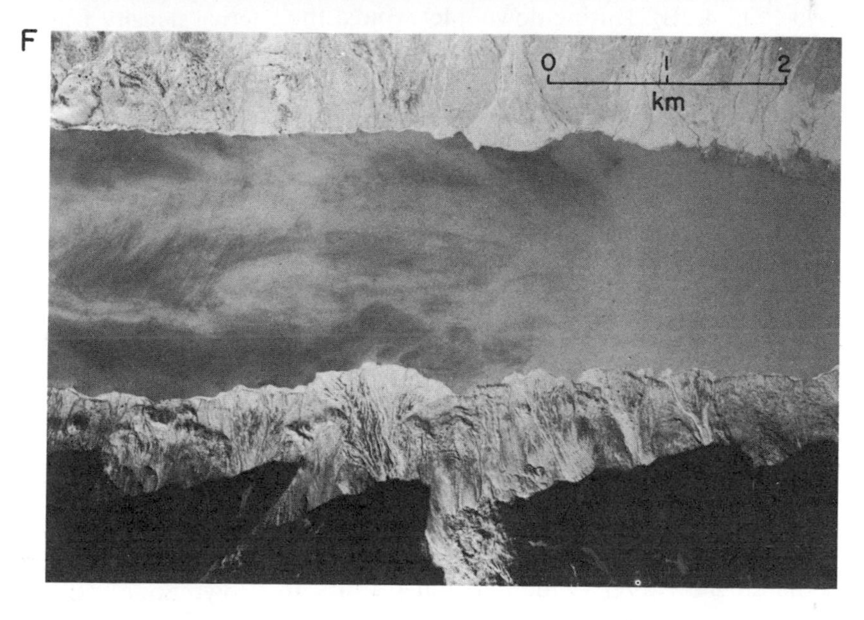

FIGURE 4.4. (*Continued*)

ness varies with $Q^{2/3}$. For wide fjords subject to wind mixing:

$$h_0 = \frac{AK\sigma^3}{[g(\rho_2 - \rho_0)/\rho_0]Q} \qquad (4.12)\,(\text{Long, 1975})$$

where A is the area subject to wind mixing; and σ is the r.m.s. turbulence velocity related to the surface friction. Thus the thickness of the surface layer will vary with Q^{-1}.

Stigebrandt (1981) provides a clear distinction between wind mixing (first term) and hydraulic control effects (second term) on surface layer thickness with:

$$h_0 = \frac{\zeta W^3 A}{Q\,g\beta S_2} + \Phi\left[\frac{Q^2}{g\beta S_2(B')^2}\right]^{1/3} \qquad (4.13)$$

where $\zeta = \psi(C_d\,\rho_a/\rho_0)^{3/2}$ and $\psi = 2k^{-1}Rf$, Rf is the Richardson number $= h_0 g(\rho_2 - \rho_0)/\rho_2\sigma^2$, and k is the wall friction constant; and β is a constant of proportionality. Freshwater thickness is then a function of river discharge as well as strong inlet winds, and these can occasionally be strongly correlated at times of hinterland storm precipitation (Farmer and Osborne, 1976).

The action of wind is twofold (Gade, 1976): (1) generation of surface waves, the breaking of which leads to a surface drift; and (2) increase of the vertical exchange (eddy viscosity) of the affected layers. During the initial stages of a wind event, the surface water is likely to obtain appreciable velocities (10% of wind velocity is not uncommon). During periods of up-inlet winds, a subsurface seaward-flowing jet may form (Fig. 4.2B). During down-inlet winds, the thickness and velocity of the surface layer will increase (Fig. 4.2B). Prolonged down-inlet winds can also remove the surface layer in a fjord (Hay, 1983a,b), or in the case of up-inlet winds, pile the surface layer up onto the fjord-head delta (Farmer and Osborne, 1976).

The tide in a fjord is predominantly a standing wave (Saelen, 1976) whose daily oscillations produce little residual flow and whose effect decreases below sill depth (Pickard, 1961). Tidal oscillations can affect the surface layer thickness and velocity, especially proximal to the river mouth (Syvitski and Murray, 1981; Syvitski et al., 1985). During flood tide, velocities are largest at the air-sea interface and decrease with depth through the surface layer (Fig. 4.2B); during ebb tide, velocity is more homogeneous in the surface layer, which may also thicken (Fig. 4.2B). Such responses to the tide are predominantly found in the upper prodelta environment (Kostaschuk and MaCann, 1983). Tidal currents can reverse the direction of the outer fjord surface layer in a complex pattern (Huegget and Wigen, 1983). This is especially true of fjords with a high tidal range during periods of low discharge.

Other buoyancy inputs that affect surface layer thickness and velocity can affect the equality of the two-layer estuarine circulation (equation 4.3). They include the exchange of intermediate and deep water (see Sections 4.1.5 and 4.4), surface heat exchange, and effects of ice (sea ice through brine rejection, icebergs and tidewater glaciers through freshwater input at depth, see Section 4.5). Surface heat exchange is given by:

$$\beta_f = gA\alpha_t H_f \rho C_p \qquad (4.14)(\text{Farmer and Freeland, 1983})$$

The consequence of surface heat exchange is to mix the surface layer downward through the water column (thermohaline convection) during the cooling phase of winter. Conversely, the warming of the surface waters during summer stabilizes the surface zone.

4.1.5 Multilayer Circulation

Imbalances to the buoyancy fluxes that drive a fjord's two-layer circulation may result in undercurrents forming, sometimes semipermanently. Such undercurrents will disturb the internal density field. These multilayered currents may involve the entire water body in the fjord, in that they are frictionally controlled and sometimes frictionally driven (Gade, 1976). For instance, Dramsfjord, Norway, was found to have a three-layer circulation in June and a five-layer circulation in December (Beyer, 1976). Dramsfjord has high summer runoff and the halocline can be as deep as 25 m. Since the fjord has such a shallow sill (8 m), these circulation cells are frictionally driven below the sill depth.

Multilayered circulation can also result from current interactions with the sill or with other buoyant inputs from outside the fjord. There are a number of examples along the Alaska and British Columbia coasts where brackish water outside a fjord forces surface flow up-fjord. Howe Sound, B.C., is a fjord that lies close to a major outside freshwater input, the Fraser River some 25 km to the south. Although the Fraser does not flow directly into Howe Sound, its plume controls the circulation in the outer parts of the fjord (Fig. 4.5A; Syvitski and Macdonald, 1982). The result is competition between the fjord-head river plume, the Squamish, and the currents driven by the Fraser River. The rough bathymetry of the outer-fjord reaches, with convergence and divergence of currents around islands and over sills, results in multilayered circulation in the inner fjord (as calcu-

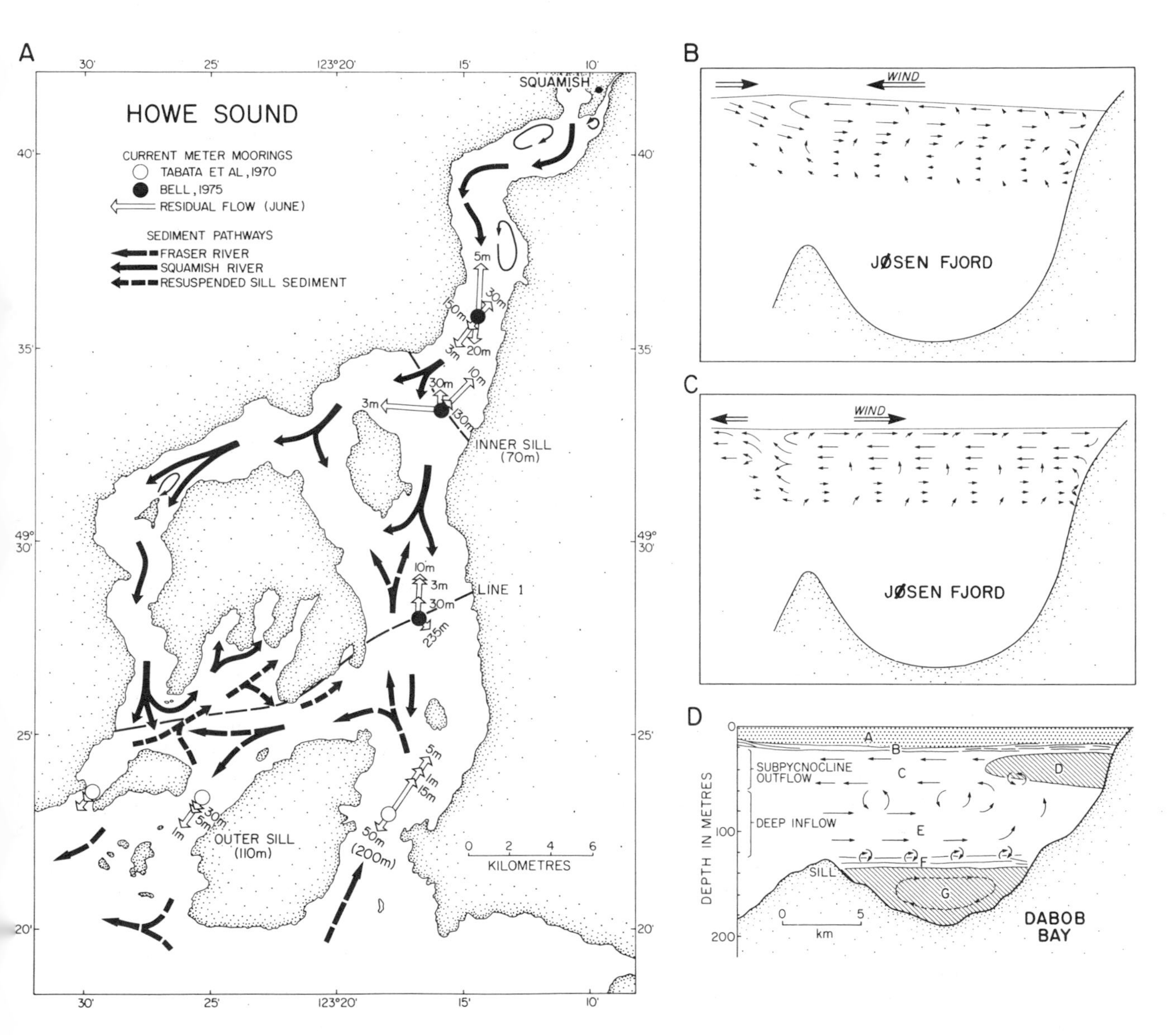

FIGURE 4.5. (A) Major sediment dispersion pathways, along with residual currents from current meter moorings in Howe Sound, B.C. South of line 1, the Fraser sediment component is obvious but is negligible north of the inner sill. The length of the arrows of residual current velocities represent relative current speed (after Syvitski and Macdonald, 1982; cf. Figs. 4.3B, C, and 4.4D). (B) Sketch of the net circulation in the upper 50 m of Jøsenfjord, June 15–19, 1974 (after Svendsen, 1977). (C) Sketch of the net circulation in the upper 50 m of Jøsenfjord, June 20 to July 1, 1974 (after Svendsen, 1977). (D) Typical circulation in Dabob Bay during late summer. Notations: A, low salinity surface layer; B, primary pycnocline; C, cooler, low oxygen content subpycnocline outflow; D, relatively static tongue of low temperature, low oxygen content water; E, warmer and more saline, higher oxygen content deep inflow; F, secondary pycnocline; and G, relatively static bottom layer of colder low-oxygen content water (from Ebbesmeyer et al., 1975).

lated from one-month velocity residuals, Fig. 4.5A).

Multilayer circulation cells can also arise from interactions of local wind stress within the fjord and wind conditions offshore. Wind-forced coastal circulation, with its geostrophic longshore currents, has a strong effect on the circulation within the fjord. These geostrophic currents control the free surface and pycnocline displacement at the fjord mouth thereby strongly affecting fjord circulation (Klinck et al., 1981). For instance in Jøsenfjord, Norway, convergent and divergent zones offshore maintain an inclined mean free surface within the fjord: a net four-layer current system results (Fig. 4.5B, C; Svendsen, 1977).

Where the inflow from outside the fjord is volumetrically greater than the seaward-flowing surface layer, a strongly developed subpycnoclinal outflow would result. An example is Dabob Bay, part of Hood Canal, Washington (Fig. 4.5D; Ebbesmeyer et al., 1975). There, the high salinity and high temperature intruding waters interleave as layers or become occluded as parcels in ambient water of comparable density. The parcels of new water enter the fjord in pulses as related to weather systems. A parcel's decay appears related to current shear.

4.2 Hypopycnal Sedimentation

Hypopycnal sedimentation is the process that describes the transport and deposition of riverine suspended sediment into a basin containing stratified water. The rate of sedimentation within many fjords can be linked directly to this form of infilling. In this section, we describe the distribution of suspended particulate matter (SPM) in fjords, both in terms of time and space, followed by a discussion on particle dynamics emphasizing marine particle interactions. Next, we examine the settling velocity of SPM and their subsequent sedimentation flux onto the seafloor. The principles and relationships outlined below are true only when two-layer estuarine circulation is well established.

4.2.1 Properties and Distribution of Suspended Particulate Matter

The sediment load carried by the river separates into two components seaward of the river mouth bar. The bed-material load settles quickly onto the delta foreset beds with the wash load being carried seaward within the river plume. The wash load is composed mostly of inorganic mineral grains of fine sand to clay size. Where hinterland glaciers are present, the suspended particulate matter (SPM) within the wash load is also referred to as glacial flour. The distribution of SPM seaward of the river mouth can be described generally using a perforated-conveyor-belt model (Farrow et al., 1983). Such a model implies that there is a rapid decrease in levels of SPM both along the surface layer and with increasing depth, as particles settle to the seafloor (Fig. 4.6A). Unlike water temperature and salinity, which are considered conservative properties of water, the concentration of SPM is nonconservative; that is, mineral grains are not neutrally buoyant like parcels of water. Conservative properties of a river plume tend to describe a linear trend out from the river mouth as the plume waters mix with the ambient basin water. Suspended particles, however, undergo settling while mixing with the ambient basin water. Data from Port Valdez, an Alaskan fjord, demonstrate these points clearly (Fig. 4.7; Sharma, 1979). Port Valdez has two main fluvioglacial inputs: the low but turbid discharge of the Shoup glacier stream and that of the Lowe River. Equally spaced surface water isotherms trend linearly with distance seaward of the river mouths (Fig. 4.7D), yet the concentration of SPM decreases more exponentially with distance seaward (Fig. 4.7B).

Based on the observation that SPM can absorb and scatter light, a variety of oceanographic techniques or instruments provide proxy information on the concentration and characteristics of particles suspended in the water. A classic method is to measure the depth associated with the disappearance (attenuance) of a standard white disc lowered into the water. The depth known as the Secchi depth is given as $1/\tau$ (τ is

FIGURE 4.6. (A) Suspended sediment concentration for summer and winter conditions in Knight Inlet, B.C. (after Farrow et al., 1983). (B) Distribution of light scattering coefficient (m^{-1}) at 546 nm in Bute Inlet, B.C., for both summer and winter conditions (after Pickard and Giovando, 1960). (C) Submarine daylight as percentage of surface light at stations from the inner to the outer central part of Hardangerfjord, Norway (after Aarthun, 1961).

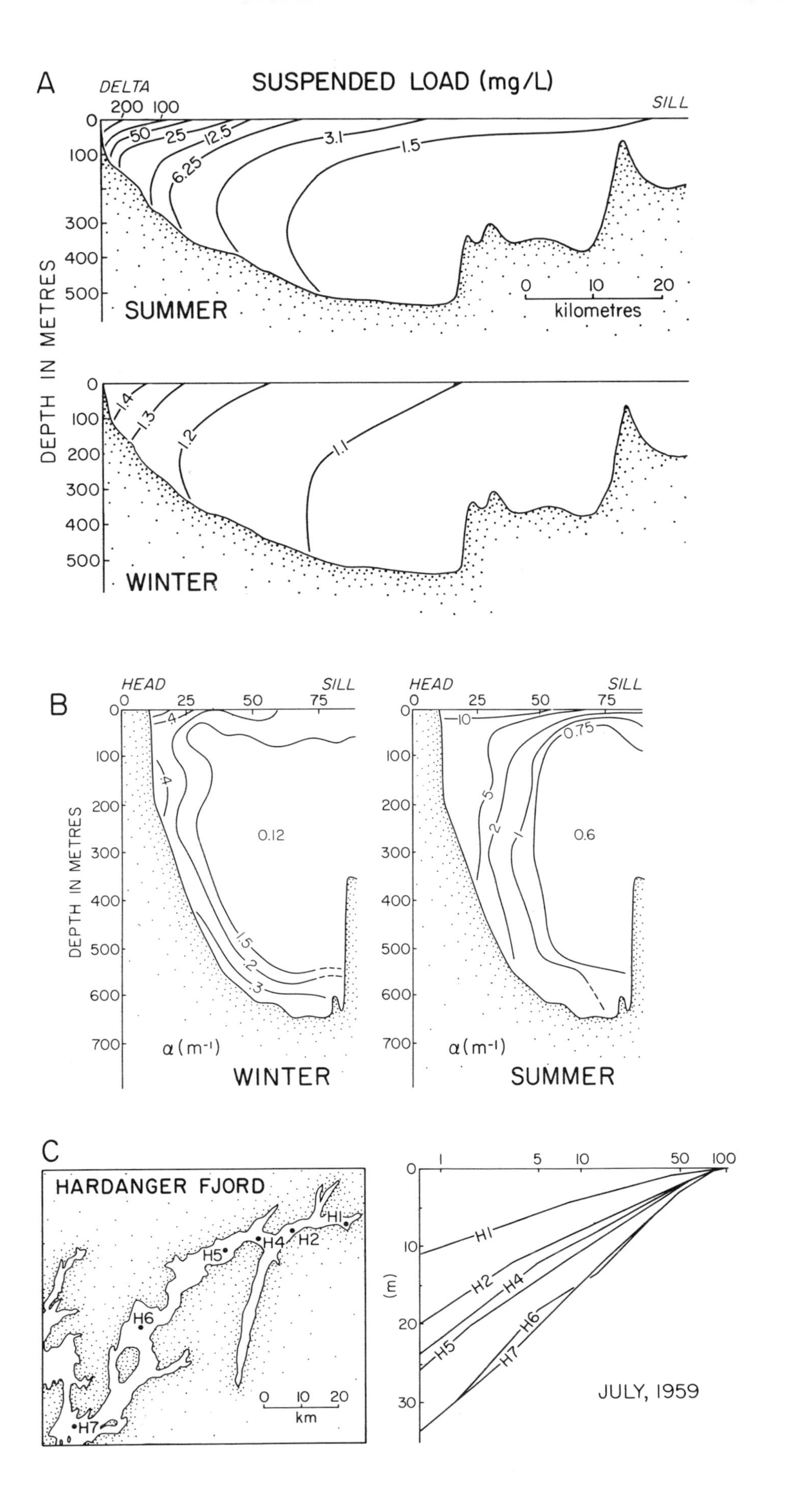
A
SUSPENDED LOAD (mg/L)
DELTA
SILL
DEPTH IN METRES
SUMMER
WINTER
kilometres
B
HEAD
SILL
DEPTH IN METRES
α (m⁻¹)
WINTER
SUMMER
C
HARDANGER FJORD
H1
H2
H4
H5
H6
H7
km
(m)
JULY, 1959

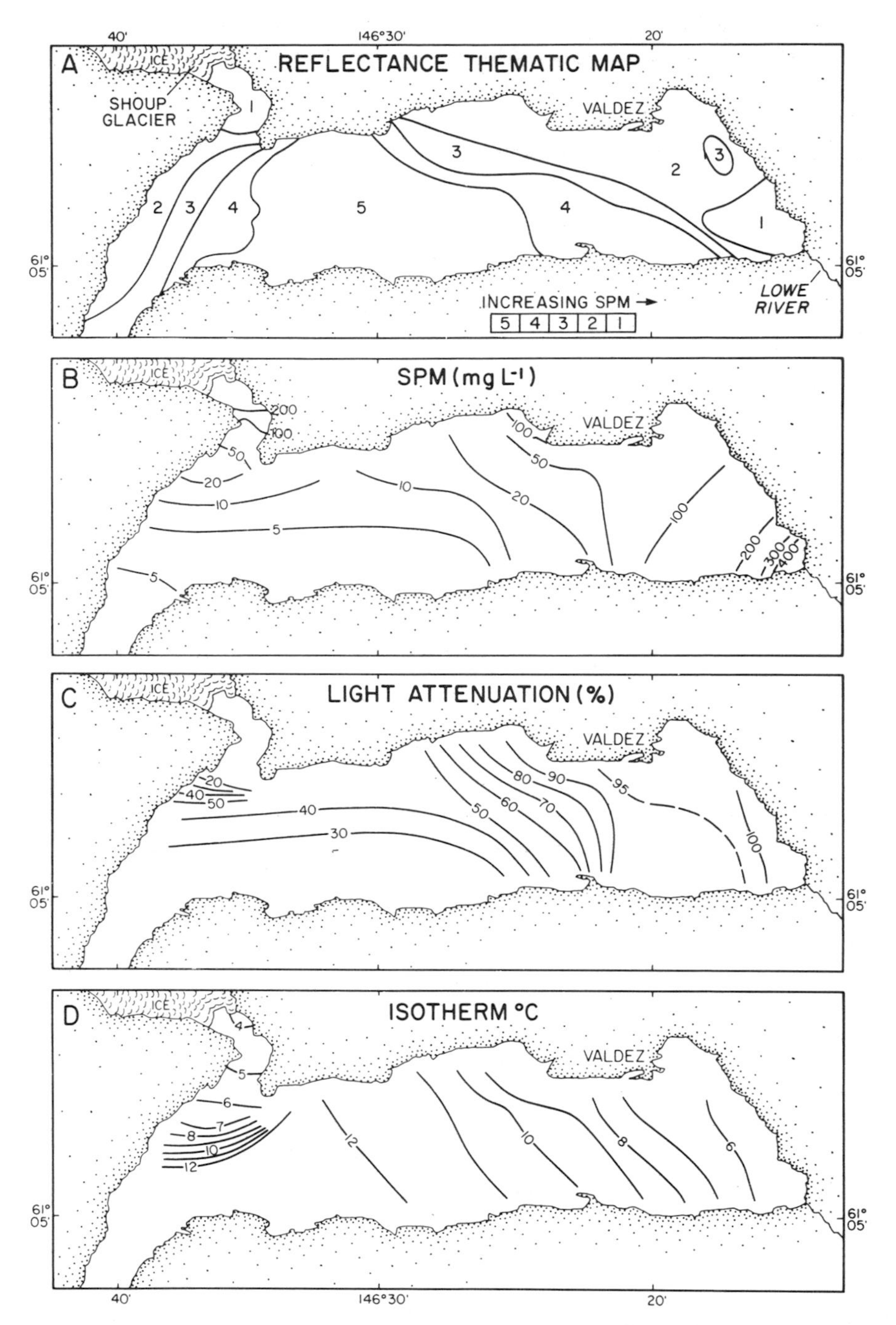

FIGURE 4.7. (A) Isodensity distribution of reflectance in satellite imagery showing relative suspended loads in near-surface water on August 15, 1973, Port Valdez (after Sharma, 1979). (B) Distribution of the surface suspended load, Aug. 2, 1972; (C) variation in light attenuation in surface waters, Aug. 2, 1972; and (D) surface water isotherms on Aug. 2, 1972, Port Valdez (after Sharma, 1979).

the level of transparency in m^{-1}). In turbid fjords of British Columbia, the concentration of SPM, C, decreased exponentially with the Secchi depth:

$$C = [0.14(l/\tau) - 0.05]^{-0.9}$$

(4.15) (Farrow et al., 1983)

The Secchi depth was found to increase linearly with distance from the river mouth.

Photometers provide a more accurate measure of water transparency, being able to profile the water column. Aarthun (1961) measured a number of submarine daylight profiles along the axis of the glacier-fed Hardangerfjord, Norway. Near the river mouth, less than 1% of the surface light could penetrate below 10 m. Further seaward, light penetrates to greater depths owing to the decrease in SPM levels (Fig. 4.6C). The more turbid fjords of British Columbia and Alaska allow less than 1% light through the first meter of surface water during periods of high discharge. During times of well-established and turbid two-layer circulation, surface light attenuation decreases linearly with distance from the river mouth (Fig. 4.7C).

Attenuance meters also measure the loss of radiant flux that is due to the process of absorption and scattering, but using a beam of light (rather than sunlight as in the above measurements) passing through a column of water, that is

$$\ln(I_z/I_o) = -\alpha Z$$
(4.16)(Jerlov, 1953)

where I_o and I_z are the incident and resultant light intensities, α is the total attenuance coefficient, and Z is the cell path length. The total attenuance, which varies with wavelength, λ, is attributed to attenuance by water, $A^{\lambda}{}_w$; absorbance by particles, $A^{\lambda}{}_p$; scatterance by particles, B_p; and absorbance by "yellow substance" or dissolved humic-like organic matter, $A^{\lambda}{}_y$, that is

$$\alpha^{\lambda} = A^{\lambda}_w + A^{\lambda}_p + B_p + A^{\lambda}_y$$
(4.17)(Winters and Buckley, 1980)

Empirical and theoretical analysis has indicated that the concentration of suspended particles can best be estimated from the attenuation of monochromatic red light (660–700 nm); and Winters and Buckley (1980) propose the following correlation:

$$C = 3.6\,(A_p{}^{680} + B_p)^{0.91} \qquad (4.18)$$

Lower monochromatic wavelengths are affected increasingly by the concentration of dissolved organic matter. The distribution of light attenuation, given in units of m^{-1} and using $\lambda = 546$ nm, is presented in Figure 4.6B for Bute Inlet, British Columbia (Pickard and Giovando, 1960). Such patterns of light attenuation during summer and winter are very similar to those for SPM concentrations given in Figure 4.6A. Further details of light conditions in fjord waters can be found in Aas (1976).

Another technique to estimate the concentration of SPM is through the use of remote sensing from aircraft or satellite. Simultaneous surveys of large areas of surface waters provide a greater understanding of the time-space variability in SPM concentrations. Multispectral scanners aboard aircraft and satellites detect surface radiation at a number of wavelengths (usually green, red, and infrared). Data products include false color composites or single band monochromatic prints (see examples in Fig. 4.4), as well as computer compatible tape with spectral data (for details see Amos and Alfoldi, 1979). The images may provide information on circulation patterns and qualitative information on the concentration of SPM (see Wright et al., 1973, and Gatto, 1976, on Cook Inlet, Alaska; and Folving, 1979, on Søndre Strømfjord, Greenland). Figure 4.7A is an isodensity reflectance map from a Landsat image of Port Valdez, Alaska. It shows the Coriolis-influenced circulation pattern and the decrease in isodensity reflectance values out from the river mouths (Sharma, 1979). Digital spectral data must first be corrected for variations in sun angle, sea-state surface and subsurface reflections, and atmospheric attenuation, among other parameters, before the multispectral information can be correlated with the SPM concentration.

Syvitski et al. (1985), building on the theoretical framework of hypopycnal sedimentation described by Syvitski and Murray (1981), developed a predictive model for prodelta sedimentation in fjords. The change in the concentration of SPM, C, at various depths within the zone of estuarine circulation, h_i, and with distance from the river mouth, x, revealed a two stage log-linear function for each depth:

$$C_i = (\overline{C}_o - b_1 h_i)\, x^{-1/2} \text{ for } 0 \leq x \leq x_b \qquad (4.19a)$$

$$C_i = (\overline{C}_o x_b - b_2 h_i)\, x^{-3/2} \text{ for } x > x_b \qquad (4.19b)$$

where $\overline{C}_o$ is the average discharge concentration just seaward of the mouth bar; b_1 and b_2 are empirical constants related to vertical flux of SPM through the zone of estuarine circulation; and x_b is the position of the upper-lower prodelta break (Fig. 4.8A). The position of x_b will change with the intensity of discharge, and during low discharge events x_b goes to zero and the upper prodelta zone is absent. At extreme discharge lows (winter), when only fine clays in low concentration are debouched into the fjord, the

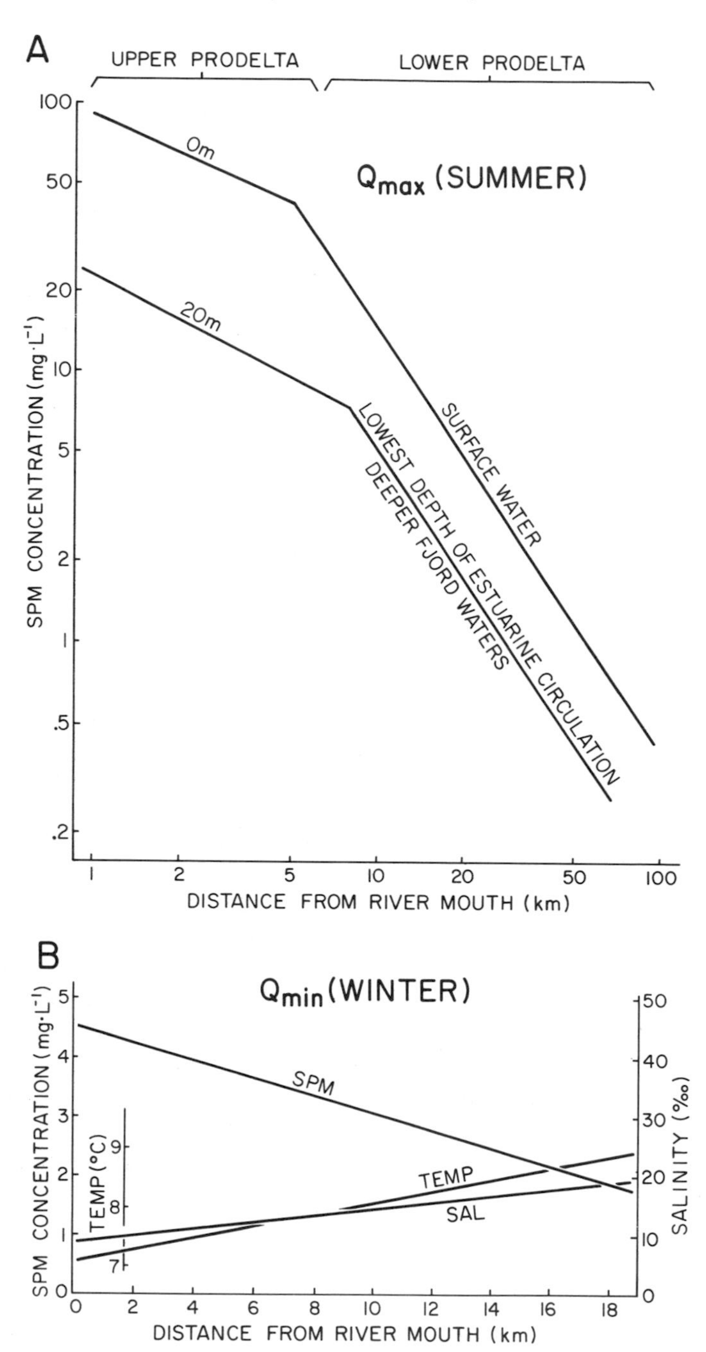

FIGURE 4.8. (A) Change in suspended particulate matter (SPM) concentration both with depth and distance down Bute Inlet, B.C., during high summer discharge. Note the change in profile between the upper and lower prodelta (after Syvitski et al., 1985). (B) Change in SPM concentration, temperature, and salinity down Howe Sound, B.C., during minimum winter discharge period. Note the conservative mixing of all three water properties (after Syvitski and Murray, 1981).

concentration of SPM behaves as a conservative water property (Fig. 4.8B).

In the deep basin waters, the concentration of SPM remains relatively low; between 5 and 10 times less than that found within the zone of estuarine circulation (Fig.4.9). Of course, deep water events such as gravity flows (Section 4.3, 5.2), renewals (Section 4.4), and storm events (Section 4.6) can substantially increase the concentration of SPM in these deep basins. In the absence of such events, the response time between the surface waters and the deep waters,

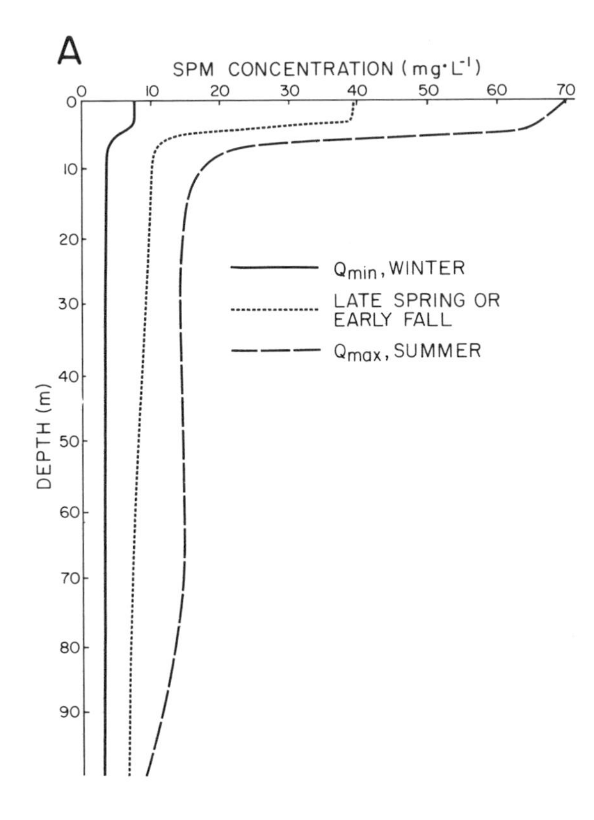

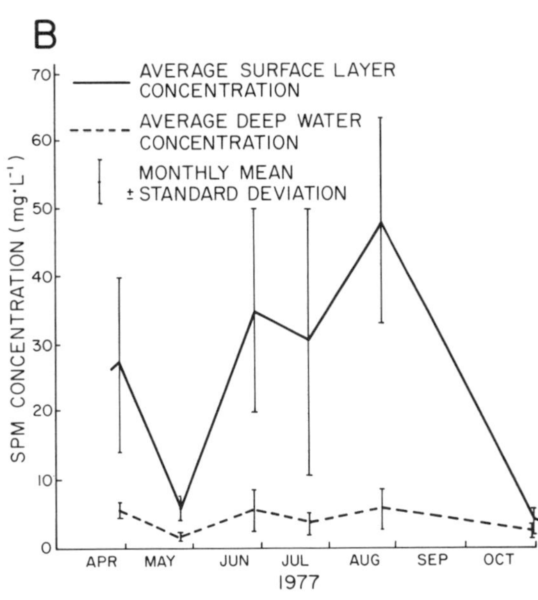

FIGURE 4.9. (A) Seasonal variations in the depth profiles of suspended particulate matter (SPM) concentration in Howe Sound, B.C. (after Syvitski and Murray, 1981). (B) Seasonal variations in the monthly averages in SPM concentration in the surface layer and deep water of Howe Sound, B.C. Note the spring storm peak (after Syvitski and Murray, 1981).

in terms of the sedimentation rate of SPM, is on the order of days (Fig. 4.9; Hoskin et al., 1976; Syvitski and Murray, 1981).

In addition to spatial dependence on the distance from the river mouth and the depth within the water column, variation in the concentration of SPM is highly time dependent. Concentration variations on the scale of seconds appear to reflect surface plume instabilities including fluvially induced turbulent eddies. Variations on the scale of minutes can result from: (1) continuous shifting of the thalweg of the plume (i.e., line of maximum surface layer velocity, Fig. 4.1D); and (2) sampling near spatially ephemeral fronts (e.g., Fig. 4.1B, C). Variations on the scale of tens of minutes to a few hours may reflect: (1) variations in river discharge dynamics (Fig. 3.3C; 4.10A); (2) changes in the state of the tide (Fig. 4.10B); and (3) radical changes in the state of the wind (Syvitski and Murray, 1981). Variations on the scale of days are more directly a function of discrete discharge events that are meteorologically controlled (Fig. 3.3B). Variations of SPM concentrations on the scale of months can be related directly to the seasonality of higher latitude river discharges (Fig. 3.3A).

Variations in the concentration of SPM, as discussed above, can be related to mixing processes and to processes that control the surface layer velocity (see Section 4.1.4). Mixing exchanges deeper, less turbid, marine water with turbid fresh water. Thus the variation of SPM decreases with depth, and additionally, the seaward transport of SPM decreases. Similarly, variations in the velocity of the surface layer will affect the competence of the surface layer to carry an increased sediment load containing higher settling velocity particles (i.e., sand). Given a constant source, the concentration of SPM in the surface layer increases linearly with the current velocities. Compared with other estuary types, stratified fjords can transport riverine particles for long distances, sometimes outside of the fjord. In well-mixed estuaries, sediment is deposited inside the estuary.

Exceptions to the above model of SPM distribution in fjords will depend on a number of complicating factors. These include (1) river discharge via a large number of small stream outlets that can cover an arctic sandur, whereby the SPM pattern is highly complex (Fig. 4.4F); and (2) side-entry fluvial or fluvioglacial inputs that can be more turbid than the main river

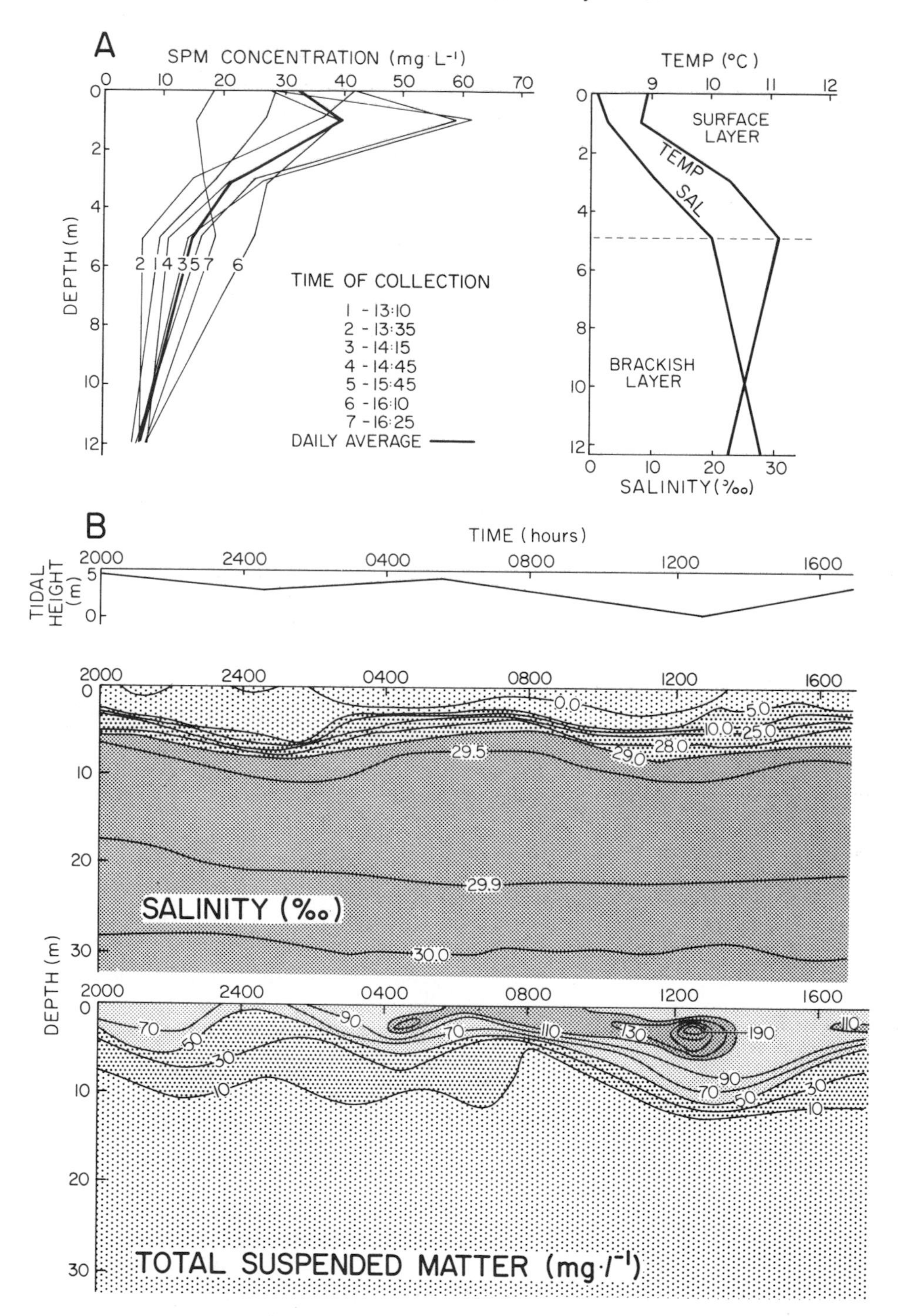

FIGURE 4.10. (A) Variability in the concentration of suspended particulate matter (SPM) during a 3-h period of ebb flow with the surface and brackish layers of Howe Sound waters (after Syvitski and Murray, 1981). (B) Time series data of surface water properties of salinity and SPM within the upper 30 m of water at an anchor station, 1.4 km seaward of the Homathko River mouth, Bute Inlet, B.C. Note the dependence of these properties on the stage of the tide (after Syvitski et al., 1985).

plume (examples include Baffin Island fjords), or less turbid (examples include the British Columbia fjords, Fig. 4.4E).

4.2.2 Particle Dynamics

Concentration maps of four discrete size fractions of SPM reveal that coarser fractions have mostly settled out in the upper prodelta (Fig. 4.11). As expected, the lower prodelta contains a higher percentage of the fine particles. The more horizontal concentration contours of the fine fractions in the upper prodelta suggest predictably a slower response between the surface and deeper waters. Patterns in Figure 4.11 can be partly modeled by:

$$\bar{d}_i = 0.018\,[\exp(0.35\,h_i)]C + 1.9 \quad (4.20)\text{(Syvitski et al., 1985)}$$

where $h_i \leq 10$ m, $\bar{d}_i$ represents the mean grain size, and $\bar{C}$ is given by equation (4.19). Equation (4.20) cannot be used for predictions beyond a

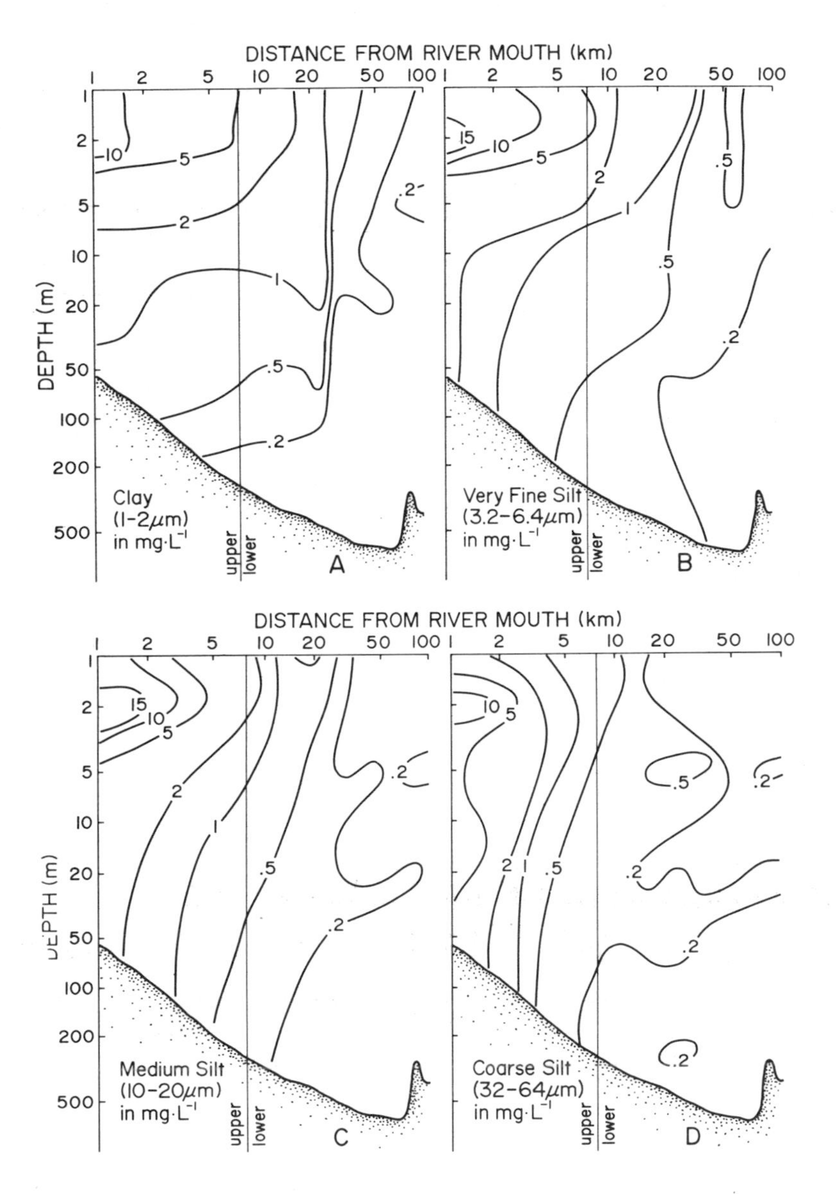

FIGURE 4.11. Longitudinal profiles of particle concentration for various dispersed size fractions: (A) clay, (B) very fine silt, (C) medium silt, and (D) coarse silt. Profiles represent a high-discharge period in Bute Inlet, B.C. (after Syvitski et al., 1985). Also shown is the approximate break between upper and lower prodelta dynamics.

seaward distance or depth that is marked by a SPM dominance by plankton cells or other non-riverine particles (see Chapter 6). In Bute Inlet, B.C., equation (4.20) is applicable for water having less than 50% organic particulate matter.

Particles suspended in a water column sink under their own weight, or in combination with other particles, that is, flocculation, agglomeration, and pelletization, and as affected by vertical diffusion. According to Stokes' Law, single particle settling is defined by:

$$w_e = \frac{2}{9}\left(\frac{\rho_s - \rho_f}{\eta}\right) gr^2, \text{ for rigid spheres of } <50\ \mu m \qquad (4.21)$$

or with the more excompassing empirical model of Gibbs et al. (1971).

$$w_e = \frac{-3\eta + 9\eta^2 + gr^2\rho_f(\rho s - \rho_f)\,(0.015476 + 0.19841r)^{1/2}}{\rho_f\,(0.011607 + 0.14881r)} \qquad (4.22)$$

where w_e is the equivalent spherical settling velocity and r is the particle radius. Equations (4.21, 4.22) have many boundary conditions (see Syvitski and Swinbanks, 1980), but are useful for most mineral grains that settle as discrete particles. Particle shape can affect a grain's settling velocity, but even nonspherical platy clay minerals can be predicted within a factor of two of equation (4.22).

Flocculation is that process that holds particles together in spite of repulsive electrostatic forces that are part of the natural chemical make-up of soil particles (Verwey and Overbeck, 1948; Whitehouse et al., 1960; Hahn and Stumm, 1970; Edzwald and O'Melia, 1975). Ions within a saline solution, however, neutralize the repulsive forces, allowing Van der Waals binding to occur. Once particles have joined together, the resultant settling velocity of the flocs is usually greater than that of the individual components (Sakamoto, 1972; Kranck, 1975).

Agglomeration is the attachment of organic detritus to mineral grains by surface tension, and organic cohesion resulting from biological activity. The process also includes: (1) mucilage coating by plankton (Lewin and Mackas, 1972); and (2) bacterial colonization (Kane, 1967; Johnson, 1974).

Zooplankton in fjord waters are mostly indiscriminate filter feeders and ingest particles based on size and not on the composition of the particle (Syvitski and Lewis, 1980). Particles are later egested in the form of fecal pellets, which can sink rapidly. The rate of pelletization depends on SPM concentration, the number and type of zooplankton, and rate of feeding (Lewis and Syvitski, 1983).

The process of diffusion eliminates gradients within transient properties. As there is a strong gradient of SPM concentration in the fjord water column, the process of diffusion moves particles away from the turbid surface layer to the clearer waters beneath.

The vertical flux, Z, for a particle of size χ can be defined by

$$Z(\chi) = C(\chi) \cdot w(\chi) \pm \left(K_z \frac{\partial C}{\partial z}\right) + P$$

(4.23) (Macdonald, 1983a,b)

where K_z is the vertical eddy diffusivity, P accounts for fecal pelleting, and z is depth downward. Based on K_z of 0.5 $cm^2\ s^{-1}$in the surface layer and 0.1 $cm^2\ s^{-1}$ within the pycnocline (Smethie, 1980), diffusion does not significantly affect particle settling in fjord waters (Macdonald, 1983a,b). Similarly, pelletization usually accounts for less than 10% of the overall vertical flux of particles, although local areas may be dominated by pelletization (Syvitski and Murray, 1981).

The in situ settling velocity, w_o, of particles of radius, r, taking into account the processes of flocculation, agglomeration, pelletization, and diffusion, was empirically found to approximate

$$w_o = 100\ w_e\ r^{-2.1}, \text{ for } r \leq 5\ \mu m \qquad (4.24a)$$

$$w_o = 10\ w_e\ r^{-0.6}, \text{ for } 5\ \mu m < r < 50\ \mu m \qquad (4.24b)$$

$$w_o = w_e, \text{ for } r > 50\ \mu m$$

(4.24c) (Syvitski et al., 1985)

where w_o is given in (m day^{-1}) for the near zone of Bute Inlet, B.C. Also, the settling velocities of the various size particles increase with depth. A 1-μm particle settles at 30 m day^{-1} by the time it has reached 5-m water depth and 50 and 100 m day^{-1} by the time it has reached the 10-m and 30-m depths, respectively (Syvitski et al., 1985).

Particle dynamics and dispersal within a fjord may be summarized as follows. Particles enter a fjord usually as single entities. Exceptions in-

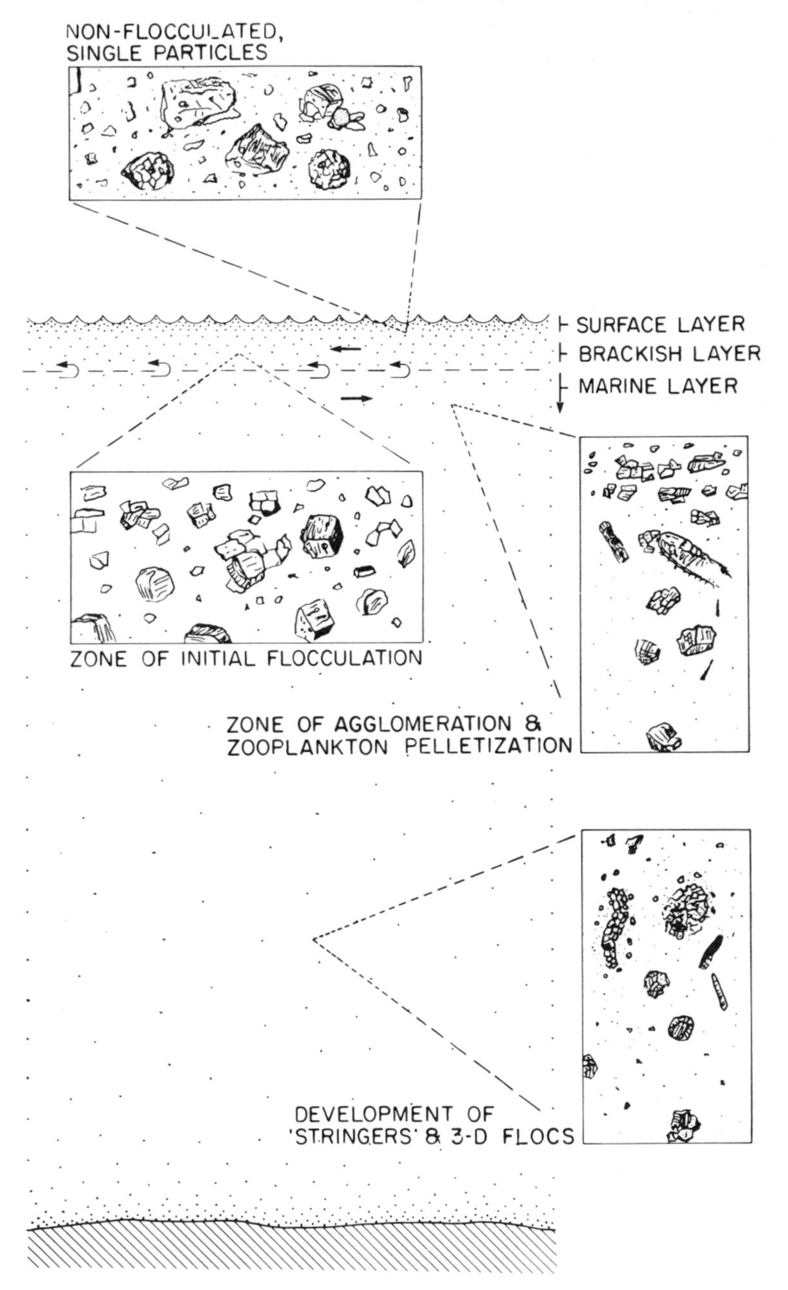

FIGURE 4.12. Schematic of the change in microtexture of suspended sediment particles as they settle through the fjord water column. Initially the particles are as nonflocculated single particles in the surface fresh water. As they move through the brackish layer they begin to flocculate, only to be consumed by zooplankton in the marine waters. With the growth of organic matter, deep water flocs are large and three dimensional.

clude: (1) coarser silt with attached hydrous oxides, organic coatings, and/or freshwater microflora (Fig. 4.12, 4.13A); and (2) clay clasts stripped from raised marine terraces now undergoing fluvial erosion (Fig. 4.13B). As the surface water mixes with the marine water, the suspended load begins to settle out. Flocculation of silts and clays begins to occur within the halocline at salinities < 3 to 5‰ (Fig. 4.12, 4.13C). Nonplaty minerals may continue to fall as single entities, especially near the river mouth. In the marine water, organic cohesive/adhesive forces allow the attachment onto the flocs of organic detritus and biogenic debris (mostly diatom frustules in various stages of mechanical destruction; Fig. 4.13D). Zooplankton, just under

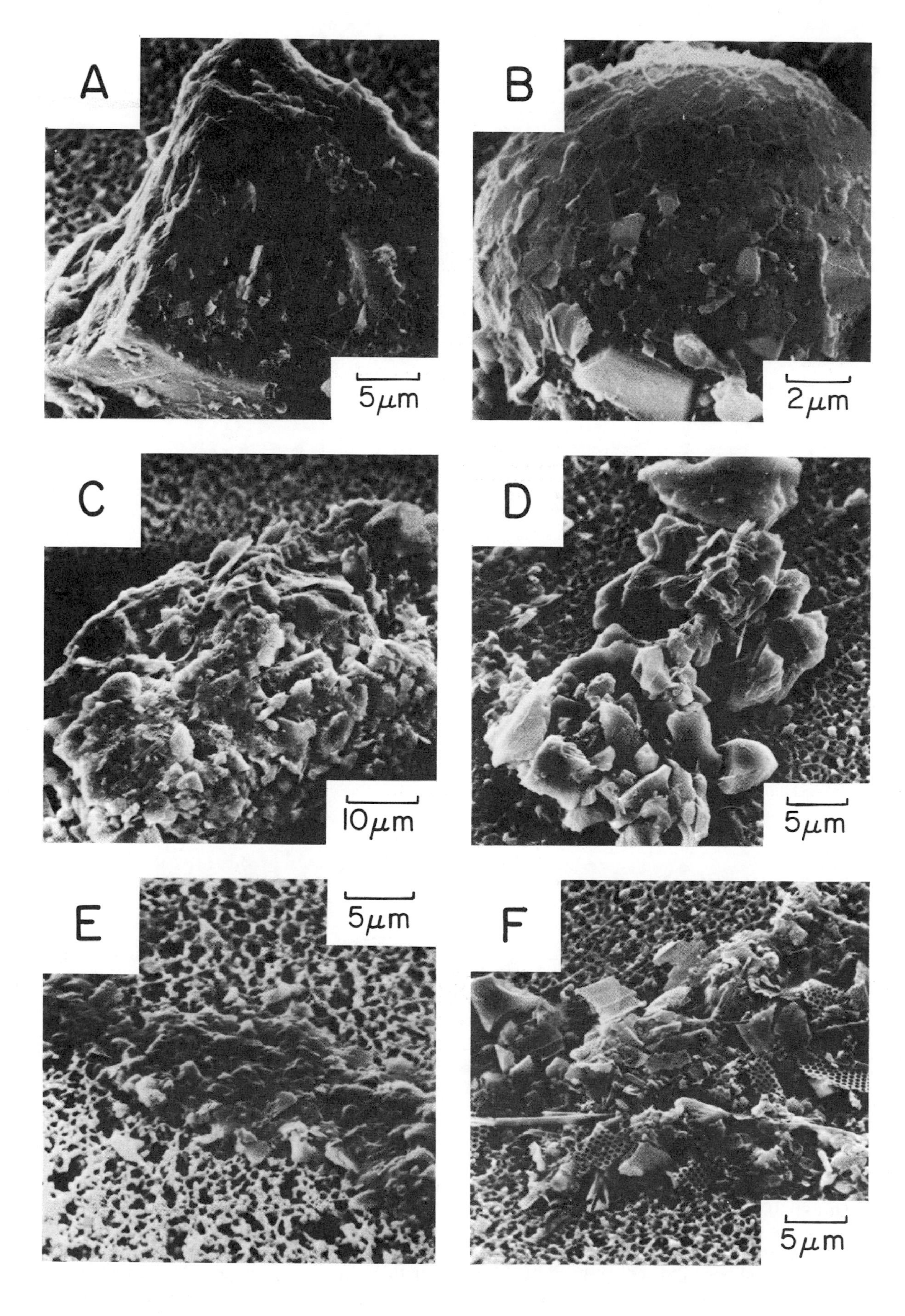
A
5μm
B
2μm
C
10μm
D
5μm
E
5μm
F
5μm

the halocline, begin to graze on these flocs and agglomerates subsequently producing mineral-bearing fecal pellets (Fig. 4.12, 4.13E). Many of the larger flocs and single grain particles, because of their size and settling velocity, escape the grazing by zooplankton; the flocs may continue to increase in size, eventually developing into mucus-coated particles (Fig. 4.12, 4.13F).

Mucoid filaments (also known as streamers or stringers), long, thin and delicate, are found suspended vertically within stable water masses (i.e., those with negligible internal shear, Syvitski et al., 1983b). Their size and concentration increases with depth and distance seaward, and they eventually become coated with suspended debris (Syvitski et al., 1985; Fig. 4.12). At depth, the filaments form delicate interconnected webs. The filaments may form from bacterial growth outward from decaying planktonic fecal pellets.

Particle settlement within the near zone of a fjord is also affected by the fluvial and tidal stage (Hoskin and Burrell, 1972; Hoskin et al., 1976, 1978). For example, the conditions of salinity and SPM concentration over one full tidal cycle are given in Figure 4.10. The concentrations of particular SPM size fractions during that time are represented in Figure 4.14. The clay and very fine silt fractions are well stratified and confined mostly to the surface layer. However, the medium and coarse silt fractions are able to breach the stratification. The coarse fractions then, are more event dependent, that is, on the tidal stage or discharge dynamics.

The size frequency distribution from surface samples within the upper prodelta are best described as lognormal distributions. Deviation from lognormality increases with water depth, so that samples collected below the halocline are strongly skewed toward the fine end of their size distribution (Fig. 4.15). The changes in the shape of the size frequency distribution indicate the increasing ability of coarser particles to preferentially escape the outward flow of the surface layer. By the 10-m depth, the size distribution of SPM is a truncated exponential curve.

FIGURE 4.13. Scanning electron micrographs of suspended particulate matter. (A) Coarse silt with attached freshwater microflora; (B) clay clasts ripped from raised marine terraces; (C) floccule of silt and clay particles in water of 10‰ salinity; (D) agglomerate with high biogenic component; (E) zooplankton fecal pellet containing mostly mineral grains; (F) agglomerate with mucoid coating.

4.2.3. Sedimentation under Hypopycnal Flows

The down-fjord sedimentation rate appears to decrease exponentially with distance from the river mouth (Hoskin et al., 1978; Relling and Nordseth, 1979; Smith and Walton, 1980; Syvitski and Murray, 1981; Bogen, 1983). The sedimentation rates reflect the exponential decrease in SPM concentrations with distance from the source (equation 4.19). The path of a floccule has been modeled and the residual descent path was found to be near vertical once the particle escaped the surface layer (Syvitski and Macdonald, 1982). Thus changes in SPM concentrations within the surface layer will affect the rates of sedimentation. Farrow et al. (1983) noted two separate stages within the exponential decay in sedimentation rates with distance from a river mouth. Apparently these two stages depict the separate conditions of sedimentation that exist under the upper and lower prodelta portion of the sediment plume.

Given knowledge of the surface layer velocity field, the vertical flux of SPM, Z, beneath the surface layer can be approximated by:

$$Z = \lambda Q_s (B_o u_o)^{-1} e^{-\lambda/u(x)} \qquad (4.25)$$

where λ is a first order removal constant in units of time $^{-1}$, Q_s is the suspended load discharge, u_o is the stream velocity at the river mouth, B_o is the channel width at the river mouth, and $\lambda Q_s(B_o u_o)^{-1}$ is the maximum rate of sedimentation at the river mouth. The removal rate is directly a function of the settling velocity of marine particles.

Sedimentation flux at the river mouth may be empirically established through the use of sediment traps suspended within the water column (see Syvitski, 1978; Hargrave and Burns, 1979; Gardner, 1980). Sediment traps positioned near the discharge outlet of five turbid fjords (Table 4.1) provide a time-averaged approximation of Z_o. The period of low discharge has concomitant low values of Z_o, about two orders of magnitude lower than rates measured at the freshet maximum. There is also a close relationship between seasonal fluctuations in SPM levels within the surface layer and that of sedimentation flux, $\underline{Z_o}$ and mean grain size of captured sediment, $\underline{d}$

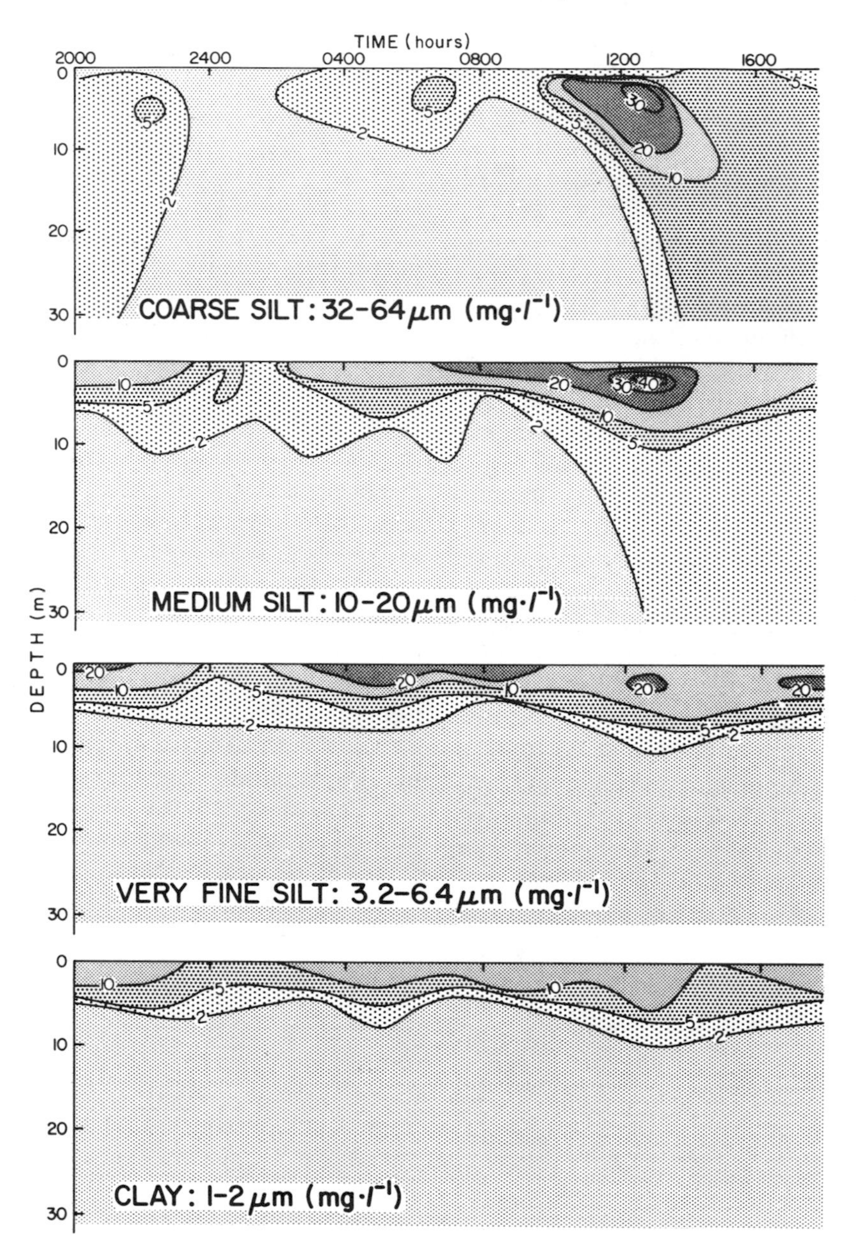

FIGURE 4.14. Time-series response of particle concentration for various size fractions (clay, very fine silt, medium silt, and coarse silt) during a 24-hr period and for the surface waters in Bute Inlet, B.C., near the river mouth (cf. Fig. 4.10B for further details; after Syvitski et al., 1985). Note the fallout pattern of the coarser particles.

(Fig. 4.16B; Syvitski and Murray, 1981). Glacier melt during late August is responsible for the highest levels in water turbidity, sedimentation flux, and largest sedimented particles.

As expected from equation (4.20), the exponential decrease in sedimentation flux away from a source is associated with a concomitant decrease in the size of particles that settle out (Fig. 4.16A, Gaupnefjord, Norway: Relling and Nordseth, 1979). The size frequency distribution effectively changes from one of a coarse size mode with a fine-grained tail nearest the river mouth, to one of a fine size mode with a coarse-grained tail farthest from the source. In other words, fallout is dominated by single component sand nearest the outlet with an increasing com-

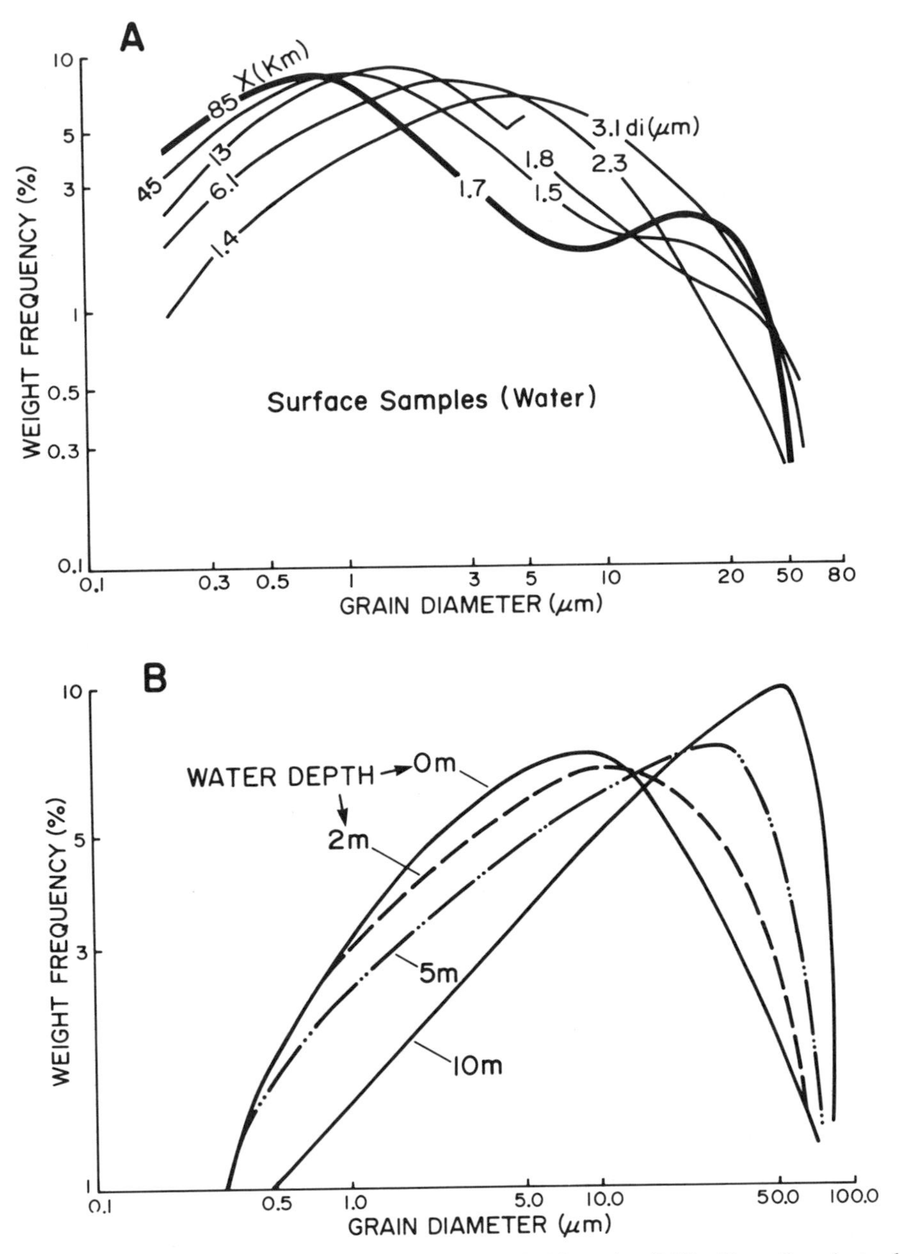

FIGURE 4.15. (A) Log-log plot of the size frequency distribution of surface water SPM where X is the position (in km from river mouth) where the sample was collected and di is the mean grain size in μm. Data from Bute Inlet, B.C., during summer freshet (after Syvitski et al., 1985). Note the phytoplankton mode (20 μm) appearing in samples seaward of 45 km. (B) Log-log size frequency distributions of samples collected in the upper 10 m in Bute Inlet water 1.4 km from the river mouth (after Syvitski et al., 1985).

ponent of silt floccules farther out (Fig. 4.16A).

Selected samples of surficial sediment from the fjord seafloor of Howe Sound also demonstrate the exponential decrease in grain size out from the river mouth (Fig. 4.16C). New sediment sources, however, especially from sediment gravity flows, can completely alter the size character of the seafloor sediment as laid down from turbid river plumes.

The net accumulation, a (in units of L T^{-1}), of sediment onto the seafloor can be given as

$$a = Z \rho_s^{-1} + B - E \qquad (4.26)$$

where ρ_s is the density of the surface sediment

TABLE 4.1. Sedimentation flux for turbid fluvioglacial fjords as measured by sediment traps near the discharge outlet.

System	Seasonal ranges (kg m^{-2} day^{-1})	Reference
1. Gaupnefjord, Norway	< 0.1 to 2	Relling and Nordseth, 1979
2. Knight Inlet, B.C.	0.04 to 4	Farrow et al., 1983
3. Howe Sound, B.C.	0.03 to 1.3	Syvitski, 1980
4. Queens Inlet, Alaska	< 1.2 to 16.8	Hoskin et al., 1976
5. Blue Fjord, Alaska	0.02 to 0.53	Hoskin et al., 1978

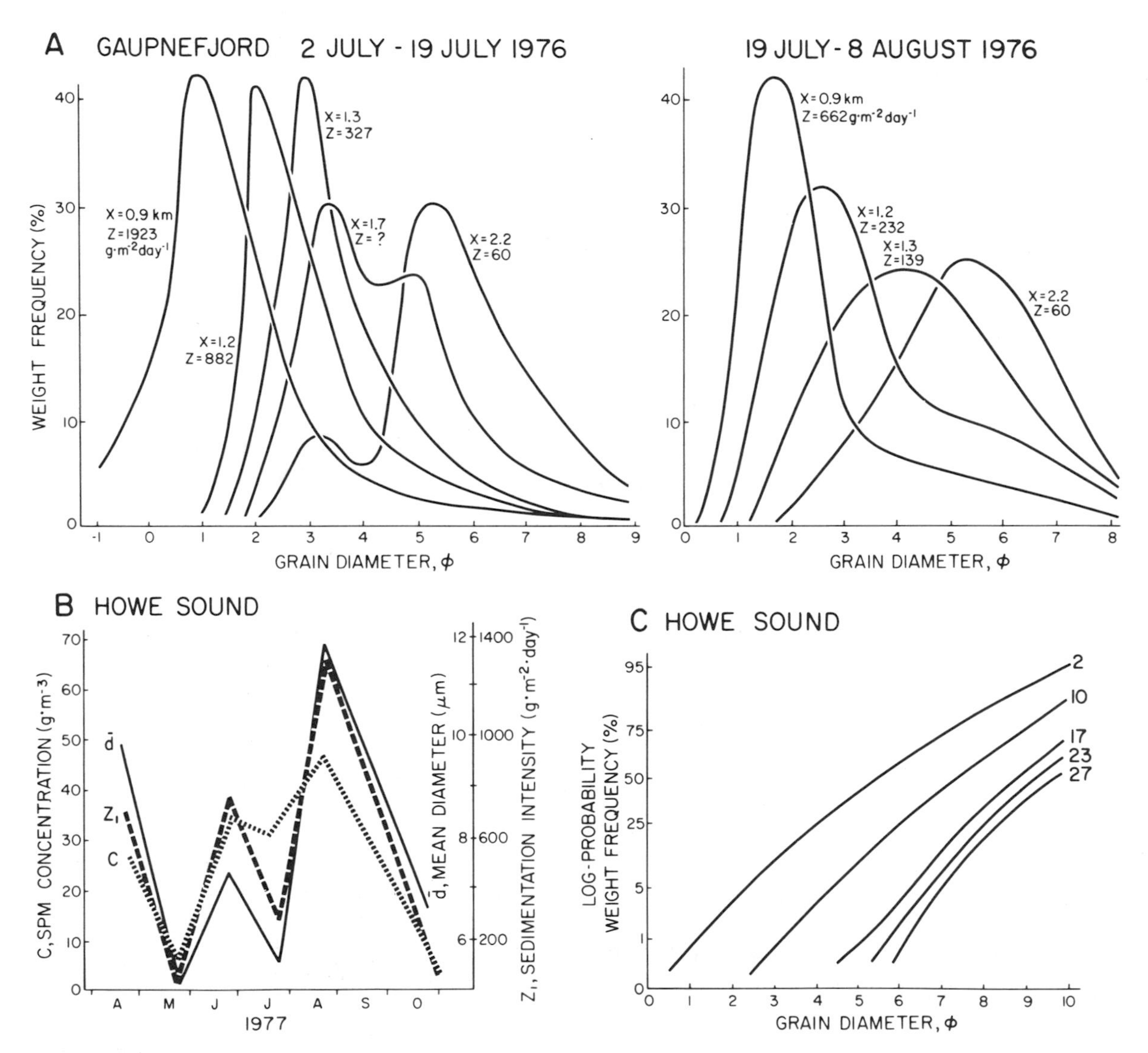

FIGURE 4.16. (A) Size frequency distributions from material collected by sediment traps anchored above the seafloor in Gaupnefjord. Shown are responses for early and mid-summer; X is the distance in km from the river mouth and Z is the sedimentation intensity in g m^{-2} day^{-1} (after Relling and Nordseth, 1979). (B) Seasonal variations in SPM concentration, sedimentation intensity, and mean grain size of sedimented material as observed in Howe Sound, B.C. (cf. Fig. 4.9B; after Syvitski and Murray, 1981). (C) Log-probability plots of the size distribution of seafloor grain size collected down the axis of Howe Sound, B.C. (after Syvitski and Macdonald, 1982). Distances are in km from the Squamish River mouth.

TABLE 4.2. Accumulation rates within fjords.

System	Accumulation (mm yr^{-1})	Method	Reference
1. Northeast Gulf fjords, Alaska (3)	450 to 2000	Bathymetry	Molnia, 1979; von Heune, 1966
2. Muir Inlet, Glacier Bay Alaska	2000 to 9000	Bathymetry	Powell, 1983
3. Boca de Quadra and Smeaton Bay, Alaska	2.7 to 4.6	^{210}Pb, ^{137}Cs	Sugai, 1985
4. Bute Inlet, B.C.	1 to 300	SPM models	Farrow et al., 1983
5. Douglas Channel, B.C.	0.2 to 2	Variety	Macdonald, 1983a, b
6. Saguenay Fjord, Quebec	0.4 to 23	^{210}Pb	Smith and Walton, 1980
7. Bedford Basin, N.S.	3	Variety	Piper et al., 1983
8. Makkovik Bay, Labrador	0.7 to 3	^{14}C	Barrie and Piper, 1982
9. Sondre Stromfjord, Greenland	25 to 35	Seismo/biostrat.	Larsen, 1977
10. Igaliko Fjord, Greenland	0.15	^{14}C	Herman et al., 1977
11. Spitsbergen fjords (3)	0.1 to 100	Variety	Elverhöi et al., 1983
12. Oslo Fjord, Norway	$<$ 0.6 to 2.6	^{14}C	Richards, 1976
13. Korsfjorden, Norway	4	Biostratigraphy	Aarseth et al., 1975
14. Norwegian fjords (6)	0.5 to 10	^{210}Pb	Skei, 1982a
15. Lock Striven, Scotland	5	Biostratigraphy	Deegan et al., 1973
16. New Zealand fjords (3)	0.8 to 4	^{14}C, SPM model	Glasby, 1978a, b
17. Milford Sound, New Zealand	1	Inference	Pantin, 1964

used to convert sedimentation flux, Z, to units of L T^{-1}, B is sediment accumulation (in units of L T^{-1}) that is due to input from nonhypopycnal sources and can include aeolian, bed-load transport, mass movement and sediment gravity flows, ice rafting, flotation and deep water exchanges; and E represents loss of sediment (in units of L T^{-1}) during periods of net erosion. Methods used to calculate a include bathymetric changes (Molnia, 1979), isotopic dating (primarily ^{210}Pb: Smith and Walton, 1980; ^{14}C: Andrews and Jull, 1985), seismo-stratigraphy (Larsen, 1977), and biostratigraphy (Aarseth et al., 1975). If we use SPM models (as outlined above or given by Farrow et al., 1983; Macdonald, 1983a,b; Glasby, 1978b) to estimate the vertical flux, and if we further assume that our reference location undergoes no period of erosion, then B (nonhypopycnal sources) can be estimated knowing a and Z. However, B is rarely calculated, and components of the accumulation rates given in Table 4.2 are seldom differentiated. One exception is the investigation of Douglas Channel, B.C. (Macdonald, 1983a,b) where estimates of Z ρ_s^{-1} were found to equal calculated values of a, that is, B and E were considered negligible during the accumulation period under investigation. In other words, the sedimentation flux to the seafloor accounts for the net accumulation rate.

The Saguenay Fjord is another system where evidence suggests much of the seafloor sediment accumulates primarily from the vertical flux of riverine particles. In addition, high sedimentation rates and high organic matter loadings with a concomitant reduction in ambient dissolved oxygen content in bottom and interstitial water can result in reduced rates of bioturbation (see Chapter 6) and enhanced resolution of stratigraphic events (Smith and Walton, 1980; Schafer et al., 1980; Smith and Ellis, 1982; Schafer et al., 1983). The lack of bioturbation has allowed marine varves to form, that is, organic-clay winter layers alternating with sandy silts composing the summer layers (Fig. 4.17A). As ^{210}Pb is absorbed onto the surfaces of clay minerals, its activity can be inversely correlated with discharge events attributed to the associated inputs of ^{210}Pb deficient silts and sands (Fig. 4.17B). Precision ^{210}Pb analysis on sediment cores collected from the Saguenay Fjord has allowed the age determination of each varve. Sediment characteristics within the summer layers can then be compared with historical discharge records of the main fjord-valley river. The results satisfactorily show that sand modal size can be used to predict freshet discharge levels (Fig. 4.17C). The possibility exists, then, for accurate hindcasting of paleoclimate before the advent of stream gauging.

The distribution of a given mineral within a fjord influenced by hypopycnal flows depends largely on the unique size frequency distribution of that mineral (Fig. 4.18C). It is important to remember that in higher latitude fjords—especially glacial fjords—minerals are less weathered so that, for example, one can get primary amphibole in clay-sized fractions and not neces-

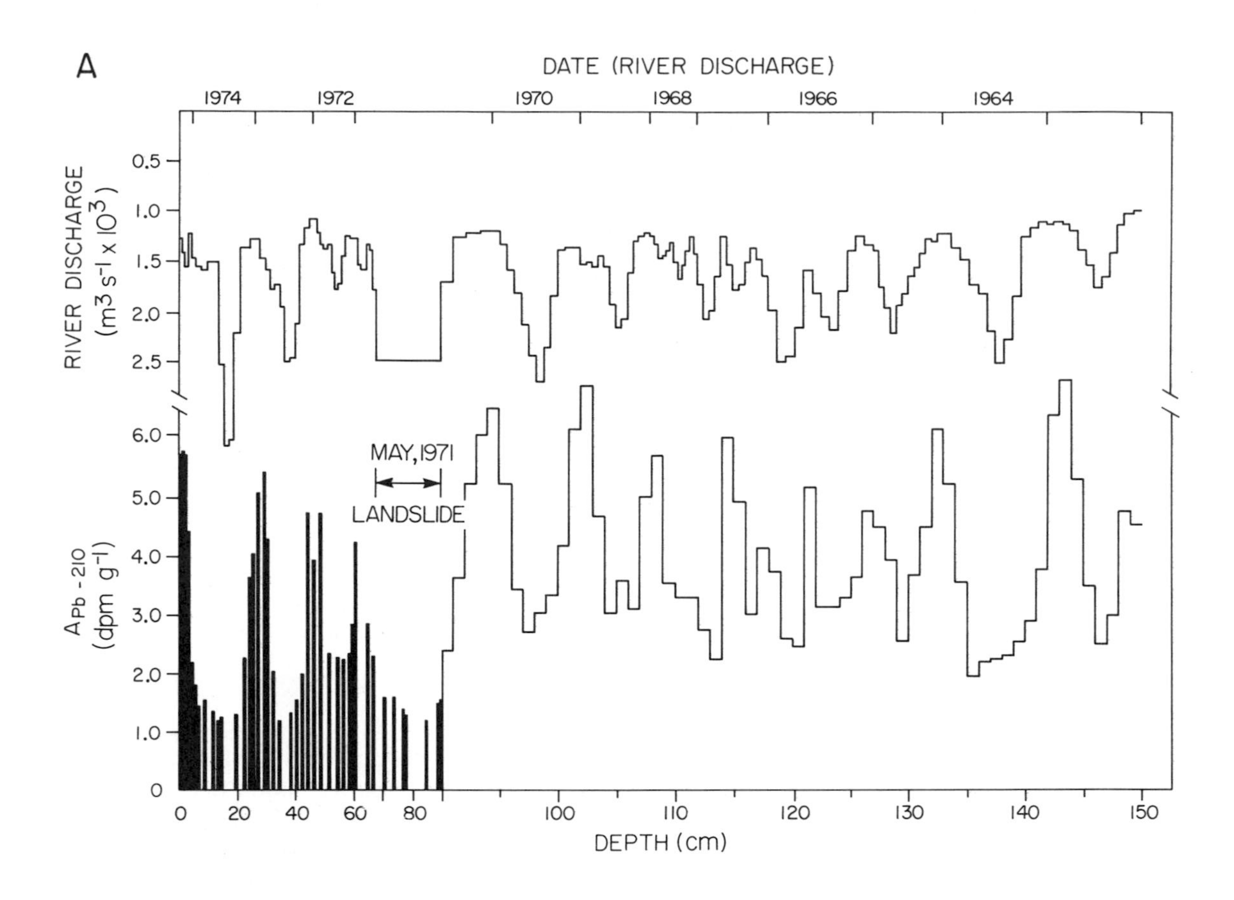

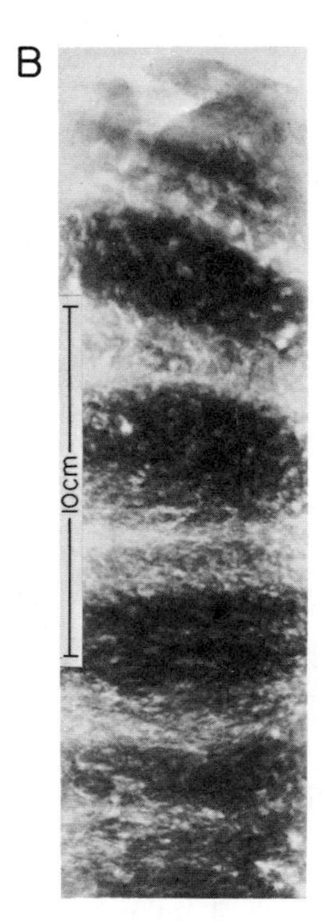

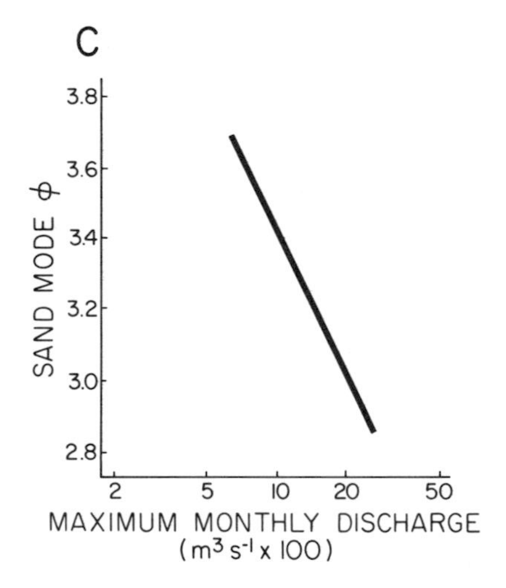

FIGURE 4.17. (A) ^{210}Pb activities as a function of depth in a core collected 300 m from the Saguenay River mouth and in a water depth of 40 m (after Smith and Ellis, 1982). An inverse correlation between ^{210}Pb activities and the river discharge rates provides excellent time-stratigraphic resolution. (B) X-radiograph of the upper 24 cm of a core collected in 74 m of water depth and approximately 2 km from the Saguenay River mouth, Quebec (from Schafer et al., 1983). The core is an example of marine varving. (C) Relationship between annual maximum monthly river discharge and phi (ϕ) mode of the sand fraction in the core shown in (B) (after Schafer et al., 1983).

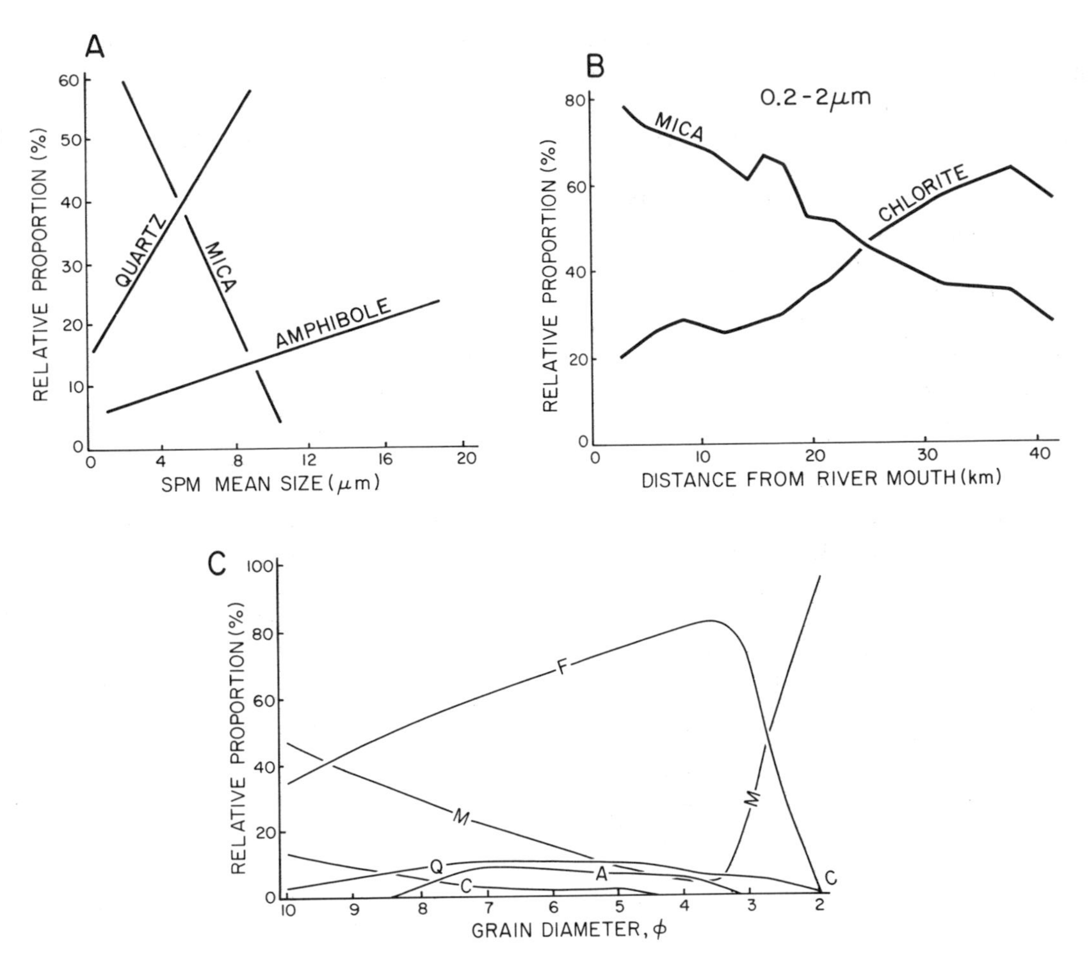

FIGURE 4.18. (A) Mineral variability within suspended particulate matter (SPM) as a function of mean grain size (Bute Inlet, B.C., data). (B) Percentages of clay-sized minerals within surficial bottom sediment as a function of distance from the Squamish River mouth, B.C. (after Syvitski and MacDonald, 1982). (C) Size frequency distribution of five major minerals found in Bute Inlet, B.C. (after Syvitski et al., 1985). M = mica, Q = quartz, A = amphibole, C = chlorite, and F = feldspar.

sarily all chlorite. It is therefore expected that a particular mineral's distribution will vary widely from fjord to fjord. If a mineral has a coarse size mode, hydraulic sorting will ensure that it reaches the seafloor near the river mouth. Some minerals are bimodal in their size distribution (see mica, Fig. 4.18C), and their seafloor distribution will thus reflect near and offshore maxima. Figure 4.18A documents the change in SPM mineralogy as a function of mean grain size within the surface water sediments of Bute Inlet, B.C. (Syvitski et al., 1985). Mica appears better represented with decreasing grain size of the suspended particles, an opposite trend to that of quartz and amphibole. In Howe Sound, B.C., the clay-sized mica content of seafloor sediments decreases linearly with seaward distance relative to clay-sized chlorite, which increases (Fig. 4.18B). The result is important as mineral dispersal patterns can effectively delineate surface plume pathways (Hein et al., 1979).

4.3 Hyperpycnal Flow

River discharge, if of sufficient density, can enter a standing body of water at depth as a hyperpycnal flow, that is, as an inflow or an underflow. This commonly occurs within freshwater lakes (see Section 3.3.1, Fig. 3.7), the physics of which have been described by Hamblin and Carmack (1978). In the case of marine basin water, concentrations of suspended sediment must be very high in order to overcome the density of salt water. For basin water of 30‰ salinity and a temperature of 3°C, the suspended sediment concentration must reach 38 g L^{-1}

(Table 4.3). Glacial meltwater streams contain high sediment concentrations, but even those seldom exceed concentrations needed to submerge the inflowing plume. With the exception of very high concentrations associated with high melt rates in ice-proximal environments (Powell, 1981a; see Section 4.5.1) or with rare or short-lived discharge events (Gilbert and Shaw, 1981), normal suspended sediment concentrations that enter into present-day fjords are several orders of magnitude less than necessary to overcome the buoyant effect of sea water (Gilbert, 1983). Rare events may include rare precipitation storms (with a several hundred year return interval) or jökolhlaups that are associated with high suspended load levels (Church, 1972). Side-entry fluvial inputs with steep valley slopes may offer the best location for hyperpycnal flows, especially during storm events. Their steep entry angle will not allow an easy separation of bed-load particles from those suspended in the turbulent flow.

Although hyperpycnal flows may not be historically significant in fjord environments, they may have been important, if not the norm, during periods of proximal ice-sheet sedimentation. Periods of ice-sheet melting and retreat (see Section 3.3.3) are known to be associated with one to two orders of magnitude higher discharge and vertical flux events (see Church, 1978; Macdonald, 1983a,b; Gilbert, 1985). These large discharges over a relatively long period (i.e., up to 3000 years) may have also reduced the salinity of the paleofjord environment, thus further increasing the likelihood of hyperpycnal flows.

Of more recent significance is the discharge of mine tailings into fjord environments (e.g., Ranafjord, Norway: Carstens and Tesaker, 1972; Marmorilik, Greenland: Asmund, 1980; Rupert Inlet, B.C.: Hay, 1982). The tailing is introduced as a slurry that enters the sea through a pipe outlet placed some tens of meters below the pycnocline, the slurry is introduced as a hyperpycnal flow.

TABLE 4.3. Concentrations of suspended sediment of density 2.70 g ml^{-1} necessary to overcome the density difference between fresh water and salt water.[1]

Salinity (‰)	Density at 3°C (g ml^{-1})	Sediment concentration g L^{-1}
10	1.008	12.7
15	1.012	19.1
20	1.016	25.4
25	1.020	31.8
30	1.024	38.1
33	1.026	41.6

[1]After Gilbert, 1983.

The greatest effects of hyperpycnal flows, especially if they are heavy enough to sink to the seafloor, is to increase both the turbidity and current regime of the seafloor, and to lower the salinity of the bottom waters. By doing so, the benthic community may become critically inundated with particulate sediment (see Chapter 6) or harmed by the lowered salinity. Many have noted the deleterious effect of particulate loading on fjord benthos (Hoskin et al., 1976; Farrow et al., 1983), and still others have noted that ice-proximal fjord sediments contain limited or no benthic life (Vilks et al., 1984) as do seafloors receiving tailings discharges (Ellis, 1982; p. 264).

4.4 Flushing and Deep Water Renewal

One of the most discussed aspects of silled fjords is the periodic flushing or renewal of their basin waters with adjacent coastal or shelf waters. Such interest may reflect: (1) the direct influence of a renewal on fjord circulation and sedimentation; (2) the results of basin stagnation: anoxia and the destruction of sea life (see Chapter 6); and (3) the importance of renewal on the biogeochemical environment of basin waters and sediments (see Chapter 7). The present discussion focuses first on the dynamics of renewals, then on the frequency of the various forcing mechanisms, and finally provides some insights into the effect of flushing on sedimentation. Details can be found in the excellent reviews by Gade and Edwards (1980) and Farmer and Freeland (1983); the physics of renewals have been discussed in Gade (1970) and Edwards and Edelsten (1977).

If the adjacent shelf water outside the sill is denser than the resident water of the fjord basin, and if it can be "lifted" over the sill, then a density current will develop and attempt to replace the basin water (Fig. 4.19A). Density currents are seldom continuous but behave intermittently on a variety of time scales depending on the circumstances. Renewal events may be triggered by tidal motion, weather systems (land/sea breezes, atmospheric pressure), and seasonal events (monsoonal winds, runoff variations).

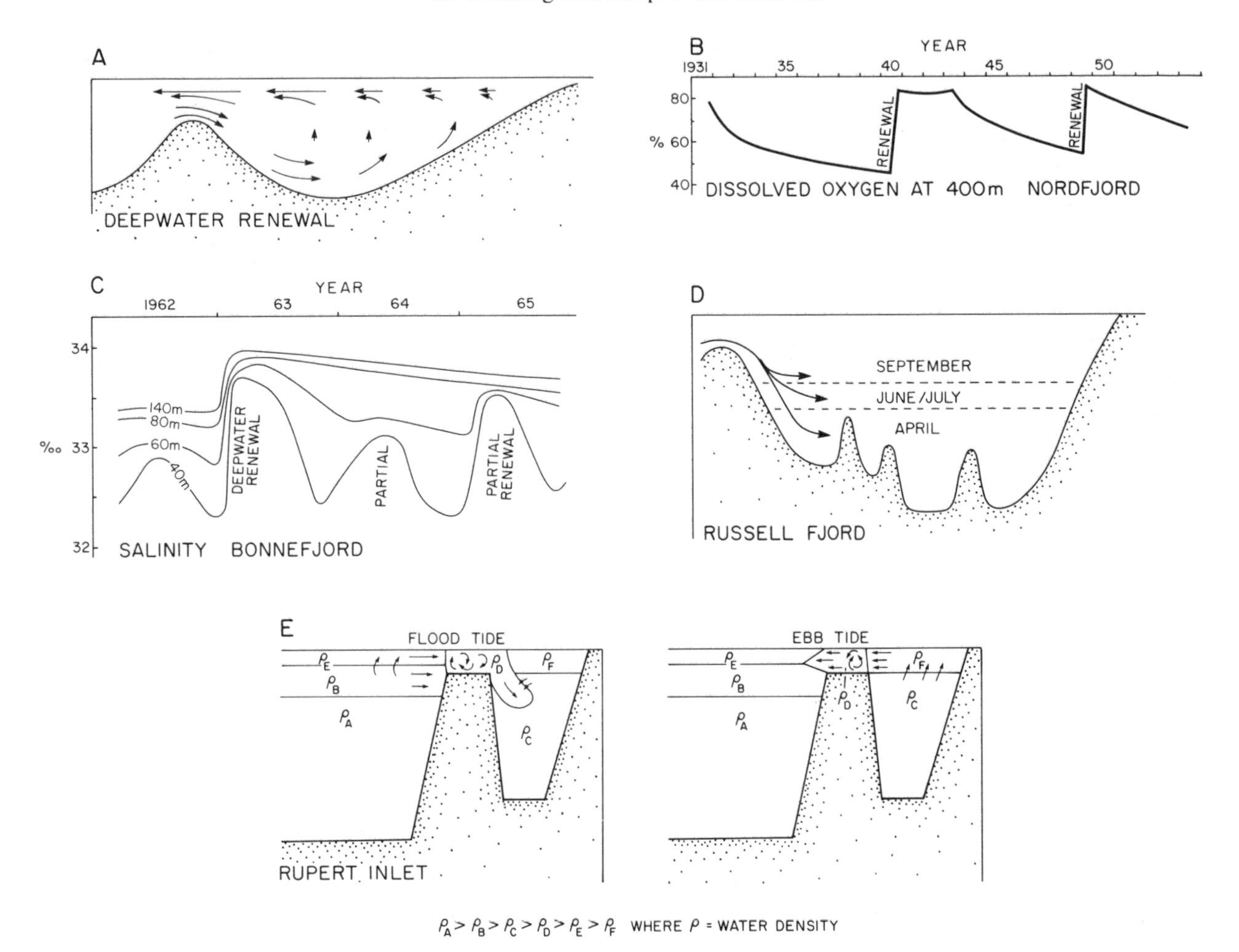

FIGURE 4.19. (A) Simple schematic of a complete fjord deep water renewal (after Gade and Edwards, 1980). (B) Schematic of the low frequency cycling of renewal in Nordfjord, Norway, based on the dissolved oxygen content at 400m (after Saelen, 1967). (C) Variation in salinity of different depths within Bonnefjord, Norway, showing both complete and partial renewal events (after Gade and Edwards, 1980). (D) Conceptualization showing dependence of depth of renewal on season in Russell Fjord (after Muench and Heggie, 1978). (E) Conceptualization of sill dynamics and renewal with the tidal cycle in Rupert Inlet, B.C. (after Drinkwater, 1973).

Having entered the basin, the heavier renewing water will sink toward the basin bottom as a turbulent density current or plume. The downward acceleration can be related to the force of gravity but modified by the density contrast, a pressure gradient resulting from the changing plume thickness, bottom drag, and entrainment characteristics of the flow (Gade and Edwards, 1980). The plume density may alter with an increase in density as a result of resuspended sediment particles and the entrainment of less-dense resident basin water (Edwards et al., 1980).

If the plume's inflow velocity is intially slow, gravity will accelerate the flow (Gade and Edwards, 1980). If, however, the initial plume velocity is high, then increased entrainment will slow down the flow. The level of entrainment is known to increase with increasing bottom slope (Edwards et al., 1980). Propagation velocities at the head of these gravity currents may range up from 20 to 70 cm s^{-1} (Geyer and Cannon, 1982; Helle, 1978; Edwards et al., 1980), corresponding to lower time-averaged velocities for the main body of the plume, for example, 1 to 10 cm s^{-1} (i.e.,Lafond and Pickard, 1975; Cannon, 1975).

Not all renewals allow for the replacement of the entire water mass of the fjord. The renewal may be considered partial under three conditions (Gade and Edwards, 1980): (1) The intruding shelf water, not sufficiently dense to replace the deepest water in the fjord, will sink to some more appropriate level, spread out, and begin to fill up the available space from that level up. (2) The intruding plume, being sluggish, may soon use up its available energy (for instance overcoming bottom friction) and come to rest without stirring up much of the resident basin water. (3) Although all density conditions for complete replacement are met, the renewal event may be of such short duration that the fjord basin water is not completely overturned. In addition, those fjords with multiple basins tend not to have deep water renewal as a single process but as a series of events that may or may not affect all the basins within the fjord (e.g., Loch Etive: Edwards and Edelsten, 1977). The "source" water may use up much of its initial energy in deep water renewal of the outermost basin (as a consequence of interlayer friction and dilution).

The shallower the sill, the more likely that flow associated with tides, wind, and meteorological disturbances will switch off inflow if directed out of the fjord and augment it if directed inward (Gade and Edwards, 1980). Therefore, renewals tend to be "pulsed" over shallow sills, rather than the more continuous flow common over deeper sills. Tides may influence renewal on a semidiurnal or diurnal time scale (e.g., Saanich Inlet, B.C.: Anderson and Devol, 1973). Tidal-influenced renewal is, however, more common on the spring-neap cycle (e.g., Puget Sound, Washington: Geyer and Cannon, 1982). The propagation of internal waves over a sill is another high-frequency mechanism for renewal. In the case of the St. Lawrence Estuary, internal waves propagate with amplitudes up to 60 m, bringing up dense water, semidiurnally, to the sill level of the Saguenay Fjord (Seibert et al., 1979).

Winds can: (1) cause baroclinic flow, which in the sill region can import renewing water; and (2) change the water level in the fjord and allow barotropic currents to diminish or augment inflow over the sill according to their direction (Gade and Edwards, 1980). Howe Sound is an example of the first case where some renewal events have been related to strong down-channel winds (Bell, 1973). Examples of the second case are more common and are related to changing coastal longshore winds affecting offshore convergence or divergence (e.g., Byfjord, Norway: Helle 1978; Skjomen, Norway: Skreslet and Loeng, 1977).

Although water on the seaward side of the sill may be of sufficient density for renewal, mixing with lighter water during its passage over the sill may constrain the renewal event. The renewal rate, then, varies with the mixing rate or supply of light water. Periods of high discharge can lead to well-developed baroclinic and barotropic currents at the sill. Such well-stratified waters would limit the dilution of the compensating current and thus promote renewal (McClimans, 1978b; Ebbesmeyer and Barnes, 1980). Mid-depth penetrations of shelf water into Howe Sound have been correlated to large discharge periods (Bell, 1975). Apparently with lowered energetics at the Howe Sound sill, increasing freshwater runoff will accelerate the baroclinic currents and the supply of renewing water.

Renewals that have a yearly return period are related commonly to the density structure offshore. In Norwegian and Alaskan fjords, major exchanges are associated with the monsoonal nature of the major wind field: sustained northerly winds in the spring and southerly winds through the winter (Gade, 1976). This results in onshore convergence through the winter, and shallow upwelling along the coast during the periods when the wind reverses (spring-early summer). During the winter, the salinity of the shelf surface waters (100 m) increases; in the spring, the salinity of the deeper shelf waters increases. The consequence is deep water renewal during winter for shallow-silled fjords (< 40 m) and continuous renewal during winter through summer for deep-silled fjords (> 100m).

Various patterns of exchange are illustrated in Figure 4.19. Dissolved oxygen content of the basin water of Nordfjord, Norway, shows a 9-year period for complete renewal (Fig. 4.19B, Saelen, 1967). Biological activity over time will deplete the oxygen from the water until the next renewal event replenishes the fjord with oxygenated shelf water (see Chapter 6 and 7 for details). Bonnefjord, Norway, had a complete renewal in early 1963 and subsequently two partial renewals: the first at 40 m in 1964 and a deeper renewal (60 m) in 1965 (Fig. 4.19C; Gade and Edwards, 1980). Russell Fjord, Alaska, had a partial renewal to intermediate depths during the summer, a shallower renewal during the fall, and a deep renewal during the spring (Fig. 4.19D; Muench and Heggie, 1978). Rupert Inlet is a dynamic example of renewal on the time scale of the diurnal tide (Fig. 4.19E; Drinkwater and Osborn, 1975).

Very little is known about the effect of deep water renewals on fjord sedimentation. It is known that deep basin surficial sediments just landward of a sill are swept clean of silts and clays (Fig. 4.20A; Gade and Edwards, 1980; Tunnicliffe and Syvitski, 1983). In an interesting set of experiments near a sill in Loch Eil, Scotland, Edwards et al. (1980) found that little sedimentation occurred during neap tides. However during spring tides, a time of known deep water renewals, local sedimentation increased by an order of magnitude. They considered that the increase in sedimentation was due to local resuspension during the renewal event and to the sediment load carried by the renewal water. The latter point has some interesting implications. If the renewal occurs during the summer when the fjord waters are more likely to be turbid compared with adjacent waters, then a renewal event may result in the removal of a significant quantity of sediment from the fjord system. The opposite is considered true for the winter, when a deep water renewal event may import resuspended storm sediment from the adjacent shelf at a time of low surface layer turbidity levels within the fjord (Winters et al., 1985). Such a hypothesis awaits field testing although similar observations based on particulate organic carbon support this relationship (Therriault et al., 1980).

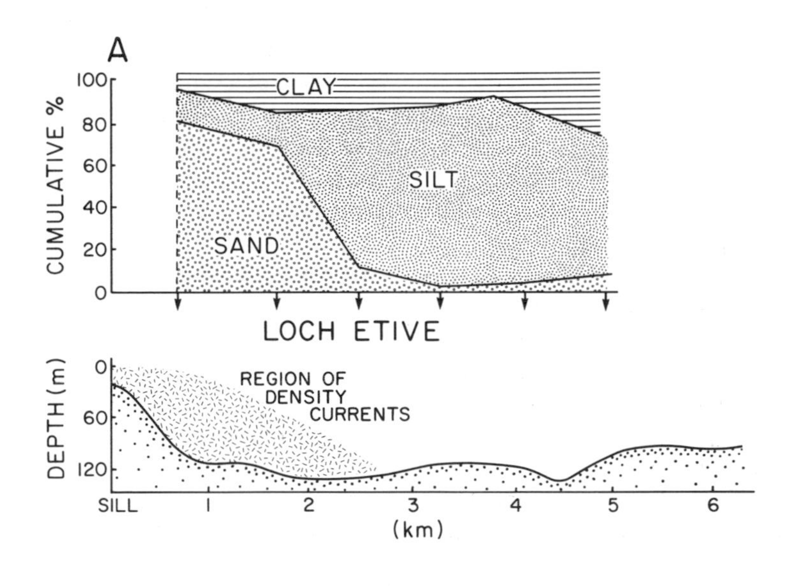

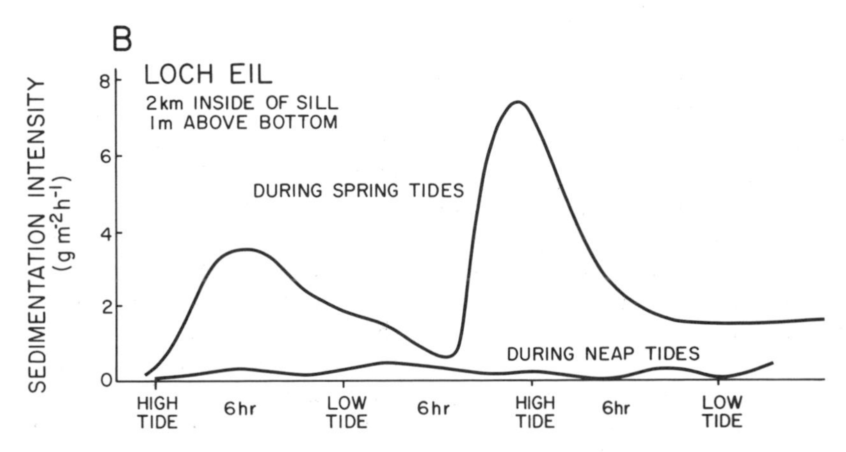

FIGURE 4.20. (A) The effect of renewal density currents on grain size fractions of surficial sediment in Loch Etive (after Gade and Edwards, 1980). (B) Sedimentation flux notably higher during spring tide in Loch Eil (after Edwards et al., 1980).

4.5 Ice Influences

Almost 70% of all fjords are influenced presently by glacier ice and or sea ice. The influence extends from effects on sediment transport and erosion to controls on water circulation. Further, by the very nature of fjord formation, all classical fjords were at one time ice dominated, and some for a very long time. For instance, the Alaskan glaciation has been in progress for the last 5 to 30 million years (Molnia, 1979). This section first reviews the formation and breakup of sea ice and the subsequent effects on circulation and sedimentation. A discourse on movement of the tidewater glaciers follows, documenting the process of iceberg calving, rafting, and scouring, and the effects of glaciers on marine sedimentation and circulation.

4.5.1 Sea Ice and Circulation

The surface layer of a fjord may freeze up sometime in the fall or winter. The period of freeze-up can be affected by air temperature, wind, waves, and pack ice conditions, amount of snowfall, and stage of the monthly tidal cycle (Taylor and McCann, 1983). New ice begins to form in the intertidal zone when the air temperature falls below the freezing point of sea water (−1.6°C at a salinity of 29.5‰, and −1.8°C at 34‰). At this stage, the pore water within the intertidal sediments freezes and a thin glaze of ice forms on the sediment surface. During calm cold days or nights, sea ice forms farther offshore (down-fjord), but preferentially near freshwater outlets where the salinity of the surface water is somewhat diluted. Near shore, ice accumulation proceeds by the deposition and consolidation of ice slush, stranded floes, ice cakes, and ice boulders, and by the freezing of wave swash, spray, and snow (Taylor and McCann, 1983). The net result is the development of an icefoot, that is, a solid bulwark of ice frozen fast to the shore. Freeze-up may be interrupted by periods of higher temperatures and wave action: in South Georgia, for example, only the sheltered fjords undergo freeze-up (Hansom, 1983).

Joyce (1950) describes various modes of icefoot formation within Neny Fjord in Graham Land, Antarctica. Some of these include: (1) the tidal platform icefoot, formed by tidal action between high- and low-water lines; (2) the storm icefoot, built up above the high-water line by spray from breaking waves; (3) the drift icefoot, fabricated from drift ice and consolidated by freezing sea water that surges through the tide cracks; (4) the pressure icefoot, formed by overriding sea-ice slabs emplaced by an onshore movement; (5) the stranded-floe icefoot incorporating beached bergy bits.

By midwinter one observes a typical sea ice zonation along most Greenland fjords: a well-developed icefoot, a bridge or transition zone of tidal cracks and leads where ice floes contract and expand with the tide, and mobile floating sea ice (Petersen, 1977). Typical first-year ice floes can reach 2 m in thickness in the cold arctic. In Cook Inlet, Alaska, sea ice is frequently in motion throughout the winter. Sea ice interactions between other ice floes and with the shoreline are common and not surprising considering the > 7-m tides and currents that may exceed 3 m s^{-1} (Blenkarn, 1970). Other less-energetic arctic fjords develop a complete ice cover by midwinter.

The development of a winter ice cover eliminates wind-induced mixing and may also dampen tides and tidal currents (Gilbert, 1983). Ice growth also leads to the establishment of a homogeneous surface layer as a result of the process of salt rejection, whereby salt is released from the freezing ice mass (Gade et al., 1974). Gilbert (1983) has calculated that during the formation of a 2 m thick ice sheet, 50 kg m^{-2} of salt is released, leaving the sea ice with a salinity from 3 to 10‰. The thickness of the mixed surface layer increases throughout the winter owing to both salt released by ice growth and by vertical mixing of deeper waters (Fig. 4.21; Gade et al., 1974; Perkin and Lewis, 1978). During ice formation, heat is lost continuously from the surface waters, and a temperature maximum migrates downward under the surface (Fig. 4.21A). As salt rejection continues, vertical mixing reaches constantly increasing depths, eventually leading to gravity flows to the middle and lower layers (Lewis and Perkin, 1982). In the shallows (i.e., near the fjord head), convection reaches the seafloor and restricted circulation can lead to an excess accumulation of expelled salt. If the density of the water in the shallows exceeds that of the basin water, it will sink down the slopes (Fig. 4.21B). The onset of

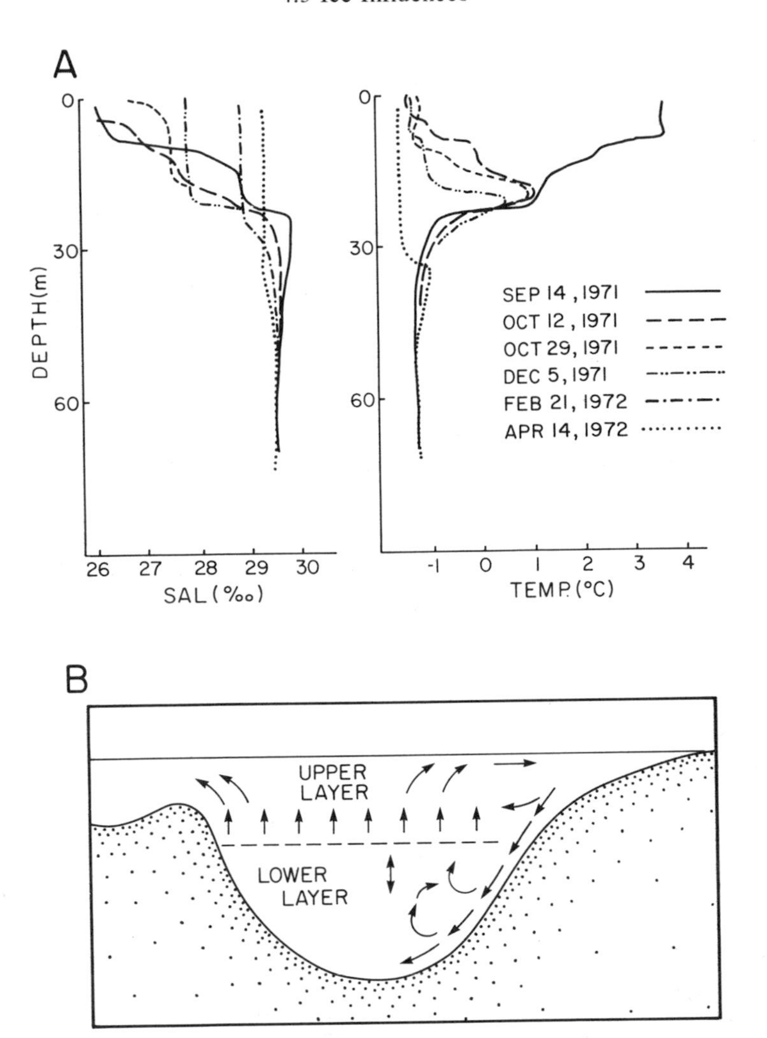

FIGURE 4.21. (A) Seasonal changes in water column temperatures and salinities at Cambridge Bay, Canadian Arctic (after Gade et al., 1974). (B) Schematic of circulation in Cambridge Bay during fall and winter caused by a flow of dense water from the shallows penetrating the main pycnocline (after Gade et al., 1974).

spring will cause a cessation in ice growth and vertical circulation will drastically decrease until ice breakup (Lewis and Perkin, 1982). The major sources of energy during the winter months include: salt-driven circulation; water movements induced by tides or atmospheric pressure changes; internal waves and internal seiche events that sometimes produce violent vertical mixing; and, where applicable, upwelling produced by melting at depth of icebergs and glaciers (Perkin and Lewis, 1978; Lewis and Perkin, 1982). Long-period waves, possibly induced by large-scale weather systems, may raise the ice cover upward along the series of tidal cracks lying a few meters offshore (Lake and Walker, 1976). These tidal cracks form a flexible hinge-connection between a narrow band of shorefast ice and the main ice sheet covering the fjord surface.

Duration and thickness of the ice cover depend on a variety of oceanographic and meteorologic conditions, but both increase generally with latitude. The higher latitude fjords may even be "frigid," in that the ice cover is permanent. Frigid fjords along the northeast coast of Ellesmere Island are complicated further by having their mouths enclosed by the Ward Hunt Ice Shelf (Fig. 4.22A). One of these fjords, Disraeli (lat. 83°N) has been investigated by Keys et al. (1969) and Keys (1978). Under the pe-

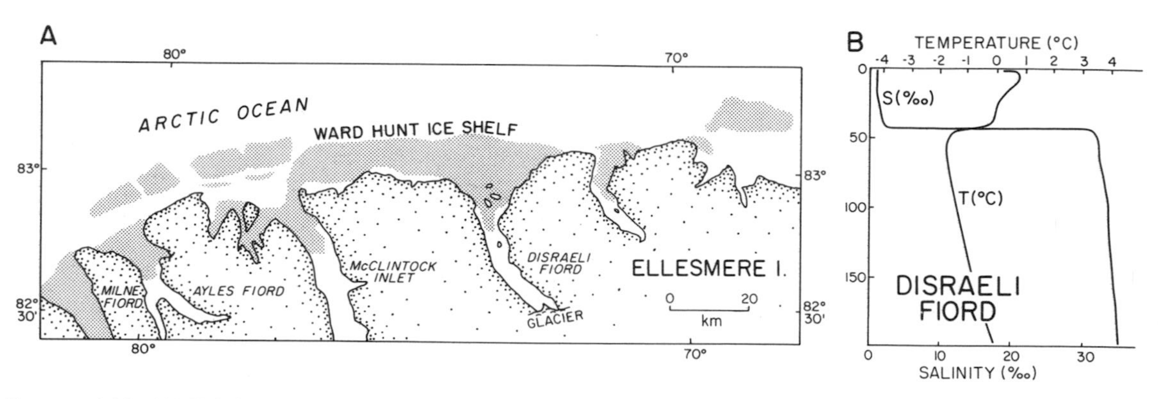

FIGURE 4.22. (A) Frigid, high arctic fjords along NW Ellesmere Island that are affected by continuous ice cover and the Ward Hunt Ice Shelf. (B) Unique salinity and temperature profiles of Disraeli Fiord as influenced by the Ward Hunt Ice Shelf (after Keys, 1978).

rennial ice cover, the upper 44 m of surface water is composed of fresh water because of the complete enclosure of the ice shelf (Fig.4.22B). This freshwater lake lies on top of 155 m of oceanic water. The pycnocline between 44 and 45 m has salinity increasing from 5 to 32‰, while temperature decreases from −0.6 to −1.6°C. Fresh water enters the fjord in the form of a glacial meltstream that sinks to the pycnocline and spreads seaward. The exact origin of the ice shelf remains speculative, but presently ice is growing from below as a result of the freshwater runoff from the nearby fjords, and from above from snow-ice accumulation (Lyons et al., 1971). Frigid fjords are noted for their small energy inputs and weak currents (i.e., d'Iberville Fiord, 80°N: Lake and Walker, 1976).

Breakup of sea ice is affected by temperature, land drainage, sediment content of the ice, wind and tidal conditions, and the position of major offshore leads (Taylor and McCann, 1983). The process begins with the melting of the snow cover on the ice, the enlargement of the tidal cracks in the littoral zone, ice breakup at river mouths owing to the influx of fresh water during the nival freshet. The ice in the intertidal zone breaks up first and melts before the offshore ice because of its fractured condition (the result of repeated groundings) and because of its decreased albedo resulting from the incorporation of dark sediment (Taylor and McCann, 1983). Ice then melts progressively from the head of the fjord to the sea, principally as a result of turbulence in the near surface (under ice) waters from the freshet discharge (Brochu, 1971: Maricourt Fiord, Quebec). The ice at the fjord mouth can disappear under the influence of tides if large (Brochu, 1971), or as a consequence of a major storm (Lewis and Perkin, 1982: Qaumarujuk Fjord, Greenland). During the latter case, major water movements can occur while the ice-free water adjusts to the sudden wind-energy input. On the other hand, the melting of sea ice may contribute to the freshwater cap and initially stabilize the surface stratification (Gilbert, 1983). Once free of their ice cover, many high-latitude fjords may soon be invaded by drifting ice from the shelf. Although not a continuous cover, pack ice will greatly reduce wind-induced mixing as well as continuing to contribute freshwater during its meltdown.

4.5.2 Sea Ice and Sediment

An important sedimentological consequence of sea ice is its ability to raft sediment. Fjord sediment can accumulate on or within sea ice by one or more of the following processes: (1) wind action, (2) stream discharge, (3) rock fall, (4) seafloor erosion, (5) wave and current washover, and (6) bottom freezing. Examples and details are outlined below.

The deposition of aeolian silts and sands transported from a sandur surface and deposited on ice is an important winter process for some fjords (Gilbert, 1980a,b, 1983: Maktak Fiord, Baffin Island). Nival melt can occur prior to the melt of sea ice and even before shoreline leads have had an opportunity to develop. Stream waters loaded with sediment may flood across the ice at high tides; high river discharges can deposit fluvial sediment a considerable distance over the still-frozen fjord surface (Knight, 1971: Ekalugad Fiord, Baffin Island). Rockfalls, slides, and dirty avalanches, released from the fjord walls by hydrofracturing during intervals of frequent freeze-thaw cycles (spring), supply colluvium to the ice surface along the entire length of a fjord (Gilbert, 1983). This process may also be an important means of adding sediment to drift ice blown in from the shelf during the late summer. Drift ice may become embedded with sediment at its base when dragged over intertidal flats with the rise and fall of the tides. Contemporaneously, waves and currents can wash considerable sediment onto the top of ice floes trapped on the intertidal flats, especially during breakup (Gilbert, 1983: Pangnirtung Fiord, Baffin Island).

Freezing of sediment to the base of ice in meso- and macrotidal environments has been recognized for some time (e.g., Sverdrup, 1931). At low tide and within the littoral zone, a 1 to 2 mm thick layer of fine sediment with its interstitial water may become frozen to the bed of the ice. At high tide when the ice floats free, clean sea water freezes to the base of the ice in a 2 to 4 mm layer. The result is a banding of dirty and clean ice (Gilbert, 1983); the number of bands range from a few to dozens. Fine sediment appears to be removed preferentially by this process (Rosen, 1979: Makkovik Bay, Labrador), whereas gravelly environments are less affected (Taylor and McCann, 1983). That is not to say that sea ice does not have the buoyancy capacity to lift large particles: stones up to 6 m in diameter may be floated by a 2 m thick ice mass.

Large boulders are more likely to be pushed instead of rafted (McCann et al., 1981), and in a seaward direction during coastal emergence. In some fjords, broad intertidal flats fringe the walls, some attaining widths of more than 600 m (Fig. 4.23). In Moraine Fjord, South Georgia, the flats originate from a lateral moraine and have recently developed into a boulder pavement: flat, tightly packed mosaics of ice-smoothed and striated boulders that have been forced into the substrate by the pressure of grounding ice blocks (Hansom, 1983). The flats of Pangnirtung Fiord are thought to have formed during the period of stable sea level during the past 5000 years with ice erosion near shore and deposition in the offshore (Aitkin and Gilbert, 1981). The Pangnirtung mud flats are characterized by random ice-rafted or pushed boulders and by a prominent boulder barricade: only 5% of the boulders move on an annual basis (Aitken and Gilbert, 1981). The boulder barricade occurs at the low-low water line between the planation flats and the steeply sloping subtidal zone. The occurrence of boulder barricades has been also noted for Frobisher Bay fjords (McCann et al., 1981), fjords of the Barents Sea (Tanner, 1939), and Labrador fjords (Rosen, 1979).

Considering the progression of ice melt during breakup (Section 4.5.1), it is expected that much of the ice-rafted sediment is deposited reasonably close to the point where it came to rest on the ice surface: sea ice melting within a fjord shows little mobility (Gilbert, 1983). Drift ice brought into the fjord from the shelf may be responsible for a more random redistribution of ice-rafted debris onto the seafloor. Quantitative assessments of the volumes of material ice rafted onto the basin floors remain unknown. Where the tidal range is high and broad, intertidal flats supply abundant sediment, and ice rafting may supply a major component of the material to the seafloor. Low tidal environments with coarse-grained talus shorelines are unlikely sources for the rafting process (Gilbert, 1983). The presence of ice-rafted boulders within hemipelagic sequences in arctic cores is ubiquitous, although unpredictable (Fig. 4.24). If a quantity of ice-rafted sediment is released at one time, the resulting deposit may be graded because of the separation of grain sizes during settling. Ice-rafted gravel fragments may distort the horizontal layering of previously deposited sediment at their point of impact. These are recognized as dropstones (Fig. 4.24).

An important but indirect sedimentological consequence of sea ice is to reduce offshore-generated wave energy before it can impact on the littoral sediment zone. Stewart and England (1983) examined the abundance of driftwood in strandlines between the present sea level and the marine limit in Clements Markham Inlet, Ellesmere Island (83°N). They interpreted driftwood variability as an indication of climatically induced changes in summer sea ice presence: 6000 to 4200 years BP was a time of summer ice-free conditions; 4200 to 500 years BP was a time of more or less continuous ice cover; and between 500 years BP and the present, the sum-

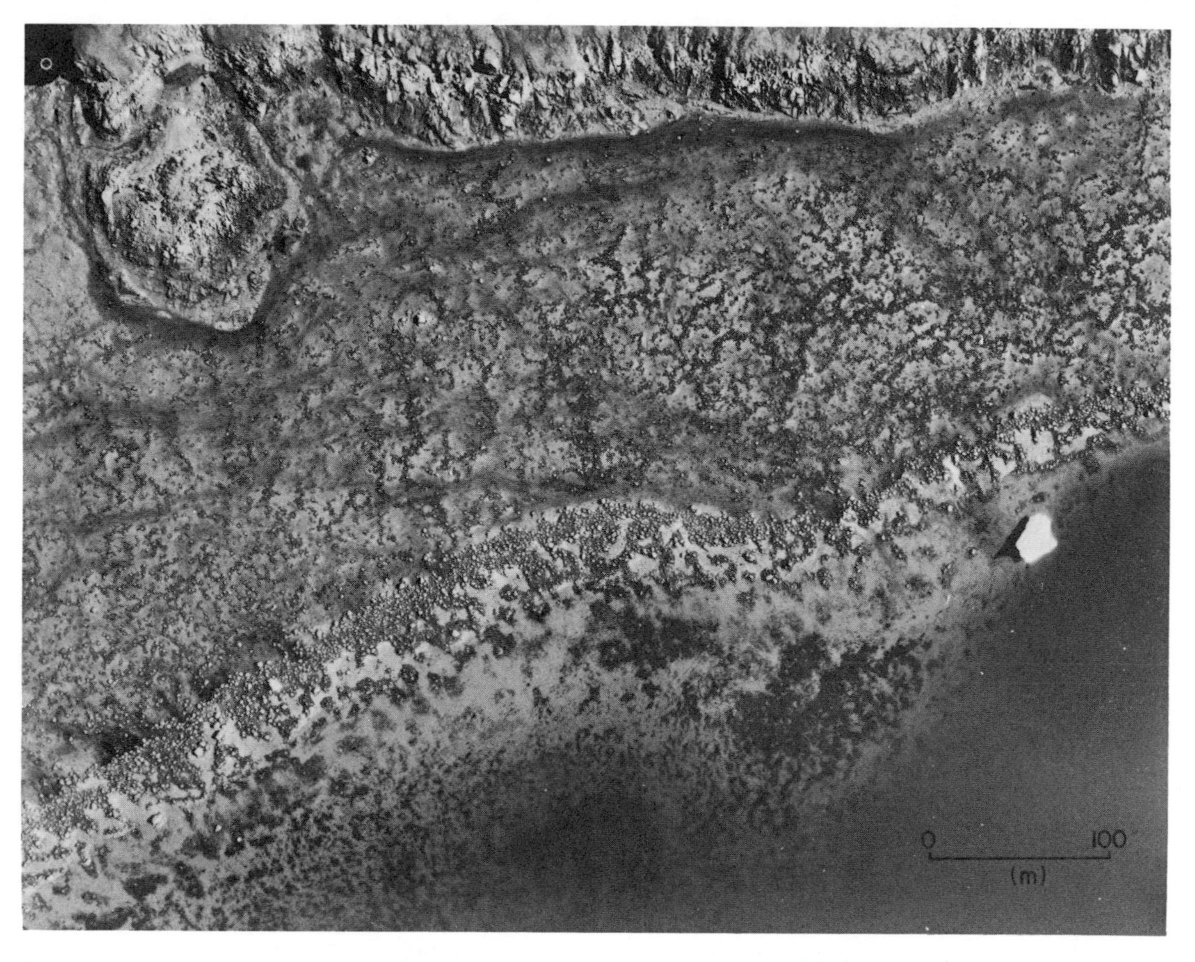

FIGURE 4.23. Air photograph of the intertidal flats in Pangnirtung Fiord, Baffin Island, showing the prominent boulder barricade near the low-tide line.

mer months saw a return to ice-free conditions. Similar conclusions were reached based on the presence or absence of wave-generated storm ridges on exposed isostatically raised terraces. Verification of this hypothesis remains to be addressed; one indicator may be found in the abundance of ice-rafted material occurring down-section in well-dated basin cores.

4.5.3 Fjord Glaciers and their Movement

Most glaciers that presently dip into fjord waters have advanced or retreated during the past 100 years, some significantly (examples are given in Weidick, 1968: W Greenland; Vorndran and Sommerhoft, 1974: SW Greenland; Hattersley-Smith, 1964: Ellesmere Island; Gilbert, 1980a,b: Baffin Island; Lavrushin, 1968: Spitsbergen; Cooper, 1937, and Field, 1947: Alaska; Hansom, 1983: South Georgia). Glaciers are known to respond to changes in the firn limit, which represents the mass-balance equilibrium between glacier accumulation and ablation. The firn limit may be defined as the highest elevation on a glacier to which the snow cover recedes during the ablation season. When a subaerial fjord glacier responds to a lowering in the firn limit, the glacier terminus advances because the confines of the fjord walls do not allow it to widen and thus increase its area of wastage. However, the terminus of a tidewater glacier is always at sea level, and if the firn limit falls, the vertical distance between the firn limit and the terminus will decrease (Mercer, 1961). Thus, the ablation area of the tidewater glacier will be less effective per unit than before, and will continue to advance. Glaciers that advance into fjords in re-

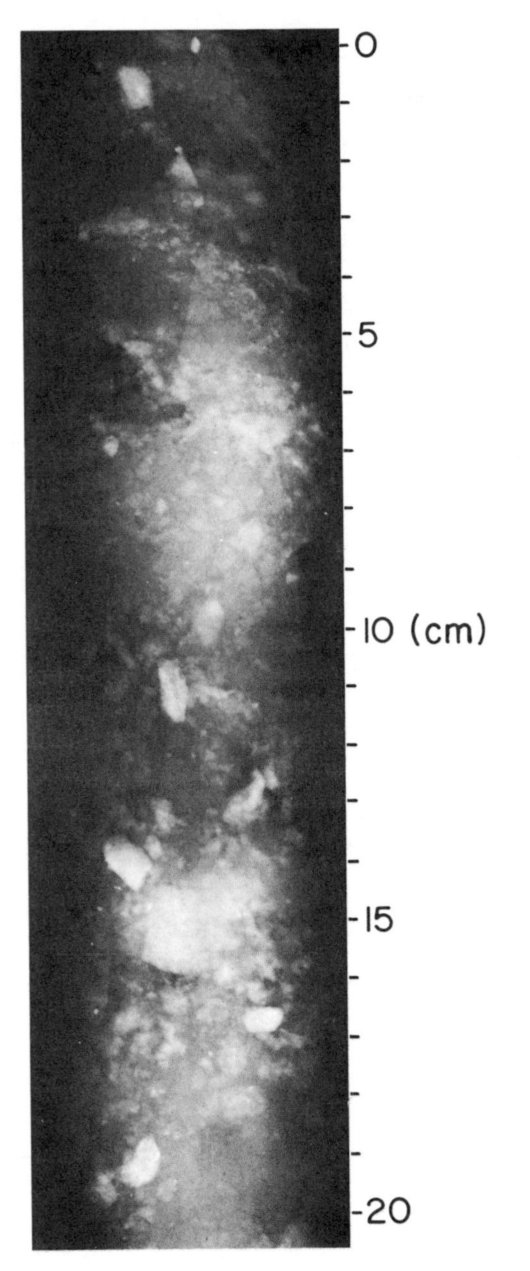

FIGURE 4.24. X-radiograph of core from Tingin Fiord showing ice-rafted gravel component within an otherwise fine-grained matrix.

sponse to a drop in the firn line normally only reach standstill positions at the fjord mouth or at pronounced points of increase in the fjord width (Fig. 4.25A; Lliboutry, 1965). If the walls narrow, or if two tributary glaciers merge to occupy a single channel, the advance will be especially rapid (Fig. 4.25A). Glaciers normally advance many tens to hundreds of meters per annum, although the calving of icebergs can significantly reduce the net advance. Goldthwait et al. (1966) provide good documentation for the Neoglacial ice advance in Glacier Bay, Alaska, from 4800 years BP to the Little Ice Age climax at 250 years BP (circa 1700 AD; Fig. 4.26). During that time, the average net ice advance was 30m yr^{-1}. The maximum extent of the glaciers in historical time is commonly seen in the form of a trim-line zone, that is, a zone marked by fresher-looking rock surfaces, and, in many places, devoid of vegetation. The 1700–1750 ice advance appears synchronous around the globe from Alaska to Greenland, although polar glaciers (i.e., N Greenland, Bylot Island) have a slower reaction time to global climate fluctuations compared with temperate glaciers (Weidick 1968, such polar glaciers have reached their maxima in only the last few decades).

Some glaciers exhibit periodic flow behavior alternating between long periods of quiescent flow and stagnation and sudden surge flow that may last for a few years. When a surge occurs, the downstream part of the ice-field reservoir begins to swell and a large bulge of ice propagates toward the terminus throughout the duration of the surge. As a result, the glacier surface is left with a chaotic crevasse pattern that is due to the increased flow velocities (Meier and Post, 1969). A surging glacier will flow both by internal deformation and by basal slip (Thompson, 1980). Surges have been observed in Alaskan fjords (Field, 1969) and in the Canadian arctic (Hattersley-Smith, 1964), but are most common to the fjord glaciers of Spitsbergen. Surge advance rates vary greatly. In Otto Fiord, Ellesmere Island, the glacier has advanced 3 km in less than 9 years. In Kongsfjorden, Spitsbergen, a 2 km surge occurred sometime within a 12-year interval (Elverhöi et al., 1980). Liestøl (1969) discusses the Negri Glacier surge in Spitsbergen where the advance reached 12 km yr^{-1}.

In one sense, glacier retreat is the opposite of glacier advance; that is, an apparent warming will raise the firn limit, and to retain equilibrium, the ice front will retreat by calving and melting. The retreat may not necessarily alter the size of the ablation area, and once the terminus begins to retreat it will continue in that direction contrary to minor changes in the firn limit (Mercer, 1961). If the fjord widens toward the head, the retreat may be rapid; between narrows the gla-

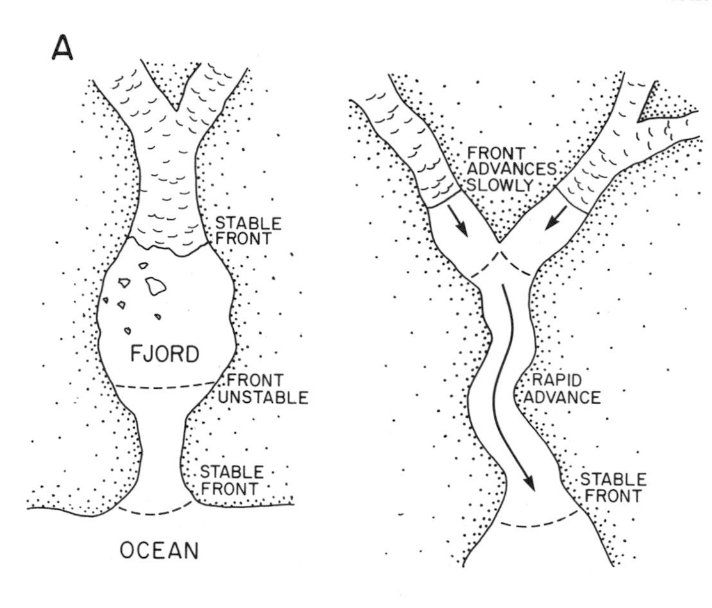

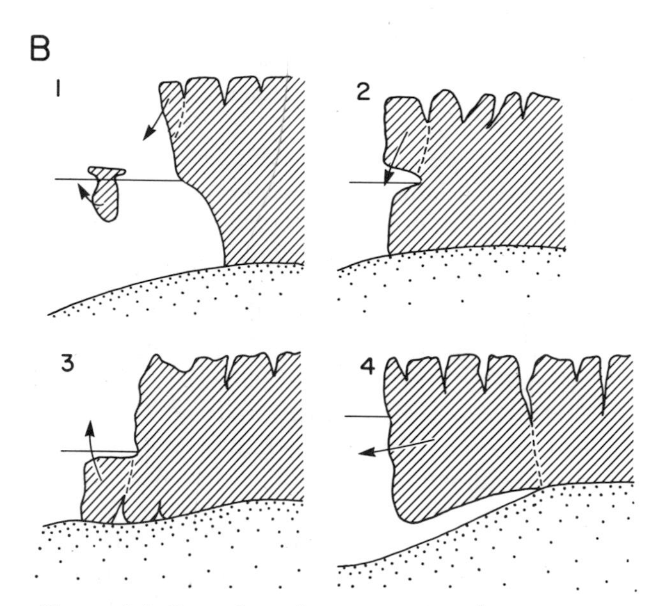

FIGURE 4.25. (A) Stability of fronts during a glacier advance within a fjord (after Lliboutry, 1965). (B) Different modes of iceberg calving from a glacier terminus: (1) subaerial launch of serac; (2) separation of berg along joint intersecting an ice tunnel; (3) subaqueous launch; and (4) separation along deeply incised crevasse (after Lliboutry, 1965).

TABLE 4.4. Controls on ice retreat rates.[1]

A. Climatic factors
- A1: Worldwide climatic change
- A2: Local climatic variations (altitude, rate of precipitation)
- A3: Sea level adjustment (eustatic component)
- A4: Seawater temperature

B. Channel factors
- B1: Bedrock factors including the size, shape, and depth of the channel
- B2: Tectonism (faults and earthquakes)
- B3: Sediment deposition
- B4: Sea level adjustments (isostatic component)

C. Glacial factors
- C1: Ice thickness, surface slope, velocity
- C2: Ice stresses (i.e., crevasse systems)
- C3: Stream discharge

[1]After Powell, 1980.

cier terminus often maintains a tenuous stability. Controls on the rate of ice retreat are listed in Table 4.4. Powell (1980) argues that channel geometry and position of crevasse systems are important controls on retreat rates. Very rapid retreat rates have been recorded from Glacier Bay, Alaska (Fig. 4.26D), and vary from as high as 4.83 km yr^{-1} in the lower basin to between 1.6 and 2.4 km yr^{-1} in Muir Inlet (Fig. 4.27; Goldthwait et al., 1966). Elsewhere, retreat rates are lower: Aarseth (1980) has calculated a 0.4 km yr^{-1} Pleistocene ice retreat from Sognefjord, Norway; in Spitsbergen modern rates of ice retreat have varied from 0.01 to 0.17 km yr^{-1} (Lavrushin, 1968). As detailed in Section 3.3.2, emergence of the land is associated with ice re-

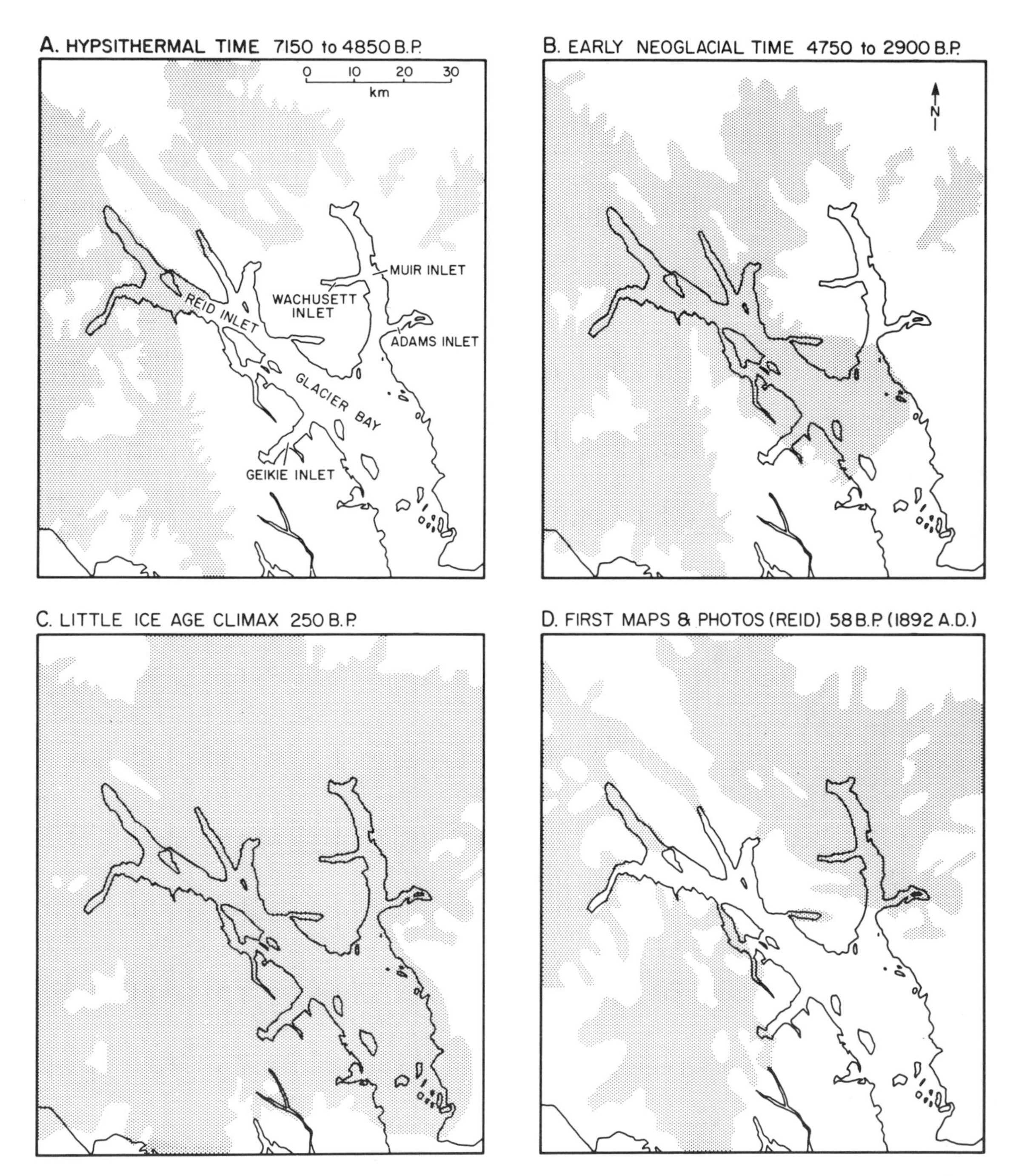

FIGURE 4.26. Maps showing the probable extent of glacier ice in the Glacier Bay area, Alaska, and adjacent Canada, between 5200 BC and 1892 AD (after Goldthwait et al., 1966).

treat. In Glacier Bay, the maximum emergence rate is > 4 cm yr^{-1} with average rates of 2 cm yr^{-1} (Haselton, 1965; Goldthwait et al., 1966; Matthews, 1981).

An important aspect of ice retreat in water is the process of calving icebergs off the glacier terminus. Lliboutry (1965) notes four different modes of iceberg production from a tidewater glacier (Fig. 4.25B): (1) subaerial jointing of ice blocks (seracs); (2) tidewater jointing of ice blocks from ice caves either tidally induced or from ice-tunnel discharges; (3) subaqueous jointing as affected by buoyant forces; (4) detachment of large icebergs along a major crevasse system, an action possibly induced from bottom slope changes. In regard to the latter,

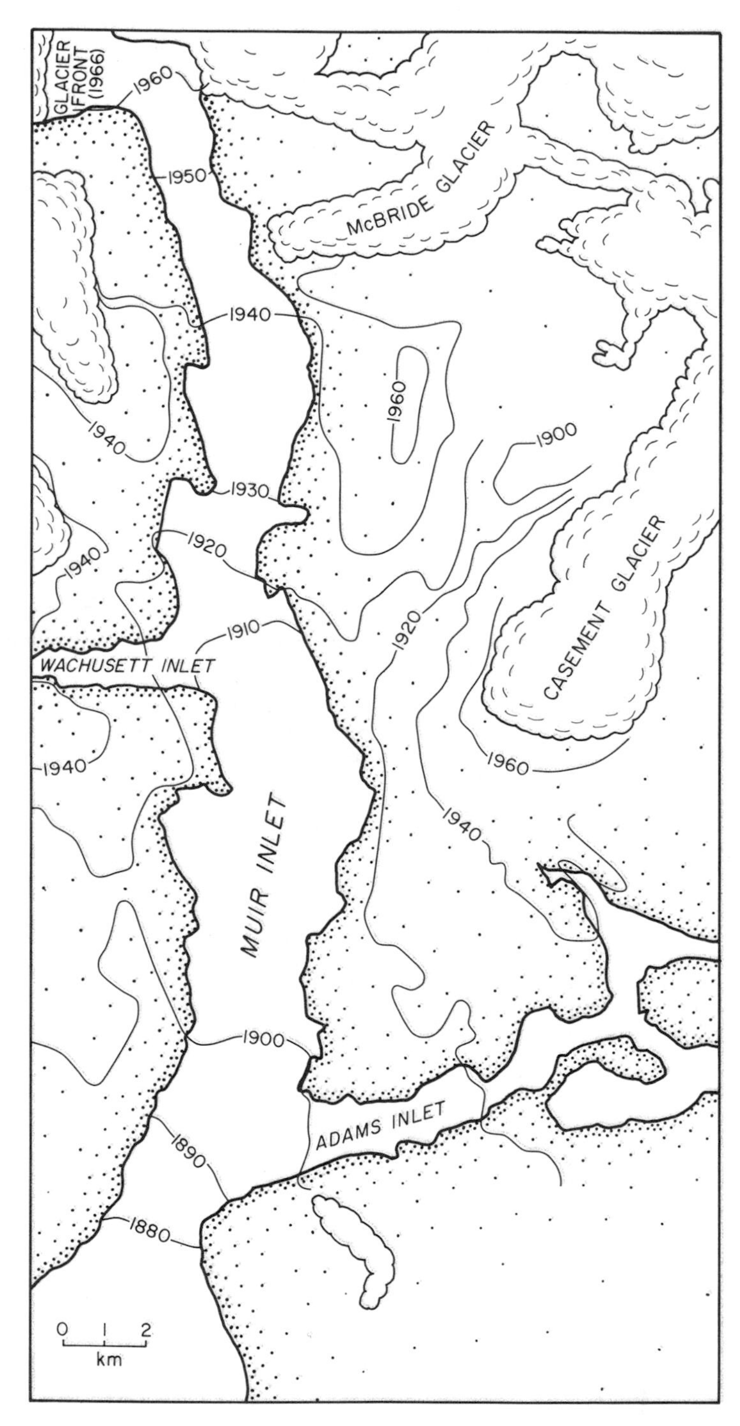

FIGURE 4.27. Ice retreat positions within the Muir Inlet, Alaska, fjord system between 1880 and 1966 (after Goldthwait et al., 1966).

we know that not all fjord glaciers are in contact with the seafloor (i.e., tidewater), some are in the form of a floating ice tongue (e.g., d'Iberville Fiord, Ellesmere Island: Hattersley-Smith et al., 1969; Skelton Inlet, Antarctica: Crary, 1966). The causes for floating ice tongues are not well understood, although the depth of water is an important factor. Powell (1980) has calculated that the paleoice front in Glacier Bay would tend to float at depths greater than 590 m. Elverhöi et al. (1983) estimated negligible shear stresses at the base of a surging glacier and "suggested that" the terminus may float in its outer part.

Increased calving is known to be associated

with transverse crevasse fields, which commonly occur between channel constrictions. Overhang of the upper ice cliff can also influence the calving rate: the amount of overhang is dependent on the ice velocity profile and the subaqueous melting rate (Powell, 1980). Ice calving also tends to be more frequent at subaqueous discharge outlets. There, rapid loss of ice from thermal erosion tends to make the glacier compensate by ice flowage toward the discharge area (Powell, 1980). Greenland is the world's largest producer of icebergs, with individual fjords having typical ice production between 10 km^3 yr^{-1} (Nordvest-fjord: Olsen and Reeh, 1969) and 30 km^3 yr^{-1} (Umanak Fjord: Carbonnel and Bauer, 1968). In a typical calving scenario for these fjords (Olsen and Reeh, 1969), ice advances for 1 month in the early summer with little calving activity. The greater velocities in the central part of the glacier cause the front to assume a curved convex shape. Calving of this central advanced part of the front may produce a few megabergs (0.1 to 0.2 km^3) with the old glacier surface in an upright position. In addition, several other bergs would rotate or break up upon calving during this time. Shortly afterwards, calving would occur from the margins of the glacier, which now are advanced owing to the major calving from the central parts of the glacier.

4.5.4 Glaciers, Icebergs, and Circulation

Glaciers and icebergs, by melting below the waterline, can inject significant amounts of fresh water at depth. Upwelling around icebergs can increase mixing and, where the number of icebergs is large, can destabilize estuarine circulation. Upwelling near the wall of a tidewater glacier may arise from one of four processes (Horne, 1985; Greisman, 1979): (1) subglacial streams of fresh water that rise owing to buoyancy; (2) wind-induced upwelling caused by offshore katabatic winds draining cold air from interior ice caps; (3) presence of sea water supercooled relative to its freezing point thus creating a conditional stability (i.e., freezing at the water-ice interface); and (4) melting below the waterline of glaciers terminating in the sea. Melting is perhaps enhanced by the release of air bubbles trapped in the glacial ice; ice-dominated fjords are known to have a very high dissolved oxygen content (Berthois, 1969). The volume flux of upwelling is dependent on the sea temperature some distance from the ice front: Muir Inlet, Alaska, has a six times greater flux than the colder, high arctic d'Iberville Fiord (Greisman, 1979). Melting near the wall of a tidewater glacier continues year-round although the rate will depend on the seasonal temperature changes in the adjacent water masses. Melting along ice fronts, up to a few hundred meters deep, will contribute to the maintenance of surface stratification although the circulation cell is many times larger than normal estuarine circulation (Fig. 4.28; Matthews and Quinlan, 1975). The size of the circulation cell is largest during the winter and smallest in the summer (Fig. 4.28). Dunbar (1951) discussed upwelling brown zones in the front of glaciers in Greenland fjords (Fig. 4.28B). The zones are saline and high in SPM, nutrients, and planktonic crustacea and are maintained as open water at the glacier face throughout the winter. The upwelling was considered to result primarily from thermal melt with contributions from ice-tunnel discharge.

As discussed in Section 3.3.1 in regard to lakes, subglacial discharge may enter a fjord as an overflow, interflow, or underflow depending primarily on the relative density of the stream water and the basin (ocean) water. There is evidence for interflows from Muir Glacier, Alaska, from Powell (1980) who represents the penetration of a turbulent interflow plume as:

$$z/Y_f = 1 - \cos\left(\frac{x/Y_f}{F\cos\theta}\right) + F\sin\theta\left(\frac{x/Y_f}{F\cos\theta}\right) \quad (4.27)$$

where: $Y_f = (\rho_2 - \rho_0)\left(\frac{d\rho_2}{dz}\right)^{-1}$,

is the modified Froude number;

$F = u\,(\rho_2 - \rho_0)\,\rho_0^{-1}\,Y_f\,g^{-1/2}$,

is the densimetric Froude number;

θ is the jet injection angle from the horizontal, and u is the jet exit velocity. Here, Y_f, gives the characteristic distance above the entering jet at which the seawater density equals that of the stream efflux.

Matthews (1981) notes that fjords that undergo

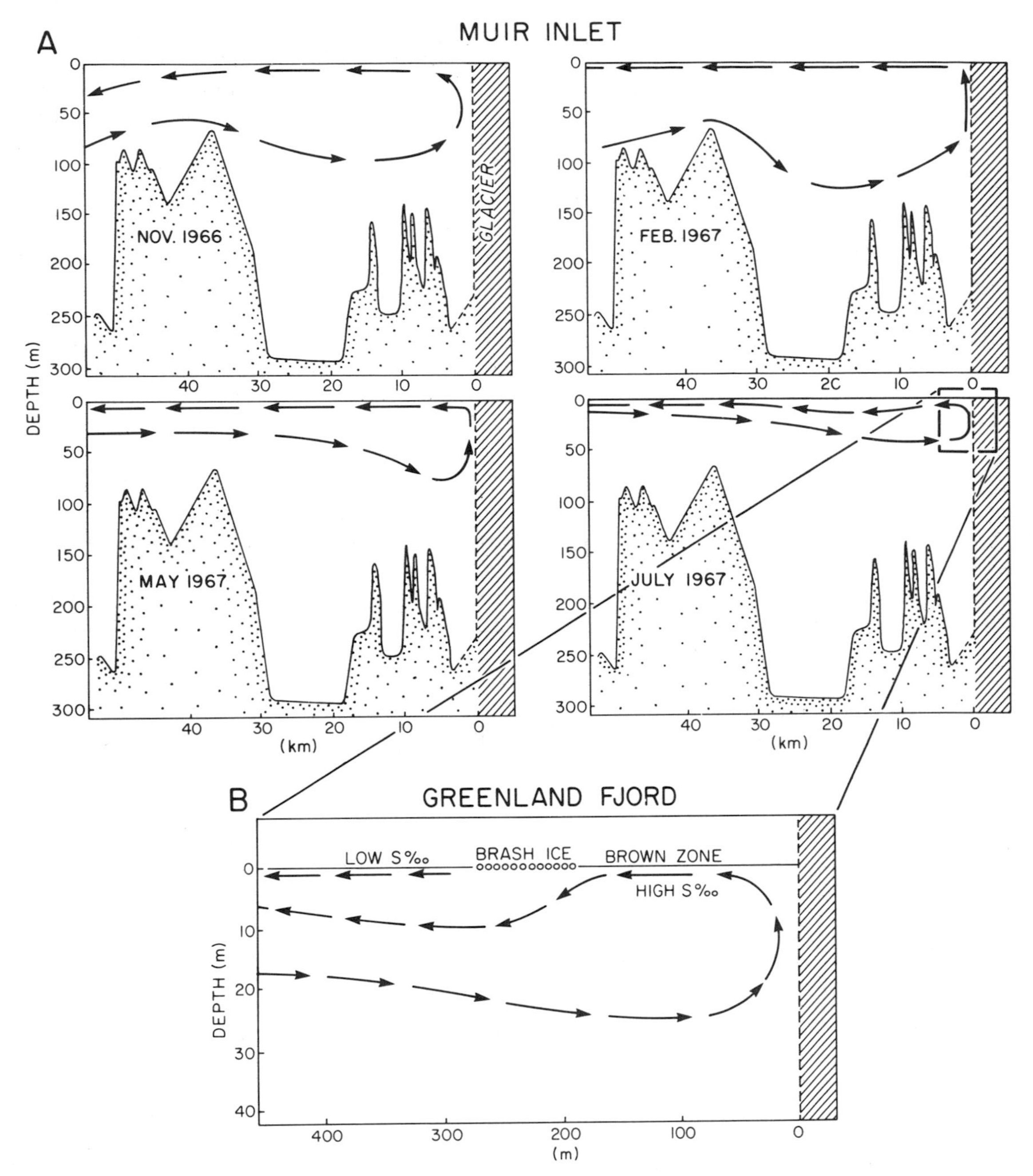

FIGURE 4.28. (A) Seasonal circulation fronting the Muir Inlet glacier, Alaska (cf. Fig. 4.27; after Matthews and Quinlan, 1975). (B) Detail of circulation fronting an ice terminus in a Greenland fjord showing prominent brown (upwelling) zone (after Hartley and Dunbar, 1938; Dunbar, 1951).

a rapid clearing of glacial ice will have continually changing circulation patterns. Mean influx of fresh water to the fjord will be increased greatly, at least initially. Year-to-year variations in upwelling will also reflect the abundance of icebergs. Land emergence may also affect sill depth and thus tidal velocities. For instance, the tidal velocities have increased by 50% in Glacier Bay since its deglaciation (Matthews, 1981).

4.5.5 Iceberg Rafting and Scouring

This and the following section are aimed at summarizing the deposition of glaciomarine sediment—a sediment type that forms the bulk of Quaternary infilling in most fjords. There are dozens of definitions for glaciomarine sediment, but we accept that of Andrews and Matsch (1983), who define it as mixtures of glacial detritus and marine sediment deposited more or less contemporaneously. The glacial component may be released directly from glaciers and ice shelves or delivered to the marine depositional site from those sources by gravity, moving fluids, or iceberg rafting. The marine component, although it may include any of the genetic types of marine bottom sediment, is comprised mainly of terrigenous and biogenous sediments. In this segment of the glacial story we concern ourselves with the contributing and modifying effects of icebergs.

The first sediment-berg interaction is at the time of calving. Icebergs of average size can either touch the seafloor or come close enough to create turbulence that will disturb the seafloor (Powell, 1980). Disturbance may take the form of load structures where bed-form layering is altered and compressed, or seafloor liquefaction leading to gravity flows. In addition, huge impact waves may wash material far up above the high-tide line upon impact with the shore. The phenomena, known as "Tagsaq," may occur several times a day, and every few years particularly high waves appear with catastrophic effects as far as 50 to 100 km from the edge of the glacier (Petersen, 1977: Greenland fjords).

Typically an iceberg maintains one position for several days while continuously releasing sediment from the submerged portion of the berg. The more rapid melting below the waterline eventually changes the center of gravity and the berg overturns and fragments. During these times, sediment that has been concentrated on the berg's surface from meltout will be released (for details see Ovenshine, 1970). The seafloor deposit from berg rafting will be similar to sea-ice rafting with a resultant graded bed but with a more till-like character. Another distinguishing characteristic may be the presence of glacier-formed pellets of sediment formed by solute rejection from pure ice crystals (Ovenshine, 1970).

As glacial sediment is not evenly distributed vertically through an ice sheet or glacier, but tends to be concentrated as basal or supraglacial sediment, it appears probable that most of the glacial debris is released within the fjord (Andrews and Matsch, 1983). Nevertheless, the contribution of berg-rafted sediment is highly variable for glacial fjords: in Kongsfjorden, Spitsbergen, the component is volumetrically minor (Elverhoi et al., 1983), as in Coronation Fiord, Baffin Island (Gilbert, 1980a,b); in Greenland fjords the component is more significant (Berthois, 1969). Andrews and Matsch (1983) give five controls on the rate of berg rafting: (1) the disposition of glacial debris within the icebergs; (2) the rate of iceberg production; (3) the rate of iceberg drift; (4) the temperature difference between the water and the iceberg; and (5) the amount of wave action. Obviously the residence time of a berg within the fjord will also be important. Escape times vary from less than one year to 10 years in the Greenland fjords. Large icebergs may not be able to exit the fjord because of a shallow sill (terminal moraine) at the fjord mouth (e.g., Vorndran and Sommerhoft, 1974; Boyd, 1948; Blake, 1977).

When icebergs impact upon the seafloor, they may create furrows (scour marks) with parallel berms (rims, levees, lateral embankments) of displaced sediment. The dimensions of these "ice scours" depend on the velocity of the berg at impact, the dimensions of the iceberg keel, and seafloor geotechnical properties (Syvitski et al., 1983c): widths typically vary from 10 to 30 m, depths range from 0.5 to 6 m, and slopes of berms from the furrow floor range from 6 to 60°. The furrows may be straight or sinuous, continuous or a series of impact pits (Fig. 4.29). Once grounded and during its period of immobility and ablation, an iceberg's continuing contribution of sediment to the seafloor may result in mounds (15 m × 15 m) of coarser grained sediment (Syvitski et al., 1983c).

Polar waters tend to be cold and icebergs may travel thousands of kilometers before they melt down. During their travels, icebergs scour and raft sediment to the offshore areas of continental shelves. Where deeper troughs cut the shelf area, onshore winds may redirect bergs into the mouths of fjords. This is especially true of Baffin Island where fjords are known to shelter icebergs calved from the Greenland coasts. Where the threshold depths at the fjord mouth are shallow, the bergs are excluded from entry and iceberg scouring is limited to the outer approaches to the sill. Where the sill depths are considerable, allochthonous icebergs may enter the fjord

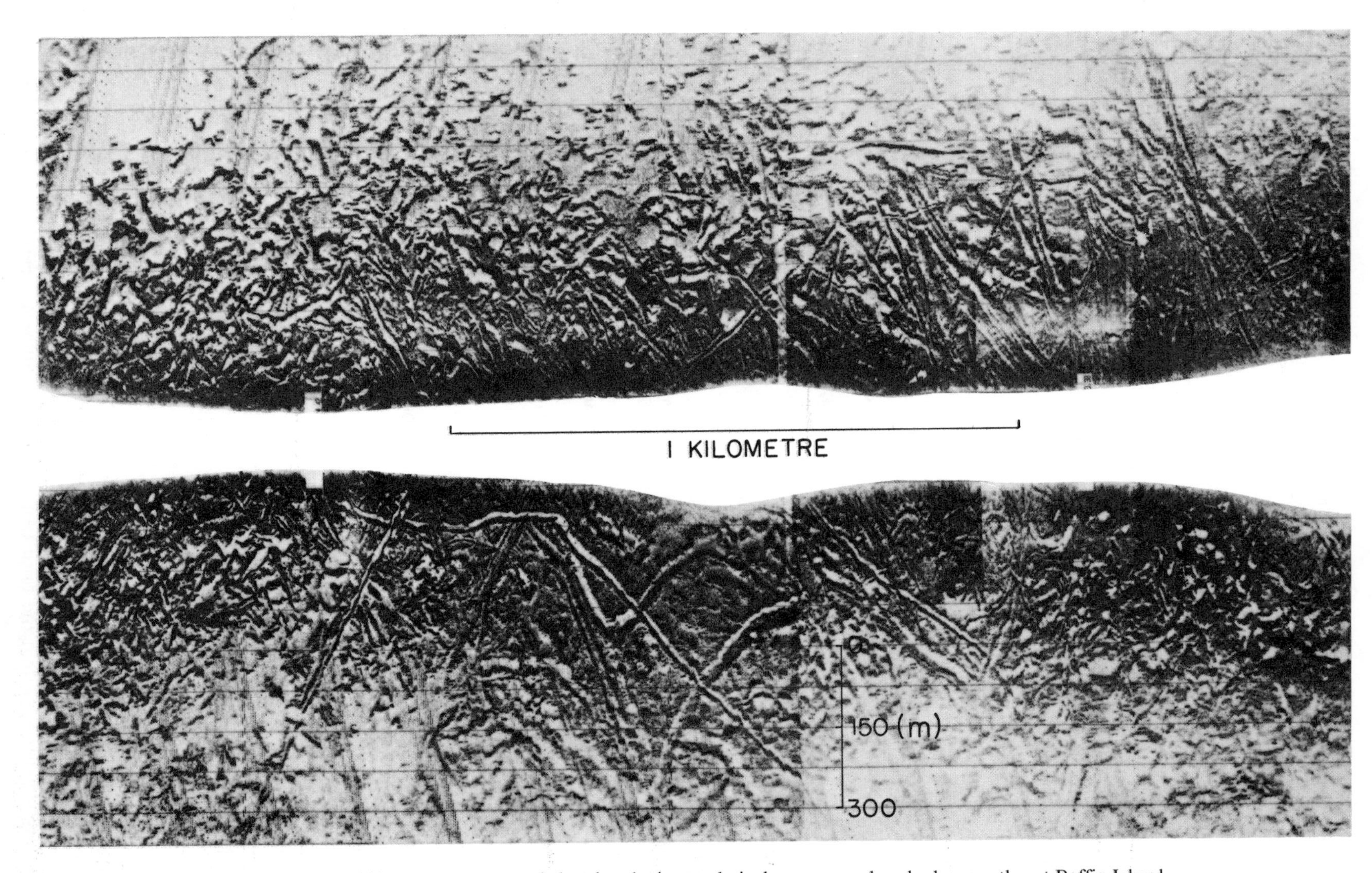

FIGURE 4.29. Sidescan sonar record showing the intensely iceberg-scoured seabed on northeast Baffin Island Shelf (from Syvitski, 1984).

and ground along the fjord margins or at the fjord-head delta (Syvitski et al., 1983c).

4.5.6 Glacier Front Sedimentation

Sources of sediment near the ice terminus are many and varied (Fig. 4.30) and include: (1) supraglacial material (subaerial and subaqueous slumping off of medial and lateral moraine till, Fig. 4.31A, supraglacial streams); (2) englacial materials (crevasse fills, englacial streams, and other englacial sediment); (3) basal material (lodgment till, waterlain till, push and surge deposits); (4) iceberg-rafted sediment (Fig. 4.31B); (5) aeolian sand and loess, blown off ice surface and along kame terraces (Fig. 4.31C); and (6) lateral (kame) deltas. Rates of sedimentation are highest immediately proximal to the ice front. For instance, seismo-stratigraphic units were mapped within Glacier Bay, Alaska, to ascertain accumulation rates since the time of deglaciation, that is, over the last 125 years (Carlson et al., 1979; Powell, 1980). The thickness of the lower unit above the bedrock floor indicates an average subglacial till formation at 0.02 m yr^{-1}. The upper unit of ice-proximal sediments had accumulation rates from 0.2 to 4.4 m yr^{-1}. Variation in the accumulation rate for individual Glacier Bay fjords is less extreme: for example, 1.0 to 1.7 m yr^{-1} for Tarr Inlet; 0.2 to 0.5 m yr^{-1} for Geike Inlet; 0.4 to 0.6 m yr^{-1} for Rendu Inlet; 0.8 to 2.7 m yr^{-1} for Muir Inlet; and 0.8 to 1.3 m yr^{-1} for Waschusett Inlet. These accumulation rates are high when compared with other world fjord types (Table 4.2), especially when compared with nonglacial fjords.

There are a number of models of ice-proximal sedimentation (e.g.,Lavrushin, 1968; Powell, 1981a, 1983; Elverhöi et al., 1980, 1983; Molnia, 1983; Andrews and Matsch, 1983; Gilbert, 1985), but all are based on limited data. For simplicity we will only review facies associations as given by Powell and by Elverhöi et al. Powell (1981a) describes one facies associated with a rapidly retreating tidewater glacier and an ice front actively calving in deep water (Fig. 4.32A, cf. Muir Inlet, Fig. 4.27). Near the ice front, subglacial stream gravel and sand overlap the subglacial till deposits and interfinger with supraglacial dump debris. Occasionally push moraines may front the ice terminus, an indicator of a recent winter advance. Farther from the ice front, iceberg-zone mud interleaves with sandy or muddy turbidites. This association features rapid summer ice retreat through calving, but tagsaqs (calving-induced waves) do not influence the seafloor, which is very deep. Exposed subglacial till may be reworked by tidal currents on stream underflows and sediment gravity flows. Large fans are not produced because of the mobile ice front.

Powell's second facies is associated with a slowly retreating tidewater glacier with an ice front actively calving into shallow water (Fig. 4.32B, e.g., Coronation Fiord, Baffin Island: Gilbert, 1980a,b; Syvitski et al. 1983a). It may feature ice-proximal morainal banks with ice contact lateral- and central-fan deltas. Gravity flow deposits are found seaward of the banks and intertongue with the iceberg-zone mud. Subaqueous stream discharges have sediment concentrations occasionally high enough to form underflows capable of transporting sand onto an otherwise muddy seafloor. Often the position of the subglacial discharge vent changes from one meltwater season to the next, developing a series of overlapping subaqueous fans. In the more distant areas of the ice front, sediment accumulation is high (i.e., 4 m yr^{-1}) thus producing thick mud layers.

Powell's third facies is deposited by a slowly retreating or advancing tidewater glacier calving into shallow water (Fig. 4.32C; e.g., Lituya Glacier, Alaska). The association consists of iceberg-zone mud deposited close to the ice front and interleaved with stream discharges that introduce gravel and sand over large subaqueous fan deltas. The association is common to protected embayments where ice loss by surface melting may be of similar or greater magnitude than calving or subaquaeous melting. During a re-advance, a glacier can override the lateral-fan deltas, causing sand to interleave with pebbly mud. The sand changes laterally into muddy sand and finally into interlaminated sand and mud. Distal muds contain thinly bedded turbidites.

The destructive or erosional power of an advancing glacier depends on its average basal shear stress, τ_b, which can be determined from:

$$\tau_b = h\rho g \sin(\theta) F \qquad (4.28)$$

where h is the ice thickness, ρ is the density of ice, g is the acceleration due to gravity, F is the valley shape factor, and θ is the slope. When

FIGURE 4.30. Major source inputs at the front of a tidewater glacier in a fjord.

FIGURE 4.31. Field photographs of the Coronation glacier, Baffin Island (after Syvitski et al., 1983a): (A) supraglacial medial moraine till; (B) bergy-bit with ice-rafted sediment component; (C) ice-cored lateral morainic material; and (D) wind shadow loess mound fronting glacier.

A
B
C
D

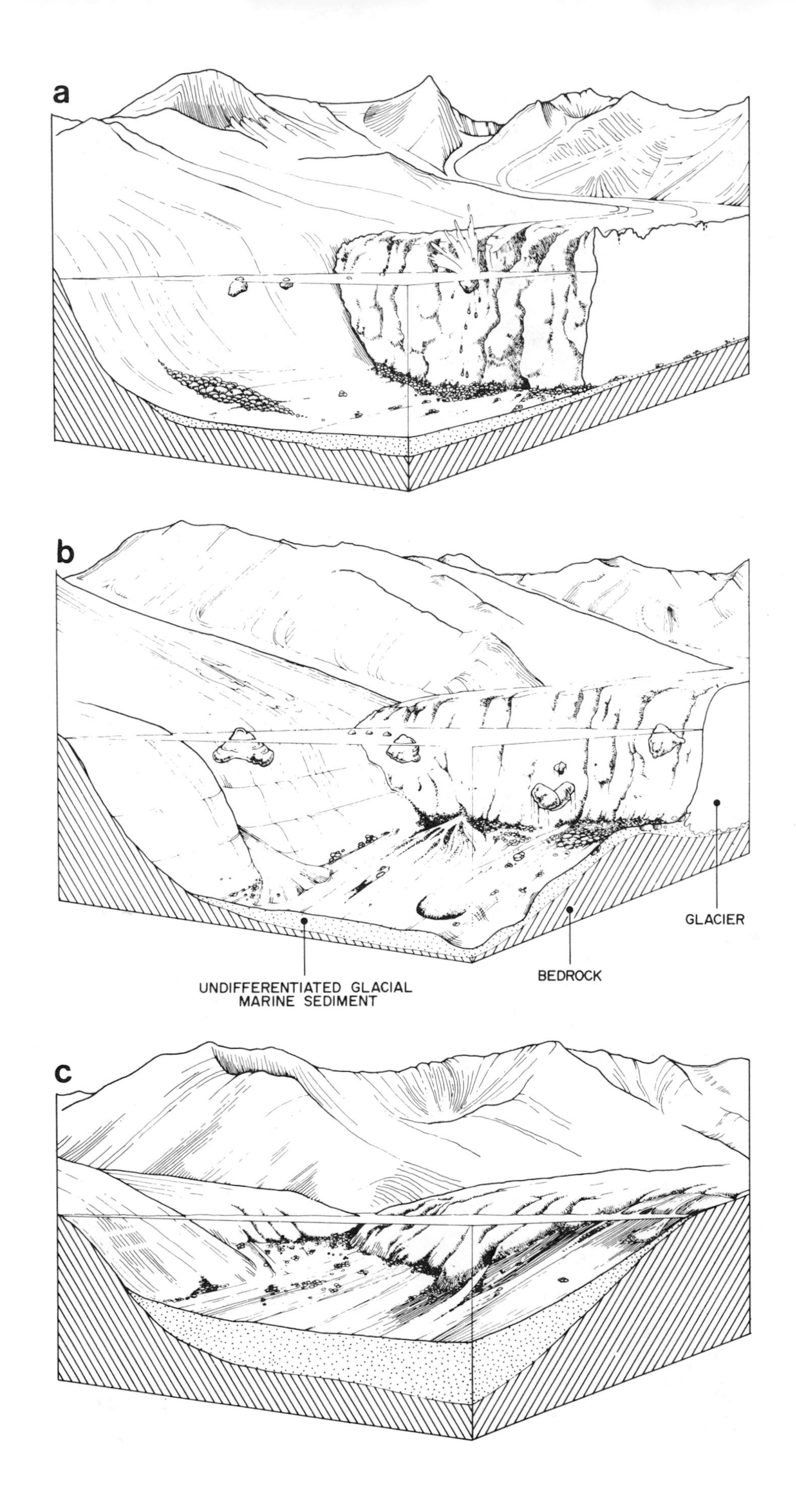
a
b
GLACIER
BEDROCK
UNDIFFERENTIATED GLACIAL
MARINE SEDIMENT
c

the shear stress is low, glaciers may slide across their beds of glaciomarine mud or simply deform the underlying muds (Andrews and Matsch, 1983). In deep fjords, glaciers will have low basal shear stress and in places the glacier might simply float across the interface with little erosion.

Elverhöi et al. (1980, 1983) provide one of the very few studies of glaciomarine sedimentation fronting a surging glacier in Kongsfjorden, Spitsbergen. Sedimentation is considerably lower than the Alaskan examples and ranges from 0.1 m yr^{-1} at the ice front to 0.001 m yr^{-1} in the outer fjord areas. The surging glacier produces glacial stream meltwater year-round. The meltwater flows out of a large tunnel into which sea water appears to penetrate for some unknown distance and thus mix with the sediment-laden fresh water (Fig. 4.33B). Sandy sediments are confined to the glacier front near the meltwater outflow where muddy and sandy layers alternate. Farther seaward, deposits are characterized by homogeneous mud with scattered ice-rafted pebbles. Chute and slump features are also observed near the ice front. The sedimentation of thick meltwater deposits outside the ice front during a surge may be a result of: (1) increased meltwater discharge owing to frictional heating during a surge; (2) radical changes in the subglacial meltwater channel; or (3) erosion of glaciomarine sediments reworked by the surging glacier.

4.5.7 Fjord Polynyas

A polynya is a space of open water in the midst of ice. It results from turbulent mixing common to tidal action over shallow sills (Makinson Inlet, NWT: Sadler, 1980; Pangnirtung Fiord, Baffin Island: Gilbert, 1984). Ground-water discharge through the seafloor is another possibility (Sadler and Serson, 1980; Hay, 1983a,b). At Cambridge Fiord, Baffin Island, a small kettle lake drains year-round through its lake bottom of glacial till. The ground water flows through a highly porous and permeable sandur so that, although in a zone of continuous permafrost, drainage is possible. The fresh water discharges into sea water at a depth of 40 m as a buoyant plume that rises to the sea surface. Mixing and upwelling of warmer water from depth is such that a polynya is formed.

FIGURE 4.32. Three ice-proximal facies associations after Powell (1981a): (A) Rapidly retreating tidewater glacier with an ice front actively calving into deep water; (B) slowly retreating tidewater glacier with an ice front actively calving into shallow water; (C) slowly retreating or advancing tidewater glacier calving into shallow water.

4.6 Mixing Processes and the Seafloor Environment

Deep waters in fjords may have sluggish currents, and in some situations may be advectively isolated, yet mixing processes remain an integral part of the shallower water regions. Oceanographers have long been interested in the contribution of shallow water turbulence to the deeper basin waters: their studies include aspects of the supply of energetics to vertical eddy diffusion. Sedimentologists and benthic biologists are also interested in boundary turbulence attributable to shear at the sediment interface, especially over sills and the approaches to basins including prodelta slopes. In particular, they have identified end-member fjords that are dominated by the action of tides or waves. In fjords where the sill is deep or even absent, tidal currents may winnow or erode bottom sediments. For the shallow end-member fjords, especially when exposed to open ocean swells, wave reworking of the shoreline margins may result in major contributions of sediment to the deeper basin areas.

This section reviews briefly the mixing processes resulting from tides, gravity waves, seiches and double diffusive effects, and their sedimentological consequence. This discussion is meant to complement the kinetic (mixing) energy sources previously discussed: wind and tidal effects on fjord-estuarine circulation (Section 4.1); underflows (Section 4.3); deep water renewals (Section 4.4); thermohaline-driven circulation (Section 4.5.1); ice-front upwelling and iceberg calving processes (Section 4.5.4)—and those mixing processes documented in Chapter 5: tsunamis, giant landslide waves, debris and turbidity currents.

4.6.1 On Tides, Internal Waves, and Fine Water Structure

The tide in fjords is predominantly a standing wave in which the sea level moves up and down

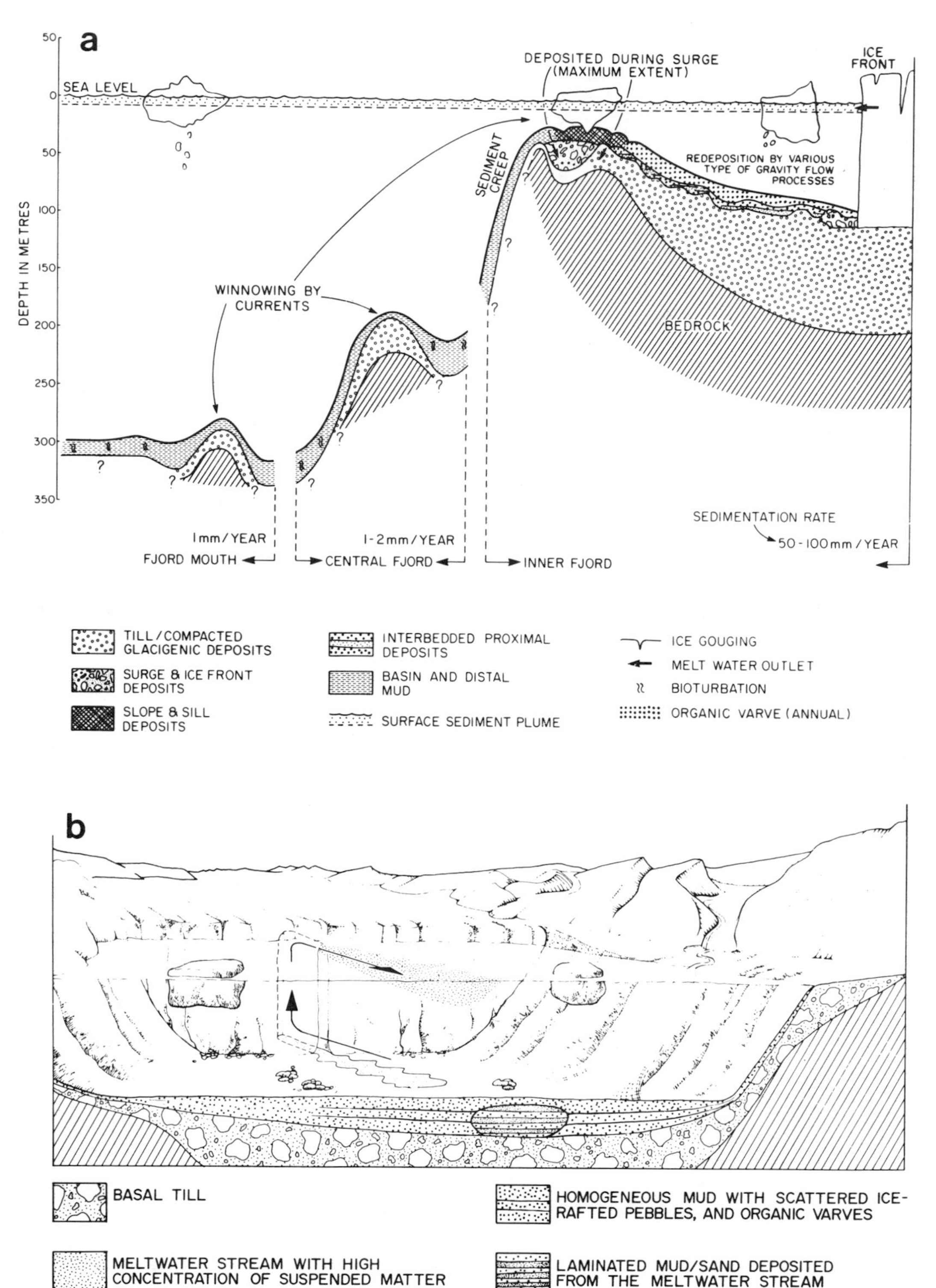

FIGURE 4.33. (A) Schematic cross-section illustrating sediment types and processes in Kongsfjorden, Spitsbergen (after Elverhöi et al., 1983). (B) Diagrammatic section of the inner part of Kongsfjorden. Meltwater is transported through the tunnel and deposits coarser grained, laminated sediment near the glacier front (after Elverhöi et al., 1980).

synchronously. Technically, the tides do propagate down the fjord in the form of a Kelvin wave. Farmer and Freeland (1983) note that the tidal amplitude, a′, will vary along the fjord as:

$$a' = 1.582 \cos (2\pi x/\lambda) \qquad (4.29)$$

where

$$\lambda = M_2 (gH)^{1/2}$$

and x is the distance from the head of the inlet, λ is the wavelength, M_2 is the main semidiurnal tidal period, and H is the water depth (taken as constant). Equation (4.29) is applicable only for low and subcritical Froude numbers, $Fr < 1$, at the sill, where $Fr = u_0 (gH)^{1/2}$, u_0 is the discharge velocity at the sill and $(gH)^{1/2}$ is the velocity of tidal propagation. In certain bays and inlets tidal amplification can be extreme because of a combination of convergence and resonant effects (for details see McCann, 1980a,b). In the Bay of Fundy, the spring tide is amplified from 3 m at its mouth to 15.6 m at the head of Minas Basin, the largest spring tidal range anywhere in the world. Cook Inlet, Alaska, is a large glaciated inlet: tidal amplification is 2.2 times with a diurnal range of 4.6 m at its mouth and 10.0 m for Turnagain Arm at its head (cf. Fig. 3.22).

For supercritical flows at higher Froude numbers ($Fr > 1$), sill choking can produce a decrease in amplitude and equation (4.29) no longer holds (McClimans, 1978b). For instance, the channel that connects Nitinat Lake, Vancouver Island, to the ocean so constricts the passage of water that the 3.3-m tidal range on the coast is only 0.3 m in the fjord (Thomson, 1981).

Tidal currents are the horizontal motion of water associated with the tidal wave form. In the deep basins, tidal currents simply oscillate back and forth with the tide. They produce little residual flow and their effect decreases below sill depth (Pickard, 1961). Slack water occurs at times of high and low phases of the tide, and maximum velocities occur close to times of midtide (variations depend on inertial effects). In the shallow reaches of approaches or long sills, tidal currents may become turbulent tidal streams, that is, well mixed from the sea surface to the seafloor and with well-defined boundary layer flow. Tidal streams tend to follow local bathymetry. Eddies, whirlpools, and upwelling domes may be generated in narrow channels by these swift currents and as induced by topographic vorticity shedding and opposing currents (Fig. 4.34A; Thomson, 1981, p. 64). Tidal streams are influenced by bathymetry, friction, the Coriolis force, river runoff, winds, inertia, and momentum. Current speeds decrease near the channel sides and seafloor because of drag on the macrorelief (boulders, irregular bed rock, bed forms) and microrelief (benthos down to individual sand grains). Where stratification is well developed, channel currents associated with the flood will tend to be strongest close to the seafloor. In models developed by McClimans (1978b), short tidal inlets with a channel length-to-depth ratio less than 50 may be treated as frictionless constrictions. For long and shallow channels, with a length-depth ratio > 100, friction dominates. Where friction is well developed, tides may be significantly dampened (Fig. 4.34C: Borgenfjorden, Norway: Glenne and Simensen, 1963). Under these circumstances, the tidal stream is driven mainly by the hydraulic head (barotropic influence) caused by the variation in tidal heights between the coast and the fjord basin (Fig. 4.34B), and slack water occurs at midtide rather than the normal periods of high and low tide.

For thresholds that are dominated by a shallow sill, a variety of tide-related oceanographic features may develop (Long, 1980; Huppert, 1980): flow separation, lee waves, hydraulic jumps, jets, bores, and internal waves. They are rather complicated features and the descriptions given below should not be considered rigorous. The discussion is partly based on observations of tidal dynamics over the sill in Knight Inlet, B.C. (Pickard and Rodgers, 1959; Smith and Farmer, 1980; Farmer and Smith, 1978, 1980a, b; Gargett, 1980; Tunnicliffe and Syvitski, 1983). When fluid moves toward a topographic high, the flowlines converge over the threshold and become compressed. If the fluid velocity is high, the associated turbulent boundary layer may break away from the boundary at a point where the substrate diverges from the direction of the mean flow. This phenomenon is known as flow separation (for details see Middleton and Southard, 1977, pp. 3.23–3.28). Flow separation over a fjord sill would occur close to the crest and, in the case of Knight Inlet, just after slack water with the shear layer evolving as a sequence of growing vortices (Fig. 4.35A). Two possible scenarios may occur next. First, for

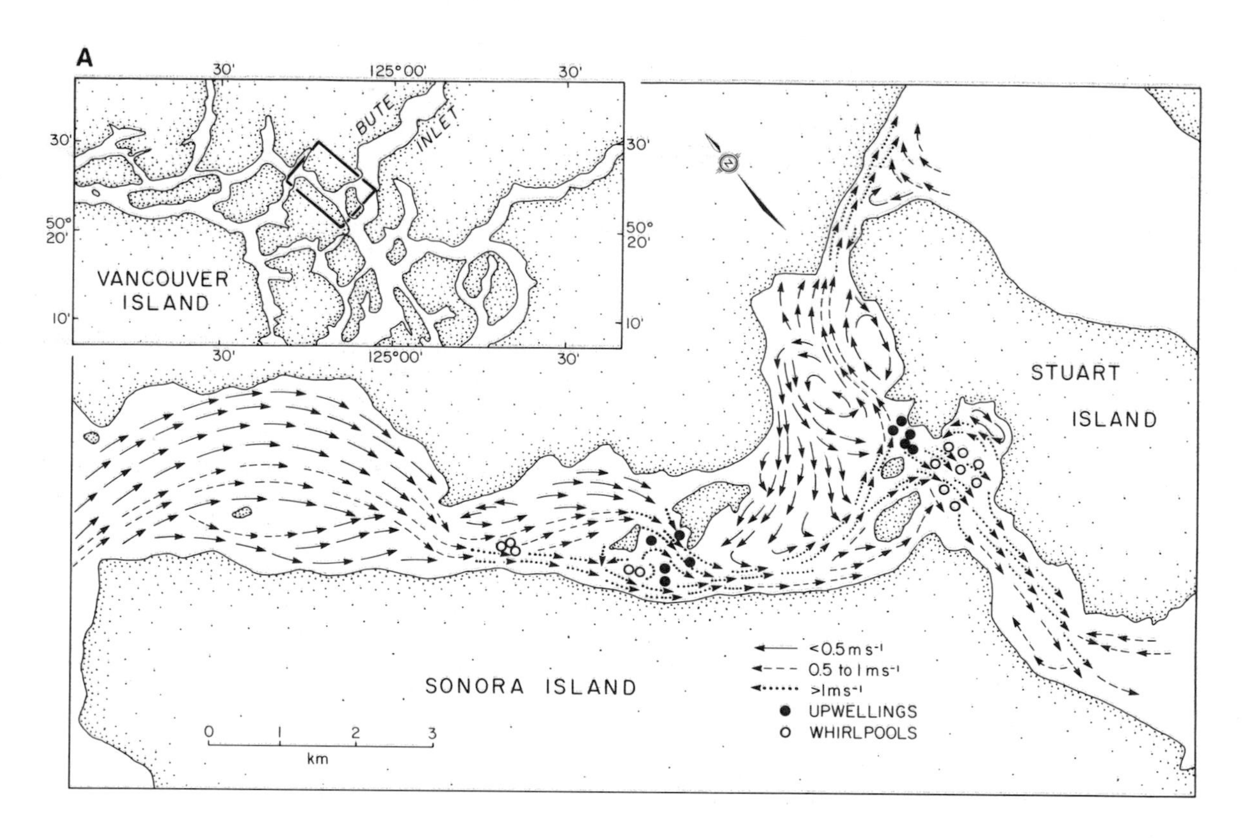

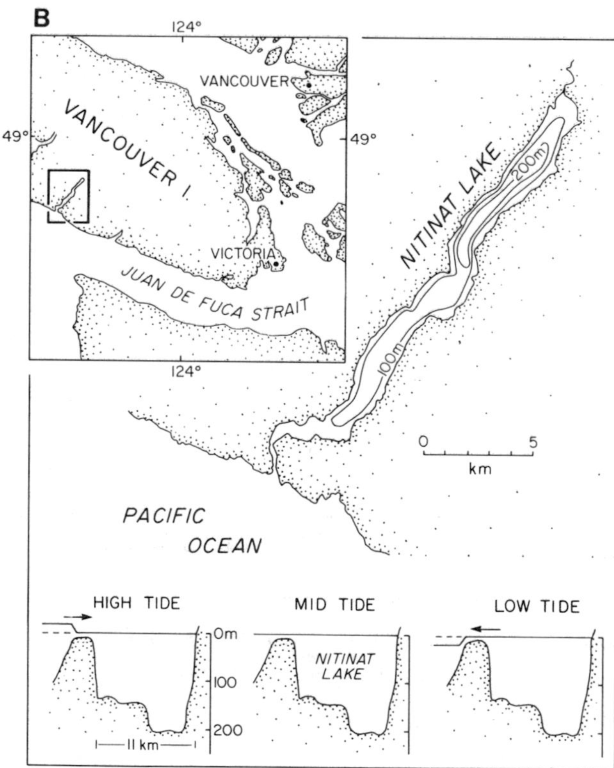

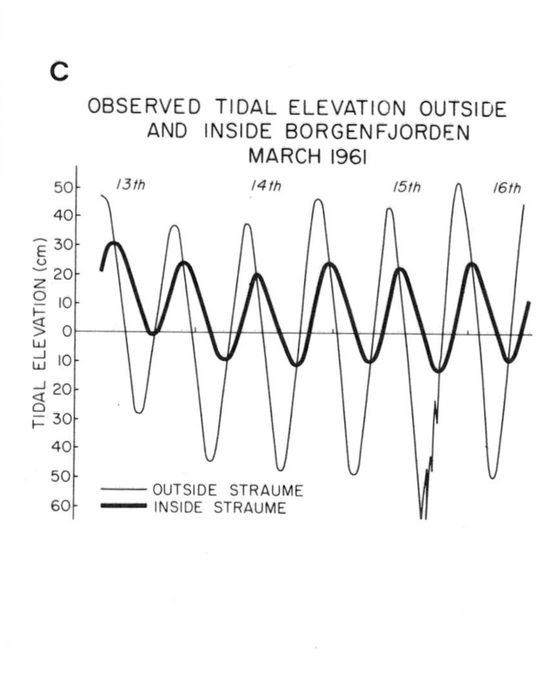

FIGURE 4.34. (A) Tidal stream 3h before the turn to ebb in Cordero Channel region near entrance to Bute Inlet, B.C., showing dangerous whirlpools and violent upwelling zones (after Thomson, 1981). (B) Relationship of oceanic tide to flood and ebb streams into and out of Nitinat Lake, B.C. Bathymetry is given in meters (after Thomson, 1981). (C) Water levels outside and inside the tidal inlet for Borgenfjorden, Norway (cf. Fig. 4.37A; after Glenne and Simensen, 1963).

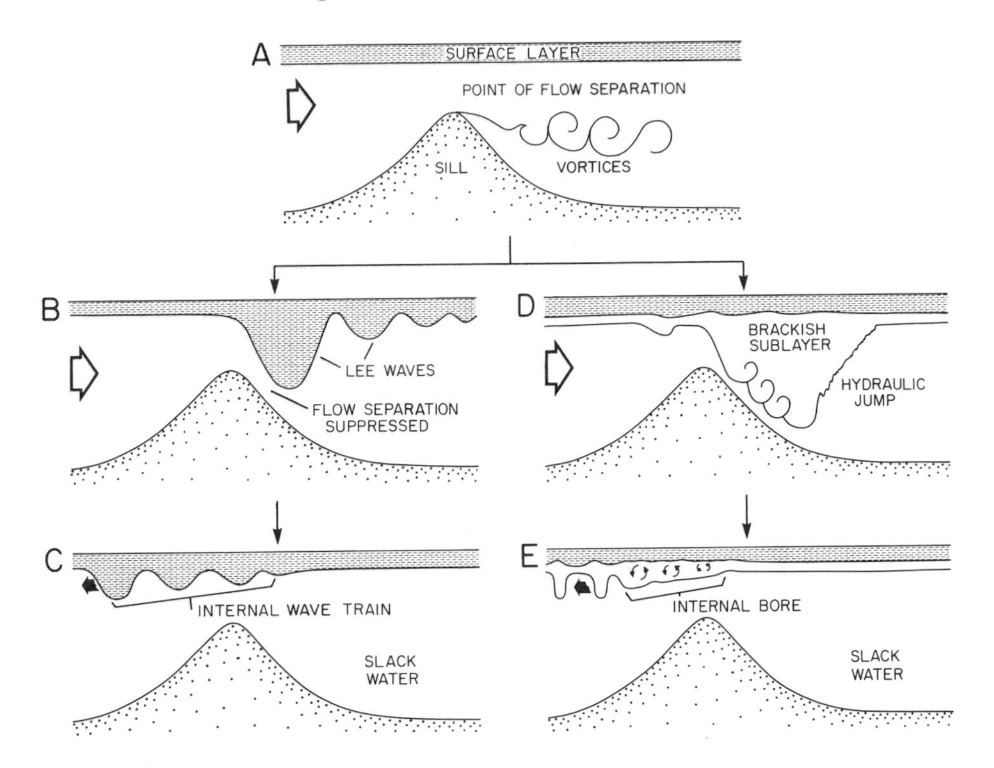

FIGURE 4.35. Dynamics of mixing over Knight Inlet sill, B.C., based on detailed observations (after Farmer and Freeland, 1983). (A) Increased tidal velocity may result in flow separation followed by: (B) and (C) the generation of lee waves and an internal wave train when tide slacks; or (D) and (E) the generation of a hydraulic jump and an internal bore when the tide slacks.

subcritical flows, large lee waves may form just before times of maximum flow, suppressing the flow separation (Fig. 4.36B). Lee waves can be thought of as vertical oscillations at the base of the surface layer (Huppert, 1980; Farmer and Smith, 1980b). Initially the lee waves are almost stationary with their upstream wave velocity balanced by the downstream tidal velocity (stream direction would depend on the stage of the tide). As the next slack water approaches, the lee waves begin to escape upstream as a train of solitary internal waves (Fig. 4.35C). If the lee waves are large, they may break upon impact with the sill; breaking of lee waves on Knight sill can cause vertical velocities up to 38 cm s^{-1}. The second scenario would have the lee wave development go from subcritical to supercritical with the subsequent development of a hydraulic jump (Fig. 4.35D; Smith and Farmer, 1980). A hydraulic jump is the extremely rapid change in the depth or thickness of a water layer (in our case the brackish layer) owing to an extreme change in boundary conditions: channel width, depth, buoyancy conditions, and water flux (Long, 1980). Eventually the hydraulic jump collapses as the tide begins to slack and an internal bore travels upstream (Fig. 4.35E). Under both scenarios, the process can be repeated, but in the opposite direction as the tide changes.

With normal supercritical flows, a turbulent jet will develop over the sill (Stucchi, 1980: Rupert-Holberg Inlet, Vancouver Island). As the jet develops, the salinity of the sill water becomes intermediate between the denser ocean water and the lighter fjord water. As a result, the jet becomes negatively buoyant during flood tide and positively buoyant during ebb tide (Fig. 4.19E). With the increase in the tidal range during spring tides, the Froude number increases and the flood jet is able to penetrate increasingly deeper into the fjord basin.

The hydraulic conditions over sills are therefore crucial to the energy exchange processes within fjords. The energy source is the barotropic tide and the sink is the divergence of mean potential energy flux where internal waves, bores, and jets form an intermediate stage for distributing energy in the fjord (Gade, 1976; Stigebrandt, 1976, 1980; Freeland and Farmer, 1980). Internal waves are gravity waves like surface waves but because they are found within the water column along pycnoclinal surfaces, the magnitude of the buoyancy-restoring forces is about 1/1000 of those of surface waves. Internal waves thus tend to be longer (e.g., $\lambda = 50$ to 200 m) and to move much more slowly (e.g., u = 50 cm s^{-1}) than their surface counterparts. Internal waves may be temporarily stationary (e.g., a standing wave), or may propagate as wave trains with the leading edge of the packet occasionally in the form of a steep-faced internal bore. Internal waves lose energy at a proportionally more rapid rate than surface waves of comparable period and are thus incapable of traveling long distances. They dissipate: (1) through collapse (Kelvin-Helmholtz shear); (2) through absorption by local currents; (3) when they break on sloping bottoms; and (4) when their amplitudes become too large to support their weight (Thomson, 1981). Breaking of internal waves is an important process contributing to vertical mixing within many fjords. Internal waves show some ability to refract around sinuous fjord walls (Farmer and Freeland, 1983), although much energy is released to walls (Stigebrandt, 1976, 1980). Vertical mixing may be intense in the boundary layer of gradually sloping seafloors, such as the fjord-head prodelta, where internal waves are likely to break (Stigebrandt, 1979).

Turbulent and diffusive processes within a stratified fluid, such as within a fjord, may lead to the development of layered instabilities within the water column, known as "fine structure." Fine structure may be present within regions of pycnoclines and is characterized by a series of discrete layers of well-mixed fluid separated by sharp density gradients. In the lower St. Lawrence Estuary, Quebec, the fine structure had 3 m thick pycnoclines in the order of $\Delta\sigma_T \approx 0.18$ m^-1, overlying 15 m thick mixed layers, $\Delta\sigma_T =$ 0.001 m^{-1} (Syvitski et al., 1983b). There are a number of possible origins for fine structure: (1) the interleaving of water masses; (2) partial collapse of internal waves (Gade, 1976); and (3) double-diffusive mixing (Farmer and Freeland, 1983). The latter case may develop when the vertical gradients of both temperature and salinity have the same sign, allowing one of the properties to be unstably distributed. Differences in the molecular diffusion rates of salt and heat allow the release of potential energy from the component that is unstably distributed, even though the mean distribution is hydrostatically stable (Farmer and Freeland, 1983). In fjords, double-diffusive effects can be expected to occur when temperature increases with depth, because salinity always increases with depth. The convection is driven by the larger vertical flux of heat versus salt through sharp interfaces allowing zones of thermal shear to exist (Syvitski et al., 1983b). The thickness of the convecting layers is determined by the viscosity, thermal diffusivities, buoyancy flux, initial density gradient, and a critical Rayleigh number for the layer at which instabilities set in (Farmer and Freeland, 1983). A sedimentological consequence of increased shear associated with fine structure is the reduction in size of the larger floccules and agglomerates of suspended sediment (Syvitski et al., 1983b).

4.6.2 On Wind and Waves

As discussed previously (Section 4.1.4), there are two time scales for wind development in fjords: (1) the diurnal land-sea breezes, and (2) those associated with large-scale weather systems. Wind speed will vary along the channel, increasing with decreases in channel width and slightly decreasing around curved portions of the fjord (Freeland, 1980). Since most fjords are the coalescence of tributary channels and valleys, winds that are funneled from the interior to the sea generally increase along the length of the fjord. In contrast, sea breezes may steadily decrease toward the interior as the inflowing air splits into tributary valleys. The daily land-sea breezes are a product of solar heating in the daytime producing rising air over land, a compensating inflow from the sea, and return flow aloft. The strength of the land breeze will depend on: (1) the distribution and elevation of hinterland topography (hypsimetric integral); (2) the distribution and size of lakes; and (3) the size and position of ice and snow fields. Where ice and snow fields are abundant, the sea breeze serves to hold back a bank of cold air that once released will rush seaward in the form of a short-lived but dramatic down-fjord katabatic wind. Along the NE Baffin Coast, the fjords have the largest portion of their ice fields along their sidewall mountains (see Syvitski and Schafer, 1985). Mariners there have experienced katabatic winds up to 150 km h^{-1} that have blown across the width of the fjord. Polar winds are commonly forced down through fjords during the winter. In British Columbia, these winds may reach 130 km h^{-1}; in the arctic fjords of Greenland and Baffin, winds can exceed 250 km h^{-1} (70 m s^{-1}). Such bursts of high wind can promote a seemingly disproportionate amount of mixing (Farmer and Freeland, 1983). Wind energy is rapidly dissipated with depth, hence its effect depends very much on the degree of stratification.

Wind turbulence, especially downdrafts, will promote surface gravity waves in the form of deep water waves: in fjords the water depth nearly always exceeds one-fourth of the wavelength of surface waves, except over the intertidal regions. The size of deep water waves for a given storm depends on the fetch, wind velocity, and a minimum duration time at that wind velocity, and there are a number of nomograms that may be used to predict wave dimensions (e.g., Bretschneider, 1952). For instance, given a fetch of 60 km, and wind velocities of 40 m s^{-1} whose duration may exceed 3 h, the significant wave heights may reach 6 m: such conditions exist in a number of fjords found along the SW coast of Greenland. Most other fjords seldom attain significant wave heights greater than 3 m with an associated wave base of 20 m. The wave base is the depth at which the seafloor begins to affect the shape and phase velocity of the surface wave, and where orbital velocities may act on seafloor sediment with enough shear to cause their resuspension.

A number of fjords, open to the full ocean fetch, may experience large shoreward-traveling swells. Swells are waves that have escaped the generating storm area where sea waves are found. Swells are relatively long-period waves that have taken a form where little energy is expended during their travel. Where a fjord mouth fronts onto a deep channel that dissects the continental shelf, swells may travel directly up to the fjord with little energy loss. Such conditions are common to the Greenland and Baffin shelves where wide troughs with depths between 400 and 1000 m front a number of fjords. The swells may first impact on the sill, if present, and once they are within the confines of the fjord the margins will dissipate the remaining wave energy. Wave diffraction will seldom allow ocean swells to remain coherent enough along sinuous fjords that they might affect the prodelta seafloor at the fjord head.

Wind drift of the surface layer can lead to changes in the depth of the pycnocline, which may lead to the formation of a standing wave (i.e., a seiche). More dramatic is the actual removal of the surface layer from the head of a fjord to its mouth (Trites et al., 1983). When the wind relaxes, an internal gravity surge is released that may travel up-fjord as a train of internal waves. These waves may break with a considerable release of energy on the prodelta slope (Hay, 1983a).

4.6.3 Seafloor Response

Although predictive equations exist for the erosion and transport of noncohesive sediment affected by wave orbital shear and bottom current shear (Madsen and Grant, 1976; Grant and Madsen, 1979), there has been virtually no field examination of their use under fjord conditions and serious doubts remain concerning their applicability to cohesive sediments. There are few current measurements near the sediment-water interface of fjords and we have little understanding of boundary-layer conditions along the steep (20–60°) fjord walls or near sills. Exceptions include: (1) deployment of a multiprobe instrument package 1m above the seafloor in Halifax Harbour, N.S., to ascertain the importance of shallow water wave erosion on fjord basin sedimentation (Delure, 1983); (2) current measurement 1 m above the seafloor in Rupert Inlet, B.C., to document the possible erosion of sediment by deep water renewal processes or turbidity currents (Johnson, 1974); and (3) current measurements and dye experiments to ascertain boundary-layer conditions over the Knight Inlet sill (Tunnicliffe and Syvitski, 1983). Consequently, the following discussion relies heavily on geologic observation and stratigraphic evidence.

One way of detecting resuspension events of bottom sediment is through the temporal observation of near-bottom maxima in the concentration of suspended sediment. The mechanism responsible usually has the occurrence of the maxima with a discernible cause-and-effect frequency. Daily or monthly frequencies tend to relate to tidal forcings (Fig. 4.36A; Pelorus Sound, N.Z.: Carter, 1976). Maxima that occur during and just after storms may relate to wave erosion (Fig. 4.36B; Piper et al., 1983), or deep water renewal events (Johnson, 1974: Rupert Inlet, B.C.), or possibly sediment gravity flows (Burrell, 1983b: Boca de Quadra, Alaska). More

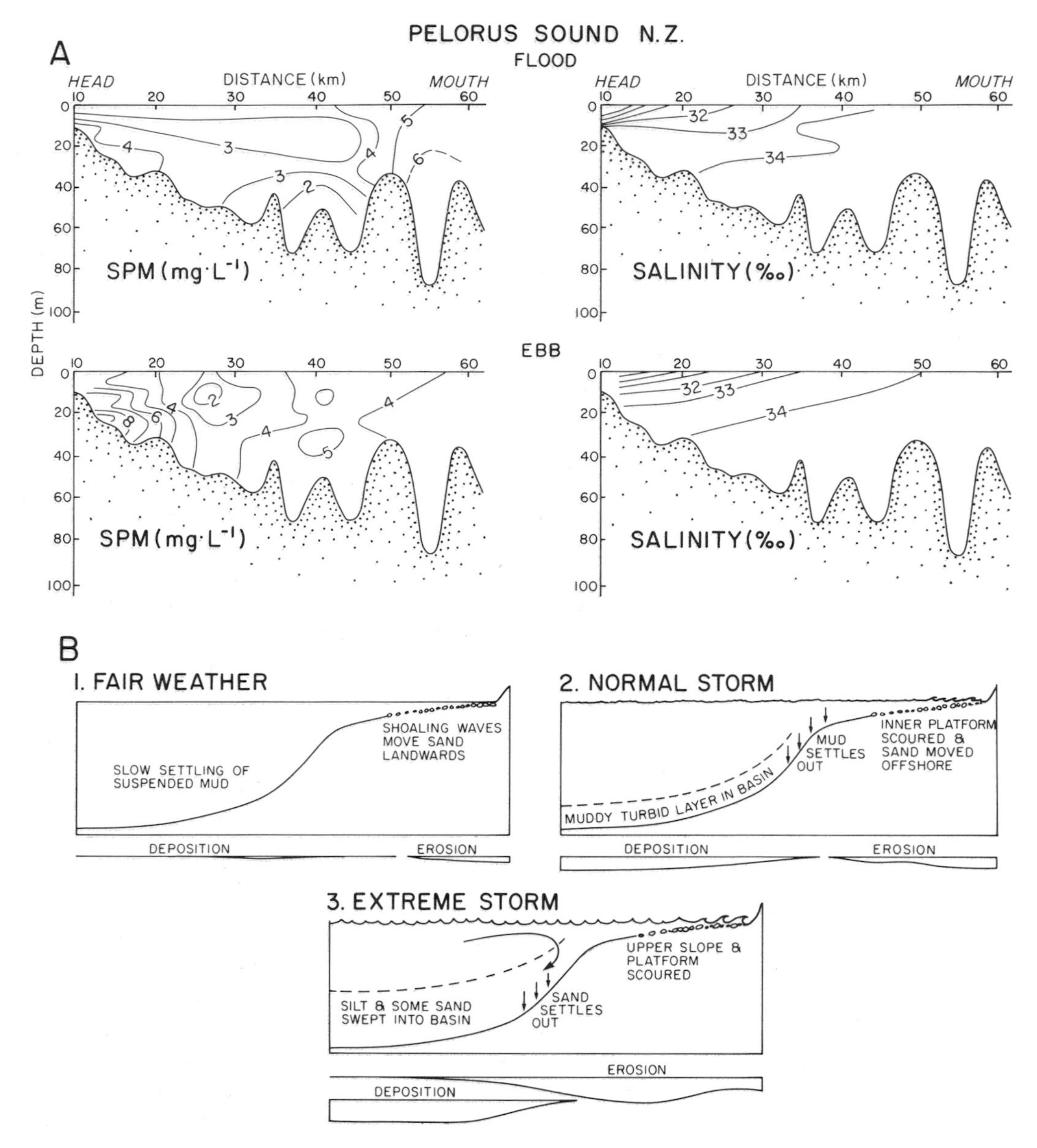

FIGURE 4.36. (A) The generation and tidal mobility of turbidity maxima in Pelorus Sound, N.Z. (after Carter, 1976). (B) Process models for low-sediment, wave-dominated fjords showing: (1) fair weather; (2) normal windy; and (3) exceptional storm conditions (from Piper et al., 1983).

constant levels of near-bottom turbid water may relate to the action of benthic bioturbators (Syvitski et al., 1983b; see Chapter 6).

In tide-influenced fjords, turbidity maxima may develop with the entrapment of particles by the tidal circulation (d'Anglejan and Smith, 1973). Turbidity maxima (i.e., localized zone of dirty water) tend to be located in less than 50 m of water and seaward of the river mouth, and in many cases, are directly related to the resuspension of recently settled material. The turbidity maxima found within the St. Lawrence Estuary, however, have more to do with advection of suspended particles than local resuspension; as such, the turbidity maxima occur mid-depth within the water column and correspond to the end of the ebb flow (d'Anglejan and Ingram, 1976). In fjordlike estuaries, tidal resuspension tends to be depth controlled: as the current energy decreases with depth, the current shear will become no longer critical to the threshold movement of sediment grains. The critical depth may change with the stage of the tide and through the spring-neap tidal cycle. When the fjord channel constricts, or where tributaries join the main channel, tidal flow will increase and thus lower the critical erosion depth. Below the zone of erosion there will exist an associated zone where seafloor sediments do not undergo erosion, yet suspended particles may not be deposited. In Cook Inlet, Alaska, the near-bottom turbidity maxima occur over thresholds, near the shallowing fjord head, and along the fjord walls (Feely and Massoth, 1982). The position of the maxima may also be dependent on the stage of the tide (Fig. 4.36A).

If the crest of the sill lies above the critical erosion depth, sediment deposited on the sill during periods of slack tide will be resuspended eventually and transported into or out of the fjord basin. Where the currents are especially strong, the sill might be mantled with a gravel lag, or even consist of exposed bed rock (e.g., Gilbert, 1978; Syvitski and Macdonald, 1982). The tidal jet generated over the outer sill in Borgenfjorden, Norway, is reflected in the coarser sediment as compared with finer grained basin sediments (Fig. 4.37A: Strömgren, 1974). Borgenfjorden is an excellent example of the close relationship between grain size and bathymetry. In general, decreasing grain size reflects decreasing current velocity with the increasing width and depth of the inlet. An interesting reversal in this "normal" sediment hydrodynamic relationship was observed on the Knight Inlet sill using submersible techniques (Tunnicliffe and Syvitski, 1983). The sill crest, at a depth of 65 m, is covered in a well-sorted pebble/cobble lag (10- to 80-mm grain size), with pronounced armoring and imbrication. Also scattered over the crest are large boulders up to 2 m in diameter, some being tightly lodged in the underlying till. Growing on top of these boulders are large fan corals (Gorgonacea), as large as 3 × 3 m and facing into the prevailing current. Tidal currents, although frequently in excess of 1 m s^{-1}, are not strong enough to transport boulders of this size. However, the drag forces on the coral fans allow for the preferential removal of the attached boulders over the smaller pebbles. The boulders collect as large rock piles on either side of the sill crest.

Turbulence as generated from hydraulic jumps at the sill may create zones of erosion where the basin sediments abut with the sill. Bornhold (1983b) provides an excellent example of both conformable winnowing (unit A sediments on Fig. 4.38A) and erosion of some ten meters of basin sediment (unit B on Fig. 4.38A). Such features may relate to tidally controlled hydraulic jumps over the Maitland Island sill in Douglas Channel, British Columbia. Where tidal streams are proximal to a sediment source, zones of erosion and selective deposition may grade with zones of deposition: scour channels and stratigraphic wedging of units may result (Piper et al., 1983). Along the approaches to Makkovik Bay, Labrador, selective tidal stream erosion and winnowing of Holocene mud result in the formation of many of these features (Fig. 4.38B: Barrie and Piper, 1982). With the availability of coarser sediment (i.e., sand and gravel), powerful tidal currents may form an assortment of bed-form groupings (for details on deep water bed forms see Amos and King, 1984). Where the basin is deep, bed forms may be found along the fjord walls, even on slopes of 30° (e.g., St. Lawrence Estuary: Syvitski et al., 1983b).

In Cook Inlet, Alaska, the sill is dissected and powerful currents rework the seafloor. Current velocities are generally high in the narrower upper Cook Inlet: poorly sorted glacial sediment that once blanketed the seafloor has been redistributed by the winnowing out of sand, silt, and clay. This has left behind a coarse-grained lag deposit and allowed deposition of sand in the wider areas down-fjord (Sharma and Burrell, 1970; Hampton et al., 1978a). In lower Cook Inlet, four major groups of bed forms have been observed (Bouma et al., 1977, 1978): (1) small asymmetric ripples ($\lambda \leq 30$ cm, $h \leq 7$ cm; where λ is the bed-form wavelength and h is the form height); (2) small sand waves ($\lambda \leq$ 10's of meters, $h \leq$ few meters); (3) medium-sized sand waves and small dunes ($\lambda = 50$ to 150 m, $h < 5$ m); and (4) large sand waves, dunes, and sand ridges ($\lambda = 350$ to 1000 m, $h = 5$ to 10 m). Several of the classes are superimposed on each other. Sand grains either roll over the bottom or form sheet flows, the latter moving as much as 30 cm s^{-1} over the crests of the larger bed forms. Small current ripples are the basic mode of sand transport, and except in sand-free areas they are ubiquitous. The boundaries between the bed-form fields can be very distinct.

The flotation of sand is another tide-related but wave-limited transport process operative over intertidal flats. Sand will be picked up and

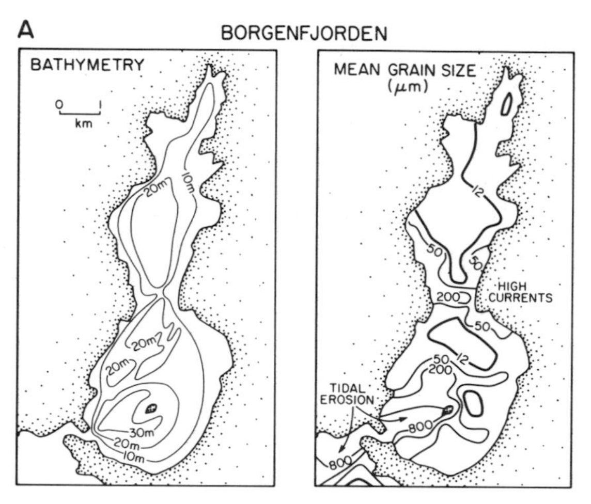

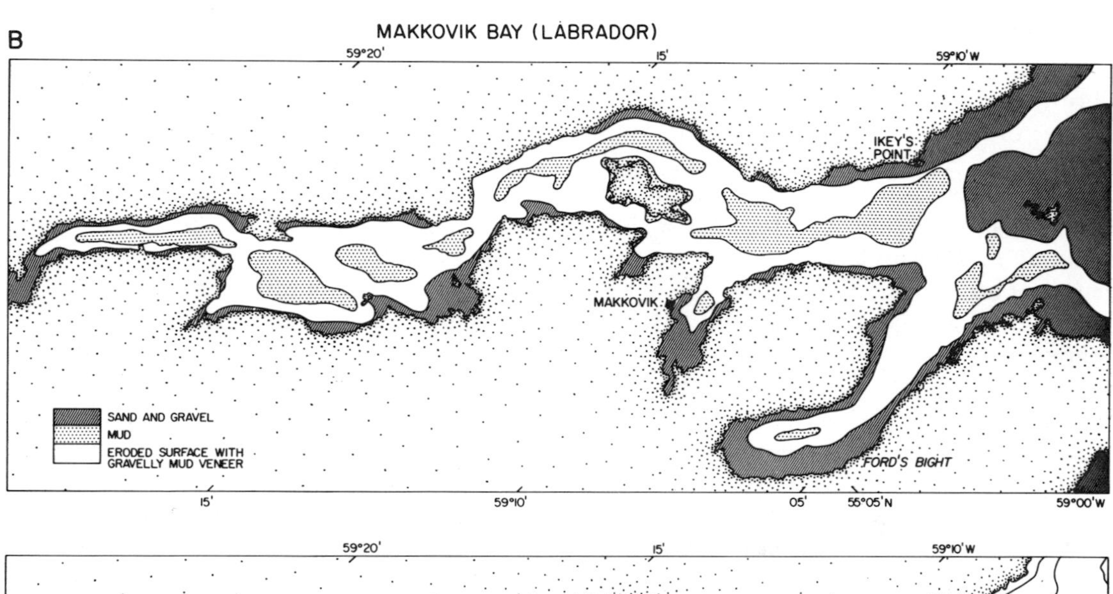

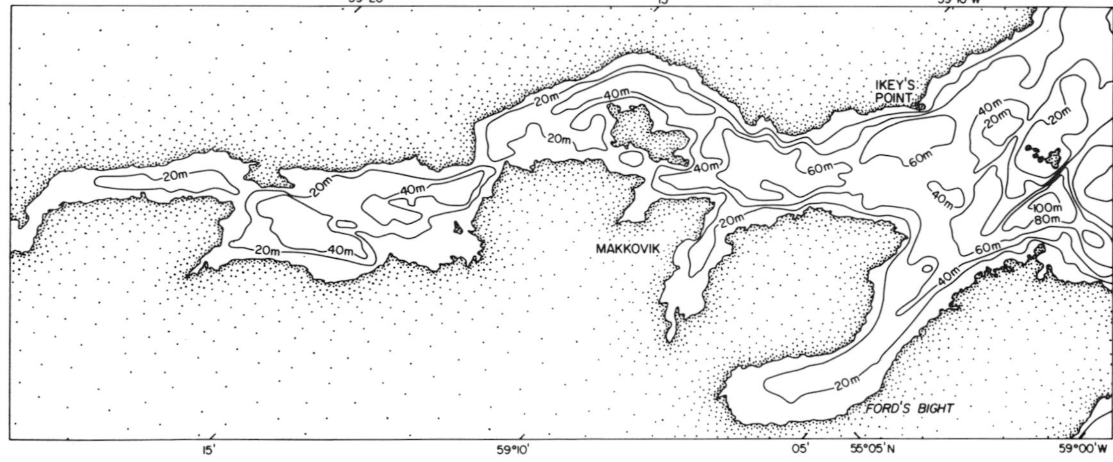

FIGURE 4.37. (A) Bathymetry and mean grain size (m) for Borgenfjorden, Norway (after Strömgren, 1974). (B) Bathymetry and facies distribution in Makkovik Bay, Labrador (after Piper et al., 1983).

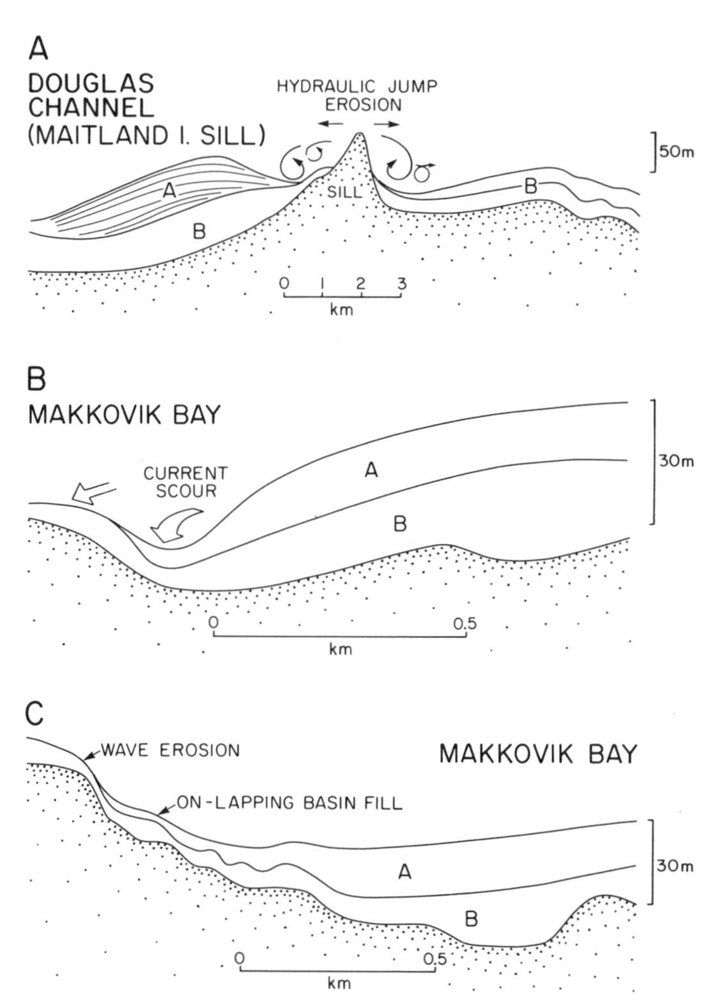

FIGURE 4.38. Schematics of seismo-stratigraphic sections that show contemporaneous scouring and/or winnowing of basin sediment. (A) Hydraulic jump erosion near Maitland Island sill, Douglas Channel, B.C. (after Bornhold, 1983b). (B) Tidal current scour in the approaches to Makkovik Bay, Labrador (after Barrie and Piper, 1982). (C) Onlapping basin fill units as a result of wave erosion within Makkovik Bay, Labrador (after Barrie and Piper, 1982).

floated on the sea surface with each rising tide, dependent on three key environmental factors (Syvitski and Everdingen, 1981): (1) proper atmospheric conditions (no fog or precipitation); (2) rising water with intact surface tension (no surface turbulence); and (3) appropriate floatable sediment for the incoming water velocity. The appropriateness of the floating sediment is a function of grain size, grain shape, grain surface texture and surface coating, and grain density. The sand, in patches as large of 100 m × 100 m can float seaward as the tide begins to fall or under the influence of gentle land breezes. The annual tonnage of sand moved seaward will depend on the intertidal area that meets the above conditions, but is typically of the order of 10^5 to 10^7 tonnes for macrotidal fjords.

Wave-dominated fjords consist of one or more silled basins, only rarely more than 100 m deep, the exception being the deeper basins of Newfoundland's fjords. Although most fjord coasts have a few examples, the fjord coasts of Nova Scotia, Newfoundland, and Labrador are predominantly of this type (Piper et al., 1983). Other examples include Flensburg Fjord, Baltic Sea (Exon, 1972), sea lochs within the Firth of Clyde, Scotland (Deegan et al., 1973), and the Kamchatka fjord coast, Russia. Wave-dominated fjords receive only 5 to 25% of their Holocene basin fill from rivers, almost all the re-

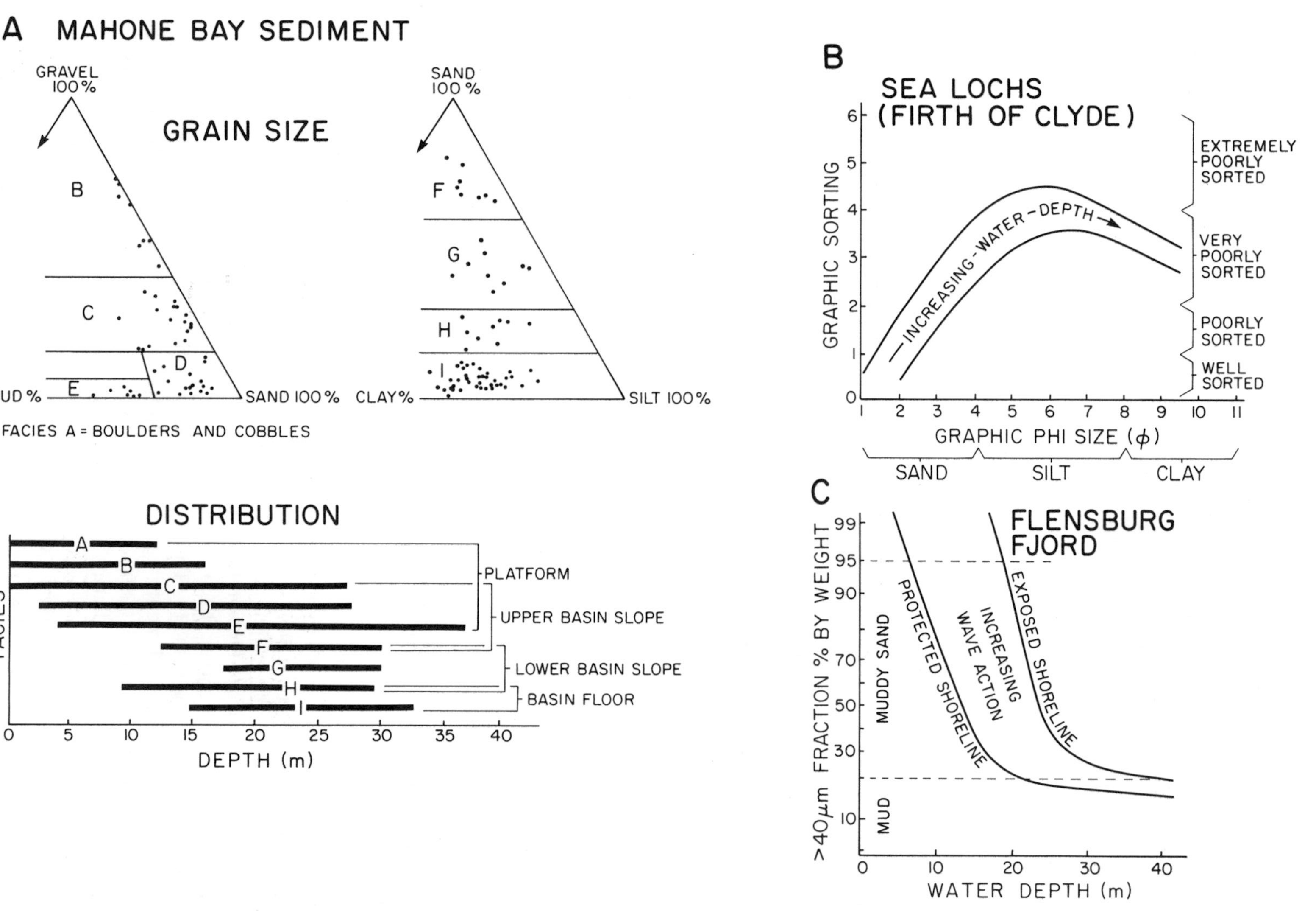

FIGURE 4.39. (A) Mean grain size and distribution characteristics of sediment facies in southwestern Mahone Bay (after Letson, 1981). (B) Relationship between graphic phi size and graphic sorting for wave reworked sediment of the sea lochs of the Firth of Clyde (after Deegan et al., 1973). (C) The decrease in the > 40-μm fraction for wave-reworked sediment samples collected from decreasing water depths in Flensburg Fjord, Baltic Sea (after Exon, 1972).

mainder is derived from waves reworking older marine sediment or glacial till along the basin margins. Sediment accumulation rates range from 0.5 to 3 mm yr^{-1} (Piper et al., 1983; Piper and Keen, 1976). The highest sedimentation rates occur in deep basins adjacent to shore areas that receive the greatest wave intensity. The larger waves arrive as open ocean swells or storm waves, and in the case of the deeper Newfoundland fjords, 10-m wave heights have been reported. Cliff retreat rates can exceed 1 m yr^{-1} near the fjord mouths, decreasing to 25 cm yr^{-1} in exposed inner fjord areas.

Severe storms are capable of erosion and resuspension of sediment found in water many tens of meters deep and as much as 100 m deep in the Newfoundland bays (Slatt, 1974; Stehman, 1976). After a storm, suspended sediment concentrations remain high in a near-bottom zone (Delure, 1983; Barrie and Piper, 1982). Fine sand and silt settle out rapidly, leading to the deposition of thinly banded graded layers on the basin floors (Letson, 1981). The internal occurrence of sandy layers suggests that severe storms recur between 20 and 50 years apart. Between these major storms, the basin margins accumulate a reservoir of sediment during less severe weather periods (Fig. 4.36B: Piper et al., 1983).

Sediment grain size in fjords controlled by wave action correlates strongly with water depth (Figs. 4.37B, 4.39); nearshore zones of sand and gravel are followed offshore with a zone of lag sediment till or bed rock, and at greater depths with poorly sorted sandy to muddy sediment. Since the severity of the storms controls the critical depth of erosion, and the onset of storms follows a time-dependent probability function, the sediment sorting decreases with water depth (Fig. 4.39B: Deegan et al., 1973). The rate of downward fining depends on local bathymetry, fetch direction (coastal exposure), and the wave size (Fig. 4.39C: Exon, 1972). Commonly, wave resuspension results in onlapping basin-fill units (Fig. 4.38C; Barrie and Piper, 1982). Severe storms may also generate strong currents capable of transporting sand (Exon, 1972). Along wave-dominated continental shelves, sands may be transported landward into the approaches to fjords and possibly into the fjord basin (Piper and Keen, 1976). Along the Newfoundland coast, some fjord sills are capped with barrier beaches; during severe storms outwash sands may breach the barrier and move into the otherwise landlocked basin (e.g., St. Mary's Harbour; Forbes, 1984).

Relative sea-level fluctuations may be critical to the development of wave- and tide-dominated fjords. Nova Scotia inlets that have experienced marine transgression have developed wave-cut platforms from the erosion of till cliffs (e.g., drumlins), eventually becoming submerged by the rising sea level (Barnes and Piper, 1978). The platforms undergo winnowing and by-passing. When carried to completion, resuspension may result in the formation of gravel pavements (Stanley, 1968). In areas of marine regression, proglacial and earlier Holocene sediments are eroded as they come above the critical erosion depth with the falling sea level. The sediment distribution therein is similar to deep water inlets undergoing marine transgression.

4.7 Summary

The dispersal, deposition, or resuspension of sediment within fjords is very system dependent and may range from fjords dominated by simple estuarine circulation to fjords more influenced by action of ice, waves, or tides. The circulation within a fjord is very seasonally modulated, influenced by the time-history of fluvial discharge and the seasonal weather patterns. System perturbations, such as deep water renewal, may cause temporary imbalances to the estuarine circulation pattern and allow the formation or maintenance of deeper and less dynamic circulation cells within the fjord. Although such circulation cells may be poorly coupled to the surface layer, they can be very dependent on sill depth and oceanographic conditions of the coastal and shelf waters.

If one examines the simple case of a river flowing into the head of a fjord and it is capable of maintaining a well-developed two-layer estuarine circulation, then the rate of sedimentation and size of the sedimented particles decrease exponentially down the fjord. This simple pattern of sedimentation will occur even though the suspended particulate matter may be in the form of floccules, agglomerates, or fecal pellets. Such hydraulically controlled sedimentation will not only control the grain size on the seafloor, but also the distribution of minerals within the fjord. Occasionally, sediment-laden river water may be dense enough to sink upon entry into the marine waters of the fjord and thus travel seaward as a gravity flow. Such hyperpycnal flows, however, are considered rare.

Although much is known on the physics of deep water renewals, the number of natural factors that can combine to allow or negate the development of a flushing event has not allowed for accurate predictions on the timing of these events. Furthermore, each renewal event is unique and thus may or may not replace all the water in the first or subsequent inner basins of the fjord. The effect of renewal events on sediment resuspension and transport is even more highly speculative.

The effect of sea ice on circulation and sediment redistribution is very important for subpolar and polar fjords. The presence of sea ice both contributes to the stability of surface water stratification (by the elimination of wind mixing in the winter and the production of a freshwater cap in the summer) and the instability of surface stratification (through the process of salt rejection). Similarly, sea ice will protect beaches and coastal areas from wave erosion in one sense, yet through ice push and rafting phenomena contribute to the erosion and transport of coastal sediment.

If we accept the definition of fjords (Chapters 1 and 2) as being glacially modified submarine valleys, then the movement of glaciers and the associated deposition of glaciomarine sediment has been (or still is) a prime force in the present state of each fjord. The presence and movement of tidewater glaciers either directly or indirectly affects such parameters as sea level (and thus circulation and tidal velocities), iceberg production (and thus the creation of bow waves, the transport of sediment, and upwelling within a fjord), surface stratification (ice-wall upwelling, year-round contribution of fresh water), and the distribution of nutrients and therefore biological productivity (see Chapter 6).

Finally, it has been popular for generalists to speculate that fjords are quiet water coastal basins or sediment traps. However, some fjords are very high-energy environments. One such class of fjords, being wave dominated, has much of its basin sedimentation as a direct result of high-energy coastal erosion. Another class, that of tide-dominated fjords, has such significant reworking of the basin sediments that highly mobile bed forms cover the seafloor.

5

Subaqueous Slope Failure

Fjords are ideal environments for the study of nearly every form of submarine slide and type of sediment gravity flow. Fjords are characterized by some very steep slopes and may have rates of sedimentation that far outpace the rates of consolidation. These underconsolidated sediments may fail under their own weight or because of stimulus from earthquakes, giant waves, and man. As a consequence, there are many examples in fjords of both loss of life and extensive damage to port facilities and settlements. Thus the study of subaqueous slope failures is environmentally important, particularly in predicting their occurrence and frequency.

The chapter begins with a brief review of mass sediment properties, because the science of geotechnical engineering is partly outside the normal study of oceanography and marine geology. Next, the various release or triggering mechanisms for slope failures and their frequency of occurrence are discussed. These relationships lead into a discussion on the mechanics and constraints of the various mass transport processes. The chapter concludes with an examination of the various macro- to microscale features that result from the myriad of failure styles.

Symbol Notation

(All other symbols as defined in the text)

a_e	horizontal earthquake acceleration
a_b	friction due to interfacial shear at the top of a current
C_v	coefficient of consolidation
c	cohesion
D	water depth
e	void ratio
f	Darcy Weisbach friction factor
ff	combined friction parameter (as defined)
G_s	specific gravity
g	acceleration due to gravity
H_t	total thickness of the sediment mass

h	thickness of the sediment gravity flow
k	coefficient of permeability
LI	liquidity index
LL	liquid limit
n	porosity
M_v	coefficient of volume compressibility
OCR	overconsolidation ratio
PI	plasticity index
PL	plastic limit
Su	undrained shearing resistance
Sr	remoulded shearing resistance
t	time
U	degree of consolidation
u_e	excess pore pressure
V	volume (defined)
v_c	creep velocity
v_s	velocity of slide or slump
v_p	velocity of sediment plug in Bingham material
v	velocity of gravity flow (defined)
W	weight (defined)
x	distance from fjord head
Z	critical depth of failure or depth coordinate (defined)

Subscripts s and w, solids and water, respectively

α	seafloor slope angle, in degres
γ	unit weight
γ'	saturated or submerged unit weight
γ_w	unit weight of pore water
η	viscosity
ρ	density
σ	vertical stress (or pressure)
σ'	vertical effective stress (or overburden pressure)
τ	horizontal component of shearing stress along a potentialslide path
Ø	angle of internal shearing resistance, in degrees

5.1 Mass Sediment Properties and Subaqueous Slope Stability

There is limited information on the mass properties of fjord sediments, although seafloor instabilities and failures are well documented in fjords. This, in part, reflects a number of historical developments including: (1) the paucity of experienced marine geotechnical engineers working in fjords until very recently; and (2) the lack of suitable sampling and profiling methods with respect to the sometimes very deep (> 1000 m) nature of fjords. As a result, there have been few attempts to predict geotechnical sediment properties within the variety fjord basins. Only recently have efforts begun to explore trends and interrelationships of geotechnical properties within fjord sediments (Richards, 1976; Hein and Longstaffe, 1985).

The physical properties of seafloor sediments include: (1) weight-volume parameters (void ratio, porosity, water content, specific gravity, and unit weight); (2) particle size and shape; (3) composition of the solids (mineralogy, organic content); (4) sedimentary and grain fabric; (5) Atterberg limits (plastic limit, liquid limit, plasticity index); (6) degree of consolidation and compressibility; and (7) shear strength.

The void ratio, e, is the ratio of the volume of voids, V_v, to the volume of solids, V_s, in a given sediment volume. The porosity, n, is the ratio of V_v to the total sediment volume, V, that is:

$$n = (V_v)(V)^{-1} = (e)(1 + e)^{-1} \quad (5.1)$$

The water content, w, is the ratio of the weight of the pore water, W_w, to the weight of the solids, W_s, in a given volume. The "natural" water content, W_n, is the ratio of W_w to the total weight, W. The unit weight, γ, in a sediment volume, V, can be given as:

$$\gamma = (W)(V)^{-1} = (W_s + W_w)(V_s + V_w)^{-1} \quad (5.2)$$

or

$$\gamma = G_s \gamma_w (1 + w)(1 + e)^{-1} \quad (5.3)$$

where G_s is the specific gravity of the solids and the subscripts s and w represent the solid and water fractions, respectively. The specific gravity is a function of sediment particle density (i.e., mineralogy and organic matter content). In a saturated soil, as would be the case of marine sediment, the saturated unit weight would be:

$$\gamma_{sat} = (G_w \gamma_w + e\gamma_w)(1 + e)^{-1} \quad (5.4)$$

or

$$\gamma_{sat} = [G_s - n(G_s - 1)]\gamma_w \quad (5.5)$$

There are dozens of functional relationships between these weight-volume parameters and for details readers could refer to Das' *Advanced Soil Mechanics* (1983) or any other standard soil mechanics textbook.

The particle size distribution either controls or affects all primary physical properties of the sediment mass. Fjord sediments have a wide range of grain size and size sorting. In some localities, there exists a spatial continuum of unimodal sands to silts to clays, as is the case near major fluvial sources. Elsewhere, sediments contain multimodal size populations as a result of mixing sediment from different depositional processes, that is, mixtures of two or more sediment populations from ice rafting, wind storms, fluvial discharge, wave reworking, and slope failures (Syvitski and Macdonald, 1982). Sediment types (lithofacies), both at the seafloor and within the sediment column, may have gradational or sharp contacts. A fjord basin may contain laterally and/or vertically extensive units, or more local units such as may result from small-scale slope failures. The vertical "anisotropy" can be related to basin history and superposition of facies. Some of these large-scale discontinuities may be potential failure surfaces. The individual beds may be homogeneous in terms of sediment population, graded, or contain a considerable particle size inhomogeneity (for details see Section 5.3). For instance, a "normal" graded bed has particle sizes that fine upward from the base to the top of the bed; in a "reverse" graded bed the particles coarsen upward.

Fjord sediments contain highly angular grains, because of their proximity to the source terrain and the nature of sediment production in glacial environments. Grain rounding does occur to a limited extent during englacial fluvial activity, or in glaciofluvial channels on outwash plains or sandurs. More extensive rounding occurs during sediment storage within the intertidal areas through the action of waves and tides. Little further rounding occurs within the fjord basin itself.

The mineralogy of sediment within a fjord basin closely reflects the bedrock composition of the local hinterland with only minor alteration of minerals resulting from soil diagenesis or marine halmyrolysis. Occasionally, foreign lithologies may have entered the local drainage basin through the action and flowage of ice caps during previous glaciations. The clay-sized minerals are mostly powdered rock fragments, possibly having undergone some hydraulic sorting during the depositional process (see Section 4.2). Nevertheless, a significant amount of clay minerals (phyllosilicates) may occur within the clay-sized fraction. Phyllosilicates are geotechnically important compared with most other mineral families in that they may increase in volume as the water content increases. The volume change depends on: (1) the type and amount of clay minerals in the sediment; (2) specific surface area of the clay; (3) grain fabric; (4) pore-water composition and salinity; and (5) the atomic number and valence of the exchangeable cations.

The content of organic matter is another important mass property of sediment. Increasing the organic content tends to increase the compressibility of the sediment. Also, organic matter tends to retain moisture within the soil. With aging, buried organic matter may undergo bacterial degradation or, through low-temperature diagenesis, can be altered to a gas. Thus a decrease in organic matter with depth is normally observed in fjords (e.g., Frobisher Bay, Baffin Island: Andrews et al., 1983). However, changes in source concentration may also account for down-core variability.

The arrangement of particles within a sediment fabric can be quite complex. Sediment fabric depends on: (1) the grain size distribution; (2) grain shape; and (3) rate of deposition. For granular material, the arrangement of particles may range from loose packing and metastable conditions, to dense packing and stable conditions (i.e., whereby there is relatively little reduction in sediment strength during loading). For instance, poorly sorted sand tends to be tightly packed with the void spaces between the larger grains being occupied by the finer grains. Well-sorted sand may be loosely packed and completely lose strength (coherency) during cyclic loading. The rapid deposition of sediment near fjord-head deltas (see Section 3.4) can result in loosely packed metastable deposits with high porosity.

The packing of cohesive sediment partly depends on the dynamics of deposition. In saline water, clays settle out as floccules, agglomerates, and even planktonic fecal pellets, with random or ordered arrangements of their edges and faces (Quigley, 1980; Syvitski and Murray, 1981). In fjord lakes containing fresh water, clay particles normally settle out in a dispersed state. Interparticle structure of seafloor sediment may also be altered by: (1) benthic bioturbators, especially through the ingestion and egestion of sediment particles; (2) bacterial breakup of planktonic fecal pellets; and (3) changes in pore-water salinity. In the latter case, floccules may become less cohesive as salt water is washed out of the sediment. Even subaqueous marine clays may become successively leached of salt from below because of artesian overpressure in fjord environments. This process of salt leaching is reversible both naturally, through the production of Mg and Ca with diagenesis of unstable minerals, and artificially through salt wells (Løken and Torrance, 1971). In general, the more random and interlocking nature of flocculated clays promotes increased sediment stability compared with nonflocculated clays.

Atterberg limits are the water content limits necessary to change a remoulded clay from a liquid state to a plastic state (liquid limit, LL), and from a plastic state to a semisolid state (plastic limit, PL). The liquidity index, LI, relates these Atterberg limits to the plasticity index, PI, by:

$$LI = (w_n - PL)(PI)^{-1} \qquad (5.6)$$

where

$$PI = (LL - PL)$$

If the natural water content is equal to the liquid limit, then the liquidity index is equal to 1. If $w_n = PL$, then the liquidity index is equal to 0. Thus for a plastic state, the natural water content is less than or equal to the plastic limit and greater than or equal to the liquid limit, with LI varying from 1 to 0. When the natural water content is greater than the liquid limit, the soil is considered sensitive (i.e., sensitive to loading and thus premature failure). If the organic content is negligible, then the specific gravity of the sediment can be related directly to mineralogy and therefore to the Atterberg limits. For Oslofjord, the liquid limit was found to vary with the plasticity index by $PI = 0.751\ LL - 11.036$ (Richards, 1976). Such a high value to the regression slope is common to many fjords (Hein and Longstaffe, 1985) and is indicative of the inorganic nature of these sediments.

Figure 5.1 gives the down-core variation in Atterberg limits for a 245-cm core taken in Inugsuin Fiord, Baffin Island (Hein and Longstaffe, 1983). The sediment thickness represents approximately the last 6000 years of accumulation. Both the plastic limits and the plasticity index decrease with core depth, in part a function of grain size variations. After some initial fluctuations in the top part of the core, the natural water content decreases down-core, in part reflecting changes in the mean grain size. The fluctuations of the liquid limit closely parallel changes in the natural water content until at depth they rapidly decrease and parallel the plastic limit. The result is highly sensitive sediment near the base of the core as documented in the liquidity index. This highly sensitive portion of the core appears to be nearly nonplastic, that is, composed of silt grains.

After deposition on the seafloor, sediment begins a process of adjustment in response to the surrounding static and dynamic forces. Initial deposition usually results in surface sediment with high porosity and associated high water content, although both parameters are dependent on the nature of the depositional process. For instance, hemipelagic deposition of floccules (see Section 4.2.2) can result in higher values of porosity and water content than suspended particles deposited under the action of current shear (see Section 4.6.3). The adjustment of these seafloor sediments in response to increased loading is known as consolidation; it involves the squeezing of water from the pores and the decrease in the void ratio. The loading may take the form of sediment burial (overburden stress), current shear stress (of consequence only as a preburial stress), and wave loading (including earthquake accelerations, internal and surface water waves). Additionally, fjord environments may experience periodic loading from ice in the form of a glacier advance (see Section 4.5.3) or from iceberg scouring (see Section 4.5.5).

It is reasonable to expect the variation in water content within a sediment mass to provide proxy information on the consolidation history of fjord sediment. Figure 5.2 gives the down-core variation in water content for two Baffin Island fjords (data after Andrews et al., 1983, and Farrow, 1983). The Holocene sediments in both cores show a linear decrease in water content, with depth, that is, conditions of normal consolidation under an approximately constant

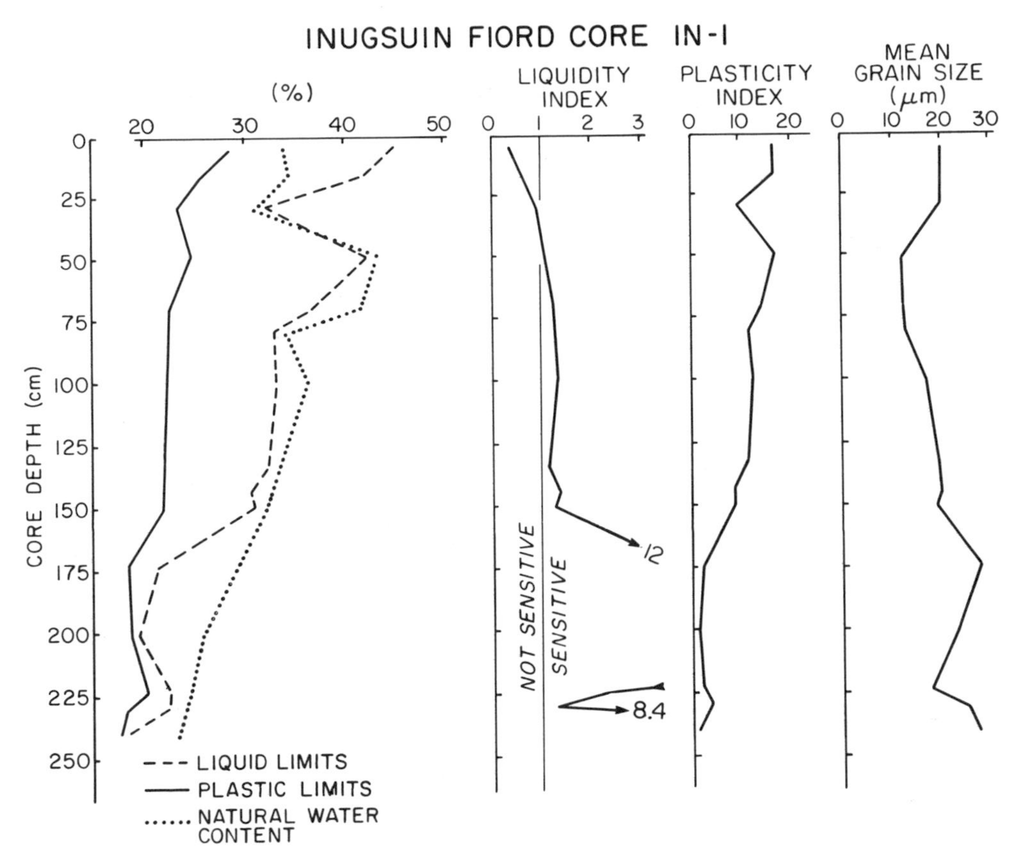

FIGURE 5.1. An example of down-core variability in Atterberg limits and mean grain size for Lehigh core IN-1 from Inugsuin Fiord, Baffin Island (data from Hein and Longstaffe, 1983). Water depth, D = 160 m; distance from fjord head, × = 5 km.

rate of sedimentation. The straight-line segment reflects virtually no mixing by local benthic bioturbators. In the Coronation Fiord core, the topmost 1.5 m of Holocene sediment is well bioturbated and this is reflected in approximately homogeneous water content values. The Pleistocene sediment shows greater and more uniform consolidation values as reflected by their relatively low water content. In both fjords there is a marked change in the water content profile at the late glacial-postglacial boundary. Some of the water content uniformity for the Pleistocene sediments of Sunneshine Fiord may reflect a history of bioturbation, and where the sediment is particularly turbated, the water content increases. Where bioturbation is not a limiting factor, as in the case of the Coronation Fiord core, the Pleistocene sediment shows a continual, albeit gradual, decrease in water content with depth.

The total vertical (overburden) stress, σ, for a submerged sediment profile is the combined unit weight of the overlying sediment (γ_s Z) minus the unit weight of the pore water (γ_w Z):

$$\sigma = (\gamma_s - \gamma_w)\, Z = \gamma' Z \qquad (5.7)$$

or

$$\sigma = \Sigma\left(\frac{G_s - 1}{1 + e}\right) \gamma_w \, dz \qquad (5.8)$$

where γ' is the submerged weight of sediment thickness, Z, within the sediment mass. A sediment may be considered normally consolidated if its effective overburden pressure can account for the total vertical stress found at that burial depth.

In underconsolidated sediment, the effective stress, σ', is:

$$\sigma' = \sigma - u_e \qquad (5.9)$$

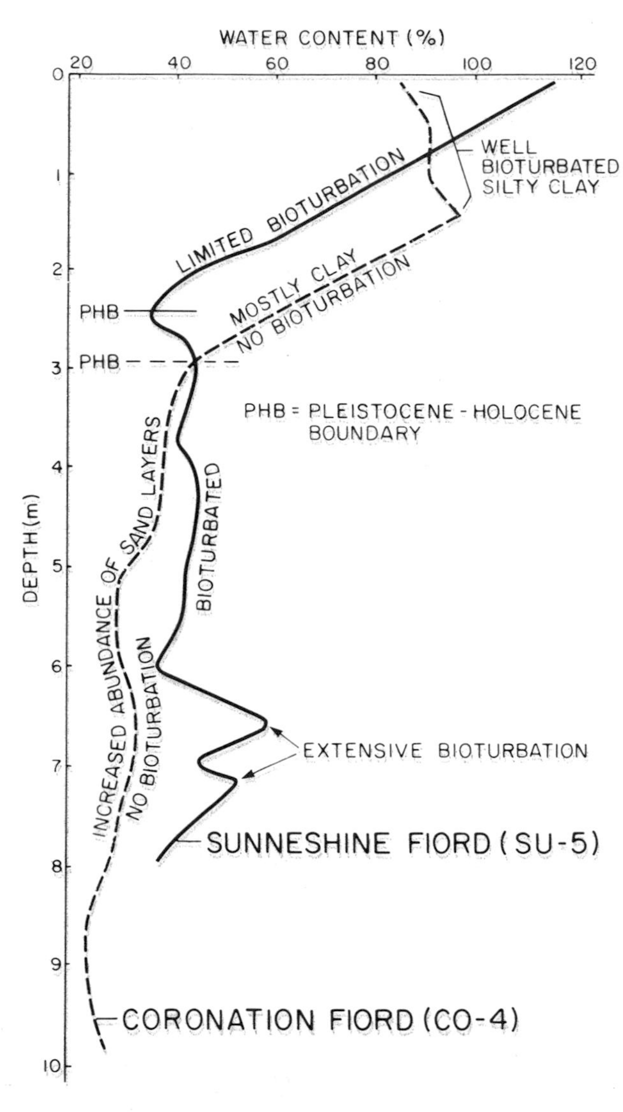

FIGURE 5.2. Down-core variability in water content for cores collected in Sunneshine and Coronation Fiords, Baffin Island (data from Andrews et al., 1983). Intensity of bioturbation data is from Farrow (1983). SU − 5: D = 155 m, × = 40 km; CO-4: D = 356 m, × = 25 km.

where u_e is the magnitude of the excess pore pressure above hydrostatic. Excess pore pressure reduces the "effective" weight of the overburden, which is normally carried by the intergranular contact pressures—the source of the sediment's frictional shearing strength. The excess pore-water pressure, u_e, at any point x above the base of a deposit is known to vary with time, t, and with the sedimentation rate ($\partial Z/\partial t$); it has been modeled by:

$$C_v \frac{\partial^2 u_e}{\partial x^2} = \frac{\partial u_e}{\partial t} - \gamma \frac{\partial Z}{\partial t} \qquad (5.10)$$

with

$$C_v = k\,(\gamma_w m_v)^{-1}$$

where C_v is the coefficient of consolidation and is in part dependent on the sedimentation rate; k is the coefficient of permeability; and m_v is the coefficient of volume compressibility. In fjords that experience rapid rates of sedimentation and where the rate of consolidation lags behind accumulation, sediments may be more than 90% underconsolidated, that is, less than 10% of the expected normal consolidation (Sangrey et al., 1979).

High values of excess pore pressure may also be associated with the in-situ production of gas. In fjord sediments, the gas is usually methane and is produced by the low-temperature diagenesis of terrestrial and marine organic matter of relatively recent age (e.g., St. Margarets Bay: Keen and Piper, 1976).

If a preconsolidation pressure, P_c, is greater than the present effective stress, σ', then the sediment may be considered overconsolidated. (Note: The value of P_c can be determined from e vs. log σ' plot using the graphical techniques of Casagrande, 1936.) For example, in Oslofjord, the upper 4 m of the seafloor is apparently overconsolidated, yet below 4 m the sediment shows normal consolidation (Richards, 1976). The exact reason remains unclear. Overconsolidated material may result from: (1) changes in total overburden stress; (2) changes in porewater pressure; (3) aging, causing a decrease in the void ratio (important for plastic clays); (4) environmental changes such as pH, temperature, and salt concentration; and (5) weathering (for instance, the submergence of weathered subaerial sediment during a recent marine transgression). Changes in the overburden stress may take the form of: (1) a decrease in wave loading (e.g., development of a protecting barrier island); (2) glaciation (ice loading); and (3) erosion of surficial sediment through slides or current action (e.g., Knight Inlet, B.C.: Farrow et al., 1983). Changes in pore-water pressure may result from: (1) fluctuations in artesian or gas pressure; (2) desiccation with subaerial drying; and (3) the effects of permafrost. In general, Pleistocene marine sediment tends to be overconsolidated, and the more recent Holocene sediment tends toward normal or underconsolidation (Molnia and Sangrey, 1979; Elverhöi, 1984).

Shearing resistance, an indicator of sediment shear strength, is the particle-to-particle resistance to a tangential stress tending to cause two adjacent parts of a sediment column to slide past one another while parallel to the plane of contact. In terms of slope stability, low undrained shearing resistance, Su, is one of the most important properties associated with underconsolidation. For underconsolidated soils, the shearing resistance is often expressed in terms of the ratio Su/σ. Sangrey et al. (1979) have noted for normal consolidation, that values of Su/σ range from 0.15 to 0.3. In general, the shear strength of glacial marine sediment is low for Holocene age material and high for Pleistocene age sediment (Elverhöi, 1984).

A slope fails as soon as the average shearing stress, τ, along a potential slide path becomes equal to the average shearing resistance along this path. At any point located in a submerged sediment mass, the strength at failure can be given as:

$$\tau = c + \sigma' \tan \emptyset \qquad (5.11)$$

where c is the interparticle cohesion and $\emptyset$ is the angle of internal friction (angle of shearing resistance). Known as the Mohr-Coulomb failure criterion, equation (5.11) is a straight-line approximation of a gentle curve. For normally consolidated clays and granular sediments, $c = 0$. For overconsolidated submarine sediments, c is greater than 0. With bioturbation and in-situ pelletization, water content may increase and soil cohesion may decrease, except where mucoid adhesion is involved. For loose silts and sands, $\emptyset$ is between 28 and 34°, and for stable clays $\emptyset$ can be as low as 9 to 13° for marine sediments. With granular soils, the angle of internal friction increases with the angularity of the grains or with the packing density.

The strength of sediment at failure is partly dependent on the moisture content. The variation of w versus log Su is approximately inverse linear for normally consolidated clays (e.g., compare Fig. 5.1 with 5.3 for core IN-1). For a decrease in water content with depth, the shear strength will normally increase (Fig. 5.3). Soil sensitivity is defined as the ratio of the undrained shear strength, Su, to the shear strength of the remoulded sediment, S_R. If $Su/S_R > 4$, the sediment is considered sensitive; when $Su/S_R > 16$, the sediment is considered quick. The sensitivity of Baffin Island fjord sediments is highly variable and reflects the thin-bedded nature of sediments: occasionally the sediments are highly sensitive (Fig. 5.3; Hein and Longstaffe, 1983). Sediment strength is also dependent on mineralogy. For a given normal stress, a soil rich in montmorillonite has lower undrained strength than one of illite and one of kaolinite, which in turn is weaker than a quartz-rich sediment (Das, 1983).

As discussed in Chapter 3, quick clays are normally thought of as subaerially exposed glacial marine clays that have been leached of saline pore water through normal weathering processes. In 1976, Richards was the first to report submarine quick clays, in relatively shallow water depths in Oslofjord. Quick clays completely lose their strength if even minor movements or loading takes place. In such cases, sliding may occur if: (1) the cumulative strains anywhere in the deposit are above some threshold value for failure; or (2) the external dynamic forces are high enough to initiate instability (Karlsrud, 1982). Submarine quick clays generally have remoulded shear strengths less than 0.7 kPa, have sensitivities greater than 16, can be poured in the remoulded state, and have Atterberg limits distinctly less than the nonquick clays that have lower sensitivities (Richards, 1976). The principal difference between submarine quick clays and uplifted quick marine clays is that the former have pore-water salinities near open ocean values.

A sediment mass with high excess pore pressure will fail at a critical depth, Z, as determined by:

$$Z = \tau (\gamma \sin \alpha)^{-1} + Zig \tan \emptyset (\tan \alpha)^{-1}$$
(Mandl and Crans, 1981) (5.12)

where Zig is the depth where the effective overburden stress, σ', deviates from the hydrostatic pore pressure gradient. Generally, $Z \approx (\sin \alpha)^{-1}$ where the critical thickness will be reached if the increase in shear stress attributable to high rates of deposition cannot be compensated instantly by a corresponding increase of cohesion and internal friction (Schwarz, 1982). Therefore, sediment failure may also be a problem of critical sedimentation rate versus the increase in excess pore pressure.

The drainage condition during shear failure is

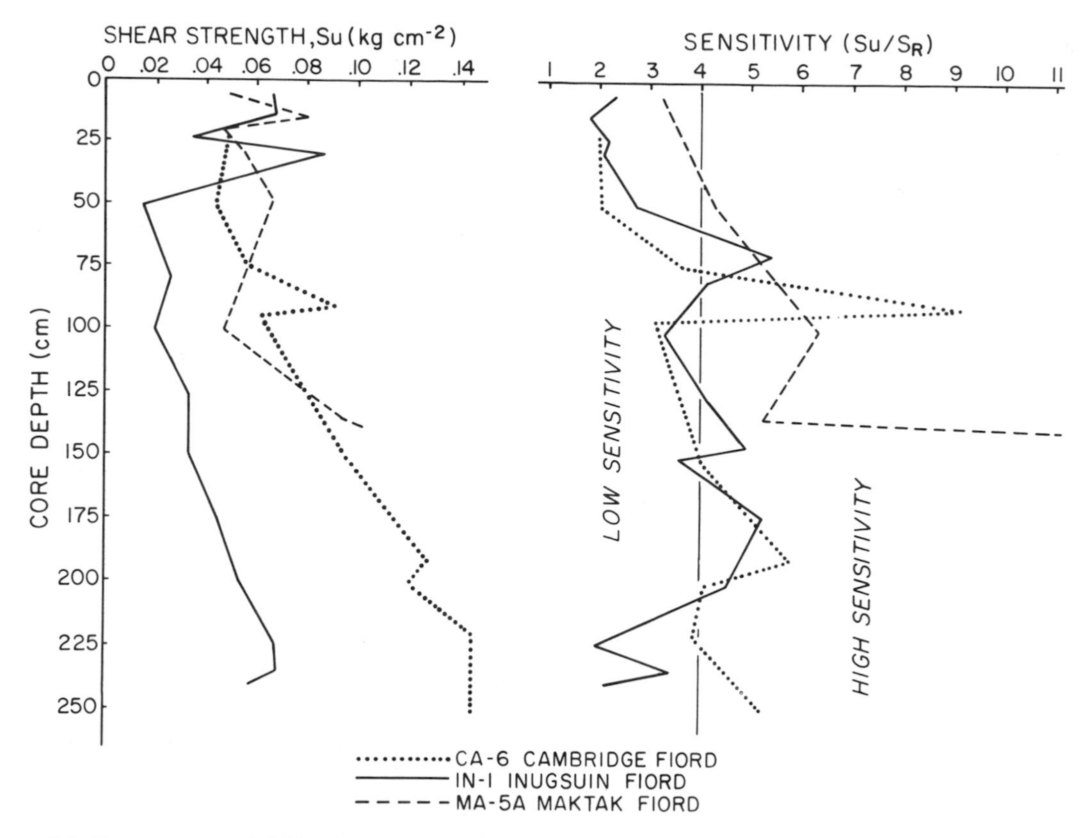

FIGURE 5.3. Down-core variability in shear strength, Su, and sensitivity (Su/S_R) for three Lehigh cores from Maktak, Inugsuin, and Cambridge Fiords, Baffin Island (data from Hein and Longstaffe, 1983). For associated Atterberg limits of core IN-1 cf. Fig. 5.1. IN-1: D = 160 m, x = 5 km; CA-6: D = 640 m, x = 75 km; MA-5A: D = 575 m; x = 33 km.

another important factor in the strength of any sediment (Morgenstern, 1967). For instance, the slope inclination at which slumping occurs is strongly dependent upon whether the initiating process induces a drained or undrained slump. The relation between the slope angle at failure under drained conditions and the properties of sediment may be given as:

$$\tan \alpha = \tan \varnothing + c(\gamma' Z)^{-1} \sec^2 \alpha$$
(Morgenstern, 1967) (5.13)

where α denotes the inclination of the seafloor slope to the horizontal and Z is the height of the sediment participating in the slump. Given a normally consolidated clay or an uncemented sand or silt, then $\tan \alpha \approx \tan \varnothing$ (Morgenstern, 1967). Slumping under drained conditions is caused commonly by depositional oversteeping such as growth of a coarse-grained delta over the steeply dipping fjord walls.

Undrained slumps may be caused by stresses induced during rapid deposition, or erosion at the base of a slope, or dynamic wave loading. For undrained slumping under a horizontal earthquake acceleration, a_e, then the undrained sediment strength may be determined by:

$$Su(\gamma' Z)^{-1} = 0.5 \sin 2\alpha + a_e (\gamma s/\gamma')\cos^2\alpha$$
(Morgenstern, 1967) (5.14)

where Su denotes the undrained strength mobilized at failure.

5.2 Release Mechanisms

Four factors control slope failures in fjords: (1) topography (Chapter 2); (2) supply of material (Chapters 3 and 4); (3) physical properties of the sediment (Section 5.1); and (4) the releasing mechanism. Release mechanisms are those natural or anthropogenic events that change one or more of the parameters that affect the strength of a sediment mass beyond some critical limit (i.e., cohesion, overburden stress, excess pore pressure, angle of internal friction). Release mechanisms include sediment loading, earthquakes, waves, sea-level fluctuations, changes in mass sediment properties, and the activities of man.

Loading may result from: (a) long-term high rates of sedimentation, although usually combined with an additional trigger event; (b) short-term heavy sediment supply to delta areas, including exceptional storm floods and jökolhlaups; (c) sediment overloading by advancing delta foresets onto underconsolidated prodelta clays; (d) oversteepening of a depositional slope up to the critical angle of failure; (e) advancing tidewater glacier; and (f) impact of sea ice or an iceberg onto sloping seafloor environments. With the partial exception of (e) and (f), these mechanisms are mostly responsible for failure in the areas offshore of the major fjord-head river deltas. Side-entry river deltas tend to be coarser grained deposits that form on the steep fjord sidewall slopes. These deltas advance with slopes near the angle of repose of the coarse-grained material. Sooner or later the shearing stresses at the base of the oversteepened material will become equal to the shearing resistance, whereupon the patch slides and the normal slope angle, $\tan \emptyset = \tan \alpha$, is re-established (Terzaghi, 1956).

In undrained deposits a critical pore pressure may be developed because of earthquakes shocks, high sedimentation rates, loss of buoyancy during an extreme low tide or the sudden change of hydrostatic pressure by wave action, and development of free gas within the sediment (Schwarz, 1982). On deltas, the overpressuring of the pore fluid permits the formation of very gently dipping slip planes (Mandl and Crans, 1981). Sudden pore pressure disequilibrium within a sediment mass composed of cohesionless grains can also result in sediment liquefaction: that is, the sudden loss of sediment strength associated with the upward movement of pore fluid, temporarily allowing sediment particles to flow. During the process of liquefaction, the sediment tends to behave as a liquid (Newtonian). During the 1964 Alaska earthquake, liquefaction-type slides incorporated sand layers exceeding 50 m in thickness (Andresen and Bjerrum, 1967).

Earthquakes are also responsible for two types of giant waves in fjords: (1) tsunamis, and (2) violent seiches. Both may result in subaerial destruction and submarine slides. The first type is a result of crustal adjustment under the oceanic surface whereby large volumes of ocean are displaced and a series of giant waves form. Tsunamis travel at high speeds (i.e., 900 km h^{-1}) with wavelengths between 100 and 400 km and periods of 10 to 60 minutes. Because their wavelengths are greater than ocean depth, they travel as shallow water waves. The nearshore effects may simply result in the gentle rise and fall of the water level or in floodlike currents (Thomson, 1981). The destructive affect of a tsunamis will depend on: (1) the shoreline proximity and "line of sight" since tsunamis are highly directional; (2) the state of the tide (more destructive at high tide); (3) natural oscillations (i.e., the matching of the natural period of the fjord basin with the tsunami's period); and (4) shoaling. If the slope of the shore is gradual over many kilometers, the wave power will be dissipated. If the shore is steep and abrupt, the water height will be that of the breaking wave. If the fjord funnels the waves, the wave height may grow. Figure 5.4A gives the water height above higher high water along the fjord coast of Vancouver Island after the 1964 Alaska Tsunamis had traveled 1700 km from the epicenter in Unakwit Inlet, Alaska. Wave amplification was strongest in Alberni Inlet where the tsunami's wave height grew to 7 m at the fjord head compared with 0.6 m at the fjord mouth (Fig. 5.4B). Closer to the epicenter in Alaska, violent seiches were produced in fjords when the natural period of the basin approximated that of the high-acceleration horizontal "Airy" earthquake phases (Goldthwait, 1968).

Giant waves have also been produced in fjords after the sudden dislodging of subaerial debris into the basin water resulting in bow wave formation. For instance in 1958, 30×10^6 m^3 of rock plunged into Gilbert Inlet, Alaska, producing a giant wave whose waters reached 543 m above sea level and traveled down-fjord at between 156 and 209 km h^{-1} (Fig. 5.5; Miller, 1960). Other examples include Tafjord, Langfjord, and Årdalsfjord, Norway, and Lituya Bay, Alaska, (Table 5.1). Large waves may similarly be generated during subaqueous landslide activity in fjords (Kitimat Inlet, B.C.; Murty, 1979).

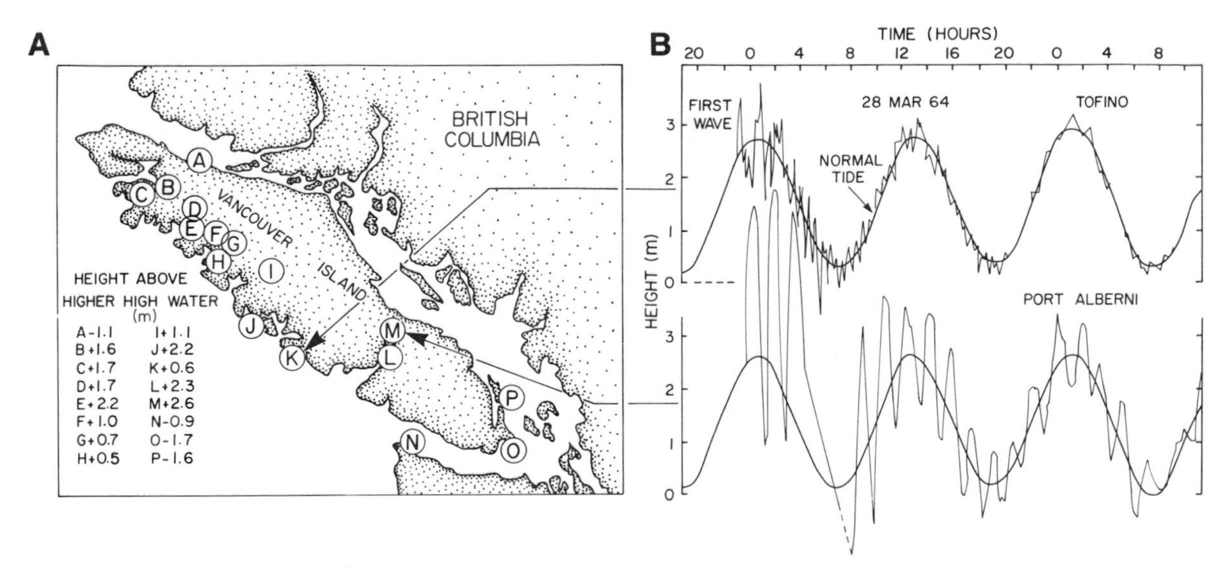

FIGURE 5.4. (A) Locations in SW British Columbia that reported significant tsunami activity in 1964. Numbers give height in meters of maximum wave crest above HHW, large tide. (B) Tide-gauge record showing the effect of amplication of the 1964 tsunami for Port Alberni, at the head of Alberni Inlet, compared with Tofino, near the mouth of Alberni Inlet (after Thomson, 1981).

The consequences of waves on slope stability result from the pressure changes on the seafloor associated with a wave's passage. The waves impose an oscillatory motion, which on a sloping ground leads to a mass transfer of sediment down-slope. In underconsolidated soft sediment, repeated reversals of shear strain in the sediment may also cause remoulding of the sediment and a reduction in shear strength (Henkel, 1972). In granular material, wave-induced cyclic shear stresses could cause liquefaction of the seafloor sediment (Clukey et al., 1980). The collapse of

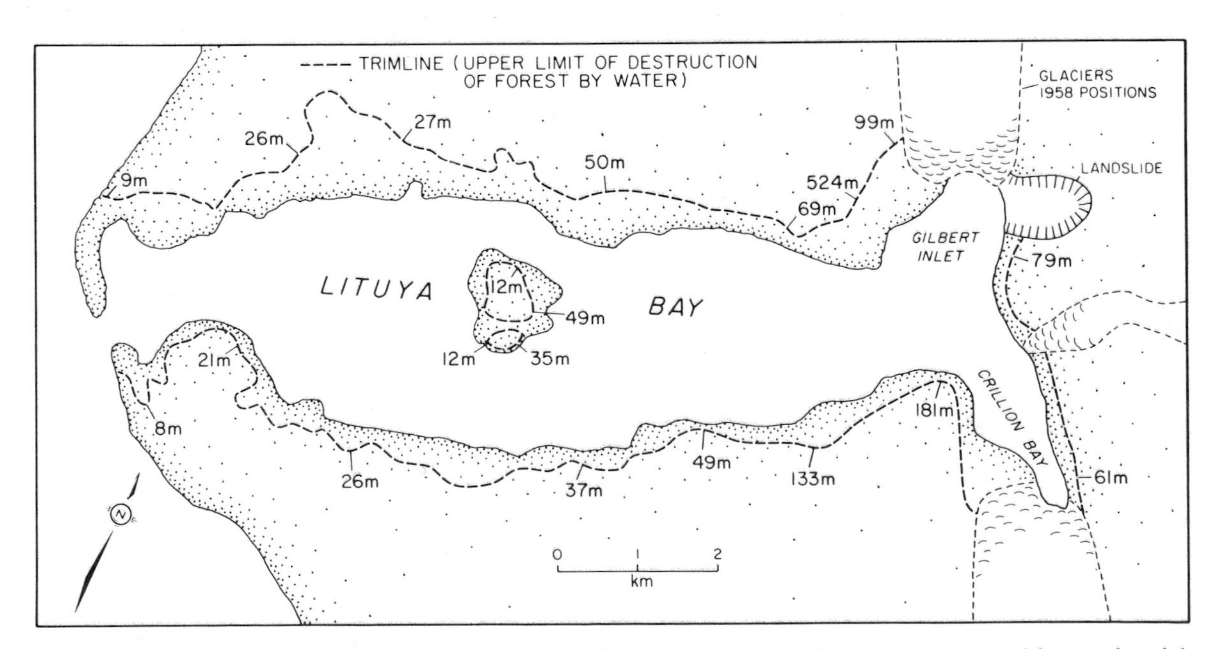

FIGURE 5.5. A trimline of destruction outlines the effect of giant waves that were generated by a subaerial landslide in 1958, Lituya Bay, Alaska (after Miller, 1960).

TABLE 5.1. Examples of subaqueous slope failures in fjords.

System	Comments	References
Norway		
Trondheimsfjorden	1950 rotational slide: one of six where shoreline damage occurred. 1888 slide: retrogressive liquefaction of mud and sand on a slope of 8 to 10°; associated giant waves caused further slides.	Skaven-Haug (1955) Andersen and Bjerrum (1967) Bjerrum (1971)
Follafjord	1952: a single block glide associated with liquefaction moved 10^6 m^3 at a velocity of 2.8 m s^{-1}.	Terzaghi (1956)
Orkdalsfjord	1930: 10^8 m^3 was released, due to a combination of earthquake and sediment loading, at a velocity of 6.9 m s^{-1} at 3 km and 2.8 m s^{-1} at 18 km from the source; upper 10 to 30 m of seafloor disturbed. Failure styles include: (1) liquefaction; (2) slide and turbidity current; and (3) debris flow.	1. Terzaghi (1956) 2. Menard (1964) 3. Andresen and Bjerrum (1967) Bjerrum (1971) Karlsrud and By (1981)
Vaerdalen	1894: mudflow due to liquefaction was transported at least 11 km at 1.7 m s^{-1}.	Terzaghi (1956)
Hardangerfjord	Slides lead to turbidity currents; spilt into deep basins from more than one source; typical turbidite sequences.	Holtedahl (1965, 1975)
Tafjord	1934: subaerial landslide due to ice fracturing resulted in 10^6 m^3 of debris entering the fjord; the resulting giant waves reached 62 m and moved at 6.1 to 11.9 m s^{-1}; 44 people were killed.	Miller (1960)
Langfjord	Heavy rains triggered a subaerial debris flow and when the 10^7 m^3 of debris entered the fjord giant waves were produced.	Miller (1960)
Fjaerlandsfjord	A slide resulting in debris flows & turbidity currents moved 10^7 m^3 of material over a slope of 25°.	I. Aarseth (personal communication, 1983)
Nordfjord	Since 1967, 10^7 m^3 volume of seafloor slid via slide and turbidity currents producing several seafloor channels up to 6 m deep.	I. Aarseth (personal communication, 1983)
Årdalsfjord	A subaerial slide of 10^5 m^3 volume produced 5 to 10 m high giant waves.	I. Aarseth (personal communication, 1983)
Eidfjord	Cores recovered from the basin floor contained turbidities.	Flaate and Janbu (1975)
Finnvika	1940: slide was followed by a series of secondary slides days later and some 5 km from the original slide.	Bjerrum (1971)
Alaska		
Port Valdez	1964: Alaska earthquake caused a slide of 10^8 m^3 to move down slopes greater than 10°.	Coulter and Migliaccio (1966)
Lituya Bay	1958: 8.3 magnitude earthquake set off a 10^7 m^3 rockslide; was 1 of 5 historical events that have caused giant waves in the fjord, reaching heights of 543 m that moved at speeds of 43 to 58 m s^{-1}; subaqueous debris flows and turbidity currents were formed: 2 people killed.	Miller (1960) Jordan (1962) Post (1974)

TABLE 5.1. Continued

System	Comments	References
Alaska Continued		
Cook Inlet	Earthquake induced fissuring and faulting of mud flats removed a volume of 10^8 m^3.	Jordan (1962)
Yakutat	1958: Alaska earthquake caused a retrogressive slide involving long distance transport. 3 persons died.	Jordan (1962)
Queen Inlet	Slide induced turbidity currents caused the formation of sinuous trenches between 3.7 and 12.7 m deep; floored with well-sorted sand.	Hoskin and Burrell (1972)
Glacier Bay fjords	A variety of subaqueous slope failures near the fronts of tidewater glaciers.	Powell (1980)
British Columbia		
Howe Sound	1955: side-entry delta failed due to oversteepening and overloading; coarse-grained deposit on slopes of 27° developed erosional chutes and channels with the action of debris flows.	Terzaghi (1956) Prior et al. (1981a, b) Mandl and Crans (1981)
Alberni Inlet	1964: Alaska tsunamis caused extensive coastal erosion.	Thomson (1981)
Jervis Inlet	Slumping occurs along the steep inlet walls.	Macdonald and Murray (1973)
Kitimat Arm	1975: two distinct slumps coalesced to form a continuous deposit 5 km by 2 km; the slide developed debris flows and turbidity currents. Involved 60 × 10^6 m^3 of sediment	Luternauer and Swan (1978) Bornhold (1983a, b) Prior et al. (1982a, b, 1983)
North Bentinck Arm	Rapid sedimentation-induced slides developed chutes up to 20 m deep that migrate across the prodelta slopes.	Kostaschuk and McCann (1983)
Rupert Inlet	Surge-type turbidity currents with frequency of 2–5 days result from discharge of mine trailings; hydraulic jump occurs when supercritical flow meets rapid change in slope.	Hay et al. (1982, 1983a, b)
Knight Inlet	Fjord walls are zones of frequent sediment failure of various forms. Prodelta area crossed with coalescing channels due to overloading resulting in development of subaqueous flows.	Farrow et al. (1983) Syvitski and Farrow (1983)
Bute Inlet	Rotational slides of retrogressive nature have developed due to overloading on prodelta.	Syvitski and Farrow (1983)
Alice Arm	A natural turbidity current was observed flowing at 0.4 m s^{-1} for approximately 1 h.	Bornhold (1983a, b)
Baffin Island		
Pangnirtung Fiord	Slumps leading to turbidity currents have eroded a single deep trench whose width averages 200 m, depth of 15 m, and side slopes of 8° with a miximum channel slope of 0.032.	Gilbert (1978, 1983)

TABLE 5.1. Continued

System	Comments	References
Baffin Island Continued		
Coronation Fiord North Pangnirtung Fiord Maktak Fiord	Due to oversteepening and overloading, slumping off the side walls and sandur front results in a wide variety of slides, slumps, and mass flows including turbidity currents.	Gilbert (1982a, b, 1983)
McBeth Fiord Tingin Fiord Itirbilung Fiord Inugsiun Fiord Clark Fiord Cambridge Fiord Sunneshine Fiord	Creep, slides, slumps, mudflows, and turbidity currents have resulted in megachannels, coalescing channels on the sandur front, contorted bedding, and gravity flow deposits and bedding.	Syvitski and Blakeney (1983) Syvitski (1984a)
Spitsbergen		
Kongsfjorden	Creep flows and slumps have resulted from overloading and oversteepening near ice fronts.	Elverhöi et al. (1983)
Hornsundfjorden	Rockslides and avalanches common.	Jahn (1967)
Adventfjord	Liquefaction and gravity flows due to overloading and oversteepening on delta front slopes 12 to 16°.	Prior et al. (1981a, b)
New Zealand		
Milford Sound	Failure on the Cleddau delta front, due to overloading and oversteepening, results in delta front area as a zone of sediment bypassing.	Brodie (1962)

bar front sediment on the Adventfjord delta, Spitsbergen, is thought to result from wave-induced cyclic loading (Prior et al., 1981b). Such a hypothesis is of special relevance to most fjord-head deltas, where isostatic emergence would continually expose unreworked prodelta sediment to the influence of waves. Giant waves, including those generated by submarine landslides, may in turn generate new subaqueous slope failures (e.g., Port Valdez, Alaska: Coulter and Migliaccio, 1966).

The emergence of marine nearshore environments, such as fjord deltas, is also associated with a loss of buoyancy of the sediment mass and an increase in the effective load upon a potential shear face (Schwarz, 1982). The isostatically exposed delta at the head of Cambridge Fiord, Baffin Island, shows a number of failures that may be attributed to this mechanism (Syvitski et al., 1983a).

With an influx of fresh water, cohesive marine sediments may be leached and shear strength reduced, possibly with microfabric breakup. In fjords, this mechanism of sediment failure is important for: (1) post-Pleistocene emerged sediments (subaerial quick clays); (2) delta front environments where foreset beds saturated with sea water in the pores become leached subsequent to a prograding salt-water–freshwater boundary (associated with delta progradation or a shift in the river mouth); and (3) the influx of fresh water into the marine sediment from beneath the seafloor under artesian pressure (e.g., Orkdalsfjord, Norway: Andresen and Bjerrum, 1967).

A common mechanism for causing subaqueous slope failures in fjords is man's direct interference. The 1955 slide of the side-entry fan delta at Woodfibre (Howe Sound, B.C.) resulted from a river channel stabilization program that allowed oversteepening and overloading to occur (Table 5.1). The disposal of mine tailings at Rupert Inlet, B.C., has resulted in surge-type turbidity currents with a frequency of 2 to 5 days (Table 5.1). The loading of waste, fill, and development of port facilities on the steep fjord walls have been responsible for a number of slope failures in Norway (Terzaghi, 1956; Bjerrum, 1971).

The frequency of slope failure is controlled

by the rate of sediment deposition and the frequency of the releasing mechanism (Schwarz, 1982). Near zones of high sedimentation, such as deltas and tidewater glaciers, the rate of deposition is most important with yearly failures probable. The time of occurrence of the initial slide frequently coincides with a period of normally low stability such as with a low tide (Terzaghi, 1956; Bjerrum, 1971). Elsewhere, the slide frequency may be on the order of tens or hundreds of years, that is, based on the local return interval of earthquakes, tsunamis, and so on. Older deposits exposed by a recent slide are more highly consolidated than the removed sequences and, consequently, have a higher resistance to shear. Under similar stress conditions no further slope failures will occur until a new sedimentary cover of critical thickness develops (Schwarz, 1982).

5.3 Mass Transport Processes

All forms of mass transport processes as defined in Table 5.2 are believed to be operational in fjords (Table 5.1). These include mass transport with an elastic mechanical behavior (rockfalls, slides, slumps, creep), transport with plastic deformation (debris and mud flows, and some grain flows), and viscous sediment flows (grain flows, fluidized and liquefied flows, turbidity currents). Prime sites of submarine failure in fjords include: (1) where sedimentation occurs over high slope "basement" topography (sidewall slopes, sills); (2) where tributary hanging valleys meet the central channel; (3) areas of active tectonic faults; (4) near areas of high sedimentation rates (such as deltas); and (5) fjord walls subject to rockfalls and avalanches.

Gravitational deformation of cohesive sedi-

TABLE 5.2. Major types of mass transport processes, their mechanical behavior, and transport and sediment support mechanisms.[1]

Mass transport processes		Mechanical behavior	Transport mechanism & sediment support
Rock fall		ELASTIC	Freefall and rolling of blocks or clasts along steep (fjord) walls.
Slide:	(a) Glide		Single block movement either tumbling, slope parallel sliding, or horizontal gliding.
	(b) Rotational slide		Block movement along a curvilinear slip face (shear surface).
	(c) Translational slide		Block movement along a slope parallel slip face.
	(d) Convolute slide (slump)		Block movement with internal folding and flow (partial plastic deformation).
Mass flow:	(a) Creep	-Plastic limit-	No visible breaks of particle contacts. Deformation is partly elastic and partly plastic.
	(b) Debris flows & Mud flows	PLASTIC (Bingham)	Shear distributed throughout the sediment mass. Strength is principally from cohesion. Additional matrix support from buoyancy. Flow is mostly laminar.
	(c) Grain flow (inertial)	-Liquid limit-	Cohesionless sediment supported by dispersive pressure in high-concentration regime.
Fluidal flow:	(a) Grain flow (viscous)	VISCOUS (Newtonian) FLUID	Cohesionless sediment supported by dispersive pressure in low-concentration regime.
	(b) Liquefied flow		Cohesionless sediment supported by an upward displacement of fluid as loosely packed structure collapses, settling into a more tightly packed framework.
	(c) Fluidized flow		Cohesionless sediment supported by the forced upward motion of escaping pore fluid.
	(d) Turbidity current		Particles are supported by fluid turbulence (autosuspension).

[1]After Nardin et al., 1979; Prior and Coleman, 1979; Schwarz, 1982.

ment always commences with macroscopically perceptible creep movement. Creep is defined as continuous yielding of the soil particles under applied stress without brittle failure. Creep of stratified sediments causes contemporaneous deformation in the form of gentle folding: an upslope area of decreased sediment thickness is balanced by an increase of thickness in a downslope alimentation area. Near the toe of a given slope, the creep displacements reach their maximum. Creep velocity depends greatly on material type and magnitude and duration of shear stress (Schwarz, 1982). The rate of creep, v_c, can be determined by:

$$v_c = [\gamma_b \cos(\alpha) \sin(2 \alpha z^2) - \cos(\alpha \tau_o z)] (4\eta)^{-1} \quad (5.15)$$

where γ_b is the buoyant unit weight of overburden, α is the slope angle, z is the distance between the creeping layer and the fixed boundary, τ_o is the shear stress below which no creep occurs and η is the viscosity (for details see Hill et al., 1982).

Creep is ubiquitous in most high-sedimentation fjords and can be found on the sidewall slopes (e.g., Knight Inlet, B.C.: Farrow et al., 1983) or on the steeper portions of the prodelta slope leading to the basin (Fig. 5.6).

Although creep may simply react to increasing overburden stress and gravity, cyclic strain from wave loading, earthquakes, or cyclic sedimentation will accelerate the creep process (Almagor and Wiseman, 1982). Such increase in strain may lead to elastic behavior in the form of breakup of feather joints in the sediment mass (Schwarz, 1982). As failure progresses, antithetic precursory faults develop (Fig. 5.7A) followed by the development of a main shear face, often with reorientation of secondary shear faces. The main shear face or glide path usually develops as a listric (curvilinear) fault. The main block slides downslope resulting in the formation of a tensional depression downslope of the scar zone, and a slide toe at the base of the slide where various fold forms and overthrust fault systems may result (Fig. 5.7B). Different types of slides are described on Table 5.2. With further available energy for deformation, and possibly the incorporation of water, the slide block may partially collapse into a convolute slide or slump where internal stratification is chaotic.

Because fjord slopes are relatively steep, acceleration of the slide mass usually results in at least partial collapse of the block system. An excellent exception can be found in McBeth Fiord, Baffin Island, where the large slide block

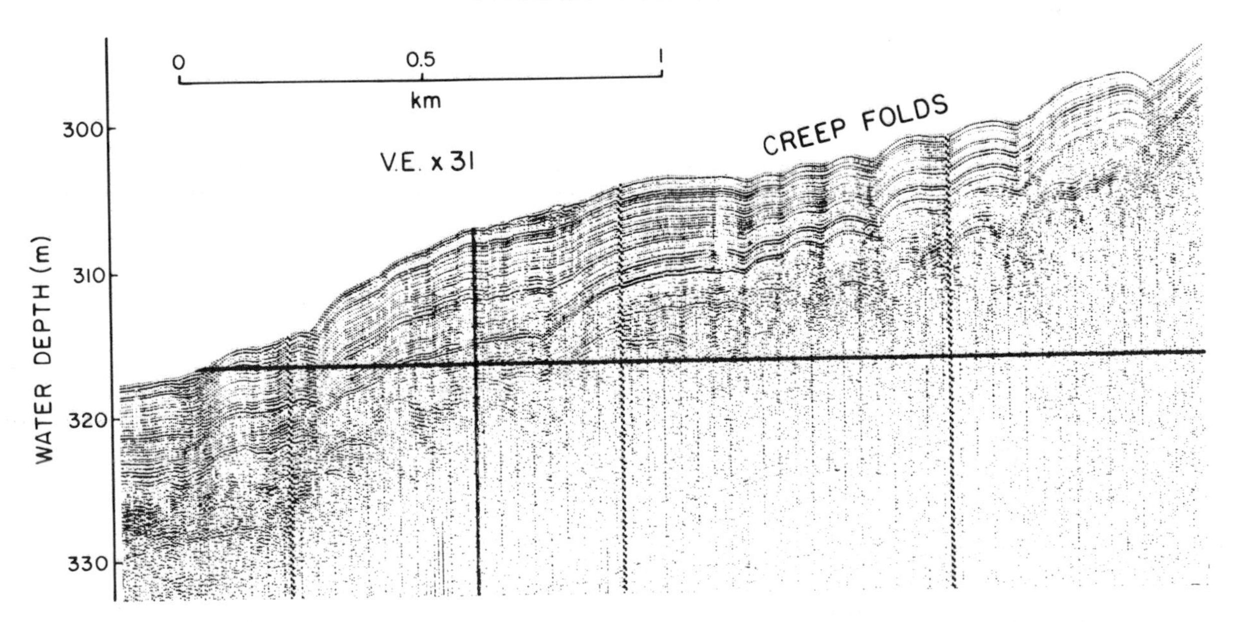

FIGURE 5.6. An example of creep folds down the prodelta slope of Tingin Fiord, Baffin Island, shown on a high-resolution reflection seismic record (30-cm vertical resolution).

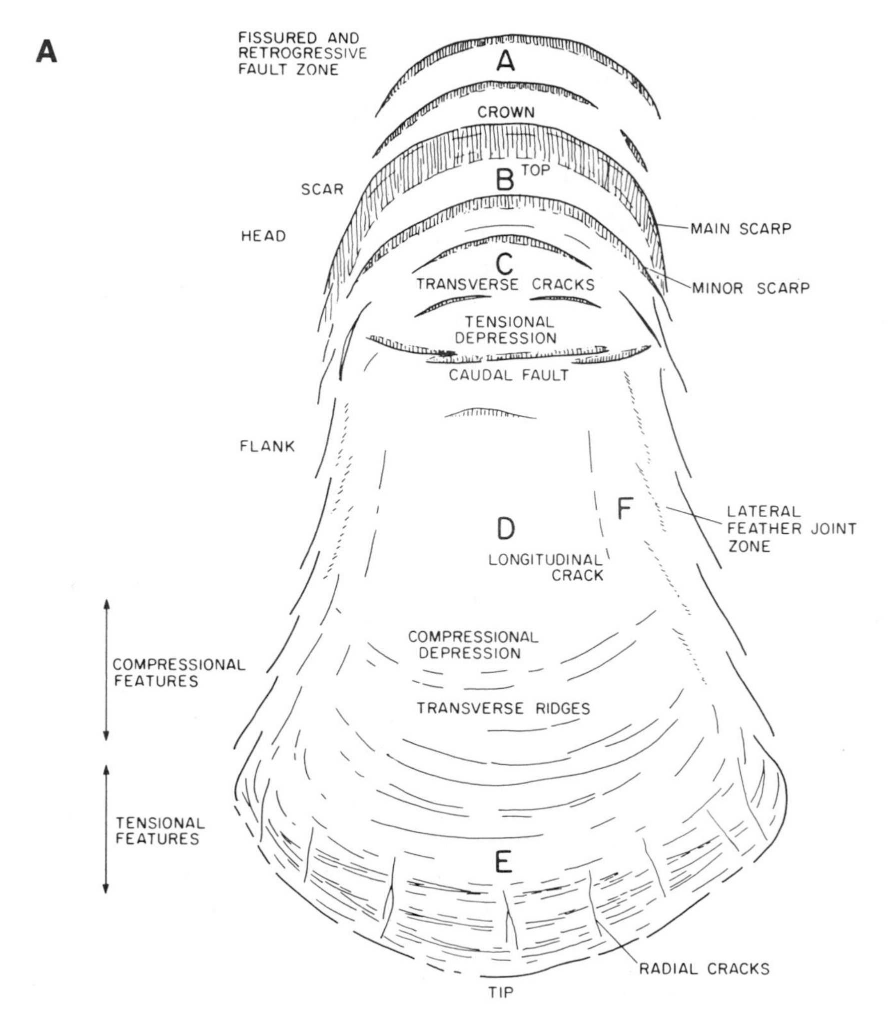
A
FISSURED AND RETROGRESSIVE FAULT ZONE
A
CROWN
SCAR
B
TOP
HEAD
MAIN SCARP
C
TRANSVERSE CRACKS
MINOR SCARP
TENSIONAL DEPRESSION
CAUDAL FAULT
FLANK
F
LATERAL FEATHER JOINT ZONE
D
LONGITUDINAL CRACK
COMPRESSIONAL DEPRESSION
COMPRESSIONAL FEATURES
TRANSVERSE RIDGES
TENSIONAL FEATURES
E
RADIAL CRACKS
TIP

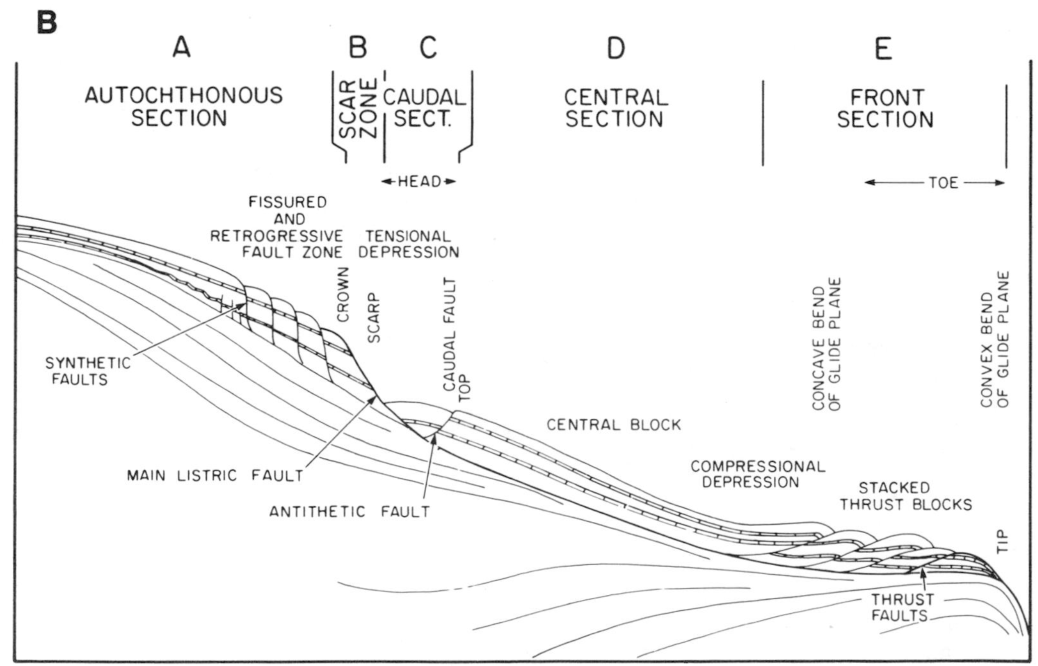
B
A
B
C
D
E
AUTOCHTHONOUS SECTION
SCAR ZONE
CAUDAL SECT.
CENTRAL SECTION
FRONT SECTION
HEAD
TOE
FISSURED AND RETROGRESSIVE FAULT ZONE
TENSIONAL DEPRESSION
CROWN
SCARP
CAUDAL FAULT
TOP
CONCAVE BEND OF GLIDE PLANE
CONVEX BEND OF GLIDE PLANE
SYNTHETIC FAULTS
CENTRAL BLOCK
MAIN LISTRIC FAULT
COMPRESSIONAL DEPRESSION
STACKED THRUST BLOCKS
ANTITHETIC FAULT
TIP
THRUST FAULTS

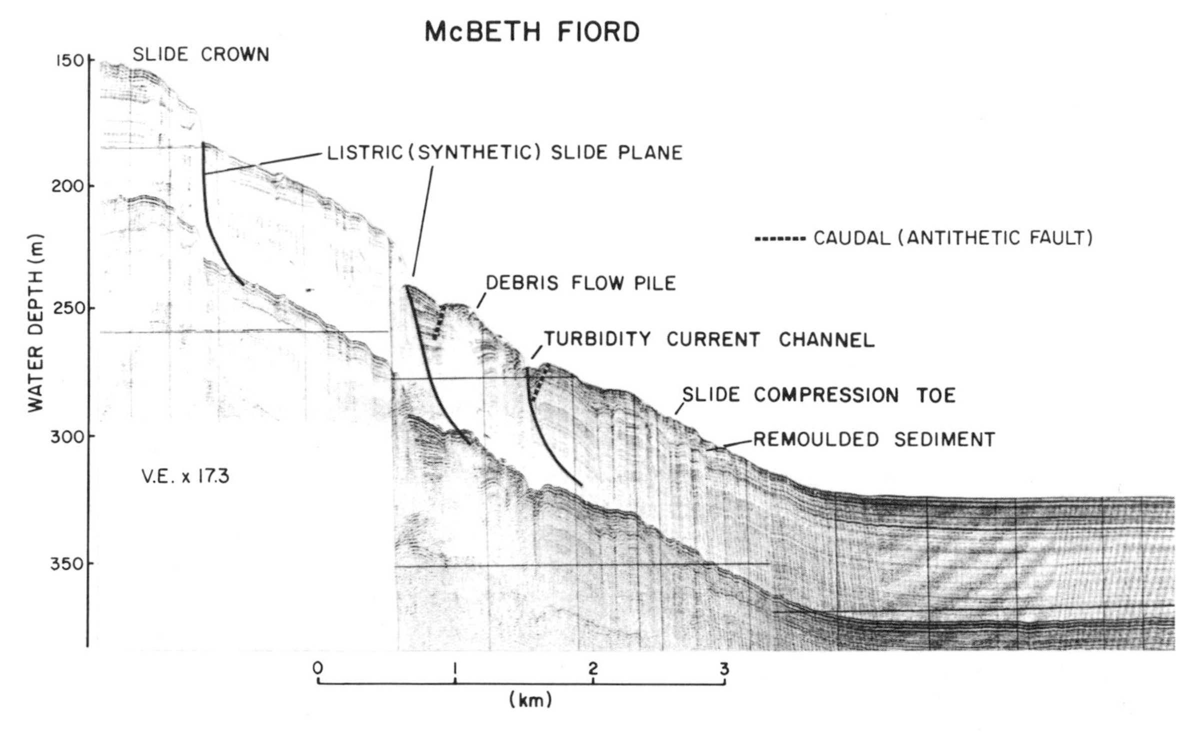

FIGURE 5.8. Internal structure and features of a slide in cross-section based on a DTS Huntec reflection seismic record from McBeth Fiord, Baffin Island. Note that the surface sediments have been partly remoulded and reworked by secondary debris flows and turbidity currents.

remained essentially intact (Fig. 5.8). The slide toe, however, underwent compression and secondary slides became remoulded and traveled farther downslope as debris flows and turbidity currents (Fig. 5.8). These secondary slides are not uncommon as the upper underconsolidated material of the slide will homogenize and develop into a gravity flow.

Morgenstern (1967) gave velocity of a slide or slump, v_s, as:

$$v_s = g\,\gamma^{-1}\,[\gamma'\sin - (\gamma'\cos\alpha^{-n})\tan\emptyset]t \qquad (5.16)$$

where g is the acceleration due to gravity, t is time, n is some number, and γ' is the effective unit weight of the slide mass. The slide velocity increases linearly, with time dependent upon the slope angle, the excess pore pressure gradient, and the density and strength of the sediment.

When the properties of the slide mass move closer to that of a plastic (i.e., Bingham material with soil yield resistance), the mass flow may develop into a debris flow where the presence of a fine-grained matrix: (1) imparts a very high viscosity with little loss of water content; and (2) supplies sufficient strength to support larger grains in the flow (Middleton and Southard, 1977). Only part of the mass may actually deform. A rigid plug may form on top of the Bingham flow for a uniform, infinitely wide flow of thickness, h, and density, ρ, down a slope with constant inclination, α. This plug will have a thickness, Tc, where:

$$Tc = K\,(\rho\sin\alpha)^{-1} \qquad (5.17)$$

(Karlsrud and Edgers, 1982)

and K is the yield strength. The steady-state velocity of the plug, v_p, is given by:

$$v_{plug} = (\eta)^{-1}(1/2)(h^2\,\rho\sin\alpha) + (K)^2\,(2\,\rho\sin\alpha)^{-1} - Kh \qquad (5.18)$$

(Karlsrud and Edgers, 1982)

FIGURE 5.7. (A) Internal structures and features of an idealized slide (A: autochthonous region, B: scar zone, C: caudal [head] section, D: central section, E: front section, and F: lateral section). (B) Internal structure and features of Kidnappers slide (Hawke Bay, N.Z.) in cross-section (after Schwarz, 1982; Lewis, 1971).

Edgers and Karlsrud (1982) developed a theoretical model useful to predict slide velocities for more laminar Newtonian fluids (i.e., linear viscous flows, grain flows, fluidized and liquefied flows). Their viscous model includes:

$$v = Fo + (v_i - Fo)e^{(t_i - t)/a} \quad (5.19)$$

$$X = X_i + Fo\,(t - t_i) + a(v_i - Fo) \cdot (1 - e^{(t_i - t)/a}) \quad (5.20)$$

where $a = (2\,\gamma' h \sin\alpha + K)(\gamma' h \sin\alpha - K)(6g\eta\,\gamma' \sin^2\alpha)^{-1}$

and $Fo = (2\,\gamma' h \sin\alpha + K)(\gamma' h \sin\alpha - K)^2 (6\eta h\,\gamma' \sin\alpha)^{-2}$

and v is the average flow velocity over slide thickness h, γ' is the effective unit weight of soil, η is the flow viscosity, t is time, and subscript i is the initial point indicator. The application of this model to numerous fjord failures indicates that viscous flows may theoretically attain very large runout distances and velocities (Edgers and Karlsrud, 1982).

Viscous flows tend to develop when the initial slide involves large quantities of saturated loosely deposited fine sand and silt. At the point of failure, particles begin to rearrange themselves into a dense configuration, and during rearrangement water is expelled (Bjerrum, 1971). Once the slide changes to a viscous liquid and begins to flow downslope, the face of the slide scar is left unsupported. This face will in turn fail and the slide will develop retrogressively slice by slice, widening the scar with each slice (Fig. 5.9A). Thus the dimensions of the final slide may be disproportionately large compared with the size of the initial slide (Bjerrum, 1971). The retrogressive motion can occur at rates of fifty to several thousand meters per hour (Andresen and Bjerrum, 1967). The sediment flow will not cease until the slide scar reaches a more dense sand deposit, or another material type, or until the rate of sediment supply to the viscous flow is greater than can be transported away. For instance, if there is a restricted exit or channel, the sediment may begin to settle out and compact within the scar area. The 1888 slide in Trondheimfjord, Norway, is thought to have resulted from this process (Andresen and Bjerrum, 1967).

Spontaneous liquefaction is another style of mass movement generated in cohesionless metastable sediments and usually initiated by a shock wave such as an earthquake. The grain size favored for this failure type is coarse silt to fine sand. The process sees the sudden build up of high pore pressures, generated by the initial sediment compression, propagated along the sea bottom in all directions to neighboring deposits of similar metastable material (Fig. 5.9B; Terzaghi, 1956). These slides may propagate at rates of 100 km h^{-1}, depending on the topography (Andresen and Bjerrum, 1967). An example of this style of failure is the 1930 slide in Orkdalsfjord, Norway, where the initial slide material was composed of nonplastic silt having a porosity of 49%. The sediment was also thought to be under artesian pressure. Another example is from the outer basin complex of McBeth Fiord, Baffin Island (Fig. 5.9C). Seismo-stratigraphic evidence includes: (1) the local disappearance of horizontal reflectors within the affected 10 m thick layer of sandy silt; and (2) the somewhat rougher seafloor compared with more deeply buried sediment layers.

Often viscous gravity flows become supercritical and turbulent and develop into turbidity currents. In many cases, turbidity currents flow within channels caused by erosion at the base of their flow and/or channels formed during the initial slide process. Assuming steady-state conditions (i.e., channelized flow) and a downstream momentum balance between friction and the downslope component of the buoyancy, the Froude number is:

$$(Fr)^2 = 2\,(\sin\alpha)(ff)^{-1} \quad (5.21)$$

(after Hay et al., 1983a)

where tan α is the channel slope, and ff is a friction coefficient incorporating the effects of friction both at the bed and at the interface between the gravity flow and the overlying water. For $Fr > 1$, the turbidity current entrains increasingly more of the overlying sea water, thus diluting the sediment concentration and slowing the flow. The average velocity, v, of the turbidity current for the main body of the flow is approximated by:

$$v^2 = (\rho_s - \rho_w)(\rho_w)^{-1}\,C'\,g\,h\,(\sin\alpha)(f)^{-1} \cdot (1 + a_f)^{-1} \quad (5.22)$$

(Harleman, 1961)

where f is the Darcy Weisbach friction coefficient, a_f is a friction term accounting for interfacial shear at the top of the current, and C′ is the concentration of the turbulent slurry.

High-density sandy currents are relatively thin

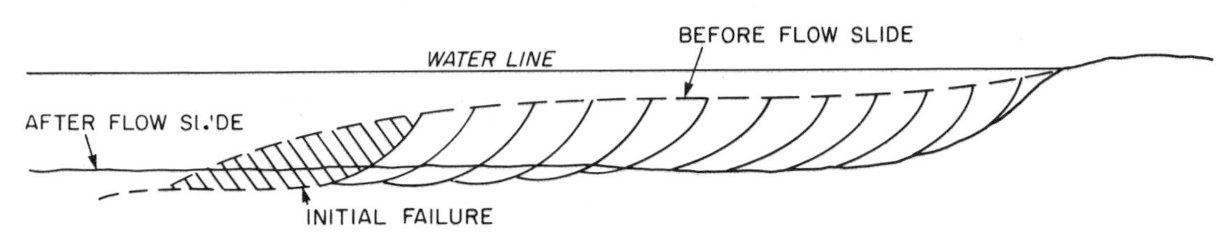

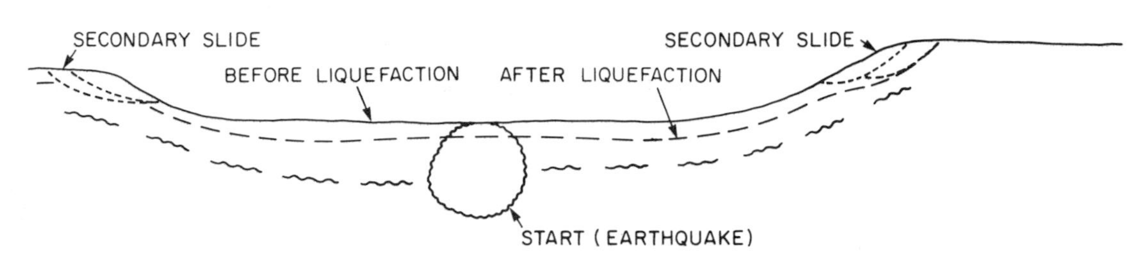

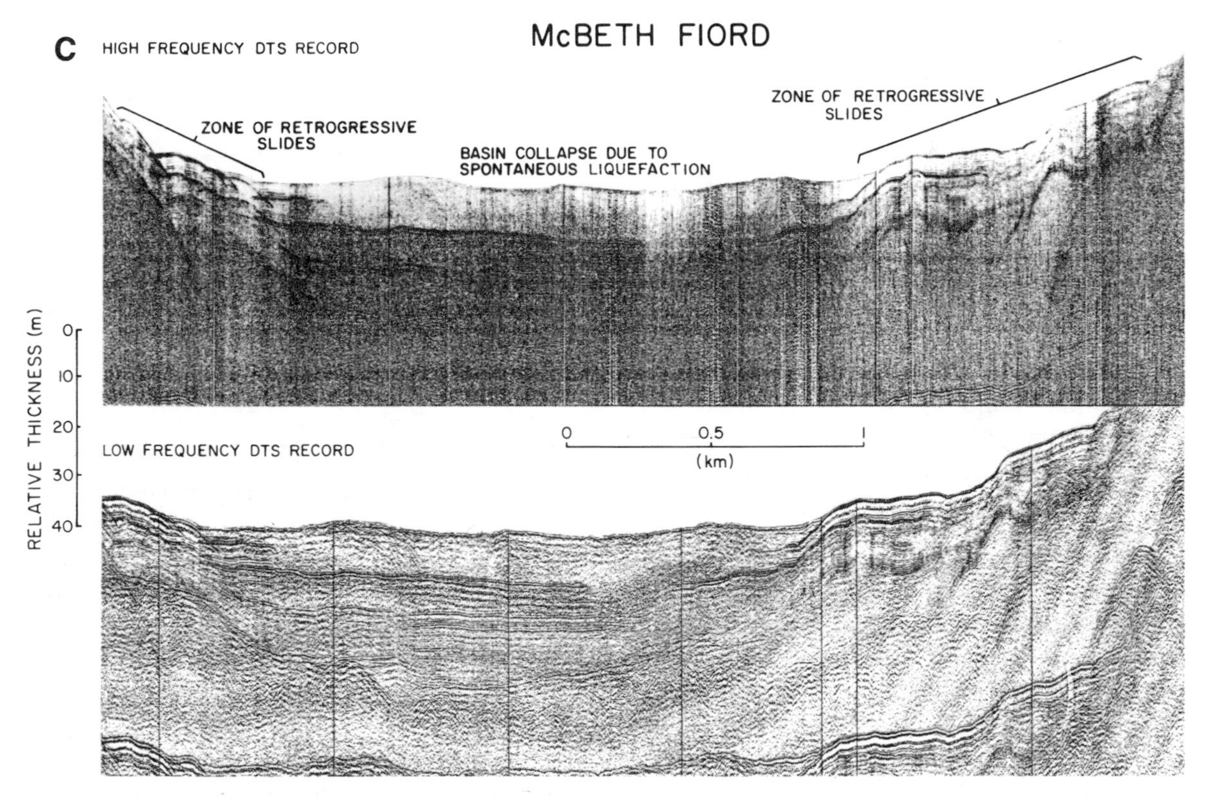

FIGURE 5.9. Idealized examples of: (A) retrogressive flow slide, and (B) spontaneous liquefaction failure (after Andresen and Bjerrum, 1967). (C) An example of progressive liquefaction slide in McBeth Fiord, Baffin Island based on a DTS Huntec seismic reflection record. Water depth ≈500 m.

and fast, whereas low-density muddy currents are relatively thick and slow (Bowen et al., 1984). Thus turbidity flows that carry coarse sediment may be confined within the channel walls and will not overtop the channel levees. If a low-density turbidity flow spills over its channel, part of the flow will be stripped away from the main body as a result of Coriolis deflection of the overbank flow. The overbank flow will undergo rapid flow spreading and sediment deposition. The channelized flow will also undergo a reduction in both the turbidity current velocity and concentration. The velocity will also be reduced with decreasing slope. As a result, the channel cross-section will decrease downslope with the decrease in the turbidity current discharge that results from overspill and deposition.

A turbidity current surge was detected in a leveed submarine channel in Rupert Inlet, British Columbia, using sounders to detect acoustic backscatter (Hay et al., 1982). Initially the turbidity current surge remained channelized, but within minutes the current spilled over the channel and levees. The event was no longer detectable after 1.5 h.

Plastic and high viscosity flows may change into a turbulent flow at major changes in the slope of the seafloor. Thus, it is not uncommon for a debris flow or a grain flow to change into a turbidity current during the downslope course of the flow. Also, muddy gravity flows that carry a substantial load of sand have properties in common with both turbidity currents and debris flows. There, the coarser grains would be supported by both fluid turbulence as well as the cohesive strength of the muddy matrix (Middleton and Southard, 1977).

During the erosive history of a turbidity current, channel walls may be undercut, initiating a new series of retrogressively developing slides. If these secondary slides add further volumes of liquid sand to the flow, the flow may be rejuvenated. If the undercutting results in the addition of plastic mud and larger mud blocks, the turbulent flow characteristic may regress to that of a debris flow or a more viscous gravity flow. When a sandy turbidity current leaves the confines of the channel walls, such as when it reaches the flat of a fjord basin, the flow slows and spreads so thinly that the sand will be deposited as a firm sand deposit (Bjerrum, 1971). The time required for this deposition increases with decreasing permeability and therefore decreasing grain size (Terzaghi, 1956). Thus, the distance the sediment mass can travel during the transition from the liquid to the solid state increases with decreasing grain size.

The runout distance of gravity flows is greater if, for at least part of the distance, the sediment mass is channelized (Edgers and Karlsrud, 1982). The runout distance, normalized with respect to the difference in water depth along the flow path, increases with the volume of sediment mass involved in the initial failure (Karlsrud and By, 1981). Variations may be caused by a slow rate of sediment feeding to the flow (i.e., slow slide retrogression), obstructions along the flow path and the tortuosity of the flow channel.

Slide volumes in fjords range from less than 10^6 m^3 to 10^9 m^3 of sediment material (Table 5.1). Slopes range from greater than 30° on the fjord walls to less than 0.1° on the flat basin floors. Slide velocities, usually calculated from the rate of rupture of underwater cables along the fjord length, range from 0.4 m s^{-1} to 6.9 m s^{-1} (Table 5.1). Runout distances are wide ranging but may exceed 60 km (Syvitski and Farrow, 1983).

5.4 The Products of Subaqueous Slope Failure

The effects of subaqueous slope failures in fjords are described in terms of seafloor morphology, sediment redistribution, and contribution to the sediment column. Examples are subdivided into three types: (1) subaqueous failures near fjord-head deltas; (2) sidewall slope failures; and (3) sediment gravity flows that generally deposit sediment in the deepest portions of the basins.

5.4.1 Fjord-Head Delta Failures

Below the low-low water line, the foresets of a fjord-head delta may dip at angles between 5 and 30° to depths between 10 and 50 m. Below this, the prodelta dips at angles between 0.1 and 5° until the flat basin floor is reached, usually at depths between 100 and 1000 m. Failure styles can be subdivided into: (1) seasonal or semi-continuous failures, and (2) discontinuous failures. The first group usually develops a well-integrated cause-and-effect response with numerous small-scale (10^3 to 10^6 m^3) failures continually adjusting to maintain maximum slope stability. Delta fronts with seasonal failure seldom experience a major catastrophe. The second group occurs at intervals of decades or centuries as solitary events or as a series of near-contemporaneous events (a number of slides within a few years). These discontinuous failures are commonly triggered by earthquakes and release anywhere between 10^6 and 10^9 m^3 of sediment.

Perhaps the most-studied discontinuous failure is the slide complex off the Kitimat Delta (Kitimat Arm, B.C.; Luternauer and Swan, 1978; Prior et al., 1982a, b, 1983; Bornhold, 1983a, b). The slide involved more than 10^8 m^3 of material, caused a 26-m deepening of the seafloor at the slide head, and resulted in the downslope deposition on the fjord floor of as much as 30 m. The slide morphology is complex (Fig. 5.10) because it results from a series of closely spaced events (i.e., 1971, 1974, and 1975). The slide geometry consists of three main segments: (1) a short and steep head on the delta front (4–7°); (2) an intermediate blocky zone with 1 to 2° slopes; and (3) an elongate depositional area more than 3.2 km long with slopes < 1° and marked by a steep toe.

Six distinctive types of seafloor morphology have been recognized (Prior et al., 1982a, b). Type I, the delta front slope, is the source area of rotational retrogressive sliding. It is highly irregular, with numerous low arcuate scarps and shallow troughs. The channel margins are marked with slightly curved tensional cracks. Some of the channels still contained blocks of relatively undisturbed sediment. Type II, the slide constriction, is an area where the 1975 slide moved down-fjord through a relatively narrow neck. Ridges, subparallel to the direction of sediment movement, indicate compression of sediment as it moved downslope and was constricted. The slide area here is superelevated above the undisturbed seafloor.

The lateral margins of the slide, type III, are well defined against the intact seafloor: very irregular and hummocky morphology contrasts

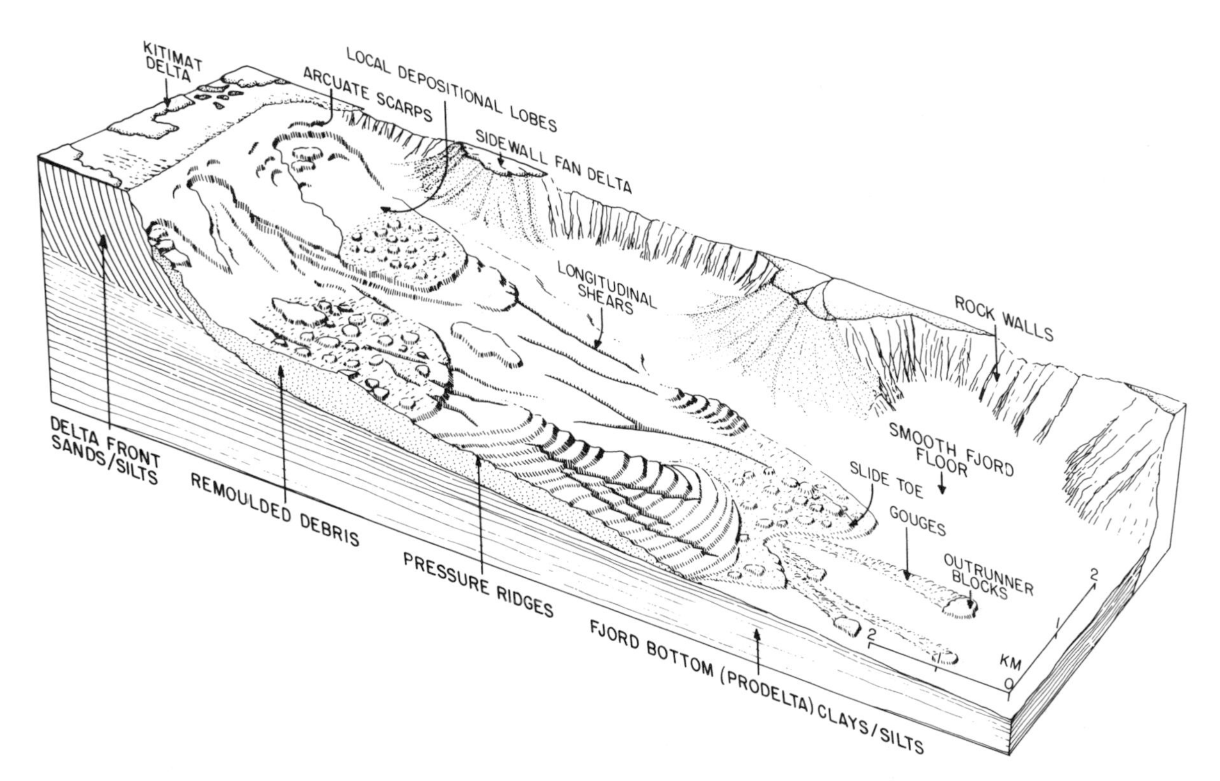

FIGURE 5.10. Schematic diagram of the 1975 and earlier slides off the Kitimat delta, British Columbia, showing the diverse surface morphology (based on high-resolution sidescan data: after Prior et al., 1983).

with smooth undisturbed sediment. At the headward region of the slide, the margins are scalloped, possibly as a result of tensional tearing. Type IV, the longitudinal shear zone, results in elongate linear patterns of low scarps arranged parallel to the slide axis. The area represents longitudinal boundary shears marking the edges of different displacement units and their differential movement within the slide mass.

Type V, pressure ridges, can be found near the slide toe with a surface relief of 1 to 5 m and widths of 5 to 15 m. Reduced motion near the slide margins, accompanied by continued sediment supply from farther upslope, gave rise to the arcuate banding (Fig. 5.10). Deformation of the ridge pattern by continued movement near the center, appears to have caused tensional stresses along the length of each ridge. The otherwise rounded crests appear somewhat segmented. The pressure ridges were deposited over and against the intact marine clays.

Type VI, the area of the slide terminus, consists of randomly arranged blocks, some > 50 m in diameter. The blocks are closely spaced and have no distinct spatial arrangement. Individual blocks appear relatively thin, rising 2 to 5 m above the seafloor. Some of these blocks have outrun the main slide mass, some by 200 m: their paths are recognized by indistinct shallow erosional swaths bound by marginal scarps 1 m high. The width of each gorge is consistent with its block dimensions. These outrunners reflect the outward extent of the slide.

There are many documented examples of seasonal or semicontinuous failures along fjord-head deltas (e.g., Spitsbergen: Prior et al., 1981a, b; Baffin Island: Gilbert, 1982a, b; Syvitski et al., 1983d; British Columbia: Syvitski and Farrow, 1983; Kostaschuk and McCann, 1983). These failures have many common seafloor characteristics: (1) channels 10 to 100 m wide and 1 to 10 m deep, cover the prodelta slope (Figs. 5.11, 5.12); (2) the channels originate from one or more arcuate re-entrants and chutes that have steep headslopes cut into the delta lip (Fig. 5.13B); (3) the channels, although not sinuous, may converge with one another or even truncate one another; (4) the channel width and depth decrease downslope until the channel form disappears: lengths typically range from 0.5 to 5

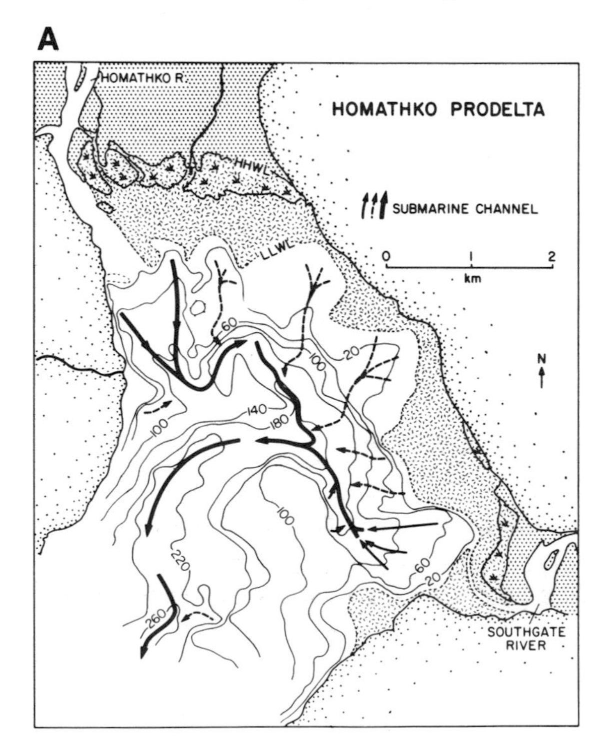

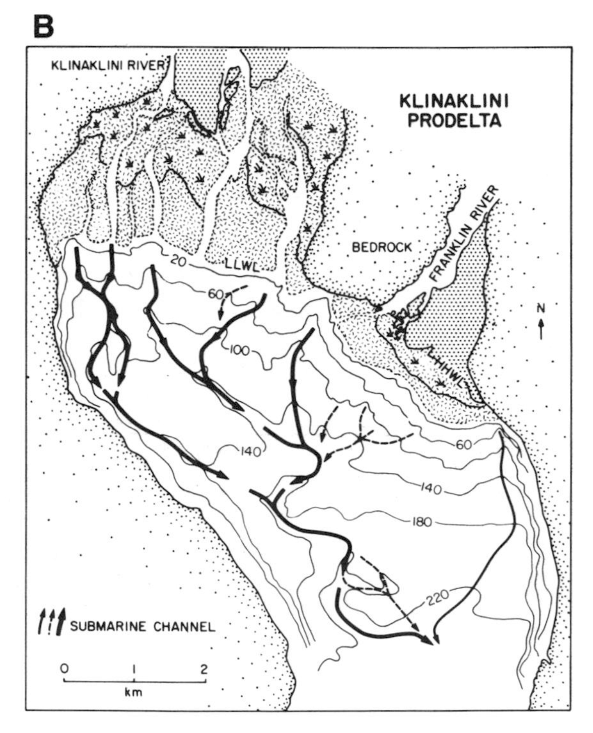

FIGURE 5.11. Prodelta bathymetry of (A) the Homathko and (B) Klinaklini (Bute and Knight Inlet, B. C.) showing submarine channels. Note that the channels of the Klinaklini prodelta line up closely with river distributaries, and the channels of the Homathko prodelta stem from arcuate scarps (from Syvitski and Farrow, 1983).

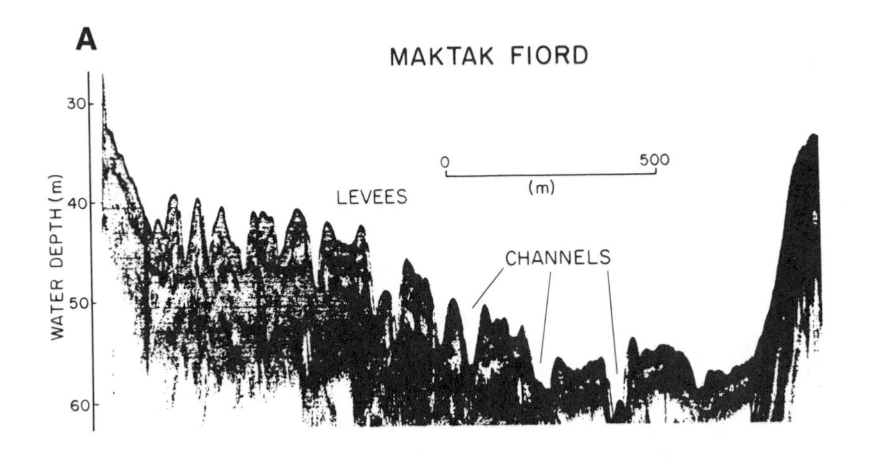

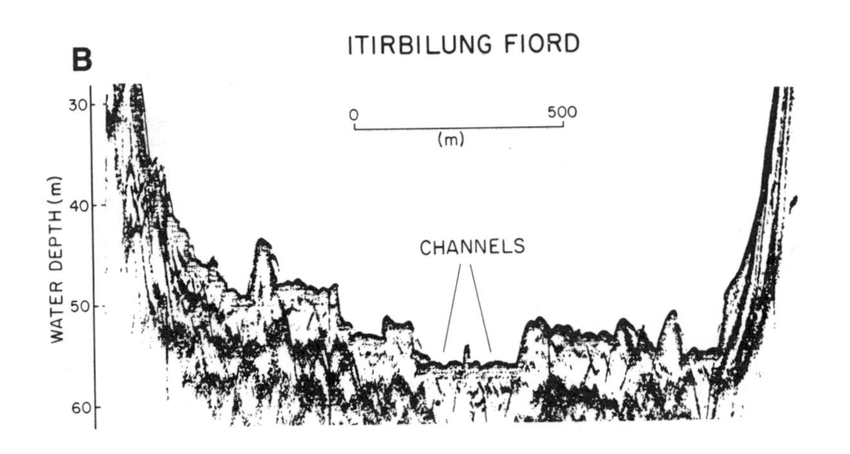

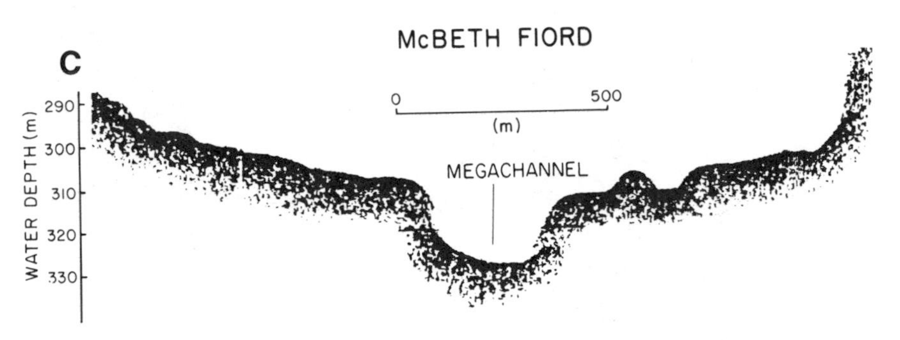

FIGURE 5.12. (A) V-shaped channels incising the Maktak prodelta (Baffin Island): note apparent levees. (B) U-shaped (flat-floored) channels incising Itirbilung prodelta (Baffin Island). (C) Megachannel that runs 10 km along the length of McBeth Fiord, Baffin Island. Note the smaller leveed channel on the right. All three records are from high-frequency sounder records run perpendicular to the fjord axis (for details see Syvitski et al., 1983d, 1984b).

km; (5) the channels contain dense, clean, and well-sorted sand and/or gravel of possible grain flow origin (Figs. 5.14, 5.17B); (6) the channels are often floored with ripples whose crests are perpendicular to the channel axis (Fig. 5.13); (7) the interchannel areas consist primarily of poorly sorted and weakly compacted very fine sandy muds; (8) the channels are commonly lined with levees when the slope falls below 2° (Fig. 5.12); (9) upslope where the channels are more numerous, the interchannel areas may be sharp crested (a result of channel migration into each

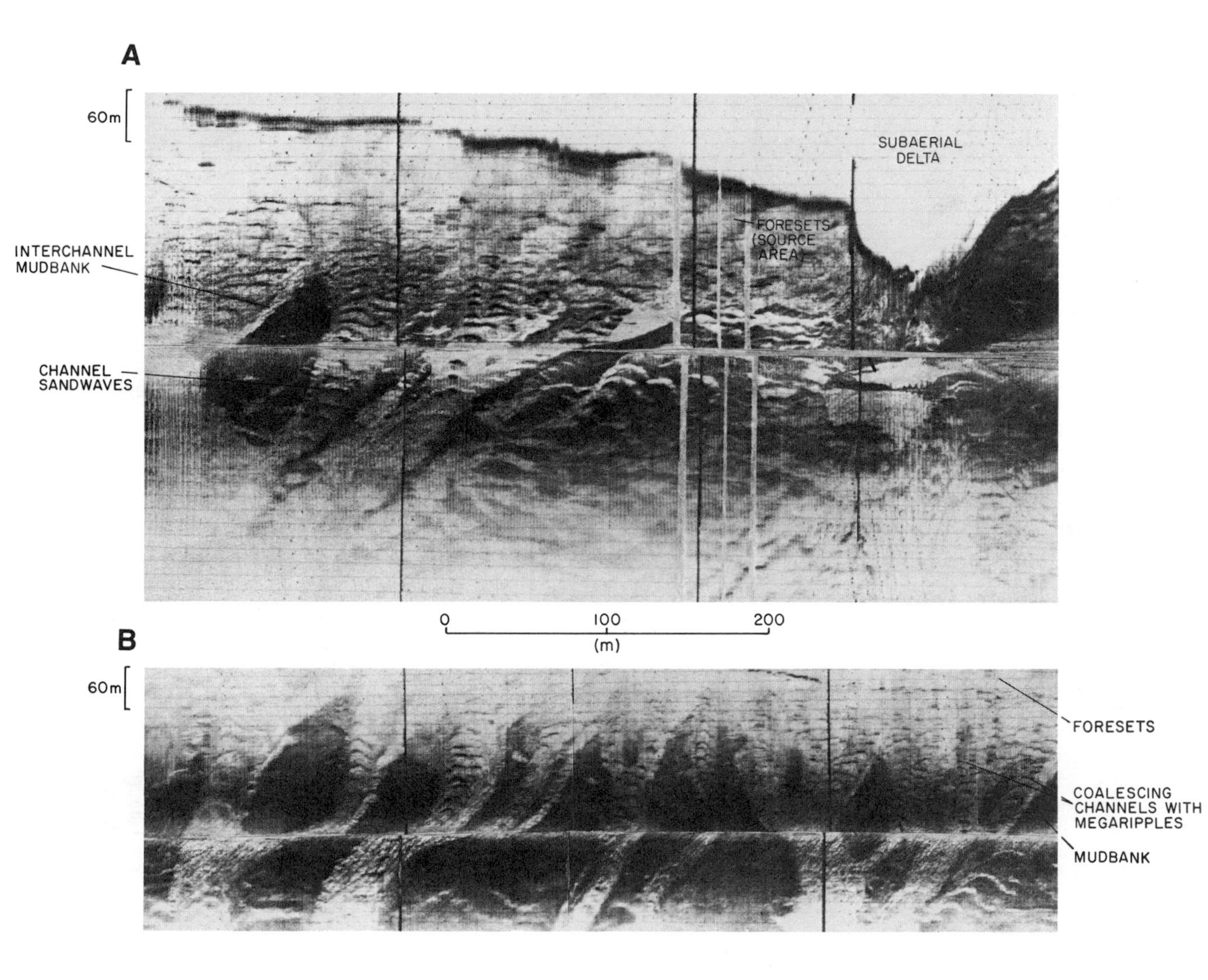

FIGURE 5.13. Sidescan sonograph of coalescing channels incising the prodelta of Maktak Fiord, Baffin Island (cf. Fig. 5.12A). (A) Note the relationships of the foreset arcuate re-entrants, chutes, and channels. (B) Note the interchannel flats and bed features that delineate the channels. (For further details see Syvitski et al., 1983d.)

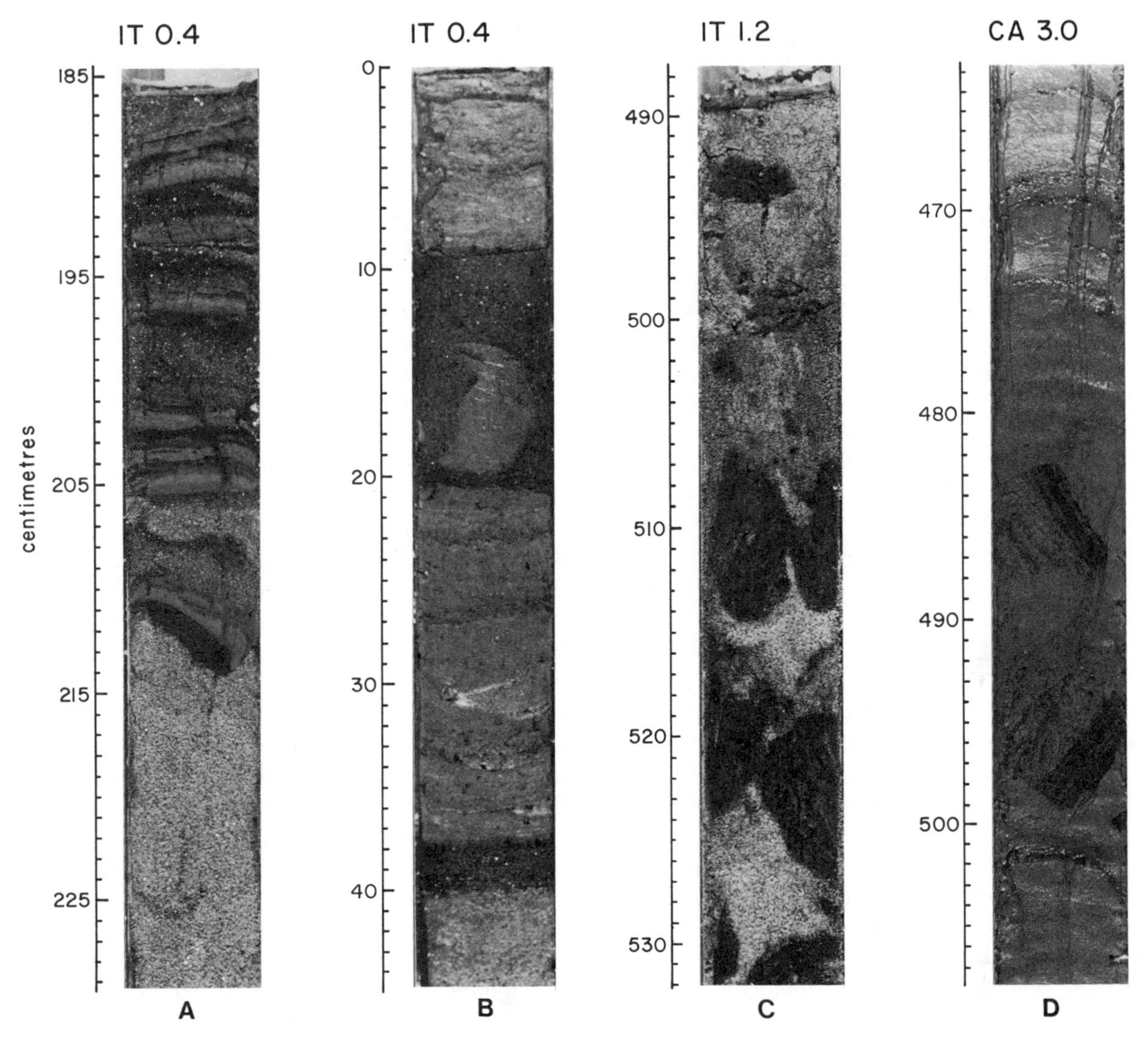

FIGURE 5.14. Photographs of core splits taken within the coalescing channel regime of the prodelta of Itirbilung Fiord, Baffin Island (cf. Fig. 5.12B); note the seasonal varving resulting from river discharge dynamics interlayered with gravity flow deposits generated from foreset failures.

other); and (10) at any time, a percentage of the channels are inactive and are being infilled.

The chutes are thought to originate from small retrogressive slides or local liquefaction that may have been generated through a combination of wave-induced cyclic loading and oversteepening after a recent period of rapid progradation. In that levees are common in the downslope channels, these mass flows must eventually go supercritical and turbulent and develop into turbidity currents.

Syvitski and Farrow (1983) compared two fjord prodelta environments that had similar tidal, wave, and discharge characteristics. The steeper river thalweg of the Klinaklini (Knight Inlet, B.C.) allows for an increase in bedload discharge with subsequent destabilization and migration of its distributary and submarine channels (Fig. 5.11A). The Homathko River (Bute Inlet, B.C.) had one stable fluvial channel and one main submarine channel that appears fed by large upslope arcuate re-entrants incised into the tidal flats and wave-built linguoid bars (Fig. 5.11B). Submersible examination of these prodelta environments revealed that both were composed of high water content (> 90%), noncohesive silt (up to 85%). Upon touching the bottom, the submersible would uncharacteristically sink up to 1.5 m and easily trigger slumping of upslope sediment.

5.4.2 Sidewall Slopes

The percentage of the fjord seafloor that is composed of sidewall slopes depends both on the duration and rate of sedimentation, as well as the original fjord geometry. When the sedimentation rate is moderate and the cumulative years of sedimentation are long, as in many British Columbia fjords with up to 800 m of Quaternary infilling, sidewall slopes account for 10 to 35% of the fjord seafloor. With higher sedimentation rates but only a recent history of infilling, as in Glacier Bay fjords (Alaska) such as Muir Inlet, sidewall slopes make up 50 to 80%. With low rates of sedimentation and a moderately long history of infilling, Norwegian and Arctic fjords have between 30 to 60% of the seafloor as sidewall slopes.

Slopes on the sidewalls range from 10° to greater than 90° (over-hangs) and average around 30°. In a detailed study of sidewall slope environments using a research submersible, Farrow et al. (1983) documented a host of common failure features (Fig. 5.15A): (1) slope-parallel grooves (5–20 cm wide) in the soft sediment that result from "microturbidity" currents set off by epibenthic activity; (2) areas of accelerating creep where burrowing crustaceans must continually re-excavate their burrow openings as the surface sediment creeps over the deeper and more stable sediments that house their permanent tunnel system; (3) exposed slide planes where the surface sediment is overconsolidated and very rough, as if the slumped surface were ripped away (Figs. 5.15B, 5.16A); (4) compressional slump toes ($\lambda > 60$ m, $h \leq 15$ m) that line the base of the side walls (Fig. 5.16C); and (5) their associated tensional joints in the upper slope regions.

The above failures are related mostly to overloading by normal suspension fallout on the steep slopes. Failure volumes range from 10^3 to 10^5 m^3 of material, although subaerial slides and flows may generate anything from small-scale failures (Fig. 5.15C) to enormous submarine slides (10^6 to 10^8 m^3). More slumps occur off the "Coriolis" wall (right-hand side facing down-fjord in the northern hemisphere, left in the southern hemisphere), a factor related to increased preferential hemipelagic sedimentation (Chapter 4). The number of slumps off the side walls also increases with the percent of seafloor consisting of side walls (see preceding discussion) and the rate of sedimentation. Therefore, in the early history of fjord infilling where rapid glacial-marine sedimentation is normal, the number and frequency of sidewall failures might be very high.

Side-entry deltas are sites of intermittent slides and slumps where the frequency is related to the development of oversteepened slopes (i.e.,where tan $\emptyset$, the angle of internal friction, is less than or equal to tan α, the slope of the delta front). The development of oversteepened slopes, in turn, can be linked to the slow or delayed switching of the subaerial distributary channels, which allows a locally oversteepened slope to build. In 1955, a slide off a side-entry fan delta (Woodfibre, Howe Sound, B.C.) altered the offshore geometry of the upper delta front slope with an increase in water depth by at least 10 m. The slide has become a classic example of slope instability involving coarse sand and gravel (Terzaghi, 1956; Prior et al., 1981a; Mandl and Crans, 1981).

The slope of the Woodfibre submarine delta had very narrow limits (26° 30′ to 28°). The delta front slopes eventually merge with the flat seafloor of Howe Sound at a depth of 200 m. The oversteepening of part of the delta lip has been linked directly to the fluvial channel confinement by man. Channel confinement resulted in increased silt content on the lateral margins of delta front. The slide occurred at extremely low tide, where lower permeability, which was associated with this siltation, interferred with the rising and falling of the water table and allowed excess pore pressure to be built up. The result of the slide was the development of large-scale scarps, disturbed hummocky blocks of sediments, and numerous discrete erosional chutes and channels. Small steplike terraces arranged roughly parallel to the bathymetric contours were also present (e.g., Fig. 5.16C). The channels were 3 to 5 m deep and 10 to 30 m wide, with levee-like features. Farther downslope the entire seafloor is composed of highly irregular and chaotically arranged blocks of sediment with a hummocky surface profile. Submersible surveys at the base of similar fanglomerate deltas in Knight and Bute Inlet, B.C. (Syvitski and

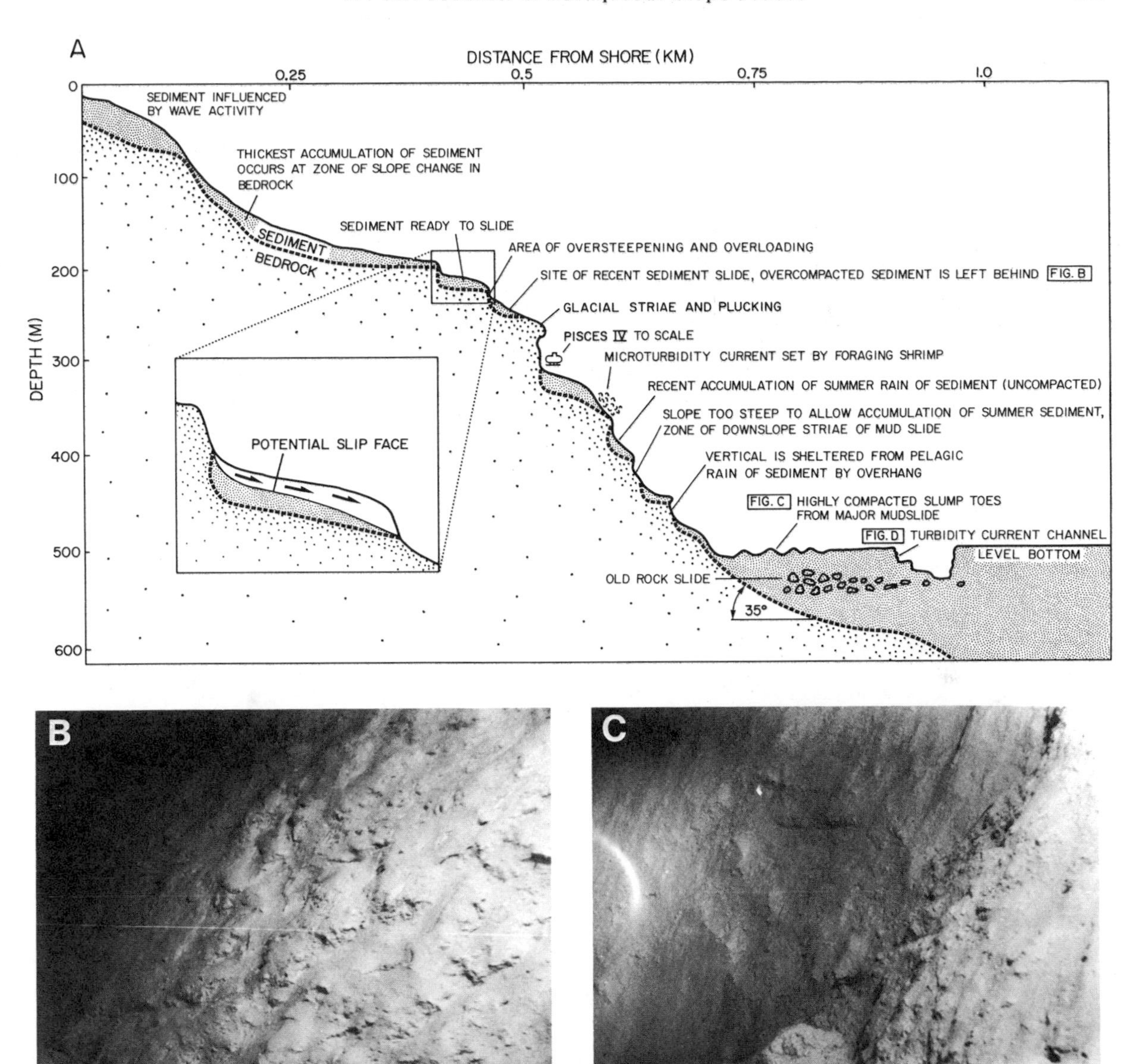

FIGURE 5.15. (A) Fjord sidewall slope environment of Knight Inlet showing sediment processes and position of the following photographs taken from a submersible (drawn to scale). (B) Over-compacted slip plane with "fluted" microrelief showing no faunal activity. (C) Recent slump covered by mudclasts. (D) 2 m high cutbank generated by turbidity current(s). Channel was located at the 500-m water depth on the fjord bottom (from Farrow et al., 1983).

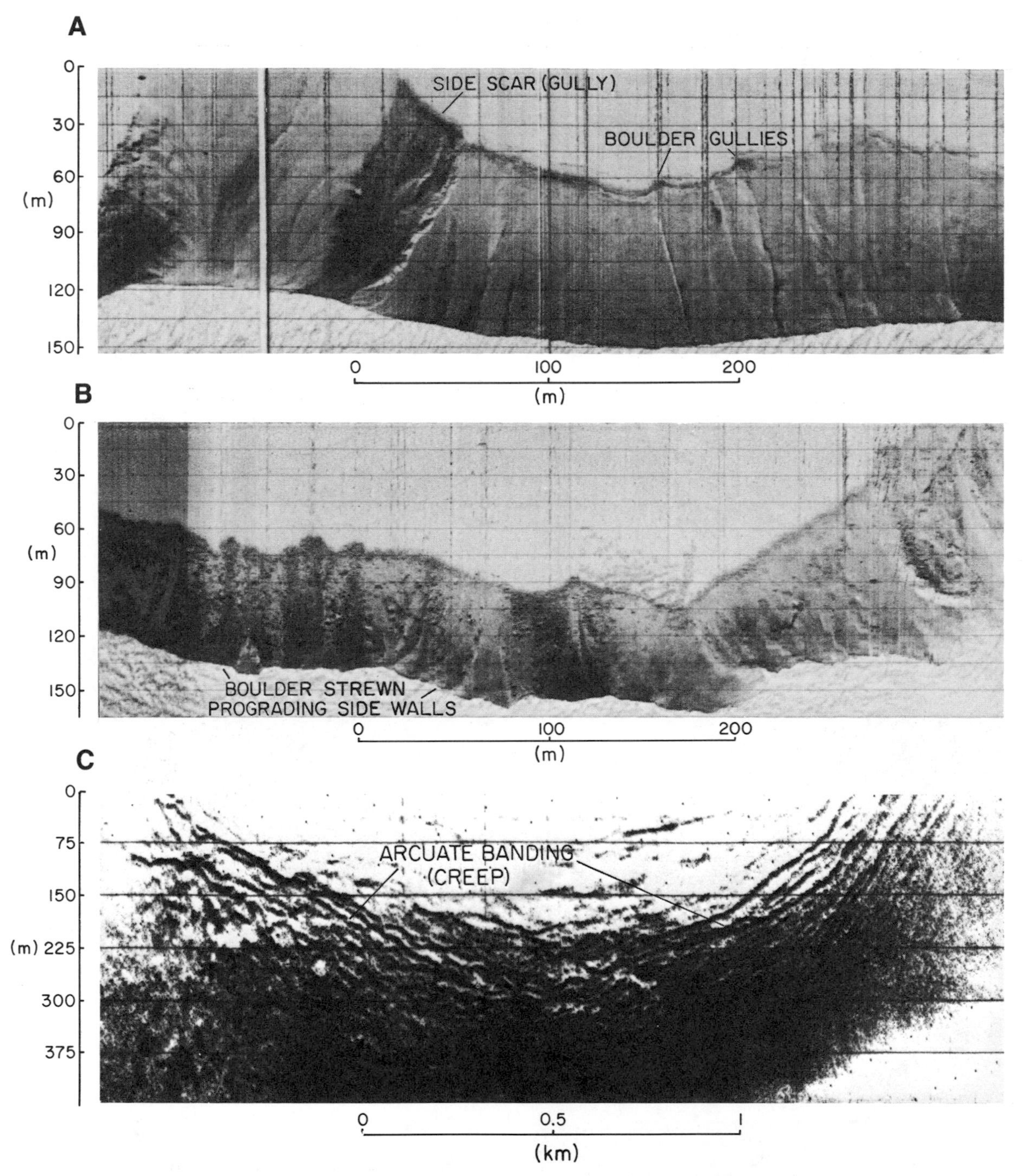

FIGURE 5.16. (A) Sidescan sonograph of fresh slide chute eroded into the sidewall slope in Itirbilung Fiord, Baffin Island. (B) Side-scan sonograph of rock debris from subaerial talus cones that have accumulated on the steep fjord walls (Itirbilung Fiord, Baffin Island). (C) Arcuate banding, possibly related to creep folding down the sidewall slope of a side-entry delta (McBeth Fiord, Baffin Island).

Farrow, 1983), observed both slump toes and glide blocks (20 to 30 m high with very steep sides, up to 80°).

Many sidewall slope failures are generated along the fjord basin where tributary "hanging valleys" meet the central channel. There, flexures and faults along steep submarine cliffs result in sediment blocks tilted and bent. As an example, where Fjaerlandsfjord enters Sognefjord, Norway, there is a bathymetric drop of 800 m over a slope of 25°: a sediment volume of 10^7 m^3 has slid, producing a plunge poollike depression at the base of the slope. Inge Aarseth (personal communication, 1983) considers the origin of these plunge pools to be analogous to snow avalanche areas on land. The depression can be many tens of meters in depth below the normal seafloor. Cores taken near the plunge pool show: (1) intensely folded laminated sequences of a slump; (2) graded beds commonly associated with turbidites; and (3) mud balls common to debris flows.

5.4.3 Deep Basin Gravity Flows and Failures

Fjord basins tend to have slopes of less than 0.1°. Many are completely level for tens of kilometers. As such, they are receiving basins for both suspension fallout and slumps or gravity flows from the side walls and upslope prodelta reaches. Within many fjords there may be a series of basins at different levels separated by bedrock or moraine sills. Sill slopes vary widely from 3 to 30°. Failure of ice-contact and other more recent glacial-marine and marine deposits plastered on these sills may also help to fill the basins.

As outlined in Section 5.3, slides or slumps may become at least partially remoulded, developing into debris or grain flows. With further incorporation of water, these flows may become liquefied or fluidized and eventually develop into viscous sediment gravity flows. If these viscous flows become supercritical, they may progress further into turbulent turbidity currents. Within this succession of possibilities, stages of flow development may be omitted or at least products marking such stages may not be preserved in the sedimentary record. Acoustic evidence of this succession commonly shows slumps, slides (Fig. 5.8), and debris flow deposits (Fig. 5.18) with channels eroded into their surfaces as the final stages of flow became turbulent.

Some fjord basins are fed sediment through one or two megachannels that have attained depths from 5 to > 25 m and widths of 100 to 1000 m (Gilbert, 1983; Syvitski and Farrow, 1983; Syvitski et al., 1983d, 1984b). These megachannels share some general characteristics: (1) the channel is commonly found in slopes less than 2°; (2) the channel decreases in depth and width with decreasing slope; (3) the channels are somewhat sinuous and may meander from fjord wall to fjord wall; (4) before a channel disappears into the flat of the basin floor, it develops levees (Fig. 5.12C); (5) upslope, where levees are not found, the channels have near-vertical walls (Fig. 5.15D); and (6) if a megachannel is still active, it contains sandy sediment compared with the surrounding hemipelagic basin muds (Fig. 5.17A). In Queen Inlet, Alaska, Hoskin and Burrell (1972) noted that its two megachannels had sediment modes of 2.3 Ø (205 μm) and 4.5 Ø (44 μm), respectively, compared with the hemipelagic seafloor muds of 6.5 Ø (11 μm).

In a classic study of Hardangerfjord, Norway, Holtedahl (1965, 1975) noted the abundance of normally graded beds in his sediment cores. He assumed that the graded layers (Fig. 5.17C) were deposited by turbidity currents and that the material gradually became finer with distance of transport (5.17D): the pathways of various turbidity currents were thus traced (Fig. 5.19). The graded beds are underlain by coarse, very poorly sorted material (Fig. 5.17C) generated from sidewall slumps (similar to the one shown in Fig. 5.18); the slumps contained littoral fauna as well as lumps of clay. The turbidites are restricted to the eroded megachannels on the flat floors. Cores taken outside the central channel did not contain turbidite layers. The vertical (within one core) and lateral grading (between cores) reflects the decreasing competence of the sediment-laden currents (Fig. 5.17C, D). Fifty percent of the sediment column within the basins of Hardangerfjord has resulted from slumps and turbidity currents, with an average accumulation rate of 0.5 cm yr^{-1}.

Rupert Inlet, B.C., is an excellent fjord to document the effect of continuous or near-continuous turbidity currents (Hay, 1982; Hay et al. 1983a, b). Turbidity currents develop because of a mine discharge outlet at depth, which allows

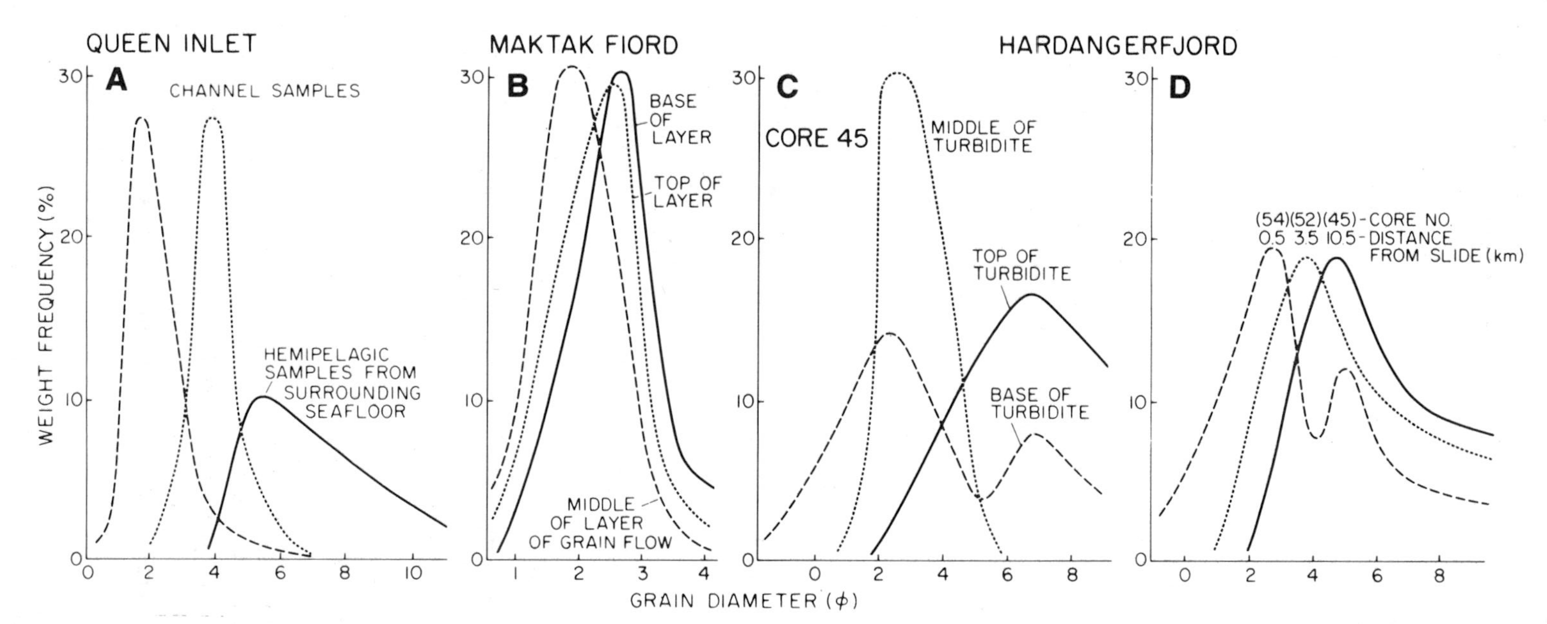

FIGURE 5.17. (A) The size frequency distribution (SFD) of samples taken within and beside megachannels in Queen Inlet, Alaska (after Hoskin and Burrell, 1972). Note the grain diameters are given in standard phi (ϕ) units equal to - $\log_{10}$ (mm). (B) SFD of samples taken within a grain flow deposit in Maktak Fiord, Baffin Island (cf. Fig. 5.12A; after Gilbert, 1983). (C) SFD of samples taken within a turbidite layer in a Hardangerfjord core (Norway). (D) SFD of samples at the base of a turbidite layer and with increasing distance from the slide source (data for C and D after Holtedahl, 1975).

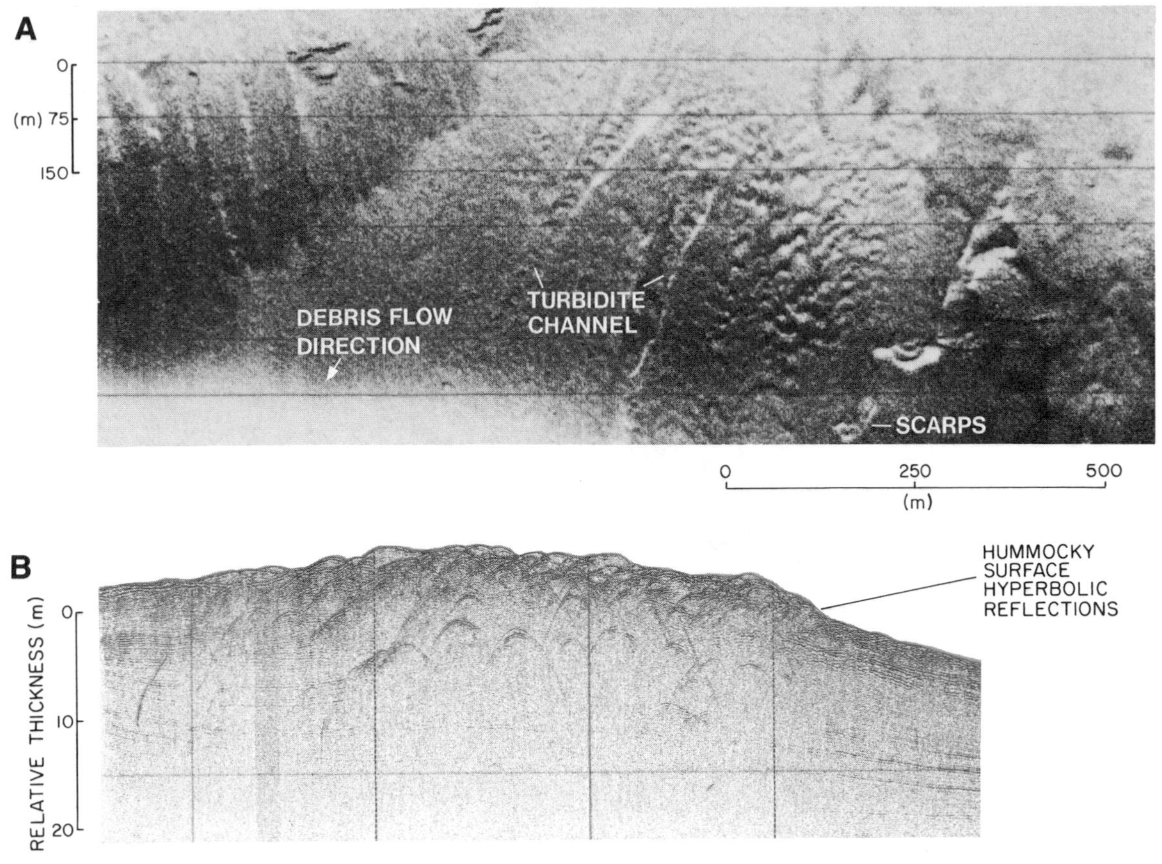

FIGURE 5.18. (A) Sidescan sonar, and (B) DTS Huntec reflection seismic image of a debris flow that had flowed off the sidewall slope of McBeth Fiord, Baffin Island. Note the two turbidite channels eroded into the debris flow on the sidescan record. Also note the hummocky surface and internal seismic hyperbolic structures common to debris flows.

the formation of leveed channels. Since its formation, the submarine channel has undergone three successive metamorphoses: (1) a meandering channel regime, (2) an apron regime, and (3) a rechannelized regime.

Leveed channels formed in Rupert Inlet in a direction parallel to the steepest accessible slopes. The levees in the proximal zone are sites of rapid deposition: the levee walls, being steep and composed of noncohesive sand and silt, in turn slump and generate the turbidity current surges. These surges, with a frequency of 2 to 5 days, deliver coarse-grained turbidites to the lower reaches of the fjord basin. Levee building by preferential deposition close to the channel from overspill flows and channel deepening by scour along the axis inhibit loss of the coarser sediment from the channel and reinforce the initial route. Meanders occur where the slope decreases below a critical threshold limit and give way to a straight channel after another slope decrease is encountered. Cores taken in the area of the meandering channel regime revealed sandy turbidite layers that increase in thickness with distance down-channel. The thickness increase relates to the progressively decreasing importance down-channel of both sediment transport and deposition from the continuous turbidity current associated with the tailings discharge and the surge-type turbidity flows.

5.4.4 Process Interpretation from the Deposit

Some lithologic or stratigraphic features are thought to be diagnostic of the type of sediment gravity flow that deposited a particular layer or layers. Given that there may be a continuum in processes for a given event, both in time and space, the characteristics of a deposit may point

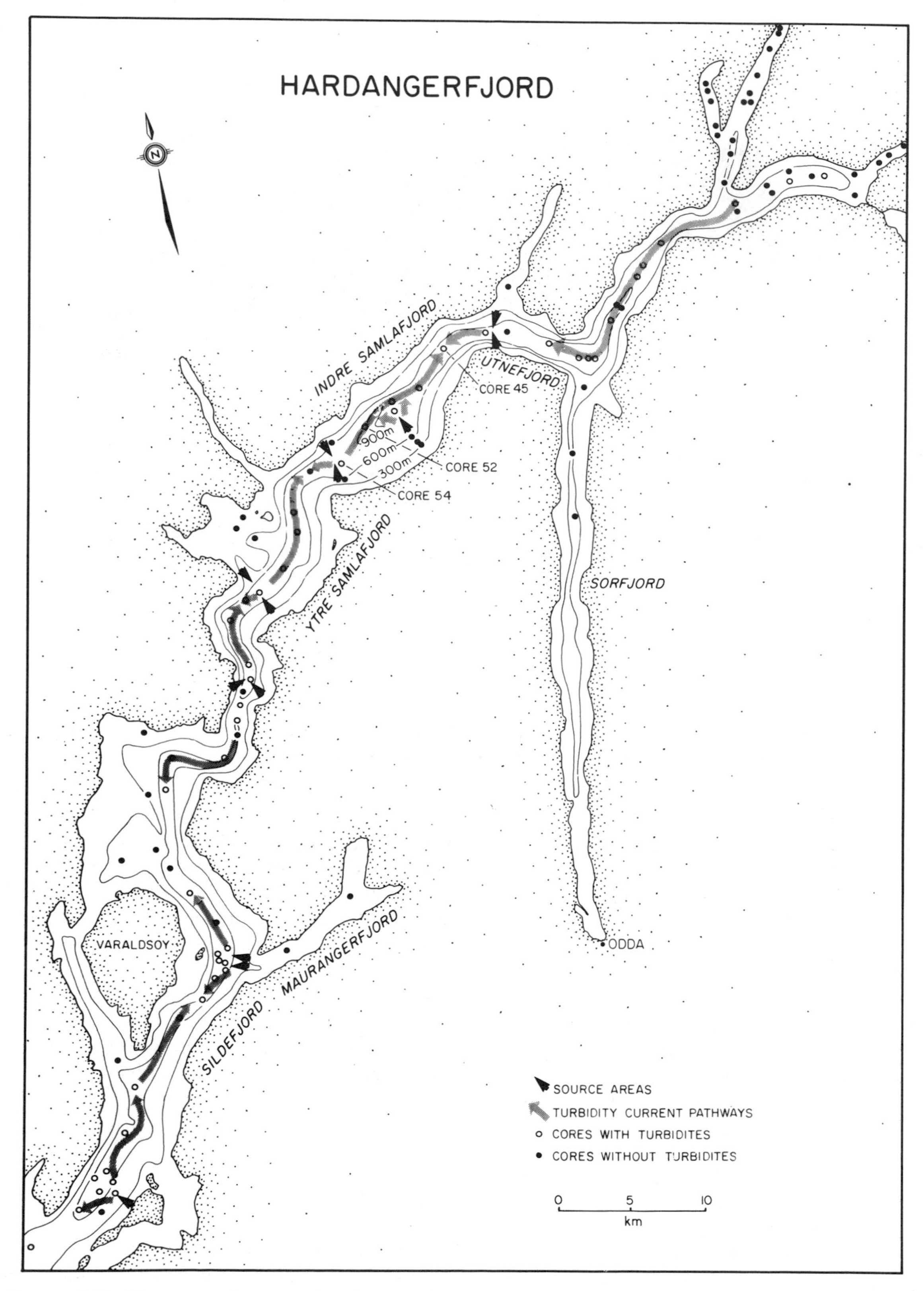

FIGURE 5.19. The source slumps and pathways of turbidity currents in Hardangerfjord, Norway (after Holtedahl, 1965).

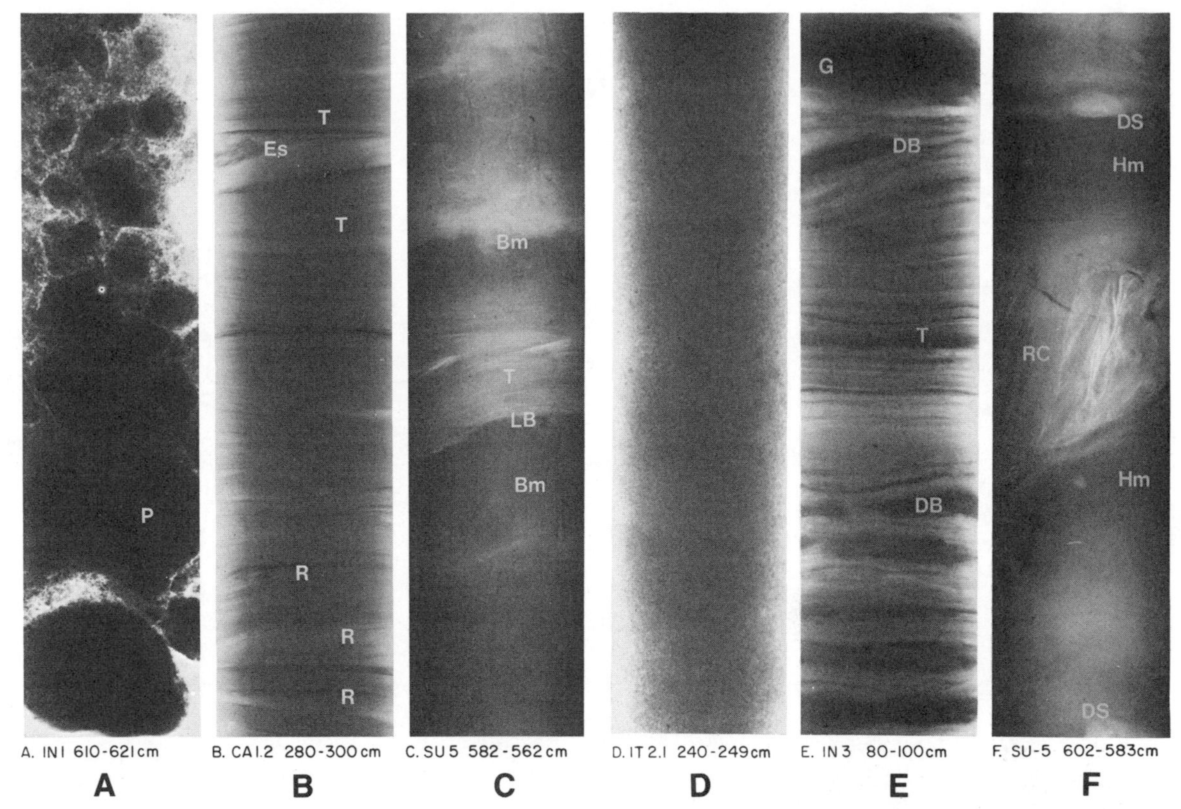

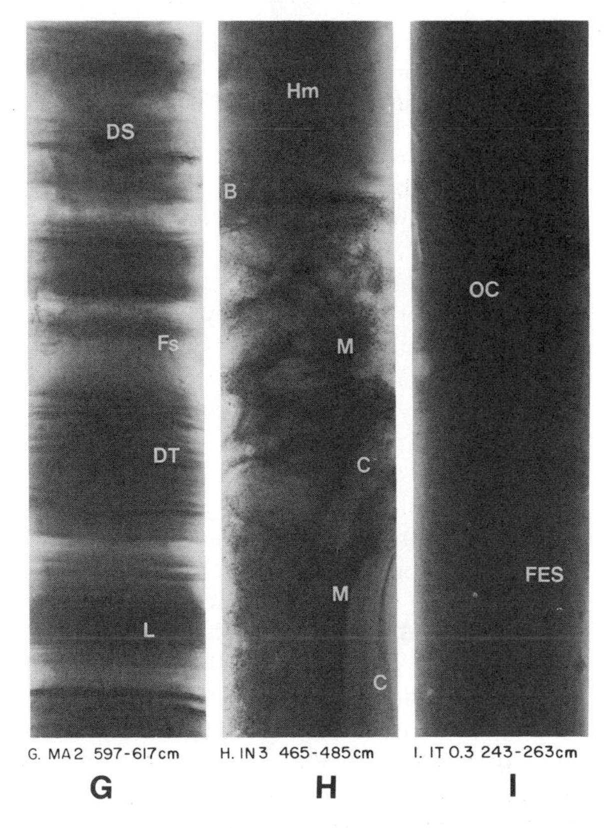

FIGURE 5.20. X-radiographs of split piston cores from Baffin Island fjords (for details see Syvitski and Blakeney, 1983, Syvitski, 1984a, b). (A) Bottom portion of 3.5 m thick (amalgamated) proximal turbidite showing fining upward of grain size from pebbles (P). Inugsuin Fiord, water depth, D = 160 m; distance from fjordhead, x = 5 km. (B) Amalgamated distal turbidite section including horizontal (T) and rippled layers (R) some with erosional surfaces (Es). Cambridge Fiord, D = 200 m; x = 3.3 km. (C) Single distal turbidite with erosive lower boundary (LB), possibly a flute mark, rippled, and horizontally layered sand (T) grading into a bioturbated mud (Bm). Sunneshine Fiord, D = 160 m; x = 40 km. (D) Massive and structureless grain flow sand with no apparent size grading. Itirbilung Fiord, D = 325 m; x = 20 km. (E) Distal turbidites (T) mixed with hemipelagic mud and grain flow layers (G). Of interest is the disturbed bedding (DB), possibly related to loading deformation. Inugsuin Fiord, D = 557 m; x = 34 km. (F) Rotated clast (RC) of rippled sand floating in hemipelagic mud (Hm), possible evidence of a debris flow. Also note the ice-rafted dropstones (DS). Sunneshine Fiord, D = 160 m; x = 40 km. (G) Amalgamated distal turbidites (DT) and liquefied flow sand layers (L). Also note the soft sediment deformation between some layers (Fs). Maktak Fiord, D = 257 m; x = 11 km. (H) A portion of a 0.5 m thick debris flow deposit with coarser synsedimentary clasts (C) near the top of the deposit. The deposit has a relatively sharp top covered by hemipelagic mud (Hm). The clasts are floating in a sandy mud matrix. Inugsuin Fiord, D = 557 m; x = 34 km. (I) Evidence for fluidization of sandy silt with clasts (OC) oriented along lines of apparent fluid escape structures (FES). Itirbilung Fiord, D = 155 m; x = 4.5 km.

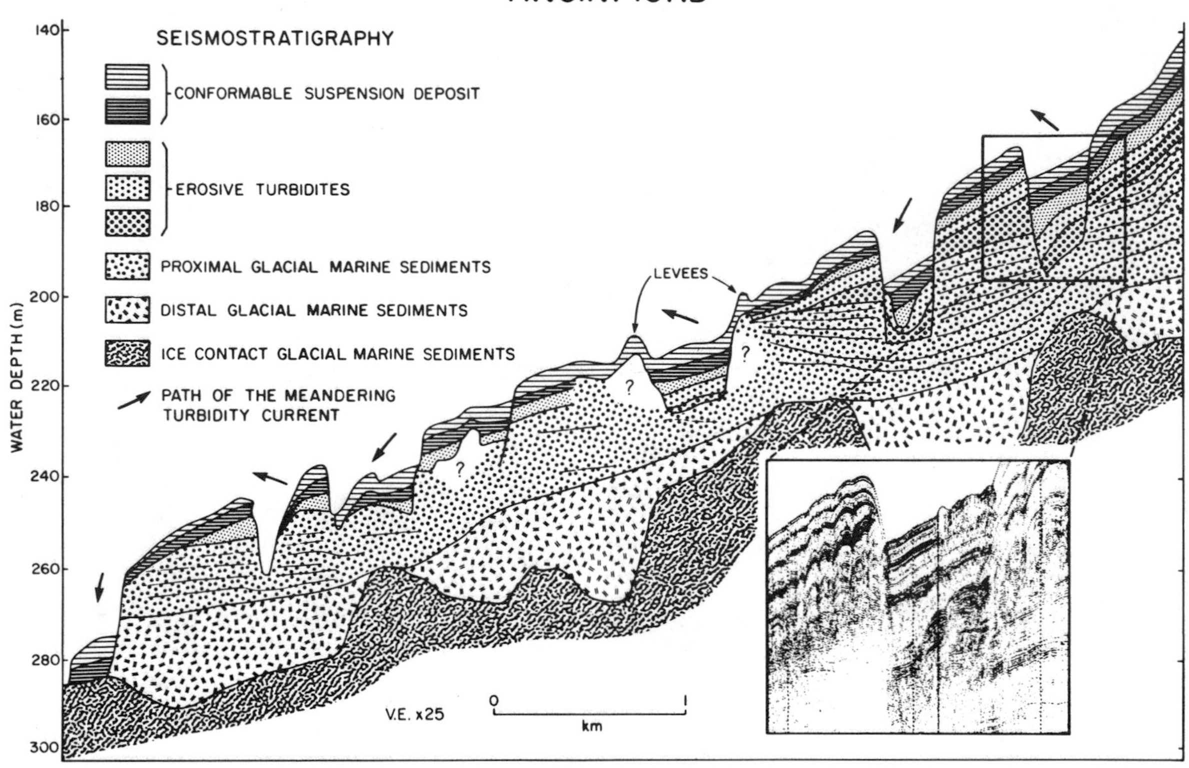

FIGURE 5.21. Interpretation of seismo-stratigraphy of the prodelta environment of Tingin Fiord, Baffin Island, with an example of the DTS high-resolution reflection seismic record. Note the erosive nature of some of the turbidite deposits and the exposure of older sediment sequences.

toward more than one contemporaneous process. Therefore, care in interpretation is stressed, especially since there are many more exceptions to the rules, than rules.

Units of poorly sorted pebbly mud are commonly interpreted as products of debris flows (Figs. 5.14D, 5.20F, H). The pebbles, either synsedimentary clasts or autochthonous gravel fragments, float within a muddy matrix. Debris flow deposits seldom undergo extensive post-depositional bioturbation, which is useful in discriminating these deposits from boulder (talus) accumulation on muddy side walls (e.g., Fig. 5.16B). Debris flow deposits tend to have the larger clasts concentrated at the top and sides of the deposit (Fig. 5.20H), compared with a more random distribution of pebbles within ice-rafted gravelly muds. The latter also contain few synsedimentary clasts. The larger clasts or boulders, within the high water content mud matrix, allow debris flow deposits to be recognized by semi-transparent hyperbolic structures in a reflection seismic record (Fig. 5.18). Some slump deposits, however, have a similar seismo-stratigraphic signature but usually with a more hummocky seafloor.

Mudflow deposits are similar to debris flow deposits except for a noticeable lack of pebbles. Grain flow deposits tend to be of a massive nature with clean and well-sorted sand with mixed grading, that is, both reverse and normal grading (Figs. 5.14A, 5.17B, 5.20D, E). Grain flows may also contain synsedimentary clasts, although the clasts are more rounded than debris flows (i.e., compare Figs. 5.14A and 5.20H with 5.14B, C). At the extreme, there may be little difference between the characters of a pebbly grain flow and a debris flow deposit, except in some arbitrary grain size value for the matrix. Grain flows are nonerosive, thus they tend to infill seafloor lows while maintaining a level surface. Amalgamated grain flows are difficult to recognize seismo-stratigraphically, since they contain no internal stratification.

Gilbert (1982a, b) recognized liquefied or fluidized flow deposits in fjords by: (1) relatively thin layers of well-sorted sand that contain almost no fine sediment; (2) no grading; (3) no internal structure; and (4) contacts above and below the layer are abrupt with no evidence of loading or current-induced deformation. A typical example is given in Figure 5.20G. Fluid escape structures (e.g., Fig. 5.20I) have also been used in liquefied or fluidized flow deposits (Middleton and Hampton, 1976).

High-concentration turbidity currents may be inferred from graded layers (Fig. 5.20A), basal load casts and flute marks (Fig. 5.20C, E), flame structures (Fs in Fig. 5.20G), and ripple sequences (Fig. 5.20B). Evidence from seismostratigraphy indicates that the basal erosive units occur within a channel-fill sequence (Fig. 5.21). Low-concentration turbidity currents will produce thin (< 1 cm) layers of clean sand or silt (Fig. 5.20G). Turbidites seldom occur as single rare events; they are more frequently found as thick units of amalgamated deposits (i.e., compare Figs. 5.20B, E, and G with 5.20C).

5.5 Summary

Fjords, with their rugged topography and dynamic basin-infilling history, are prone to a variety of slope failures. Where bottom currents are strong, either from wave activity or tides, sediment accumulates under stable conditions and slope failures are rare. Fjords that have a particularly rugged seafloor and high rates of sedimentation, however, are dominated by mass transport processes. These processes can influence both the benthic community and the geochemical environment. For instance, the local infaunal community may be limited from: (1) exposure of overconsolidated sediment after slide activity; and (2) sediment loading where the frequency of slope failure is high (Farrow et al., 1983, 1985). Scientists have yet to determine both the short-term and long-term effects of mass failures on biogeochemical cycling of elements.

Sediment slides and gravity flows may so alter the structure of the seafloor that a radically different seascape is formed. For example, the megachannel in Tingin Fiord, Baffin Island, has exposed a wide range of chronostratigraphic units, each with its unique geotechnical and geochemical properties (Fig. 5.21). So, what might have been a simple benthic or geochemical environment has been altered to an incredibly complex environment. Shells and sediment older than 20,000 years BP might easily become mixed with sediment of recent age.

6

Biotic Processes

In this chapter the important role played by the biota in fjord sedimentation processes is explored. In many fjords a substantial proportion of the sediment transported into the marine environment by rivers and other freshwater discharge is biogenic, and further quantities are produced in situ within the marine fjord (autochthonous carbon). The latter may be augmented by import from contiguous coastal areas; or, alternatively, organic material originating in the fjord may be transported out. Although much of this material is recycled within the water column—supplying the indigenous pelagic ecosystem—conditions in many fjords are conducive to significant downward transport of organic detritus to the sediments. Frequently the flux of this component of the sediment has a distinctive seasonal pattern. Degradation of the more labile organics at the benthic interface and within the near-surface sediments generates the spectrum of chemical microenvironments discussed in Chapter 7.

Organic detritus is also the major energy source—in deep fjord basins the sole energy source—fueling the benthic fauna, which in turn plays an important role in mixing, sorting, and binding sediment particles, and in defining the sedimentary chemical environment. Finally, the distinctive patterns of sedimentation and secondary transport of sediments within many fjords profoundly affect the types of benthic communities that may colonize and survive on and beneath the basin floor. These processes are summarized schematically in Figure 6.1.

6.1 Pelagic and Littoral Processes

6.1.1 Littoral Production

In relatively shallow tidal estuaries, where the ratio of intertidal to subtidal areas is frequently greater than one, intertidal and shallow subtidal primary production is generally of greater quantitative importance than is marine phytoplankton

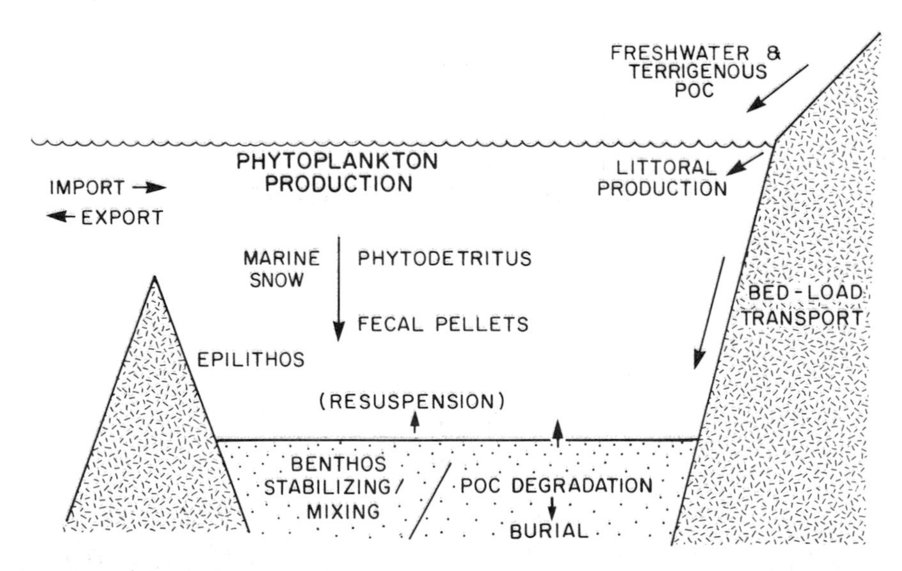

FIGURE 6.1. Sources of particulate organic carbon (POC), biogenic sedimentation, and the interaction of benthos and sediments in fjords.

production. For example, in the Nanaimo River (British Columbia) estuary, Naiman and Sibert (1979) have shown that phytoplankton production contributes only 10 to 20% of the total estuarine primary production. Frequently, the areal rate of carbon fixation by the shallow fringing plant communities may be one or more orders of magnitude greater than that shown by the phytoplankton (Seki, 1982; Mann, 1982). The production rate of eelgrass within the shallow Isefjord, Denmark, is approximately 1100 gC m^{-2} yr^{-1}, of which some 50% is exported as detritus (Rasmussen, 1973). However, because of their distinctive steep-sided topography, most fjords do not conform to this pattern: the littoral regions are characteristically much reduced compared with the marine surface area. This is particularly so in rugged terrain and where the associated rivers are relatively small. Pickard and Stanton (1980), reviewing a wide range of Pacific coast fjords, found the range of intertidal to subtidal areal ratios to be between 1:13 and 1:40.

Where present, the fringing seagrass and kelp communities of energetic intertidal and subtidal estuarine regions are highly productive. Mann (1975) compares two fjordlike inlets on the south coast of Nova Scotia. One supports extensive *Laminaria* beds in shallow water areas subject to strong wave action, and here the macroalgal productivity (1750 gC m^{-2} yr^{-1}) is nearly an order of magnitude greater than that of the marine phytoplankton. The other basin is sheltered, and seaweed production there is far less important. Relatively low kelp productivity (120 gC m^{-2} yr^{-1}) has also been recorded by Johnstone et al. (1977) from a sheltered locality within a Scottish sea loch. Burrell (1983b) has estimated that fringing community primary production in a southeast Alaska fjord is probably not greater than 15% of the mean annual phytoplankton production, and that less than 1% of this originates from *Fucus distichus* attached to the steep intertidal rock walls.

Sieburth (1969) has demonstrated that a large fraction of the gross fucoid production may be released as soluble organic exudates, the fate of which is largely unknown. In fjords having substantial intertidal areas, detritus from macroalgae and seagrass communities may be an important component of the sediments, inasmuch as this material is refractory and, unlike phytodetritus, is unlikely to be substantially remineralized within the water column. Depending on ambient temperatures, sedimentation rates, and other local factors, organic detritus from this source may, directly or indirectly, be an energy source for the deposit-feeding benthos. The more refractory constituents, together with much of the terrigenous particulate organic material, will be buried in the sediments and lost from the immediate fjord carbon cycle.

Benthic microalgae often account for a significant percentage of the total primary production of estuarine and inshore ecosystems. However, with a few exceptions (e.g., Evansen, 1974) this production source has not been extensively studied in fjords. Grøntved (1960) gives a mean value in excess of 400 gC m^{-2} yr^{-1} for fine-grained localities within the shallow Danish fjords, but Steele and Baird (1968) note that benthic microalgal production is less than 10 gC m^{-2} yr^{-1} in wave-disturbed regions of Loch Ewe, Scotland.

6.1.2 Phytoplankton Production and Grazing

In most fjords, primary production is largely due to phytoplankton, and hence the rate of growth, distribution, concentration, and diversity of these plants determines the course of all subsequent secondary and higher order production. In higher latitude estuaries there are marked seasonal cycles at all trophic levels, and the spring phytoplankton bloom—keyed to an irradiaton threshold (Stockner et al., 1979; Hegseth, 1982; Erga and Heimdal, 1984)—is the most distinctive feature of phytoplankton growth (Fig. 6.2). In the absence of intense grazing, blooms occur in water which is sufficiently stabilized to ensure that turbulence does not persistently transport cells out of the euphotic zone (Sverdrup, 1953; Wyatt, 1980). This condition is generally reached both within and outside the estuary with increased insolation and warming of the near-surface water, but blooms characteristically occur earlier within fjords (Braarud, 1974; Tett and Wallis, 1978), concurrently with spring freshets, which stratify the surface zone. Conditions favoring generation of a significant bloom within fjords located in polar regions may be delayed until late in the brief summer season, or may be absent.

Vernal blooms in fjords characteristically consist of one dominant and a few subsidiary

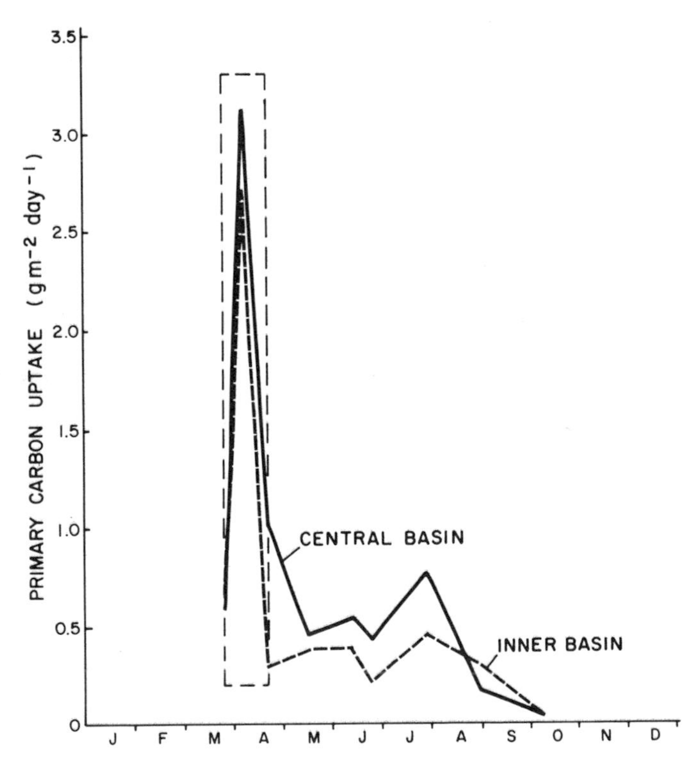

FIGURE 6.2. Integrated seasonal ^{14}C primary uptake rates (gC m^{-2} yr^{-1}) through the euphotic zone (to the 1% light level) at two localities in Boca de Quadra fjord, southeast Alaska (55°N) in 1982. Approximately 40% of the total production in this year (annual rate of 135 gC m^{-2} yr^{-1} at the central basin locality) occurred at the time of the spring diatom bloom in March-April (boxed region). Data from Burrell (1983a, in press).

diatom species, which may differ in adjacent inlets and in successive years. Many environmental factors influence the community structure. For example, *Thalassiosira nordenskioldii* appears to be better adapted to cold water (Kattner et al., 1983), and *Skeletonema costatum* has special flotation properties (Braarud, 1976) and so may be dominant in poorly stratified water. The spring diatom production and biomass maxima in fjords frequently occur below the surface layer (Burrell, 1983b; Eilertsen and Taasen, 1984), and, as the bloom progresses, the larger diatom cells tend to sink naturally in the weakly stabilized water (Platt and Subba Rao, 1970). Sinking—accentuated as the proportion of senescent cells increases—is a recognized adaption to decreasing nutrient levels (Eppley et al., 1967). Subsurface biomass maxima occur within the pycnocline and at density discontinuities (Fig. 6.3A).

Spring blooms develop within fjords in the absence of intense grazing, and are probably terminated primarily because of nutrient depletion (Sakshaug, 1978), and by self-shading and increasing densities of herbivores (especially calanoid zooplankton) in the euphotic layer. In fjords where blooms develop before marked stratification of the near-surface zone, nutrients can be replenished and need not be limiting, and the culmination may be attributable to sinking (Eilertsen and Taasen, 1984) and to increased grazing pressure. At high latitudes, there tends to be a lag between peaks in the primary and herbivore production (e.g., see Longhurst et al., 1984). Wyatt (1980) notes that a more efficient energy transfer results if the highly seasonal phytoplankton stocks are permitted to build to some optimum level. Since the trophic transfer efficiency is generally lower in arctic and subarctic regions (Conover, 1974), a higher proportion of the primary production may be directly transferred to the benthos, as has been demonstrated by Petersen and Curtis (1980). Hargrave (1980) has emphasized that the trophic transfer efficiency to benthic detritivores is increased if phytoplankton cells are directly sedimented, and this may be a particularly important energy transfer pathway in many boreal fjords. As noted below, cells and detritus originating from the spring flowering are an important component of most fjord sediments, and the major source of energy to the subtidal benthos.

Matthews and Heimdal (1980) and others have emphasized that different trophic pathways may

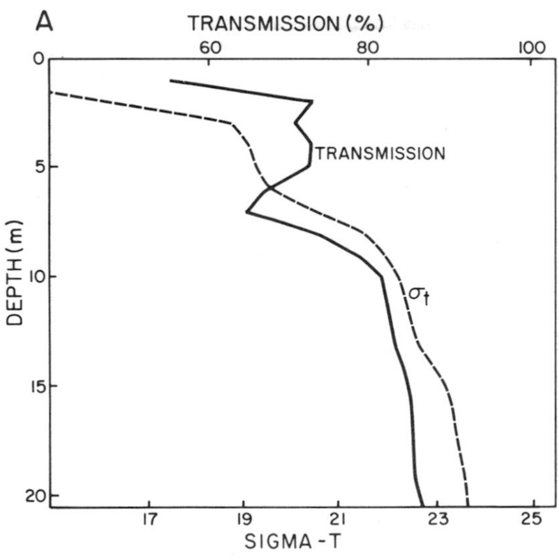

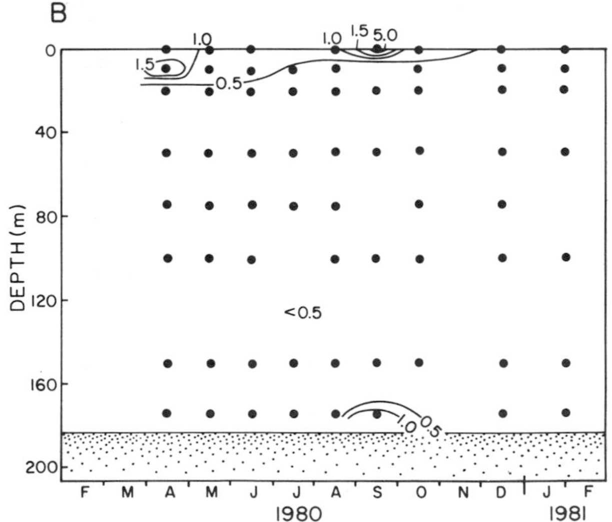

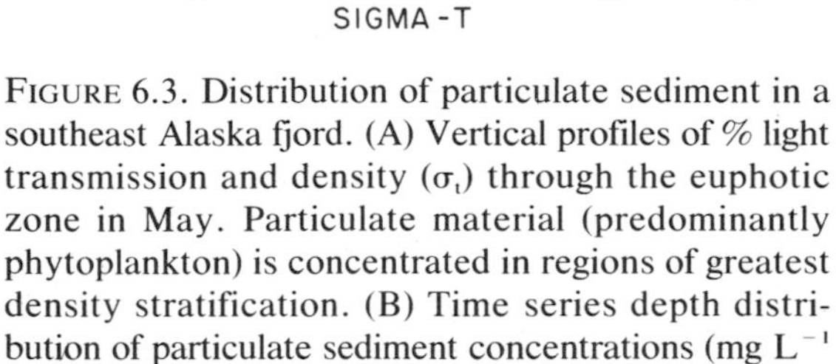
FIGURE 6.3. Distribution of particulate sediment in a southeast Alaska fjord. (A) Vertical profiles of % light transmission and density (σ_t) through the euphotic zone in May. Particulate material (predominantly phytoplankton) is concentrated in regions of greatest density stratification. (B) Time series depth distribution of particulate sediment concentrations (mg L^{-1} of particle size > 0.4 μm; closed circles mark the sampling localities). The subsurface maximum in spring is the diatom bloom; that in September marks the annual peak influx of freshwater sediment. It is only through the latter period that marked resuspension of bottom sediment is observed in this fjord. (Data from Burrell, 1983b.)

be followed in fjords, sensu stricto, compared with polls. The latter tend to lack significant populations of the larger copepods, and the spring diatom bloom, if it occurs, may not be intensively grazed (Wassmann, 1983). At least two factors are important here. In the first place, polls are generally too shallow to accommodate over-wintering populations of calanoid zooplankters. More importantly, the shallow sill depth, in relation to the spring-summer pycnocline (this is the key characteristic of polls), acts to prevent recruitment of zooplankton from coastal regions outside the fjord. An excellent example of this is provided by Fosshagen (1980).

Later in the summer, as stratification of the near-surface waters improves with increased insolation and freshwater influx, resupply of nutrients to the euphotic zone is slowed, and concentrations generally remain low (Wassmann and Aadnesen, 1984, show that this situation does not necessarily apply to polls, however). Under these conditions, the phytoplankton tends to be dominated by a more diverse flagellate flora. This commonly described species succession could partially be the result of silicon depletion (Officer and Ryther, 1980; Paasche and Østergren, 1980). The smaller phytoplankters also have a higher uptake efficiency at low ambient nutrient levels (Skjoldal and Lannergren, 1978; Erga and Heimdal, 1984; Harris, 1984), and microflagellates are generally more tolerant of euryhaline conditions. At this time of year, the phytoplankton and herbivore cycles tend to be closely coupled, and phytogenic detritus is "leaked" from the euphotic zone predominantly as fecal pellets.

The spring bloom illustrated in Figure 6.2 accounted for approximately 40% of the annual phytoplankton production. At this fjord basin locality, the euphotic zone is stabilized through most of the summer, and phytoplankton production continues at a reduced rate, largely fueled by regenerated nutrients. In the absence of intense grazing, sustained or episodic periods of increased productivity are potentially possible in any fjord during the summer, when near-surface stratification is greatest, where there is influx of "new" nutrients (sensu Dugdale and Goering, 1967) into the euphotic zone. Subeuphotic marine water is the major nutrient reservoir, and upward transport may be affected by entrainment-driven estuarine circulation (Sakshaug and Myklestad, 1973; Winter et al., 1975), by localized tidal turbulence, and by sporadic wind mixing (Takahashi et al., 1977). Fjords containing tidewater glaciers possess a further mechanism: upwelling of the buoyant freshwater plume from subglacier inflow and glacier-face melt (Greisman, 1979). Enhanced nutrient concentrations—and apparent increased productivity—in front of marine glaciers has long been advocated (Hartley and Dunbar, 1938; Dunbar, 1973), and nutrient upwelling within a narrow (< 0.5-km) zone adjacent to the ice face has recently been confirmed (Horne, 1985).

In many lower latitude estuaries, the rivers may be an important source of nutrients, and this is so also in some fjords that are supplied by mature river systems draining extensive watersheds (Sakshaug and Myklestad, 1973; Solorzano and Grantham, 1975). However, mean nutrient concentrations in the fresh water entering higher latitude fjords may be lower than in the nondepleted marine water. This appears to be the case for glacier-fed fjords, where precipitation is initially stored as snow, and generally in fjords indenting mountainous coastal terrain where the residence time of rainfall in the limited catchment areas is very short. In temperate and boreal fjords, freshwater runoff may contain high concentrations of dissolved humic material, which may locally depress marine phytoplankton production by reducing transparency (Sugai and Burrell, 1984a). Wood et al. (1973) have shown that in Loch Etive, Scotland, greater than 90% of the primary carbon uptake during most of the summer is confined to the surface 10 m. In many other fjords, secondary blooms are suppressed by increased surface turbidity during times of high precipitation and freshwater influx. Surface plumes of particulate sediment are characteristic features of glacial fjords at the time of peak meltwater discharge, usually toward the end of summer.

A single phytoplankton bloom may occur during the brief, high-summer season within polar region fjords, but Nemoto and Harrison (1981) suggest that the peak may be relatively more intense than at lower latitudes because of the zooplankton grazing lag. Polar phytoplankton is adapted to light levels rather than ambient temperatures, and is generally light and nutrient limited (Harrison et al., 1982). Ice algae may make a significant contribution to the primary production in ice-covered waters (Horner, 1976), although there are no data specifically relating to fjords. Clasby et al. (1973) and Alexander and Horner (1976) have shown that algae attached to the underside of sea ice on the Beaufort Sea shelf bloom early in the arctic summer (in May: maximum production rate of 65 gC m^{-2} yr^{-1}), suppressing submarine phytoplankton growth by attenuating light passing through the ice. The primary sources of nutrients are probably from within the ice, and from in-situ regeneration. Matthews (1983) suggests that convective circulation set up by brine drainage beneath nearshore ice (Lewis and Walker, 1970) may be an effective mechanism for transporting nutrients to epontic algae, but this seems unlikely because major brine drainage coincides with summertime disintegration of the ice, at which time production is rapidly decreasing.

Table 6.1 gives mean annual (^{14}C) phytoplankton carbon uptake data for localities within the main body of fjords and polls at various northern latitudes. In recent years it has been demonstrated that the standard ^{14}C uptake methods for determining primary productivity probably seriously underestimate open ocean rates. Welschmeyer and Lorenzen (1985) show that, in one temperate fjord, a major potential source of error (herbivore domination by microzooplankton) is much reduced, but there is increasing evidence (e.g., Albright, 1983) that bacterioplankton production is an important component in fjords. Because of large seasonal and short-term fluctuations in phytoplankton productivity, it is doubtful whether the mean annual production has been closely estimated to date for any fjord. The values listed in Table 6.1 do not demonstrate any clear latitudinal trend. This may possibly reflect a sampling bias favoring the intense, but brief, spring-summer bloom periods (in which case these data are also predominantly for "new" production). However, high arctic localities are not represented; from the data compiled by Nemoto and Harrison (1981), the mean arctic shelf primary production rate may be of the order of 15 gC m^{-2} yr^{-1}.

In many fjords (e.g., Burrard Inlet and Puget Sound of Table 6.1), high phytoplankton production values may be the result of cultural eutrophication (see also Chapter 8). However, as emphasized previously, natural production can be significantly increased at times when the stability of the surface layer is reduced, permitting an increase in the nutrient supply. Hence relatively high production rates have been recorded in association with frontal zones (Pingree et al., 1978; Parsons et al., 1981). Such conditions characteristically occur in the vicinity of fjord sills, which, depending on the constriction cross-section and the volume flux, are likely to be re-

TABLE 6.1. Mean annual phytoplankton primary productivity in fjords and polls (gC m^{-2} yr^{-1}).

Locality	Latitude (°N)	Fjord	Productivity	Source
W. Greenland	69	Godhaven Disko	90	Anderson, 1977
	64	Godthaabfjord	95	Steeman Nielsen, 1958
N. Norway	69	Balsfjorden	114	Eilertsen et al., 1981 Eilertsen & Taasen, 1984
W. Norway	60.7	Lindaaspollene	95	Wassmann, 1983
	60.3	Vaagsbopollen	180	Wassmann & Aadnesen, 1984
	60	Korsfjorden	100	Erga & Heimdal, 1984
	59.2	Boknafjorden	140	Erga & Sørensen, 1982
Alaska	61	Port Valdez	150	Goering et al., 1975
	55	Boca de Quadra	140	Burrell, 1983b
W. Scotland	57.8	Loch Ewe	100	Steele & Baird, 1972
	56.5	Loch Etive	95	Wood et al., 1973
S. British Columbia	49.5	Howe Sound	140	Stockner et al., 1977
	49.3	Burrard Inlet	390	Stockner & Cliff, 1979
Washington	47.5	Puget Sound	465	Winter et al., 1975
Nova Scotia	44.5	Bedford Basin	220	Platt, 1975

gions of enhanced vertical turbulence. Parsons et al., (1983; 1984) have performed frontal-zone analyses at the mouths of two British Columbia fjords, and have demonstrated that elevated productivity occurs when the sill is relatively shallow.

6.1.3 Biogenic Sedimentation

One or more annual intensive blooms of "net" (> 20 μm) phytoplankton, primarily diatoms, are characteristic features of fjords (Fig. 6.3B). As noted previously, many diatom species are adapted to a sinking-upwelling mode of existence, and the large senescent or nutrient-starved cells that occur during late stages of a bloom tend to sink much faster than healthy cells (Bienfang et al., 1982). Lännergren (1979) has recorded settling rates of 9 m day^{-1} following the peak of the diatom bloom in a western Norwegian fjord.

Many examples may be cited where substantial fractions of the spring bloom are sedimented to the floor of relatively shallow fjords (< 220 m: Gucluer and Gross, 1964; Davies, 1975) and seas, such as around the margins of the Baltic. Elmgren (1980) has recorded a 40 to 50% transfer of the largely undecomposed bloom in a western Baltic inlet, and Theede (1981) notes that as much as 60% of the relatively high primary production occurring in the Kiel Bight may reach the sediments. It would be expected that progressively less phytoplankton detritus would arrive at the bottom with increase in water depth because of consumption and decomposition. However, there is increasing evidence that diatoms may also be seasonally sedimented to the benthic floor under open ocean conditions (Davies and Payne, 1984), and in very deep water (Billett et al., 1983).

One important factor permitting survival of intact diatom cells to the deep fjord sediments is reduced grazing pressure. The peaks of primary producers and consumers frequently do not coincide in fjords (Section 6.1.2); especially early in the season at high latitudes (Nemoto and Harrison, 1981) and in polls (Wassmann, 1983). But survival, and seasonally pulsed sedimentation, could also be the result of sedimentation rates much higher than would be predicted given the general size and density range of diatom tests and fragments. Settling rates of diatoms were found to be around 10 m day^{-1} in a Norwegian (Wassmann, 1983) and a Canadian (Hay, 1981) fjord, in line with laboratory results obtained for tests of single species (Smayda, 1970). However, much higher rates have been reported by Bodungen et al. (1981) in the Baltic. Intact *Thalassiosira* tests have been recovered at 385 m in an Alaskan fjord within two weeks of the bloom (Fig. 6.4). Billett et al. (1983) suggests that accumulation of diatoms at density discontinuities (as illustrated in Fig. 6.3A) may lead to aggregation and subsequent sinking rates as high as 100 to 150 m day^{-1}.

Although usually lagging the peak diatom bloom, herbivorous zooplankton populations, especially the larger copepods, generally increase in the fjord euphotic zones through the spring-summer period contemporaneously with increased stratification (Hargrave and Taguchi, 1978), and phytogenic debris may be scavenged in the form of fecal pellets (Mullin, 1980). Lewis and Syvitski (1983) have demonstrated that fjord zooplankton may also actively ingest inorganic particles, which are adsorption sites for nutrients and bacteria. Fecal pelletization performs two functions of importance to the efficient downward transfer of detritus: it increases the mean particle size, and encapsulates the detritus within a protective membrane. Most of the mass flux to the sediments is in the form of large particles (McCave, 1975), and pellets from the larger copepods may settle at rates well in excess of 100 m day^{-1} (Smayda, 1969; Honjo, 1980; Bienfang, 1982). It should also be noted that sedimentation velocities are greater for elongated than for more spherical particles of equivalent volume (Lerman, 1979). Enclosing membranes delay pellet disintegration (Schrader, 1971), and Honjo and Roman (1978) have shown that membranes degrade at a much slower rate with decreasing ambient temperature, so that the flux of biogenic detritus in higher latitude fjords should be further enhanced. In any environment, however, the fecal pellet flux must generally progressively decline with depth. One reason for this is multiple reingestion by other organisms: Poulet (1976) believes that organic detritus constitutes the major energy source for some copepods. The relationship between water-column depth and the surviving fraction of euphotic zone production is not as simple in shallow coastal waters as appears to be the case in deep oceanic

FIGURE 6.4. Electron micrograph of intact diatom test (*Thalassiosira sp.*) at the sediment surface (283 m) of a fjord basin in southeast Alaska, recovered approximately 2 weeks after the termination of the spring bloom. (Hong and Burrell, unpublished data.)

regions (Suess, 1980). This is primarily because of the relatively greater importance of mixed-layer processes (e.g., Hargrave, 1973; 1980). Nevertheless, in a number of fjords it has been shown (see below) that the amount of energy-producing biogenic debris reaching the floor, and hence the level of benthic productivity, decreases with depth.

In recent years, direct observations from submersibles have shown that marine particulate matter is frequently present as fragile macroscopic aggregates: marine "snow." Because subsurface, small-scale turbulence is generally much reduced in fjord estuaries, such mucus-bound globules and filaments may be a common feature of fjords. Organic snow has been observed, for example, at all depths below the surface zone in a number of British Columbia fjords (Mackie and Mills, 1983; Farrow et al., 1983; Syvitski et al., 1985). Figure 6.5 (from Harrison et al., 1983) shows a representative sketch of a vertical profile observed in Saanich Inlet by Mills (1982). Since the downward flux is a function of particle size, Shanks and Trent (1980) hypothesize that marine snow (individual aggregates containing thousands of fecal pellets have been observed) may be the primary vehicle for transporting organic detritus to the sediments.

Nonglacial fjords may be important sites where organic carbon is lost from the biogenic cycle through sediment burial. Relatively high

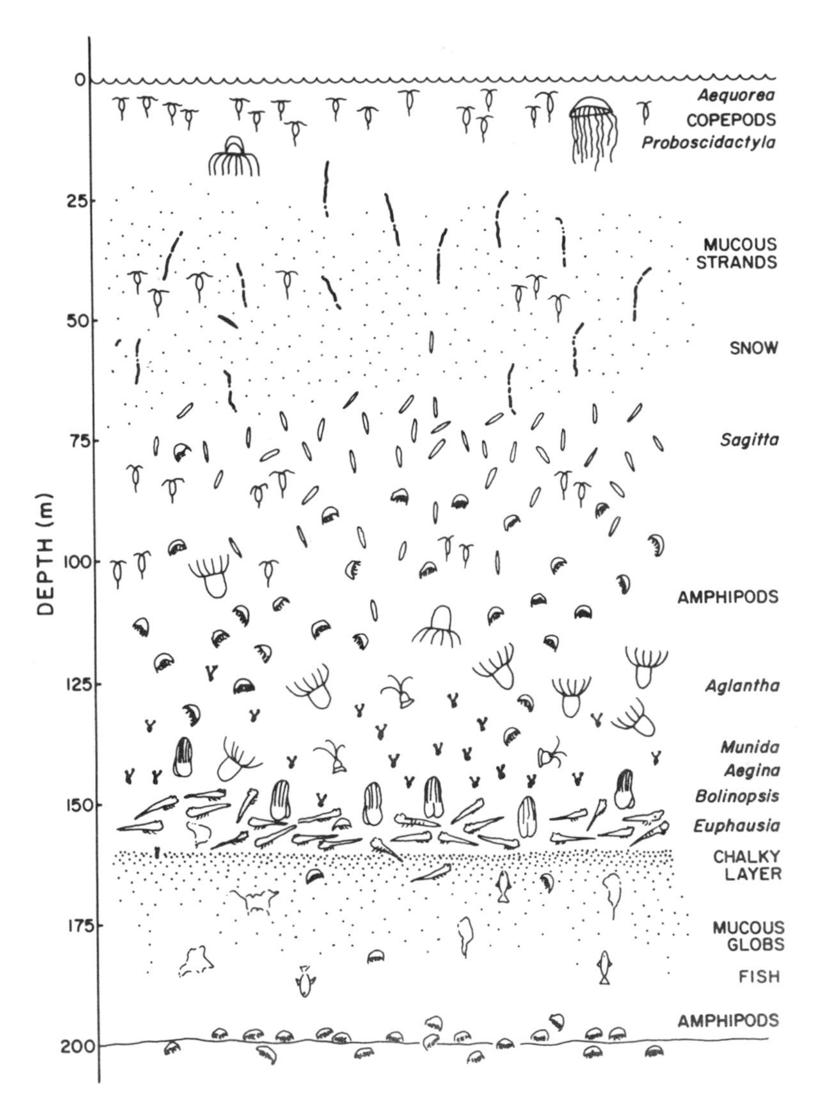

FIGURE 6.5. Composite sketch of the vertical distribution of plankton and "marine snow" observed from a submersible in Saanich Inlet, southern British Columbia. From Harrison et al. (1983) after Mills (1982).

inputs of refractory terrigenous organic material, low sediment temperatures, and high sedimentation rates favor incomplete degradation of organic detritus in the near-surface sediments (see Chapter 7). Figure 6.6 shows the mean annual carbon budget computed for a deep (circa 385 m) basin locality in a southeast Alaska fjord: here some 55% of the total carbon flux arriving at the sediment surface is buried. It has been suggested (Walsh et al., 1981) that burial of the carbon in shallow shelf and coastal marine areas may constitute an important sink for anthropogenic increases in atmospheric carbon dioxide.

6.1.4 Import-Export of Biogenic Material

The organic detritus sedimented in any particular fjord need not be predominantly autochthonous or terrigenous. Although the residence time of water in fjords is commonly considered to be much longer than in other types of estuaries, generally this would only be the case if,

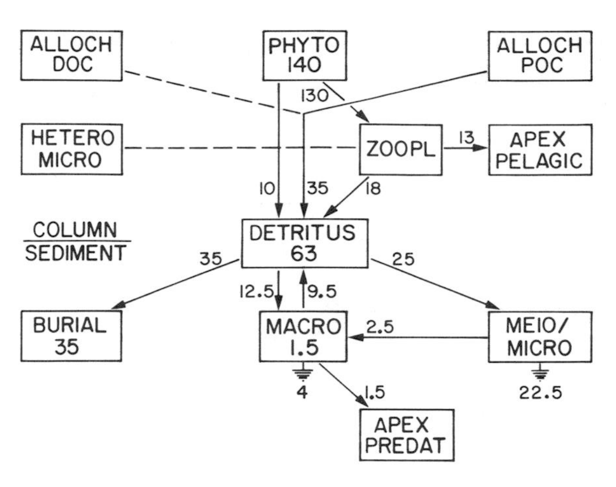

FIGURE 6.6. A partial mean annual carbon budget (gC m^{-2} yr^{-1}) for a deep basin (circa 385 m) locality in a southeast Alaska fjord (55°N). Phytoplankton and benthic macrofaunal production, benthic community respiration, POC flux out of the euphotic zone and to the sediment interface, and sediment carbon burial rates have been measured. Other values are derived assuming a 10% trophic transfer efficiency. Note that the rate of burial of organic carbon (55% of the total carbon flux to the sediments) matches the riverine input.

unrealistically, the entire volume of the fjord were involved. The layered, estuarine circulation zone may be flushed relatively rapidly. Except on time scales of the order of days (Mann, 1975; Lewis and Platt, 1982), most fjords thus cannot be considered as isolated ecosystems, and interchange of organic material between the fjord and the contiguous coastal region may be significant. Whether a particular fjord is a net importer or exporter of carbon basically depends on the relative levels of productivity inside and outside the estuary and on the water exchange characteristics (e.g., Burrell, in press), but individual evaluation is mandatory because of the multiplicity and interdependence of factors involved. Even though entrance sills may conveniently reduce the cross-sectional area of the seaward boundary of the fjord, the direct measurement of advective fluxes over a meaningful time scale is a very difficult practical operation. Platt and Conover (1971) attempted this for a small fjordlike inlet on the coast of Nova Scotia (over a 24-h period only), and determined that nearly 60% of the phytoplankton production was exported. In shallow temperate fjords, such as occur along the Danish coast, there may be major export of refractory detritus from intertidal and littoral plant communities.

As outlined above, phytoplankton blooms may originate and develop, inside or outside the estuary, when water stability, light, and grazing conditions are most favorable. But fjords, like all estuaries, theoretically are favorable environments for continuing relatively high levels of primary production through the summer because the estuarine circulation and frequently the rivers are sources of new nutrient supplies for the euphotic zone. For example, it appears that mean productivity is lower in the Strait of Georgia, British Columbia, than in many of the bordering fjords (Parsons et al., 1970). Although few comparative data are available, net export of energy would be expected from the "classic" type of fjord estuary. However, in very many individual inlets, conditions are far less conducive to sustained production after the initial bloom. In rugged terrain, freshwater inflow may be depleted in nutrients, and, where the catchment area is small, the volume flux may be relatively low, so that the resultant gravitational circulation is weak. An increase in near-surface turbidity or discoloration through the summer, limiting the euphotic depth, is also a common feature of fjords. In such cases, primary and secondary production may be greater in external coastal regions, generating the potential for import into the fjord. Stockner et al. (1977, 1979) have documented primary production in a portion of the Strait of Georgia and adjacent fjords near the outfall of the Fraser River. Phytoplankton production rates and biomass decrease up-fjord in Howe Sound with increasing turbidity, but are enhanced within nonturbid regions of the Fraser River plume (primarily because of entrainment of nutrients from the underlying marine water). Under these conditions, particulate carbon generated at the surface outside the fjord may be advected in and sedimented within the interior basins. This phenomenon also has been described for another coupled external river-fjord system by Therriault et al. (1984).

Since phytoplankters are passively transported with the near-surface estuarine water, an export flux from the fjord may be formulated (Ketchum, 1954) from mean measured flushing and reproduction rates. Lewis and Platt (1982) have shown that the interdependence of physical and biological scales can be quantified in terms of a scale length, which is a function of longitudinal dispersion and the biological reaction rate. Phytoplankton turnover rates are of the order of days or hours, and, where the rate of exchange of the surface water is relatively slow, indigenous populations may be substantially retained within the fjord. More vigorous seasonal or aperiodic circulation can generate drastic changes in both phytoplankton and zooplankton dynamics. Terminations or replacements of diatom blooms by different dominants as a consequence of episodic influxes of coastal water have been well documented for a number of fjords along the west and south coasts of Norway (Braarud, 1975; Kattner et al., 1983; Erga and Heimdal, 1984).

Recruitment of large oceanic copepods into the fjords and coastal waterways along the coast of British Columbia and southern Alaska is an important allied process. Ontogenetic migration of species such as *Neocalanus plumchrus* (Parsons et al., 1970; Harrison et al., 1983) and *Eucalanus bungii* (Krause and Lewis, 1979) result in late copepodite stages being present in shelf surface waters in the spring. The prevailing pattern of onshore convergence at this time of year may then transport these grazing stocks into the contiguous fjords. Cooney (1985) believes that this phenomenon may constitute a major source of energy to glacial fjords in Alaska, which have a paucity of autochthonous production.

6.2 The Fjord Benthic Environment

6.2.1 Subtidal, Soft-Bottom Benthos

6.2.1.1 Estuaries and Fjords

As commonly defined, an estuary is a relatively shallow body of water with marked spatial salinity gradients, usually subject to cyclic tidal disturbance. It was noted earlier in this chapter that, in this conventional type of estuary, intertidal and other fringing community primary production generally greatly exceeds phytoplankton production. Similarly, intertidal and shallow subtidal benthic production, both heterotrophic and autotrophic, is paramount. The extensive literature on estuarine benthic ecology (e.g., Wolff, 1983) therefore tends to emphasize stress effects resulting from mixohaline conditions, tidal currents, and suspended sediment dynamics. Because of the major influence of such abiotic factors, and because estuaries in general are recent and ephemeral geologic features, the estuarine benthos is usually considered to represent an early stage on the stability-time successional framework formulated by Sanders (1968).

A number of fjords—and especially polls—conform to this classic estuarine scenario, and in many others the physical-chemical environment at the head, and frequently the near-surface zone throughout the fjord, is subject to acute temporal and spatial gradients. Rosenberg and Möller (1979), for example, describe markedly different communities above and below the halocline maintained at approximately 15-m depth in fjords along the west coast of Sweden. However, the most characteristic and distinctive benthic environment of fjord estuaries is the relatively deep, subeuphotic, basin-floor substrate, and it is this environment, and the organisms adapted to it, that are of special concern here. The emphasis is primarily on the macrofauna (organisms larger than about 1 mm). Information on the subtidal meiobenthos of fjords is quite sparse, although Marcotte (1980) suggests that meiofaunal biomass may exceed that of the macrofauna in deep water environments. A general review of fjord benthic ecology (both shallow and subtidal) has recently been given by Pearson et al. (in press).

6.2.1.2 The Physical-Chemical Environment

Fjords are relatively recent geologic features, and are sites of net, and frequently very high, sediment accumulation. Away from the principal river discharge zones, especially in multisilled fjords, particle-by-particle deposition from the overlying water column is believed to be the primary mode of sedimentation. Bottom current shear is generally very low in the "quiet" basin environment, and the cyclic resuspension and secondary deposition of sediment, characteristic of shallow tidal inlets, is unimportant. Thus, in one deep Alaskan fjord basin, net sediment accumulation rates, and the mean annual flux into traps moored 5 m above the interface, were found to closely agree. However, episodic sedimentation events, such as slumping from steeply sloping walls, and ice rafting of coarse-grained material, are also common features in many fjords. The biological impacts of "catastrophic" sedimentation are addressed separately below.

An annual temperature range of at least 5°C is the norm at depth in temperate-zone fjords (e.g., Milne, 1972; Pickard and Stanton, 1980). Conversely, temperatures at the bottom of boreal and polar fjord basins will generally be considerably lower than the mean in shallow estuaries at lower latitudes, and are likely to be remarkably uniform year-round and from year-to-year. Figure 6.7A shows the limited annual temperature fluctuations at 365-m depth in a fjord in southeast Alaska (the multi-year warming trend here probably reflects El Niño conditions in the adjacent Gulf). Lewis (1983) has cited annual variation of less than 0.5°C in two British Columbia fjords. One important consequence is that the rates at which the microbially mediated degradation of organic detritus occurs will be approximately constant year-round. Seasonal fluctuations of the degradation rates in temperate coastal sediments result in vertical migrations of geochemical reaction zones (Chapter 7). This is generally absent, or occurs to a far lesser extent, in deep, higher latitude fjord sediments. Low mean sediment temperatures are a major control on the productivity of the deep fjord benthos because respiration rates decline exponentially with decreasing temperature. It has been frequently reported (e.g., Myers, 1978; Rhoads and Boyer, 1982) that "azoic conditions" may be approached at very low ambient temperatures. In fjord basins where the density of infaunal organisms is low (usually because of a limited food supply), and productivity is also low, sediment bioturbation is far less prevalent than in the more widely studied temperate estuaries. (Reduced oxygen contents and deeper water conditions, as well as low ambient temperatures, are contributing factors.) Conversely, in fjords where such conditions do not apply, rates of biological mixing as great as in other types of estuaries are possible. In the temperate and relatively shallow Loch Creran, Scotland, Pye (1980) found that large burrowing crustaceans and fish could disturb sediment to a depth greater than 50 cm; at a density of some 5 organisms m^{-1}, the mean surface area at these localities was increased by 10%. The degree of biological mixing of the near-surface sediments is an important control on the rate of vertical transport of both solid phase and dissolved interstitial water constituents as discussed in Chapter 7.

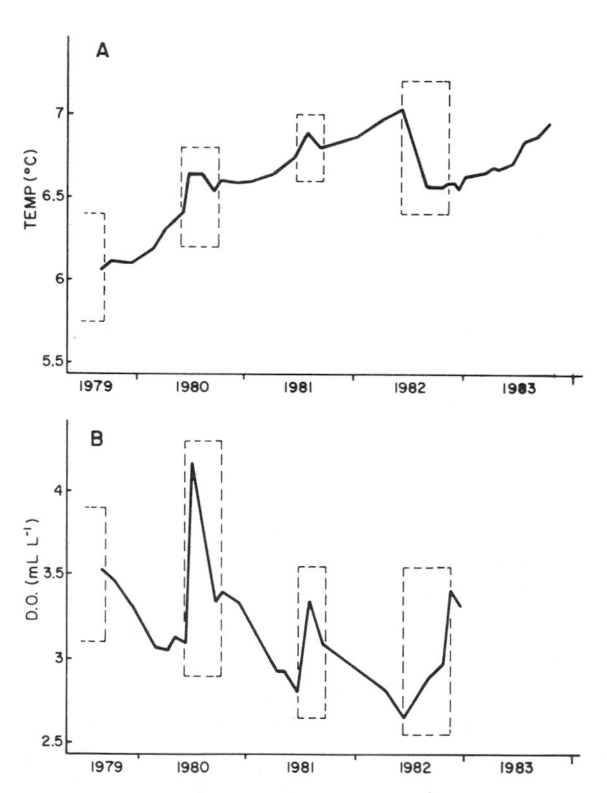

FIGURE 6.7. Multi-year variability in water properties at the bottom of a circa 385 m basin in Boca de Quadra fjord, southeast Alaska (55°N). (A) Temperature (°C) pattern. (B) Dissolved oxygen concentrations (ml L^{-1}). The boxed regions mark periods of bottom-water renewal. (Fig. 6.11 shows macro-infaunal density variability over the same period at an adjacent locality.)

Dissolved oxygen concentrations in the bottom waters of fjord basins are highly variable, depending on renewal processes and rates, and on rates of utilization in both the water column and in the near-surface sediment. Figure 6.7B shows multi-year variability in dissolved oxygen levels at the bottom of a deep fjord in southeast Alaska where the basin water has been renewed annually in the summer. Burrell (in press) has computed the wintertime utilization rate in the basin to be circa 0.75 ml l^{-1} yr^{-1} (very much smaller than values cited for fjords farther south in southern British Columbia and Washington: see Section 7.2.4), and anoxic conditions were not approached over the 1979–82 observation period. Here, progressively lower mean oxygen concentrations in each of the three consecutive winters illustrated appear to be a result of systematic, long-term changes in the character of the shelf-derived, summer replacement water. Dissolved oxygen concentrations at the bottom of fjord basins can be reduced to zero where the water is flushed at infrequent intervals, or where the flux of labile carbon, and degradation rates, are relatively high. The impact of low oxygen tensions on benthic biota is considered in Section 6.2.4.

Many fjord basin sediments contain high concentrations of organic matter, and the redox front may be within millimeters of the interface (Revsbech et al., 1980), especially where the sediments are largely undisturbed by physical and biological processes. Such an environment severely restricts the niches available to macro- and meio-infauna that are obligate aerobes. Figure 6.8 illustrates the community structure and density of organisms inhabiting 0.05 m^2 of sediment at a subeuphotic site within Sullom Voe (Shetland Islands), where the input of organic detritus is high, and bottom currents are small (Pearson and Eleftheriou, 1981). The oxygenated surface zone here is only a few millimeters deep, and the benthos is characteristically dominated by deposit-feeding species concentrated near the interface. To some extent, macro-infauna can locally extend the aerobic zone, but, because of lower faunal densities and metabolic rates, biological stirring of the sediments is generally less important in deep fjords than in most other estuarine environments.

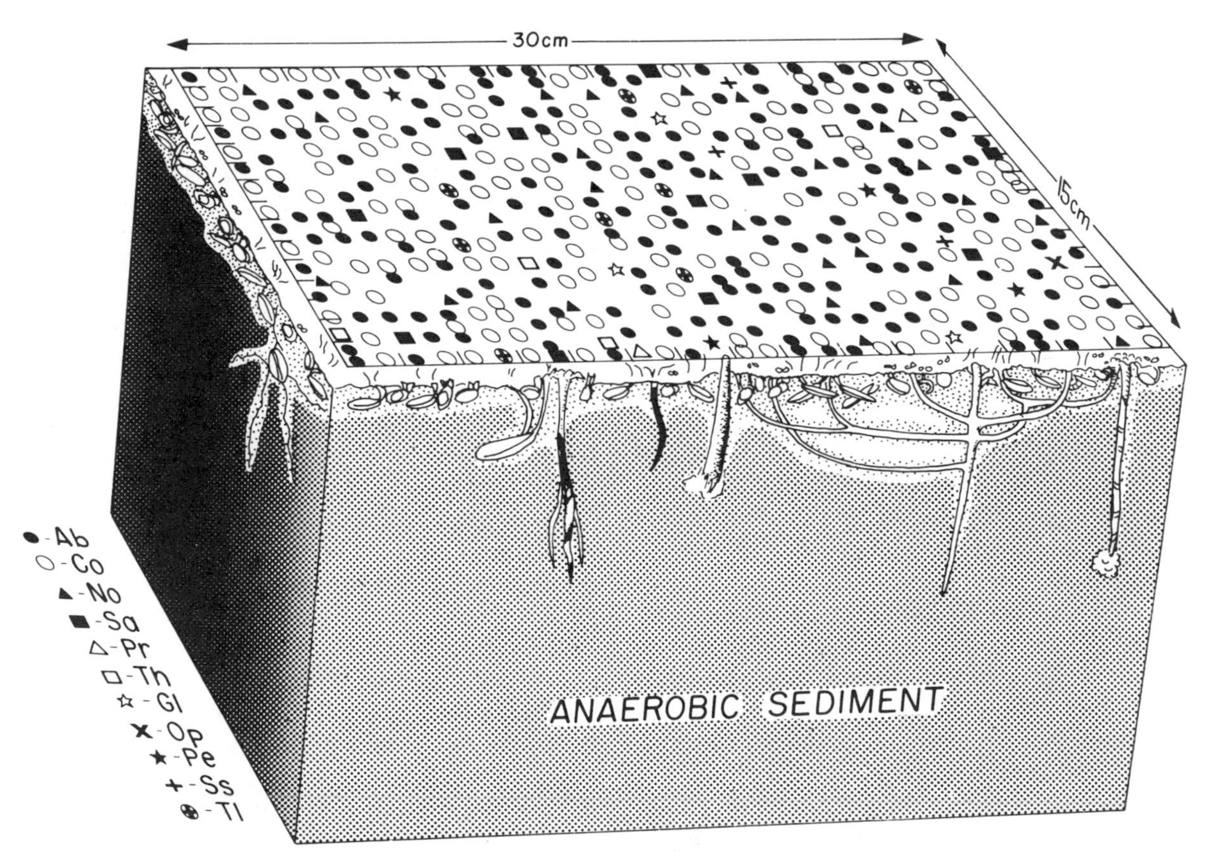

FIGURE 6.8. A representation of the community structure and density of organisms inhabiting 0.05 m^{-2} of sediment in a region of Sullom Voe, Shetland Islands, having low bottom currents and a relatively high organic flux. The oxygenated zone (light stipple) is < 1 cm deep, and the fauna is dominated by small deposit-feeding organisms. Ab–*Abra alba*; Co–*Corbula gibba*; No–*Notomastus latericeus*; Sa–*Scalibregma inflatum*; Pr–*Prionospio spp.*; Th–*Thyasira spp.*; Gl–*Glycera alba*; Op–*Ophiodromus flexuosus*; Pe–*Pectenaria koreni*; Ss–*Spisula elliptica*; Tl– *Tellina fabula*. From Pearson (1980b).

6.2.1.3 Benthic Community Structure and Organization

In marked contrast to those of shallow, temperate-zone estuaries, fjord benthic communities have been extensively examined in only a very few localities, and it is perhaps unfortunate that most of these studies have dealt with either artificially or naturally vigorously stressed systems. Consequently, relatively little is known about subtidal infaunal communities in the majority of fjord environments, either in terms of analytical structure (sensu Watling, 1975), or of trophic organization. In general, the soft-bottom, fjord-floor environment is one of physical homogeneity, both in space and time. Silled fjords of moderate depth are less subject to conditions that strongly affect the community patterns developed in shallow temperate estuaries (Boesch et al., 1976): to vigorous tidal mixing, or to marked daily and seasonal variability in temperatures and salinity, for example. Gray and Christie (1983) note that the coastal benthos below around 100 m appears to remain "stable," in terms of persistence, over long periods of time; conversely, such communities are likely to lack disturbance resilience, and are highly vulnerable to stress (e.g., Jerneløv and Rosenberg, 1976). Lie and Evans (1973) and Josefsen (1981) have recorded multi-year species persistence in deep water in Puget Sound and in the Skagerrak, respectively.

In some silled fjords, it has been shown (e.g., Burrell, 1983b) that surficial sediment blanketing the relatively flat basin floor is remarkably uniform in terms of grain size distribution and chemical composition (such as the organic carbon content). In such homogeneous and persistent physical environments, benthic communities characterized by organisms having relatively high standing crops but slow turnover rates—biologically accommodated, K-selected assemblages as defined by Sanders (1968)—would be predicted (Gage, 1972a; Pearson, 1980a; Levings et al., 1983). This has been observed in a number of localities. For example, extensive surveys along the Gulf coast of Alaska (Feder and Jewett, in press) show that the subtidal macrofauna in many fjords is dominated by a few large, long-lived organisms and is hence relatively unproductive.

It would be expected, a priori, that the infauna of physically and chemically homogeneous sediment should approximate a random distribution. Inevitably, the most thoroughly studied fjords are those that have been extensively impacted by man (e.g., Rosenberg, 1974; 1977). Nevertheless, it appears to be generally true that fjord soft-bottom macro-infauna occurs in patches having a wide areal scale range. One commonly used measure of patchiness is the sample dispersion index (I), formulated in terms of the Poisson distribution: $I = \text{variance}/\text{mean}$. Perfectly randomly dispersed samples would have an I value of unity, and $I > 1$ indicates a tendency to aggregation, depending upon the Poisson expectation for that sample. Spatial heterogeneity of the benthos appears to be the norm in all marine coastal environments. Patterns of benthic recruitment from the water column (see below) are believed to be a major contributing factor, and faunal aggregation generally appears to be a function of localized differences in the sediment and sedimentation regimes. Catastrophic disturbance of the fjord sediment, discussed in Section 6.2.4, will almost certainly result in an aggregated fauna.

Biologically adapted "equilibrium communities" are diverse assemblages of organisms; that is, they comprise many species that tend to be evenly distributed. In localities where the benthic environment has been stressed, naturally or by man, diversity is generally observed to decrease (e.g., Mirza and Gray, 1981; Rygg and Skei, 1984: see also Chapter 8). However, in a number of documented cases where the fjord benthos has been impacted by organic effluents (Pearson, 1975; Rosenberg, 1976: see Section 8.12), temporal and spatial diversity trends are confused and cannot be easily correlated with periods of pollution discharge or abatement. This is largely because the most commonly used diversity index—H', the Shannon-Wiener diversity function (Pielou, 1975)—combines two basically independent components: species richness (abundance) and "evenness." Species abundance, per se, frequently has been found to be a very useful bioassay tool (Sheehan, 1984). However, the evenness component of H' usually predominates (Josefsen, 1981, has termed H' a dominance-diversity index), and under increasingly depauperate conditions assumes overwhelming importance. Some forms of stress, such as low ambient oxygen levels (Rosenberg, 1977), may affect all species to an approximately equal extent, so that evenness tends to be preserved. However, it has been more commonly observed that, as the impact on the benthic environment increases, rarer species are disproportionately eliminated, and there is a concomitant increase in the number of species present at higher abundances (Pearson et al., 1983). Toxic chemical pollution invariably appears to differentially affect specific classes of organisms or trophic levels. For example, in fjord environments, both metal (Thomas and Seibert, 1977) and oil (Johansson, 1980) pollution have been noted to severely impact the zooplankton, permitting enhanced phytoplankton blooms. The important effects of sediment pollution on the structure and organization of indigenous organisms are discussed further in the following sections and in Chapter 8. It should be noted that, other factors being equal, benthic community diversity should naturally be lower in well-sorted sediments, rather than in more heterogeneous grain-sized sediments, which will have a greater diversity of potentially habitable niches (Gray, 1974).

Under natural conditions, community dominance by one or a few species may occur periodically as a consequence of differential larval recruitment. In most fjords, invertebrate larvae meroplankters are at a maximum in the water column in spring-fall, coupled to the primary production cycle. However, it should be noted that in high-latitude environments, the benthos increasingly employs brooding rather than pelagic larval development. Thorson (1950) has related this trend in part to reduced phytoplankton production, and to a previously noted tendency for non-synchronization of the primary and secondary pelagic production cycles. Seasonal recruitment to the benthos generates high abundances and local dominance by a few species, but die-back and re-establishment of more even distributions are likely to occur through the succeeding winter (see Fig. 6.11, which follows). Kiørboe (1979) has demonstrated an approximately 50% population reduction between summer and the following spring in a Danish fjord. In cases where detritivore-dominated communities locally destabilize the near-surface substrate (see Section 6.2.3.4), Eagle (1975) has shown that drastic year-to-year changes in community composition may occur if larvae of the dominants are unable to settle. As a generality, however, Pearson (1970, 1980a) believes that the physical confinement of fjord basins is likely to lead to "saturation colonization," and that such a mechanism may help to explain the frequently recorded lack of readily identifiable classical "discrete" communities (sensu Petersen, 1913) in fjord estuaries. A gradational, rather than a discrete community structure, would also be favored where the fjord-floor sediments were areally homogeneous.

Deposit-feeding organisms become proportionally more important with depth in the "quieter" environments of deep fjord basins

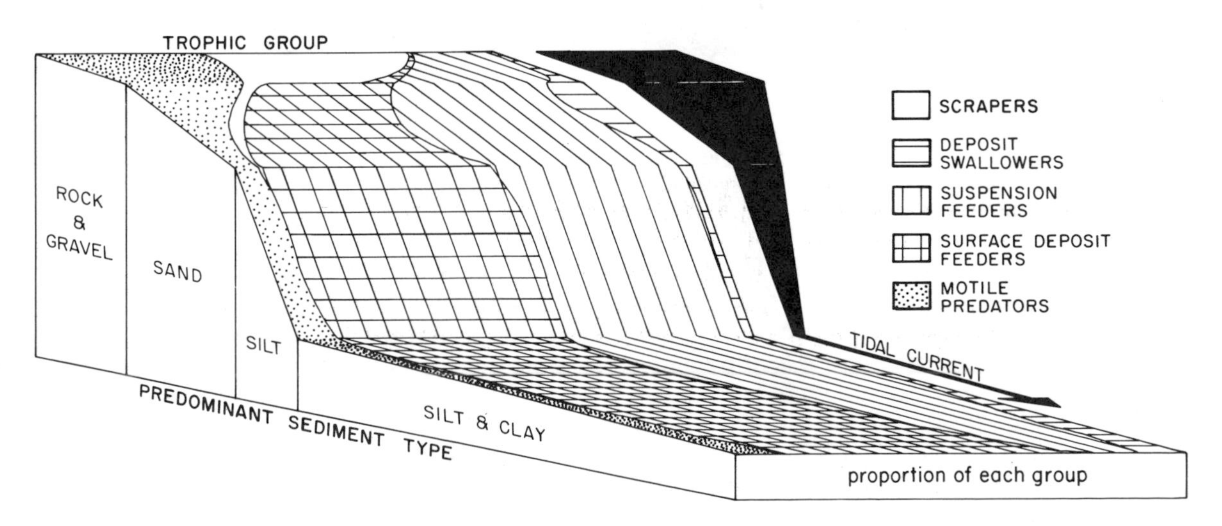

FIGURE 6.9. Schematic of proportional distribution of benthic trophic groups with depth, and in relation to bottom tidal currents and substrate categories, in Lochs Eil and Linnhe, west Scotland. Detritivores predominate at depth, displacing suspension feeders. which prefer coarser-grained sediment and higher bottom currents. From Pearson and Rosenberg (1978).

FIGURE 6.10. (A) Station IT1 in Itirbilung Fiord at 167 m. Bottom current direction (orientated vane) is sub-parallel to the grooves. Stubby anemones occur with abundant ophiuroids belonging to several generations. (B) Station IN1 in Inugsuin Fiord at 160 m (both this and the locality shown in (A) are near the fjord head). Shown is the rheotactic behavior of elongate anemones leaning into current from northeast. The compass weight has a diameter of 7.5 cm (photos after Farrow, 1983).

(Fig. 6.9; from Pearson, 1971), paralleling a similar trend in the open ocean (Jumars and Fauchald, 1977). This characteristic is related in part to the absence or diminution of bottom currents (deWilde, 1976; Petersen, 1978; Wildish and Peer, 1983). However, sessile filter feeders are abundant on rocky fjord walls, and in other localities, such as in the vicinity of sills, where flow over the sediments is enhanced (Section 6.2.2). Figure 6.10 (Farrow et al., 1985) shows epibenthic anemones and ophiuroids at localities within two Baffin Island fjords where stronger bottom currents occur.

Although there are numerous exceptions, it is probably true that fjord basin sediments are characteristically fine grained, poorly sorted, and relatively compact. Such substrate conditions should result in minimum interstitial pore volumes, and hence meiofaunal densitites lower than in shallower, more dynamic estuarine

FIGURE 6.10. *Continued*

areas. The data of Marcotte (1980) do not, however, support this, and Hulings and Gray (1976) suggest that, in general, pore space is not the major factor determining meiofaunal abundance. Such a deep fjord sediment environment would also favor infaunal organisms that live and feed close to the interface (hence decreasing the intensity of subsurface biological mixing). It is unfortunate that little is known about general trends in the functional morphology of fjord benthos. Types of fauna favored under specific circumstances—such as chemical pollution, or high sedimentation rates—are better documented, and this topic is addressed in Sections 6.2.3.4 and 6.2.4.

6.2.1.4 Biomass and Production

It is convenient to distinguish between shallow and deep soft-bottom benthic production in fjords. The shallow water regions of fjords, as in any estuary, are subject to variable physical conditions: in particular, to daily and seasonal fluctuations in currents, salinity, temperature, and turbidity. In high-latitude fjords, sea ice may be a major stress factor. On present evidence, it appears that both biomass and production in the "physically dominated" regions of fjords are as high as, or higher than, in other types of estuaries at equivalent latitudes. This is evident, for example, from comprehensive surveys of the shallow water macrofauna of inlets along the west coast of Sweden (Rosenberg and Möller, 1979; Pihl and Rosenberg, 1982; Möller and Rosenberg, 1983).

In deep water (an environment generally absent from conventional estuaries), both biomass and productivity progressively decrease with depth; except that, in high arctic fjords, biomass and species density maxima are below the winter ice impact zone. Declining standing stock as a function of depth parallels the food supply (Wildish, 1977; Gray, 1981), since labile carbon is utilized throughout the basin column (e.g., Hopkins, 1981). Figure 6.11 illustrates seasonal differences in macro-infaunal densitites at two localities within adjacent shallow (140 m) and deep (385 m) basins of Boca de Quadra fjord, SE Alaska. Macrofaunal production at the deep basin site has been estimated (Burrell, 1983b: see Fig. 6.6) to be appproximately 1.5 gC m^{-2} yr^{-1}: Parsons (in press) advocates a mean regional benthic production value for the entire Alaskan fjord province of 2.6 gC m^{-2} yr^{-1}. The macrofauna of deep fjord sediments comprise predominantly sluggish and sedentary taxa, and productivity is much reduced compared with contiguous shallow water environments characterized by higher proportions of more rapidly maturing species. Decreased metabolic rates result primarily from lower mean temperatures, and, in some cases, from very low oxygen levels. Relatively large, long-lived, low metabolic rate organisms are also an adaptation to the seasonally pulsed food supply (description follows), which is a common feature of higher latitude fjords. Benthic production in shallow and/or more temperate fjords is intermediate between that in "shallow estuarine" and "deep basin" environments: for example, Davies (1975) has given a value of 30 gC m^{-2} yr^{-1} at 30-m depth in Loch Thurnaig, Scotland.

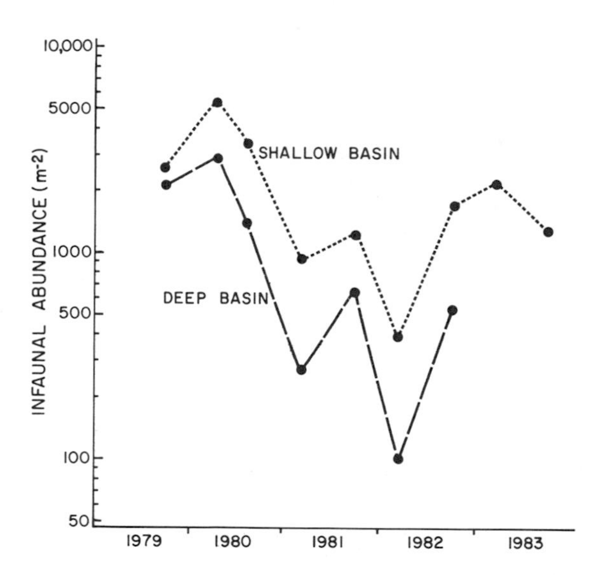

FIGURE 6.11. Seasonal and multi-year fluctuations in macro-infaunal abundances at the bottom of two adjacent shallow (innermost basin, circa 140 m) and deep (central basin, circa 385 m) fjord basins, Boca de Quadra, southeast Alaska. (Fig. 6.7 shows contemporaneous temperature and dissolved oxygen conditions.) Recruitment to the benthos occurs in the summer, followed by die-back and reduction in densities of many species each winter. Data from VTN (1984).

The contribution of microbial and meiofaunal fractions to fjord benthic production is largely unknown. Limited studies in higher latitude, intertidal, and shallow estuarine habitats (Elmgren, 1976; Ankar, 1977; Schwinghammer, 1981) have shown that the meiofauna may constitute between 5 and 70% of the total infaunal biomass, possibly increasing with depth (Marcotte, 1980), and with latitude. Since the smaller size fraction organisms are approximately 5 to 10 times more energetic than the macrofauna, recorded macrofaunal productivities may be only a small fraction of the total. However, certain characteristics of the deeper fjord substrate (for example, fine-grained and poorly sorted sediments, and the shallow redox front) may lead to reduced densities of meiofauna. Much of the bacterial and meiofaunal production is likely to be funneled through the macro-organisms, which are predominantly deposit feeders.

It was noted in Section 6.1.3 that pulses of autochthonous carbon may periodically be sedimented out of the euphotic zone, especially at the time of the spring bloom. This has been well documented (Smetacek, 1980) for shallow areas in the western Baltic where Graf et al. (1982) have also shown that labile detritus, rapidly transferred to the bottom in the spring, immediately stimulates benthic production (aided by increasing ambient temperatures at this time of year). Stephens et al. (1967) have recorded peaks in carbon input to the sediments of a British Columbia fjord that correlate with, but lag by a month, phytoproduction and riverine input maxima in the surface zone. It might be expected that this seasonality in the carbon supply to the benthos would decrease and become more diffuse as the depth of the overlying water increased, and hence that the relatively long-lived, deposit-feeding organisms, which constitute a major component of the macrofauna of most fjord soft bottoms, would not be affected by short-term variability in the energy supply. However, it is the larger sized particles, having settling rates of the order of 100 m day^{-1} (McCave, 1975; Shanks and Trent, 1980), that are responsible for most of the particulate carbon sedimenting to the basin floor. Billett et al. (1983) have demonstrated a spring peak in the organic detrital flux (and tentative metabolic responses by the benthos) at a depth of 2000 m in the northeast Atlantic. Since both autochthonous carbon production and the freshwater influx of organic material into many fjords are seasonally pulsed, corresponding maxima in the fraction arriving at the benthic interface may be the rule. Figure 6.12 illustrates a basically bimodal pattern of carbon flux at the bottom (385 m) of an Alaskan fjord, corresponding to phytogenous detritus immediately following the spring bloom, and predominantly terrigenous (elevated C/N ratio) material in the fall.

6.2.2 The Subtidal Epilithic Environment

Since fjords characteristically have steep rocky walls, the sub-tidal epilithic fauna may be quantitatively and energetically very important. However, unlike the soft-bottom benthos, these organisms have very little direct effect on sedimentation. Unfortunately, primarily because of sampling difficulties, comparatively little is known about these communities; although, in recent years, there have been a number of photographic surveys (e.g., Gulliksen, 1980; Sandnes and Gulliksen, 1980; Christie and Green, 1982) and direct observations from submersibles (Tunnicliffe, 1981; Levings et al., 1983).

Epilithic organisms are primarily suspension feeders and sedentary, and are hence particularly susceptible to sediment loading (Evans et al., 1980; Farrow et al., 1983). Figure 6.13 illustrates the longitudinal distribution of selected epilithic groups along the walls of the highly turbid Knight Inlet, British Columbia. Here, the apparent response to sediment loading varies between taxa. The density of the dominant organisms, serpulid polychaetes, decreases up-

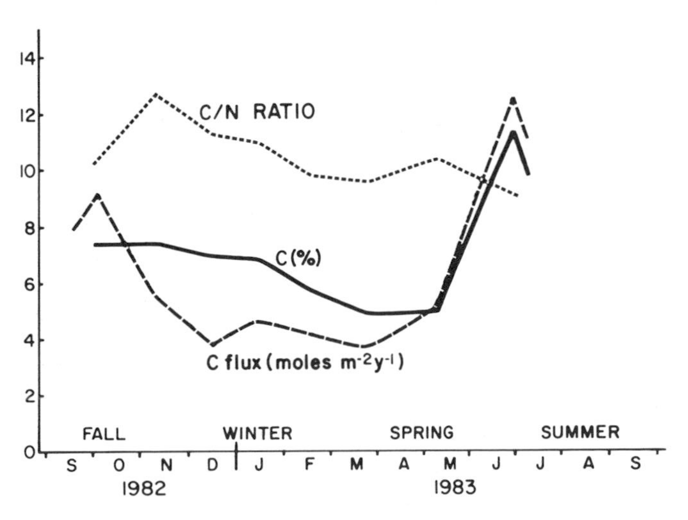

FIGURE 6.12. Seasonal variations in the character of biogenic sediment reaching sediment traps located circa 5 m above the floor of a deep (385 m) fjord basin in southeast Alaska. The carbon flux is bimodal: one peak lags the spring diatom bloom in early summer, but the major input is of predominantly terrigenous (high C/N ratio) material in autumn. (Hong and Burrell, unpublished data.)

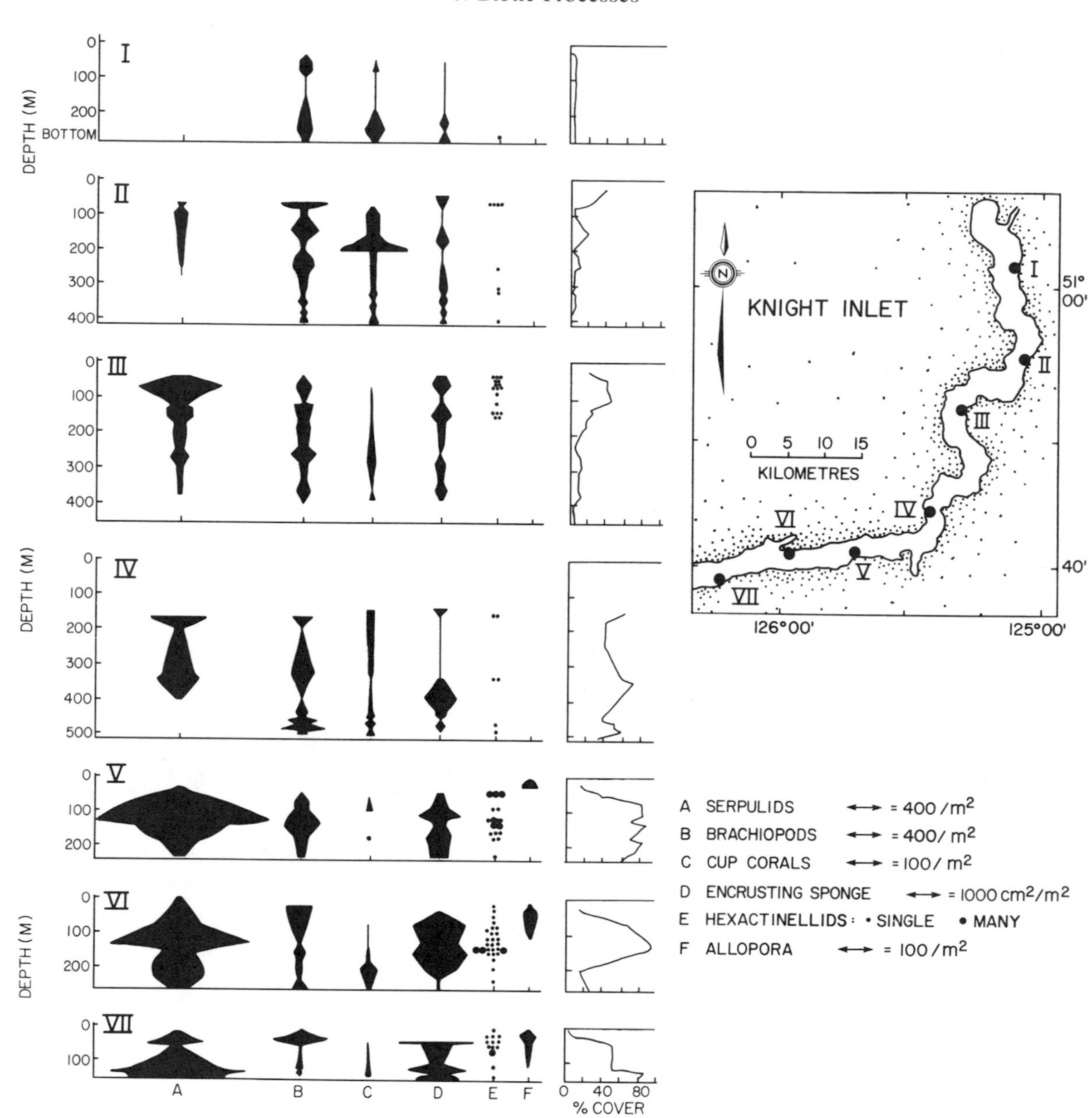

FIGURE 6.13. Vertical distribution of epilithic benthos along the walls of Knight Inlet, a turbid fjord in southern British Columbia. From Farrow et al. (1983).

fjord with increase in the sedimentation rate; conversely, encrusting sponges increase in both size and abundance toward the head, and the brachiopods appear to be approximately evenly distributed throughout the length of the fjord. A comprehensive, multi-year investigation of the rocky bottom fauna in Borgenfjorden, Norway, by Gulliksen (1980) has shown that, although total abundances fluctuated seasonally (largely because of annual settlement), faunal composition and mean densities remained remarkably stable and persistent over the nearly four-year study period.

The sill areas of fjords are generally regions of increased bottom currents where rock and boulder surfaces are likely to predominate because fine-grained sediment is not retained. Such physically dominated localities are highly suitable environments for a number of attached, filter-feeding organisms (e.g., Gage, 1972b), since the enhanced flow supplies abundant food, and removes waste products. Although there have been no detailed studies of the benthos of fjord sills per se, the ecological implications of this type of dynamic environment have been reviewed by Darnell (1979), and Gulliksen (1978) has described faunal associations in a sub-arctic coastal gully.

6.2.3 Effect of Biota on Fjord Sediments

6.2.3.1 Sediment-Biota Interactions

Benthic biota play a major role in modifying and transporting deposited sediments, and studies of these processes have greatly increased our understanding of geologic sedimentary structures and environments (Frey, 1973, 1975). Animal-sediment relationships in the sea have been extensively reviewed in recent years (Gray, 1974; Rhoads, 1974; McCall and Tevesz, 1982), and the major processes and expected diagenetic characteristics are generally well known. In temperate-zone estuaries, and in the shallow regions of most fjords, the living and feeding behavior of the indigenous macro-infauna is probably the most important factor determining the biogeochemistry and dynamics of the near-surface, soft-bottom sediments. However, this is unlikely to be the case in many deep fjord basins (which are the particular concern of this chapter), where food limitations permit only a relatively sparse fauna, and where the physical-chemical subsurface environment favors communities of shallow dwelling and epibenthic species. In other fjords, exogenous physical factors may largely dictate the functional types of benthic organisms that may best inhabit the soft-bottom fjord floor; physically accommodated communities are considered in Section 6.2.4.

Two broad categories of biogenic sedimentary processes may be distinguished: those that bind and stabilize the near-surface sediment, and those that generate instabilities. This contrary behavior has led to functional group classifications of the macrofauna (Woodin and Jackson, 1979; Woodin, 1983), which may cut across conventional taxonomic and trophic categories. Brenchley (1981) and Wilson (1981) believe that biological disturbance of the sediment is particularly a function of the relative motility of the community members.

6.2.3.2 Effects of Organisms on Sediment Transport

Of primary concern here is the stability of the interface in response to shear from flow of the overlying water (discussed in Chapter 4). Biological processes that act to destabilize the sediment decrease the critical erosion velocity, and vice versa. Theoretical and laboratory treatments relating shear stress to mobilization usually relate to noncohesive substrates, and, although a number of studies have explored the dynamic properties of cohesive sediment (e.g., Moore and Masch, 1962; Rhoads et al., 1978b; Owen, 1977; Mehta et al., 1982), it is generally not possible to predict critical shear stress from knowledge of flow conditions alone (Richards and Parks, 1976). Nowell et al. (1981) have noted that there is very little experimental data available for fine-grained (silt-clay) sediment. Current velocities of the order of 20 cm s^{-1}, 5 to 10 m above the sediment surface, may be a reasonable operational value for the critical erosion velocity in fjords (Hay, 1981; Baker et al., 1983). Fjord basin soft-bottom sediments as a class may be quintessentially cohesive, and especially the fine-grained, organic-poor material flocculated and deposited in glacial fjords. Macdonald (IOS, Sidney, Canada; unpublished observations) has noted that finely ground, inorganic mine tailings discarded in British Columbia fjords similarly form very compact substrates that resist resuspension.

The situation becomes increasingly complex with increase in the mean organic carbon content. Much of the organic detritus, especially the terrigenous fraction, may initially be manifest as "flocculant" low-density material at the interface. There is some evidence (Burrell, 1983a) to suggest that this fraction is preferentially resuspended with episodic increases in bottom current shear, and may for that reason constitute a useful tracer of such events. The enhanced agglomeration characteristically associated with organic-rich sediment (Johnson, 1974) appears primarily to be due to exudates of mucopolysaccharides by, primarily, the micro- and meiobenthos (Newman, 1974; Frankel and Mead, 1973; Hobbie and Lee, 1980). Rhoads et al. (1978b) have demonstrated in the laboratory that mucus produced by micro-organisms present in concentrations much lower than may occur naturally can effectively bind the substrate. Clay mineral particles, which have large surface areas and carry strong negative surface charges, are prime candidates for adsorption of, and binding by, mucilaginous products (Rhoads and Boyer, 1982).

Organic binding strengthens the surface sediment (Reimers, 1982) and increases resistance to erosion. But the major influence of the macrobenthos is to decrease the critical shear stress required for resuspension or transport by raising the water content of the sediment. Organisms may also increase the microtopography ("roughness") of the interface: the effects of projecting animal tubes (Eckman et al., 1981; Rhoads and Boyer, 1982) and surficial tracks (Nowell et al., 1981) are notable examples of this behavior. It is now well known (Rhoads and Young, 1970; Brenchley, 1981) that the activities of deposit-feeding and mobile organisms lower critical erosion velocities, and in several fjords, a turbid zone observed directly overlying the sediments is believed to be caused by burrowing fish and other mobile benthic organisms (Farrow et al., 1983; Syvitski et al., 1983b). Detritivores agglomerate the surface sediment through the production of fecal and pseudofecal pellets. An individual macrodeposit feeder may injest a volume of sediment of the order of 100 $cm^3 yr^{-1}$ (Rhoads, 1974), so that, in temperate estuarine environments, the surface sediment zone may be processed many times each year (Myers, 1978). In Long Island Sound and adjacent estuaries, Young et al. (1981) and Bokuniewicz and Gordon (1980) have shown that fecal pellets may comprise the bulk of the surface 1 to 3 cm, and Gray (1981) notes that much of Oslo Fjord is similarly mantled. Although fecal pellets may not, per se, be more easily resuspended than the unreworked sediment particles (Nowell et al., 1981), the resultant high porosity texture greatly facilitates particle entrainment and transport.

6.2.3.3 Bioturbation

The life styles of many macro-infaunal organisms can cause significant mixing of the upper sediment layer (Fig. 6.14, from Rhoads, 1974). The best documented examples are those vertically oriented organisms that feed at depth and defecate at the surface to generate "conveyor-belt" (Rhoads, 1974) or "convective" (Fisher et al., 1980) sediment mixing. In a number of fjord regions, and over a wide depth range, the activities of large burrowing demersal and epibenthic species (megafauna), although present at relatively low densities (of the order of one individual m^{-2}), are potentially capable of mixing large volumes of sediment to depth of 30 to 50 cm or more (e.g., Chapman and Rice, 1971). Bioturbation is an important mechanism for the vertical movement of chemical constituents within the sediment. Some of the more commonly used mathematical models are reviewed by Matisoff (1982), and this topic is briefly considered in the following chapter. Irrigation from the surface to depth is quantitatively an even greater factor (Aller, 1978) in promoting the vertical transport of chemicals within the sediment, and in changing the bulk sediment chemistry, by modifying the redox environment (Aller, 1980a; 1980b; 1982), for example. In temperate-zone estuaries, the effects of bioturbation may vary seasonally with the annual temperature cycle (Aller, 1980d; Elderfield et al., 1981; Hines et al., 1982), but such oscillations are unlikely to be a feature of deep fjord sediments.

6.2.3.4 Community Dynamics

Clearly, the magnitude of the impact that biota will have on sediment structure and dynamics depends primarily on the density of the macroinfauna. But it is also apparent that the functional behavior of the infaunal community is important: certain locomotion and feeding modes promote mixing and instability of the interface. Rhoads and Young (1970) show that detritivores, by their reworking feeding habits, can generate habitat instability, which tends to exclude other trophic groups (trophic group amensalism). Rhoads and co-workers (Rhoads et al., 1978a; McCave, 1976; Rhoads and Boyer, 1982) and others have also discussed the ecological succession that would be expected along a gradient from depauperate or azoic sediment toward an equilibrium or climax community structure. This is a very important concept with regard to fjord sediments because such envi-

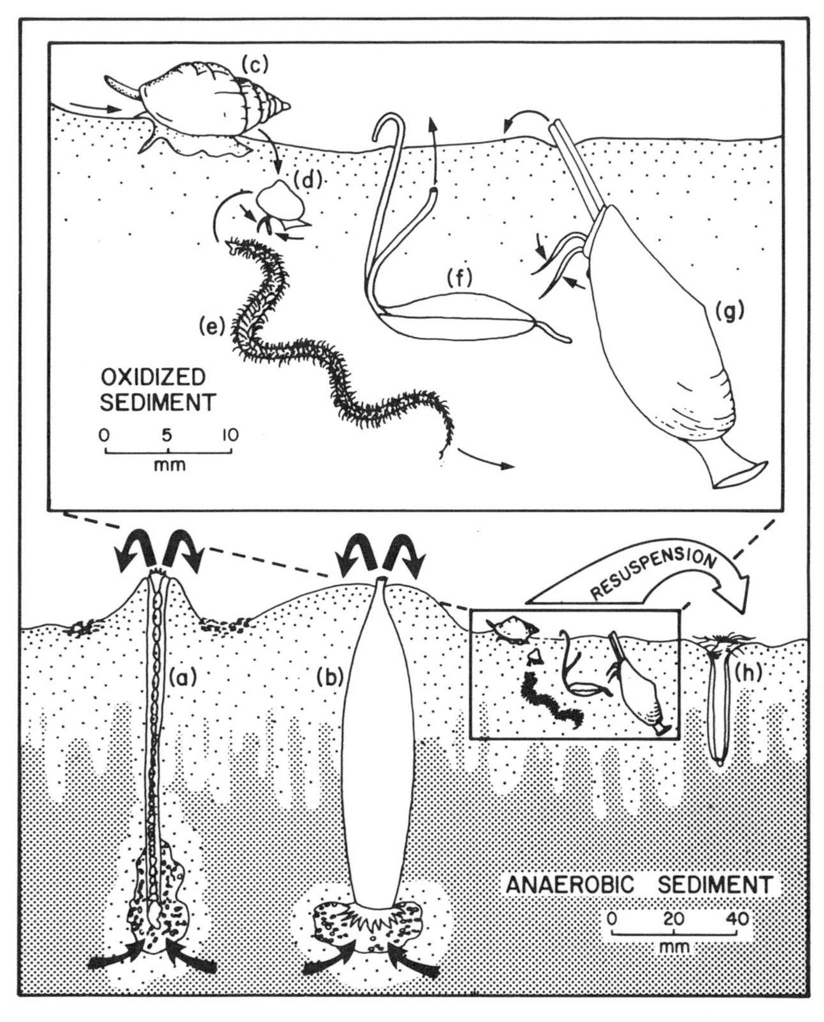

FIGURE 6.14. Mixing and resuspension of sediment by deposit-feeding organisms. (a) Maldanid polychaeta; (b) Holothurian; (c) Gastropod (*Nassarius*); (d) Nuculid bivalve (*Nucula sp.*); (e) Errant polychaete; (f) Tellinid bivalve (*Macoma sp.*); (g) Nuculid bivalve (*Yoldia sp.*); (h) Anemone (*Cerianthus sp.*) From Rhoads (1974).

ronments are susceptible to episodic physical and chemical disturbance that hampers attainment of long-term community stability. It has been shown (McCall, 1977; Rhoads et al., 1977, 1978a: Fig. 6.15) that during successional pioneering stages, surface-feeding organisms are favored. Such organisms tend to be chemically isolated from the substrate, so that physical disturbance owing to foraging and feeding below the interface is likely to be slight. Reduced bioturbation may, in turn, help maintain the redox discontinuity close to the interface: a positive feed-back favoring surface feeding (Levington, 1979).

The influence of the benthos on bottom sediments is summarized in Figure 6.16. In fjords, these effects, being primarily a function of the density of the macro-infauna, should generally decrease in importance with increasing latitude and basin water depth. In addition, where the fjord benthos is physically rather than biologically controlled (i.e., where physical-chemical disturbance is endemic), biological mixing of the sediment may be negligible. Regardless of any biotic conditioning of the interface, resuspension of deposited sediment is much reduced in fjords in comparison with shallow, tidally dominated estuaries. Near-bottom currents are usually weak, except episodically, and especially at times of deep water replacement (Skei, 1980; Stanley et al., 1981).

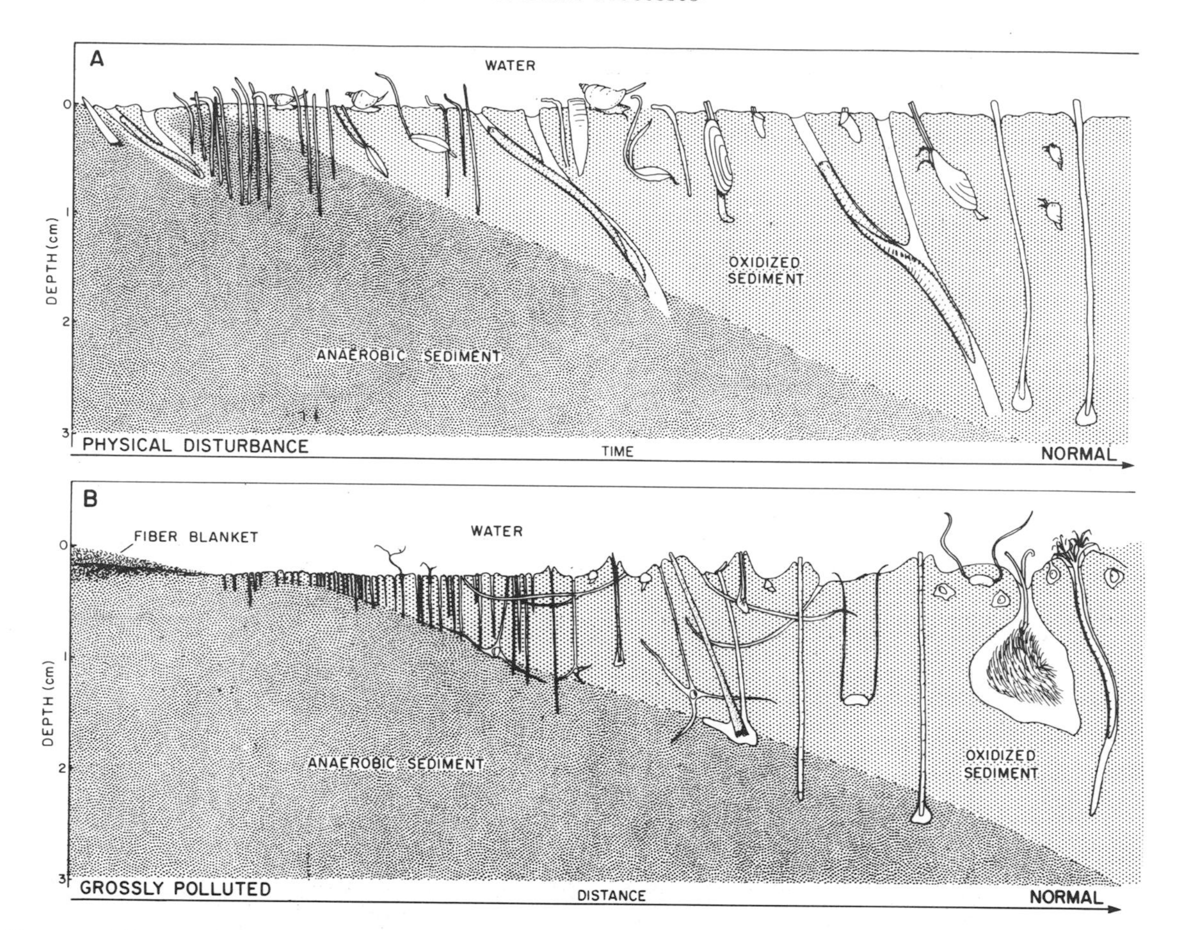

FIGURE 6.15. Development and successional trends of infauna responding to physical-chemical disturbances; the biota generates, and responds to, increasing aeration of the near-surface sediment. (A) Temporal ecological succession from a depauperate, predominantly surface-feeding, pioneering community to a more diverse "equilibrium" assemblage that includes deeper dwelling and feeding organisms. From Rhoads et al (1978a). (B) Gradient from organically polluted to "normal" sediment. Opportunistic species (r-strategists) initially respond to the high organic loading (see also Section 8.12). From Pearson and Rosenberg (1976).

6.2.4 Physical-Chemical Stress in Fjord Basins

6.2.4.1 Community Stress Effects

Earlier in this chapter it was suggested that, ideally, the subtidal, soft-bottom benthic environment of fjord basins might be expected to favor persistent, equilibrated (biologically accommodated) communities. However, many fjord basins may be periodically, or more typically, episodically physically or chemically stressed. The most important and common of such impacts are periods of depleted oxygen or anoxia, and various catastrophic sedimentation events, which may occur naturally, or be induced or exacerbated by man. The role of the resident benthos in modifying the sedimentary environment of fjord estuaries has been briefly surveyed in Section 6.2.3; here the reverse response of the biota to profound physical-chemical changes to the substrate is outlined.

The nature, spatial and temporal patterns, and periodicity of various exogenous impacts determine the responses of the resident soft-bottom benthos. Gray (1979) has distinguished between chemical stress, affecting productivity, and disturbances that destroy or remove individuals, and it is important to conceptually separate impacts which are stochastic and pervasive, and events differentially affecting particular components of the benthic biota. Pertinent ex-

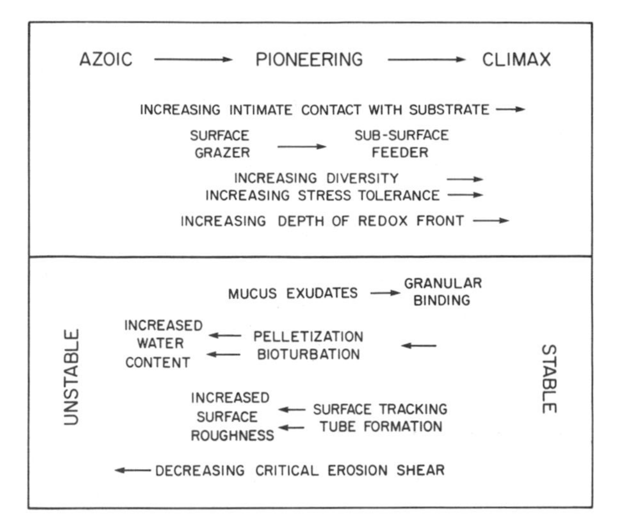

FIGURE 6.16. Soft-bottom, organism-sediment relationships.

amples of "catastrophic" events impacting fjord infaunal communities are: large fluctuations in the food supply or of sedimentation rates; periods of reduced oxygen or anoxia in the overlying water column; and ice scour. Massive or chronic anthropogenic heavy metal or hydrocarbon pollution also result in catastrophic-type stress (e.g., Rygg and Skei, 1984: see also Chapter 8).

Patterns of infaunal population changes subsequent to defaunating impacts would be expected to conform broadly to Rhoads' successional paradigm (Section 6.2.3.4: Fig. 6.15). However, Rhoads and Boyer (1982) note that this scheme may be less relevant in cases of persistent chemical stress when selectively tolerant species may assume dominance. Early colonizing organisms are typically rapidly reproducing, opportunistic species (r-strategists), responding to decreased competition, and to temporary resource abundances (Rosenberg, 1980; Thistle, 1981). As noted previously, such pioneering communities tend to be decoupled in a physical-chemical sense from the substrate, and hence their impact on the bulk properties of the sediment may be minimal.

Modes of re-establishment of the soft-bottom benthos following disturbance appear to be complex and have been studied in only a few estuaries. Pelagic larval settlement is commonly cited (e.g., McCall, 1977). The ability of invertebrate larvae to actively select the settlement substrate is well known (Meadows and Campbell, 1972; Crisp, 1974; Scheltema, 1974). For many organisms, successful colonization depends on conspecifics or other specific stimuli (processes that may naturally promote patchiness). By disrupting or eliminating these cues, physical disturbance may adversely affect larval settlement and survival (Levington and Bambach, 1970; Menzie, 1984). Pearson and Rosenberg (1978) have suggested that adult migration plays an important role in colonizing depauperate areas. Vagile species are prominent among recorded early colonizers (Rosenberg, 1972; Brunswig et al., 1976; Santos and Simon, 1980; Arntz and Rumohr, 1982), especially in the case of small-scale patches of impacted substrate. There is also some evidence (Gage and Coghill, 1977; Burrell, 1983b) to suggest that motile organisms may preferentially persist in physically disturbed environments. Surveys of naturally impacted shallow coastal sediments by McCall (1978) showed that organisms employing some form of brood protection were most resilient to physical disruption. Small-scale recolonization experiments conducted in an Alaskan fjord (Winiecki and Burrell, 1985) have similarly resulted in disproportionate colonization by such taxa compared with those having planktotrophic larval forms.

Pearson and Rosenberg (1978) observed that the soft-bottom infauna of stressed fjord estuaries are likely to comprise relatively low diversity but resilient communities. In general (e.g., Boesch and Rosenberg, 1981), communities subject to variable physical conditions are more resistant and resilient to other forms of stress than are the more diverse "equilibrium" communities adapted to uniform physical environments. However, it must be emphasized that infaunal "patchiness" is believed to be a natural attribute of marine substrates, even in cases (such as certain fjord basins) where the sediment appears to be relatively uniform, and that such frequent small-scale disturbance results in a spatial and temporal "mosaic" structure (Johnson, 1970; Grassle and Saunders, 1973). Intermittent perturbations may be the norm in a wide variety of environments (e.g., Sousa, 1979; Hughes, 1984), "climax" states may seldom or never be attained, and localized disequilibrium may be the natural state of marine communities. A number of investigators (e.g.,

Boesch et al., 1976; Pearson, 1981; Zajac and Whitlatch, 1982) have suggested that postdisturbance successions are likely to be unpredictable, depending on change factors to yield multiple community structures (i.e., the "neighborhood stability" concept; Gray, 1981).

The extent to which the community subsequently moves toward "stability" may depend upon the biological time scales of the dominant organisms relative to the frequency of the physical-chemical disturbance (Arntz, 1981; Pearson, 1981). Large, long-lived invertebrates are important components of many fjords, and, if the habitat is impacted catastrophically at frequencies of once a year or more, such species may not become established, or be re-established. As noted previously, seasonal abundance fluctuations of short-lived organisms are generally the most conspicuous feature of the fjordic benthos. Figure 6.17 is a schematic representation (see also Arntz, 1981) of both a single catastrophic event, and chronic chemical pollution, on short- and long-lived infaunal components.

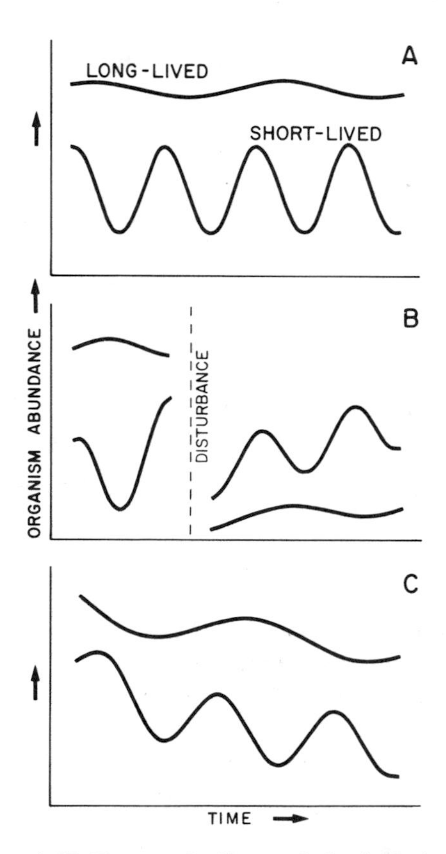

FIGURE 6.17. Temporal effects of physical-chemical stress on fjord soft-bottom benthos. (A) "Equilibrium" community of long- and short-lived organisms. (B) Effect of single catastophic disturbance. Recolonization is initially by rapidly reproducing, opportunistic species. (C) Effect of chronic, nondiscriminatory chemical pollution.

6.2.4.2 Sedimentation

It has been emphasized that high rates of sedimentation are characteristic of many types of fjords, and large-scale episodic deposition events, such as slumping from the head and sides of deep inlets, have been identified in fjords in many parts of the world (see Chapter 5). The mean range of sedimentation rates in all fjords ranges over at least five orders of magnitude. Rates of the order of 0.01 cm yr^{-1} are common in many relatively mature European fjords (e.g., Aarseth et al., 1975; Calvert and Price, 1970). But, in glacial areas, annual depositions as great as 1 to 10 m have been recorded (Hoskin and Burrell, 1972; Mackiewicz et al., 1984), and this flux may occur predominantly over a few months—or even weeks—during the summer. High sedimentation rates produce porous, noncompacted, and readily eroded substrates (Inderbitzen, 1970), and diffuse interfacial regions which constrain the type of colonizing fauna. Sessile species tend to be excluded from areas subject to high rates of sediment deposition (Jumars and Fauchald, 1977), and suspension-feeding organisms will be largely absent where poorly compacted sediment may be periodically resuspended by relatively weak bottom currents (Section 6.2.3.2). Thayer (1975) has discussed morphological adaptations of invertebrates to soft substrata. Using cluster analysis techniques, Feder and Matheke (1980) have identified two primary infaunal communities within a south-central Alaskan fjord (Fig. 6.18), associated with regions of relatively high and low glacial sedimentation, respectively. In the few fjords studied along the Alaska Gulf coast (e.g., Hoskin et al., 1978), infaunal biomass appears to be inversely related to the sedimentation rate. Gulliksen et al. (1985) have also recorded diminishing densities of macrobenthic organisms with decreasing distance from the heads of Svalbard fjords, ascribed to increasing sedimentation. As noted previously, benthic communities adapted to such conditions would be expected to be both resistant and resilient to sedimentation stress,

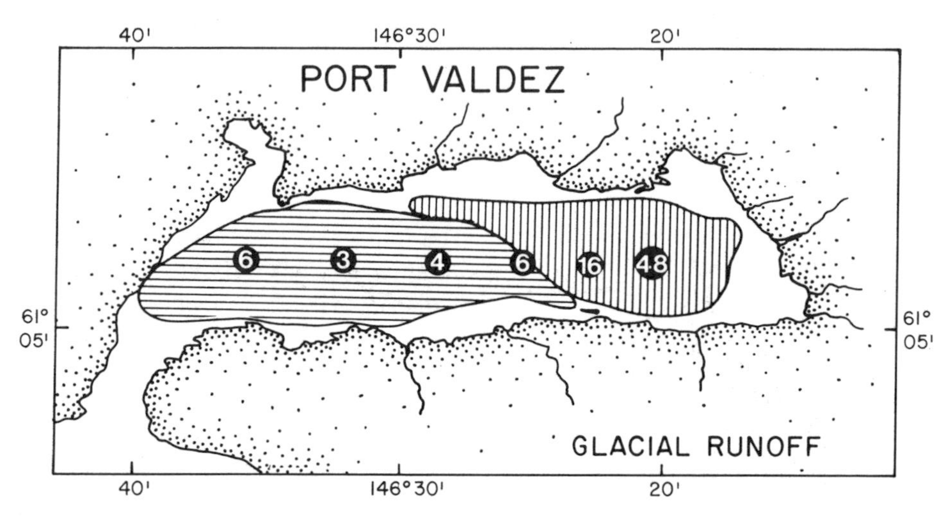

FIGURE 6.18. The deep, benthic macro-infauna of Port Valdez, Alaska, may be separated into two broad community groups. At the head of the fjord, where sedimentation rates are high, few filter-feeding organisms occur; down-fjord the communities consist of a mixture of filter and deposit-feeding taxa. Sedimentation flux values (mg cm^{-2} day^{-1}) are shown for a period in the summer. From Pearson et al. (in press).

and Feder and Jewett (in press) believe that this is so for the infauna studied over a ten-year period in Port Valdez, Alaska (see also Section 8.4). Farrow et al. (1983) suggest that sedimentation rates in excess of 20 cm yr^{-1} in Knight Inlet, British Columbia, exclude biota. However, the data of Hoskin (1977) suggest that some (predominantly macro-epibenthic) species may survive much higher rates. Hoskin et al. (1976) recovered a brittle star fragment (from approximately 64 L of sediment) near the head of a glacial fjord in Glacier Bay, Alaska, where the mean annual sedimentation rate exceeded 2 m yr^{-1}. The upper tolerance level may also be breached by catastrophic sedimentation events. Syvitski and Blakeney (1983) have recorded abiotic zones in sediment cores from Baffin Island fjords coincident with episodic turbidity flows (see also Chapter 5).

6.2.4.3 Carbon Loading

Pearson (1980a) and Pearson et al. (in press) have noted that many fjord basins tend to be eutrophic under conditions of low hydrodynamic energy input. In deep and high-latitude fjords, especially glacial fjords as discussed above, this condition is less likely since the benthos there is more typically carbon limited. High sedimentation rates of labile carbon decrease the depth of oxygenated sediment (Fig. 6.8) and lead to dominance by surface-feeding organisms. Extreme eutrophication (more commonly a feature of temperate localities) may generate periods of water-column anoxia.

The head of Oslofjord consists of two basins, circa 160 m deep, contained by a shallow (20 m) sill. It is considered likely that portions of the innermost basin may historically have become naturally anoxic at times, but this condition has been exacerbated during the present century with the progressive increase in anthropogenic organic waste discharge. Mirza and Gray (1981) have shown that the soft-bottom benthic fauna of this part of Oslofjord are dominated by opportunistic species, suggesting a community structure restrained at an early stage of the classic successional progression. It will be of interest to observe what changes occur following implementation of current plans to greatly curtail disposal of organic wastes into the upper reaches of this fjord. Catastrophic solid carbon waste dumping in fjords, especially of lumber industry effluents, is, or has been, a relatively common practice (see Chapter 8). In such cases, smothering of the biota and other limitations associated with high sedimentation rates as noted above may occur in addition to eutrophication effects. Benthic successional patterns following impacts of the latter type have been reviewed by Pearson (1975) and Pearson and Rosenberg (1978), and specific examples are described in Section 8.12.

6.2.4.4 Low Oxygen Conditions

Benthic biota in general are unable to survive prolonged periods where oxygen concentrations are below about 0.5 ml L^{-1} (Levings, 1980a; Arntz, 1981), and there have been numerous reports of total defaunation within fjord basins (e.g., Kitching et al., 1976) and of large areas of the fjordlike Baltic Sea (Tulkki, 1965; Leppakoski, 1969), with the spread of anoxic conditions. Various "critical thresholds" in the range 1 to 2 ml L^{-1} dissolved oxygen, below which organisms are rapidly impacted, have been proposed or implied (Hargrave, 1969; Rhoads and Morse, 1971; Rosenberg, 1980; Dethlefsen and von Westernhagen, 1983), although Rosenberg (1977) has shown that oxygen concentrations of less than 1 ml L^{-1} maintained for short periods (order of days) had no perceptible effect on the shallow infauna within Byfjord, Sweden (see also Section 8.13). Chronic low-oxygen conditions, or cyclic repetitions of low-oxygen stress, however, are likely to result in the establishment of communities adapted to this environment (Davis, 1975; Levings, 1980a; Tunnicliffe, 1981). It should be noted also that organisms generally exhibit increased tolerance to reduced oxygen concentrations with decrease in the ambient temperature (Theede et al., 1969). The 0.5 to 2.0 ml l^{-1} zone that constitutes an environment in which low-oxygen tolerant species (e.g., certain polychaetes: Fournier and Levings, 1982; molluscs: Theede, 1973, Rosenberg, 1980; echinoderms: Shick, 1976, Levings, 1980b; and crustaceans: Burd and Brinkhurst, 1985) may be at a competitive advantage; but invertebrate larvae are generally susceptible. Because of lower metabolic requirements, sessile organisms frequently appear to be more tolerant of low dissolved oxygen conditions (Vernberg, 1983). However, motile species may escape or avoid regions where ambient oxygen concentrations are slowly declining (e.g., Burd and Brinkhurst, 1985), as they may also survive spatially limited sedimentation perturbations within fjords, such as small-scale slumps. Low oxygen levels in the water above the interface concomitantly result in an upward migration of the redox front within the sediments (see Chapter 7), further deleting habitation niches, as described by Levington (1979).

6.2.4.5 Ice Impact

The impact of winter ice on the shallow water benthos of high-latitude fjords constitutes a particularly severe form of physical stress generally absent in temperate estuaries. Many investigations in arctic (e.g., Wacasey, 1975) and subarctic (Feder and Keiser, 1980; Cimberg, 1982; Skadsheim, 1983) fjords have shown extremely depauperate benthic fauna over depth zones subject to seasonal shore- or bottom-fast ice, or to ice scour. In fjords on Svalbard (Gulliksen et al., 1985) and Baffin Island (Farrow et al., 1985) maximum benthic biomass values have been recorded not in the near-surface zone, but at depths in the range of 30 to 85 m. Craig et al. (1984) note that the ice-impacted benthic macrofauna within Simpson Lagoon, on the Beaufort Sea coast of Alaska, appear to be maintained at an early, pioneering successional stage. It is suggested that in this environment the most successful epibenthic fauna consist of versatile and relatively motile organisms that are tolerant of a wide range of physical conditions.

6.2.4.6 Natural and Anthropogenic Impacts

The emphasis in this section has been on natural forms of stress characteristic of fjord estuaries. Although it is apparent that many types of fjords are naturally susceptible to various chronic and catastrophic impacts, these generally have been poorly studied to date. For pragmatic reasons, work has largely concentrated on industrial and urban polluted environments, and specific examples of these are the subject of Chapter 8. It is clear that changes in the structure of soft-bottom benthic communities, resulting from anthropogenic stress, may be indistinguishable from natural changes (e.g., Eagle, 1975; Gamble et al., 1977), and that unambiguous identification of most forms of "pollution" is far from simple, and may, in many cases, be impossible. Much of the extant "impact study" literature lacks essential background knowledge of short- and long-term natural variability, and consequently has limited scientific value. Finally it should be emphasized that community structure, in the sense used here, and generally by marine ecologists (Sheehan, 1984), need not be closely related to function. Physiological responses to environmental stress (e.g., Bayne, 1985) are unfortunately outside the scope of this book.

6.3 Summary

The unit area primary production of littoral and shallow subtidal plant communities in estuaries in general may be around an order of magnitude greater than that of the marine phytoplankton. However, because of their characteristic topography, production of the open marine regions in fjords is typically of greater quantitative importance. Phytoplankton production is seasonally constrained by available light levels to progressively shorter summertime periods with increasing latitude. Maximum annual production usually occurs over a short intense bloom period in the spring, which typically is terminated by limited nutrient availability as the stratification (stability) of the near-surface zone increases in the summer. (There is an increasing lag period between peaks in phytoplankton growth and the herbivorous zooplankton with latitude.) Post-bloom, summertime phytoplankton production rates depend primarily on various local means of injecting new nutrient supplies into the euphotic zone, and on grazing patterns. Hence, although mean annual primary production is a function of latitude, specific values vary widely and are site specific. Downward fluxes of carbon into the fjord basin and to the deep sediments frequently follow a regular seasonal pattern with spring-summer peaks of autochthonous phytogenous detritus, and riverine (usually more refractory) organic material predominating at the time of maximum freshwater discharge. Except in a few specific cases (such as the export of seagrass detritus from shallow temperate fjords), the import-export patterns of particulate carbon between fjords and adjacent coastal areas are not well known. Fjords are commonly thought of as poorly flushed estuaries, but where and at times when the characteristic fjord-type (entrainment-driven) circulation is well developed, exchange of material between the near-surface "estuarine" zone and contiguous waters may be relatively rapid.

Estuaries are transition zones between land and sea and the characteristic physical-chemical spatial and temporal gradients result in benthic communities best adapted to these conditions. In most fjords, this "estuarine" environment is a feature of the near-surface zone. However, the deeper subtidal regions are generally quantitatively more important and are considered to constitute the most distinctive fjordic benthic habitats. Fjords ideally contain large areas of subeuphotic lithic and soft-bottom substrates subject to uniform and temporally stable physical conditions. Whereas littoral and shallow communities are likely to be dominated by R-selected, pioneering organisms, basin communities characteristically consist of relatively low productivity, long-lived species. Such communities, which are likely to be susceptible to episodic environmental impacts, are akin to the "biologically accommodated" benthic assemblages of deep sea sediments. The fauna of basin sediments in higher latitude (especially glacial) fjords are believed to be carbon limited. The flux of labile organic detritus, generated autochthonously within the fjord, to the deep sediments may fluctuate seasonally, but resultant effects on the benthic fauna, if any, are largely undetermined. The basic macrofaunal community organization is one where more motile deposit feeders are generally dominant in deep, soft-bottom environments. Sessile filter feeders predominate on hard substrates and locally in regions, such as in the vicinity of sills, where bottom currents are greater and the sediment coarser. Although fjord basin biota may be analogous in a number of ways to that of deep sea environments, even the deep basin regions of fjord estuaries are an integral component of the land-sea estuarine buffer zone. Many temperate fjord basins are naturally eutrophic and may become periodically anoxic. Under carbon-enriched conditions, the zone of oxygenated sediment is reduced and infaunal communities tend toward opportunistic, surface-feeding organisms. Under these conditions (and also in low-carbon environments where the benthos is sparse), bioturbation of the near-surface sediment may be slight. Prolonged periods of anoxic or low-oxygen conditions will decimate or eliminate the resident biota. Many fjords, particularly those directly influenced by glaciers, have very high sedimentation rates. In such localities the benthos must accommodate to high burial rates, poorly consolidated and easily resuspended sediments and, in many cases, dilution of the food supply. Episodic, catastrophic events (such as sediment slumping and ice scour) are also common occurrences in fjords. The response to such nonsystematic impacts depends on the event frequency, but is generally likely to result in maintenance of pioneering stage benthic communities.

7

Biogeochemistry

Organic detritus is thermodynamically unstable under most natural conditions, and early diagenetic reactions in marine sediments result from the progressive oxidation of these compounds. Hence the emphasis in this chapter is on the distribution and reaction of organic material, metabolic products, and other redox-sensitive inorganic components in generalized fjord sediment environments. Fjords potentially offer a wide range of conditions for studying early sediment diagenesis. In temperate latitudes, if the fjord basins are poorly flushed, the flux of organic detritus to the benthic boundary may be very high, and the sediment is consequently anoxic to, or very close to, the interface. Under these conditions, a relatively large fraction of the more refractory organic material may be preserved and buried in the sediment. The most extreme case occurs in those silled fjords where renewal of water within the basin is infrequent and the supply of reactive organic material high, so that dissolved oxygen is depleted, allowing anoxic conditions to develop in the water column above the sediments. As shown in the previous chapter, the activities of benthic organisms are generally much reduced in cold and relatively deep sediments, and will be absent entirely where anoxic conditions persist for extended periods in the overlying water. In such localities, where sediment bioturbation is minimal, preservation of the depth zonation resulting from early diagenetic reactions provides convenient natural laboratories for studying estuarine anaerobic processes under conditions that frequently approximate steady state. At the other extreme, in other fjords—especially the glaciated estuaries of polar regions—the supply of labile organic material is likely to be very low, resulting in oxic and suboxic diagenetic reactions similar to those found in deep sea pelagic and hemi-pelagic sediment.

Although the intent of this chapter is to focus on sedimentation and sedimentary processes that are typical of the entire range of silled fjords, very few of these environments have been examined in any detail to date. In particular, considerable attention has been given to the unique geochemistry of long-term or seasonally anoxic fjords, while far less work has been done in the more abundant permanently oxygenated fjords. Fortunately, over the past decade, there have been very detailed investigations in a number of estuarine and deep sea localities having a wide range of sedimentation rates and fluxes of reactive organic material. These studies, supplementing the sparse field data available, permit prediction of generalized fjord sediment reactions and conditions.

Symbol Notation

P_M	concentration of M on solid phase
$[M]$	concentration of M in solution; e.g., [Cu], [Mn]
$[M]_{fw}$	concentration of M in fresh water
$[M]_{sw}$	concentration of M in marine water
C	concentration of solid phase organic carbon in sediments
C_o, C_z, C_∞	subscripts refer to C at sediment surface, at depth z, and at z_∞, respectively
C_s	% dry weight solid phase organic carbon concentration
$[N]$	interstitial water concentration of metabolite species
$[N]_o$, $[N]_z$, $[N]_\infty$	subscripts as for C
k	general reaction rate constant
k_1, k_{-1}	forward and backward reaction rate constants
k_d	organic decomposition rate constant
k_s	first-order scavenging rate constant
K_d	solid-solution distribution coefficient of species M
K_{sp}	ionic solubility product
D_b	sediment mixing coefficient (vertical)
D_s	bulk sediment diffusion coefficient (vertical)
K_z	vertical turbulent diffusion coefficient
J_s	particulate flux (downward) at the sediment surface
J_b	particle burial flux
J_N	flux of soluble species (vertical) in sediment
τ_M'	apparent residence time of M in solution
ω	sedimentation rate (cm yr^{-1})
r	sediment mass flux (g cm^{-2} yr^{-1})
ϕ	sediment porosity
ρ	sediment solid phase density
Ω	saturation coefficient
γ	activity coefficient
pH_{ZPC}	pH of zero surface charge

7.1 Particulate Sediment

7.1.1 Riverine Input

The geochemistry of fresh water entering fjords varies widely, but some important features are common to broad groupings of fjord types. Fjords characteristically are located in rugged terrain that tends to restrict the watershed area associated with each inlet and limits the residence time of precipitation in this reduced catchment area. The mean annual freshwater inflow into a typical mountain-range fjord is likely to be much smaller than that into a coastal plain estuary of comparable dimensions. However, because fjords are also higher latitude estuaries, inflow follows a pronounced seasonal pattern, with most of the freshwater discharge occurring through the summer. As noted in Chapter 4, this feature has important implications with regard to annual estuarine circulation patterns. In some temperate-boreal fjord provinces, the freshwater influx is characteristically bimodal: the result of release of stored precipitation (snow) in spring–early summer and maximum annual rainfall in late summer–autumn. In subarctic and arctic regions, a single freshwater inflow peak in mid-late summer is more usual. Figure 7.1 illustrates these features for two Alaskan fjords that are approximately 5° of latitude apart. Where residence times in the catchment areas are short, freshwater concentrations of dissolved chemical constituents are relatively low, and this is especially so in the case of glacial meltwater and summer freshet discharge from snow melt.

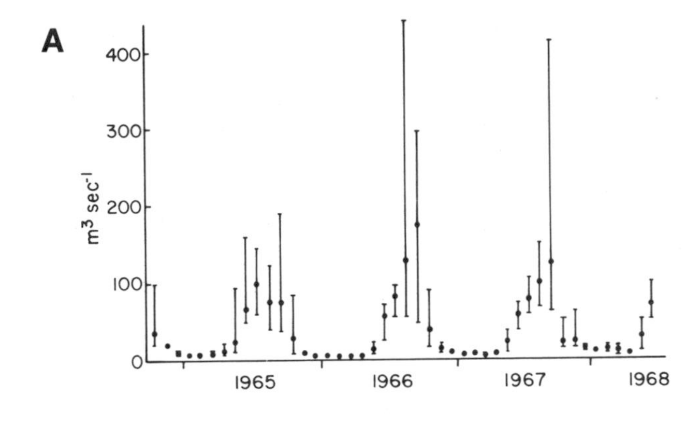

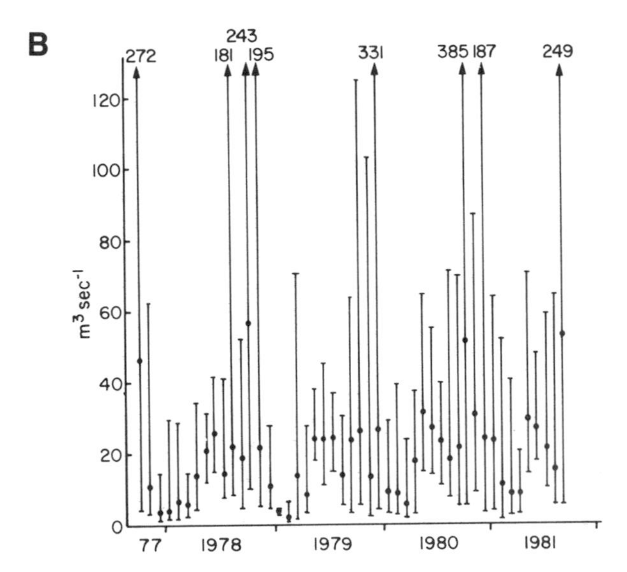

FIGURE 7.1. Seasonal patterns of freshwater discharge into two Alaskan fjords. (A) From the Resurrection River into Resurrection Bay (60°N). Runoff, including glacier and ice-field melt, peaks in late summer–autumn, the time of maximum direct precipitation. (B) From the Keta River into Boca de Quadra (55°N). Maximum rainfall discharge is in late summer–early winter. There is a secondary peak influx in the spring from winter snow melt.

The quantity of sediment moved by freshwater streams increases exponentially with the volume discharge rate (transporting competence). This means that the bulk of the annual riverine sediment discharged into the fjord—both material in suspension and as bed load—may occur over a very limited period of peak seasonal flow. Particulate load concentrations of mud-sized sediment in excess of 10 g L^{-1} have been measured (Burrell, 1972) in glacial streams traversing morainal deposits. Marine surface plumes of particulate sediment are characteristic features of estuaries having a fjord-type, layered circulation. As described in Chapter 4, Syvitski et al. (1985) have recorded that particulate load concentrations may exceed 1 mg L^{-1} some 50 km down-fjord from the head of Bute Inlet, British Columbia. In the southeast Alaska fjord system illustrated in Figure 7.1B, around 90% of the annual particulate load is transported at the time of autumn storms, but peak freshwater discharge is an order of magnitude less than that into Bute Inlet, and a distinctive turbid plume is only rarely observed at the head of the fjord.

The quantity and nature of riverine organic material is a major factor governing the geochemical behavior of soluble and particulate constituents carried into the marine environment. Freshwater discharge into fjords from a typical bog and forested hinterland may contain high concentrations of humic material (Sugai and Burrell, 1984a) in the form of complex polymers, derived primarily from biophenolic precursors, which tend to be resistant to microbially mediated degradation. These compounds contain dominant carboxyl and phenolic hydroxyl functional groups (Perdue, 1979) favoring the formation of negatively charged hydrophilic colloids in river water. Humic compounds also stabilize inorganic hydroxy colloids, chelate heavy metal ions (Reuter and Perdue, 1977; Mantoura et al., 1978), and sorb strongly onto particulate surfaces (Neihof and Loeb, 1974; Tipping, 1981). All natural surface water particulates examined to date have been shown to be negatively charged (Hunter and Liss, 1982; Tipping and Cooke, 1982), and this is believed to reflect the presence of ionized carboxylic and phenolic groups associated with virtually ubiquitous organic coatings. In some freshwater environments, high concentrations of humic substances reduce the pH and solubilize iron and manganese. Subsequent reoxidation results in iron-manganese hydroxide coatings on particles and co-precipitation or sorption of other heavy metals (Jenne, 1968).

7.1.2 Mixing Zone Reactions

Sea water is a high ionic strength medium, of remarkably constant chemical properties, and is very different from the freshwater influx. The zone where river and marine waters mix is one of strong gradients in, particularly, major ion concentrations and overall ionic strength, which profoundly affect the physical-chemical character of the inflowing constituents. One of the more important mixing zone reactions is aggregation of small particles into larger ones, which enhances sedimentation (Eckert and Sholkovitz, 1976; Kranck, 1981). Estuaries are traps for imported terrigenous material (Martin and Whitfield, 1983, estimate that less than 10% reaches the pelagic environment), and material deposited in fjord basins is physically restrained. Flocculation of river-borne organic and inorganic colloids is primarily a consequence of the reduced repulsion potential resulting from compression of the surface double layer with increasing ionic strength (Hahn and Stumm, 1970; Hunter, 1983), aided by charge neutralization. Aggregation clearly requires inter-particle contact, and is hence a function of the concentration of the particulate material. This is well demonstrated (Hoskin and Burrell, 1972) in the case of highly turbid surface plumes commonly present in stratified fjords fed by glacial streams: flocculation and rapid sedimentation occur when entrained sea water increases the salinity to around 4‰ (ppt). Krone (1978) advocates increasing "orders" of aggregation, each with progressively decreasing shear strength, which supports Kranck's (1973) concept of a limiting mean size for marine flocs. Edzwald et al. (1974) suggested that differential flocculation of specific clay minerals occurred, but it is now generally believed (e.g., Gibbs, 1977) that such effects should be masked by the properties of the surface organic coatings. It would be of interest to see whether such trends could be identified in high-latitude glacial fjords where organic concentrations may be very low.

The humic acid fraction of fresh water, together with associated heavy metals, appears to be significantly flocculated in sea water (Höpner and Orliczek, 1978; Fox, 1983). Figure 7.2 illustrates iron and manganese encrustations on biogenic debris at the head of a subarctic fjord. Sholkovitz (1976) has demonstrated that flocculation may occur over a wide (up to around 20%) salinity range. However, in most river-estuary systems studied to date (there is obviously a data bias in favor of temperate-zone examples) the organic carbon fraction is dominated by relatively low molecular weight fulvic acid components (Beck et al., 1974; Reuter and Perdue, 1977), which remain in solution on mixing with sea water (Moore et al., 1979; Hunter, 1983; Mantoura and Woodward, 1983). This is believed to be the case also in southeast Alaskan fjords studied by Sugai and Burrell (1984a) where peak riverine dissolved organic carbon concentrations (> 5 mg L^{-1}) were some twenty times greater than the coexisting particulate carbon contents.

7.1.3 Partitioning of Chemical Species

On entering the marine environment, freshwater chemical species may behave either conservatively or nonconservatively. A conservative mixing pattern is one in which the concentration versus salinity plot (i.e., between riverine and marine end-members) is linear; conversely, nonconservative trends represent either loss or gain of the constituent within the mixing zone. In this section, the focus is primarily on solid-solution transitions, which, because of the op-

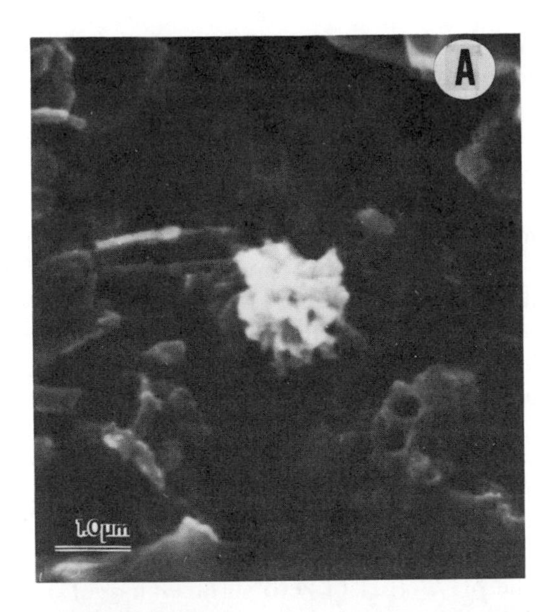

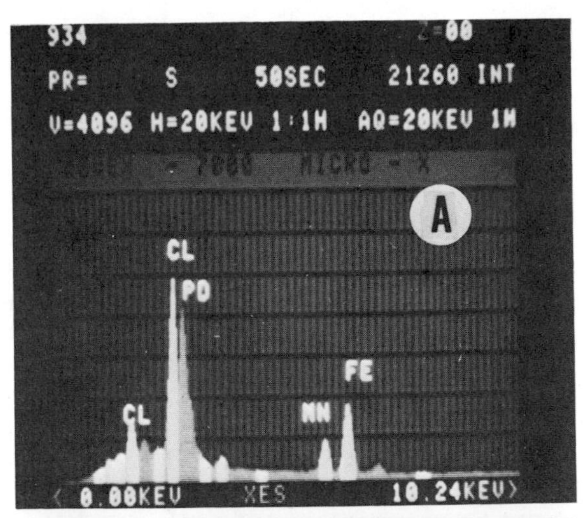

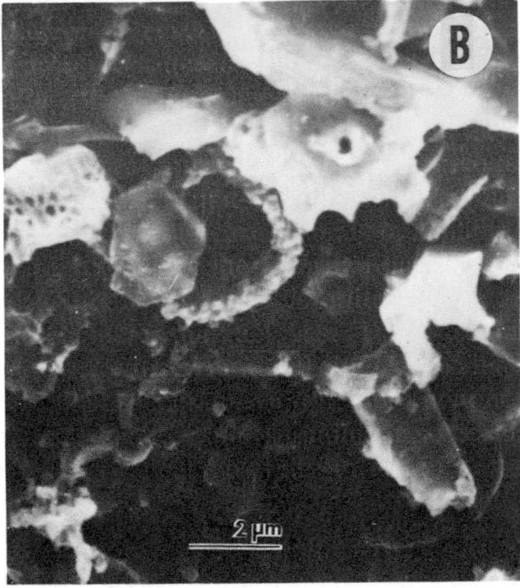
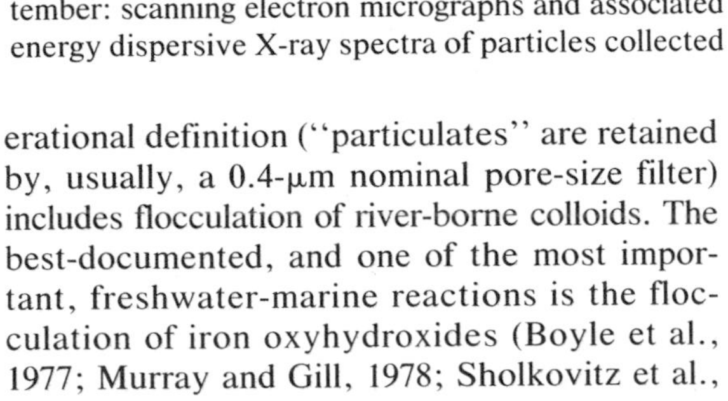

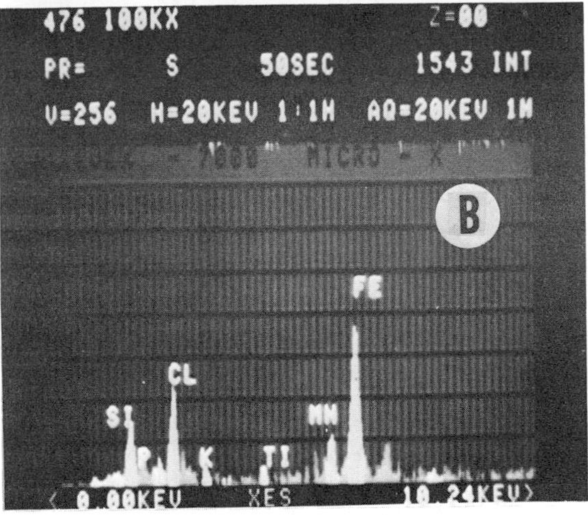

FIGURE 7.2. Iron and manganese enriched floccules and encrustations at mid-column depths at the head of Boca de Quadra fjord, southeast Alaska in September: scanning electron micrographs and associated energy dispersive X-ray spectra of particles collected on 0.4-μm pore-size filters. (A) Small Fe-Mn enriched particle (Mn/Fe = 0.5). (B) Fe-Mn encrustation on diatom fragment (Mn/Fe = 0.3; Fe/Si = 2.3). From Robb (1981).

erational definition ("particulates" are retained by, usually, a 0.4-μm nominal pore-size filter) includes flocculation of river-borne colloids. The best-documented, and one of the most important, freshwater-marine reactions is the flocculation of iron oxyhydroxides (Boyle et al., 1977; Murray and Gill, 1978; Sholkovitz et al., 1978; Hunter, 1983).

It has long been known (Krauskopf, 1956) that many chemical species exist in solution in sea water in concentrations far lower than would be predicted from the solubility products of their common least soluble compounds. It is apparent that such soluble species are "scavenged" by solid phases, and a partition coefficient can be defined (e.g., Li et al., 1984):

$$K_d = P_M/[M] \quad (7.1)$$

where P_M and [M] are the concentrations of species M associated with unit weights of solid and solution, respectively. Nonconservative removal in a body of water such as an estuary is also reflected by the relative residence time in solution:

$$\tau'_M = \frac{A_M}{\bar{J}_M} \quad (7.2)$$

where A_M is the mass of M in the estuary, and $\bar{J}_M$ is the mean flux through the system (Whitfield, 1979). At steady state, $\bar{J}_M$ is the same as the supply or removal rate. If the influx rate into the estuary is constant, and $[M]_{fw}$ and $[M]_{sw}$ are the freshwater and sea-water concentrations of M, then τ'_M is a function of $[M]_{sw}/[M]_{fw}$. Such operational residence times are relative to the flushing rates of the estuary (Morgan,1967a).

It is apparent that a relationship exists between the partition coefficient and residence time of M. Whitfield and Turner (1979) and Turner et al. (1980) have shown that:

$$\log \tau'_M = -a \log K_d + b \quad (7.3)$$

where a and b are empirical constants. Chemical species, including most common heavy metals, which have short residence times (or large K_d values), are variously termed "depleted" or "particle reactive" (Santschi et al., 1983).

Many recent theoretical and laboratory studies have emphasized the importance of the adsorption of ions onto hydrolyzed oxide surfaces (Li, 1981; Balistrieri et al., 1981). The surface groups (L—OH) behave amphoterically depending on the pH:

$$\text{L—OH} \rightleftarrows \text{L—O}^- + H^+$$
$$H^+ + \text{L—OH} \rightleftarrows \text{L—OH}_2^+ \quad (7.4)$$

The surface charge becomes more negatively changed with increasing pH, and cationic species are taken up. Considering adsorption thus in terms of reversible thermodynamic equilibria—that is, treating the solid surface group as a complexing ligand—permits determination of apparent equilibrium constants (Oakley, 1981; Balistrieri and Murray, 1983), and idealized chemical speciation models have been developed that specifically incorporate sorbed species (e.g., Vuceta and Morgan, 1978). It should be noted, however, that theoretical thermodynamic partition models usually disregard the reaction kinetics:

$$M^{z+} + \text{L—OH} \underset{k_{-1}}{\overset{k_1}{\rightleftarrows}} \text{L—OM}^{(z-1)+} + H^+ \quad (7.5)$$

where k_1 and k_{-1} are the sorption and desorption rate constants for species M^{z+}. This is potentially a serious concern in estuaries where the residence time of the water is relatively short. Nyffeler et al. (1984) have shown that time constants for the equilibration of a number of elements on particles appear to be of the order of days, and hence that certain soluble species may not reach equilibrium with the solid phase while in transit through the estuary.

Assumption of ideal thermodynamic reaction implies that, under certain conditions (notably a decrease in pH), ionic species should be desorbed (Bourg, 1983). It has been noted previously that FeOOH surfaces, as flocs and coatings on particles, are present in relatively high concentrations in most estuarine mixing zones. The pH_{ZPC} (pH of zero surface charge) of goethite appears to correspond closely to that of sea water (around 8: Balistrieri and Murray, 1979), so that weak sorption of either anions or cations would be predicted. A considerable body of field evidence supports desorption of, for example, manganese (Morris et al., 1978; Wilke and Dayal, 1982) and copper (Moore and Burton, 1978; Windom et al., 1983) from riverine sediments during estuarine mixing. Li et al. (1984) have also demonstrated in the laboratory a decrease in K_d values of radiotracers of cadmium, cobalt, manganese, and zinc with increasing salinity, implying that desorption should occur under these conditions. However, unlike typical riverine waters entering temperate and boreal fjords, these experiments utilized low-organic-content water. Sholkovitz (1976) has pointed out that the frequently cited laboratory studies of Kharkar et al. (1968), showing desorption of cations from clay particles mixed into sea water, also omitted organic phases. Since natural water particulate matter has been shown to be universally negatively charged, and abiotic uptake of anionic species has not been demonstrated within the mixing zone, it appears likely that organic matter is generally ubiquitously sorbed onto the inorganic surfaces and that exchange characteristics are primarily dictated by the major humic functional groups. In some cases, it is possible that observed enhancements of manganese and copper, in particular, in shallow estuarine waters are the result of transport out of the bottom sediments (see Section 7.2.3), or resuspension of fine-grained sediment.

As noted, the present utility of thermodynamic models for describing the distribution of chemical species in fresh and marine waters is primarily limited by researchers' inability to adequately incorporate organic ligands. Knowledge of specific organic compounds present in all natural waters is very rudimentary, so that modellers (e.g., Morel and Morgan, 1972; Stumm and Brauner, 1975) have generally been forced to incorporate simple ligands, for which stability data are known, but which do not occur, or are insignificant, in natural environments. Humic and fulvic acids are major components in many fjord systems and, although bonding with the active groups of these complexes is generally relatively weak, stable complexes with iron and, especially, copper may be important. A number of investigators (Mills and Quinn, 1981; Huizenga and Kester, 1983; Kremling et al., 1983) have shown that copper-organic complexes do not sorb onto manganese oxide surfaces. This is the likely explanation for the conservative behavior of dissolved copper (and a few other elements) frequently observed in estuaries receiving high concentrations of terrigenous humic material (Holliday and Liss, 1976; Hunter, 1983) or sewage discharge (Mills and Quinn, 1984). Figure 7.3A illustrates linear mixing of copper through the mixing zone of a fjord in southeast Alaska in October. At this time of year, the influx of fresh water (Fig. 7.1B) and humic material is at a maximum. In an adjacent fjord, Sugai and Burrell (1984a) have shown that riverine "dissolved" organic carbon peaks in autumn, and that total iron concentrations remain relatively constant, with increasing salinity, down the length of the inlet. A number of studies of the variously defined "nondetrital" fraction of fjord sediments (Loring, 1976a; Hogarty, 1985) have shown correlations between copper, iron, mercury, lead, and zinc and organic carbon concentrations, and also with C/N ratio values, suggesting that these metals may be at least partially complexed with particulate terrigenous organic material.

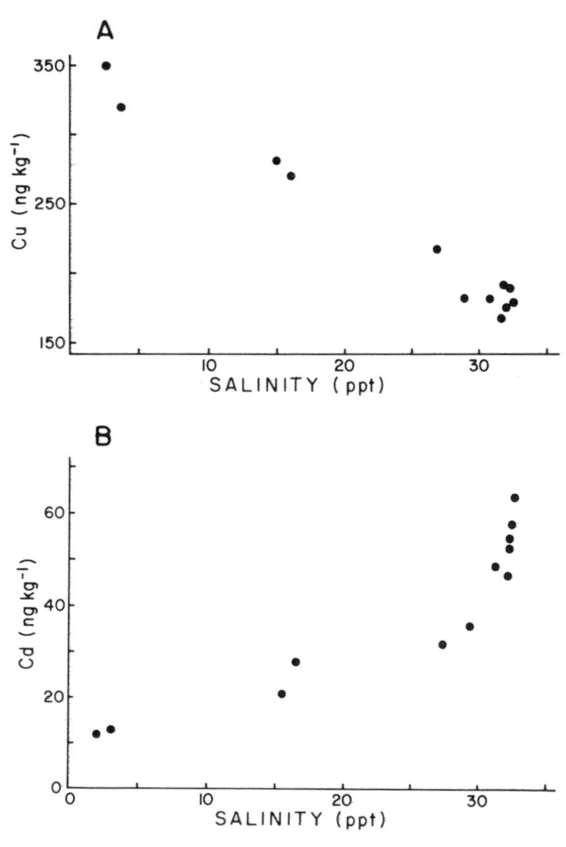

FIGURE 7.3. Concentrations of soluble (μg kg^{-1}; < 0.4 μm) (A) copper and (B) cadmium as a function of salinity in Boca de Quadra, southeast Alaska. At this time of year (October), copper appears to be behaving conservatively, and cadmium non-conservatively. Data from Erickson and Stukas (1983).

In the open ocean, biological uptake is a major removal mechanism for a number of constituents present in solution in trace quantities, and these processes are also likely to be seasonally important in most fjords. Cobalt, chromium, copper, iron, manganese, molybdenum, nickel, vanadium, and zinc are required micronutrients; but other nonessential metals, such as cadmium, are taken up also. Table 7.1 is an operational classification of biologically active metals, based on the cited literature, which attempts to identify those elements that are taken up and regenerated relatively rapidly in the near-surface region, and those that are transported in particulate form (by processes discussed in Chapter 6) to depth. These may be thought of as phosphate- and silicate-type distributions, respectively. In the latter case, in fjords, the elements are likely to be predominantly carried to the sediment floor, where aerobic remobilization may occur. This sequence is considered further in Section 7.2.3. Figure 7.3B illustrates a nonconservative distribution of cadmium at the head of a fjord, which is believed to reflect biological uptake in the euphotic zone.

7.1.4 Temporal and Spatial Distributions

Fjord particulate sediment consists primarily of aluminosilicates, erosion products from bed rock within the surrounding watershed (Sundby and Loring, 1978). The aluminosilicates entering higher latitude fjords may be less degraded than in more temperate regions. This is especially so in the case of glacial sediment. For example, O'Brien and Burrell (1970) have recorded primary amphiboles in the clay-sized fraction of sediment deposited within a tidewater glacier fjord. There is some evidence (e.g., Loring, 1984) to suggest that the non-detrital fraction of heavy metals (i.e., sorbed or otherwise weakly bound) extracted from arctic sediments is smaller than from estuarine sediments elsewhere, observations which may also reflect limited weathering and/or short residence times in the watersheds. As discussed in Section 7.2.3., particulate sediment sampled near the benthic interface is commonly enriched in manganese that has been transported out of the sediments (e.g., Bewers and Yeats, 1979).

Nepheloid zones above the benthic boundary are common features of most shallow estuaries. Resuspension of deposited sediment back into the water column is far less prevalent in fjord basins, where bottom shear is usually slight, but increased concentrations of near-bottom particulate material have been recorded at times of bottom-water renewal (Skei, 1980; Stanley et al., 1981). In deep regions of fjords, enhanced particulate loads often reflect sediment focusing within the basin (Gulliksen, 1982; Wassmann, 1984).

On a mean annual basis, less than half of the particulate organic material present in the fjord surface zone is likely to be derived from the

TABLE 7.1. Biologically active elements.

Shallow regeneration	As	Andreae (1978)
	Cd	Bruland (1980) Boyle & Huested (1983) Collier & Edmond (1984)
	Mn*	Martin & Knauer (1980) Collier & Edmond (1983, 1984)
	Ni	Collier & Edmond (1983, 1984)
Combined	Ni	Sclater et al. (1976) Bruland (1980)
	Se	Measures & Burton (1980)
Deep regeneration	Cr*	Cranston (1983)
	Cu	Collier & Edmond (1983)
	Fe*	Collier & Edmond (1983)
	Ge	Froelich & Andreae (1981)
	Zn	Martin & Knauer (1973) Collier & Edmond (1983)

*Primarily particle-reactive elements.

surrounding watershed. However, terrigenous detritus tends to be much more refractory than phytogenous material, and the allochthonous/autochthonous carbon ratio increases significantly in the sediments. Chemical examination of the organic detritus may help differentiate between autochthonous and terrestrial plant material. Characteristic C/N ratios and $\delta^{13}C$ ranges are commonly employed, and examples of this type of analysis applied to the organic component of sediment deposited within the basins of the Saguenay Fjord, Quebec, are cited in Section 8.7. Marked seasonal variations in provenance are distinctive features in many fjords. The September surface POC maximum shown in Figure 6.3B (Chapter 6) illustrates the peak seasonal discharge of terrestrial organic detritus into a boreal (southeast Alaska) fjord, with a separate maximum in April, which marks the diatom bloom. In this region, the mean annual flux of riverine particulate organic carbon is approximately equal to phytoplankton production within the fjord (Sugai and Burrell, 1984a).

At higher latitudes, freshwater inflow is reduced or ceases entirely during the winter, and the riverine flux of sediment into arctic and subarctic fjords has a large seasonal range. Since the carrying competence of the rivers is an exponential function of the flow rate, it follows that a very large fraction of the annual sediment load may be transported into the fjord over a short time interval. In polar regions, this material will be almost entirely inorganic.

It should be re-emphasized that marine "particulate sediment" is commonly operationally defined as material, collected in discrete samples, that is retained by various standard (but physically arbitary) pore-size filters. The sparse, randomly distributed larger particles (such as calanoid fecal pellets) cannot be accurately sampled in this fashion, but may be quantitatively retained in sediment traps.

7.2 Aerobic Diagenetic Reactions

7.2.1 Decomposition of Organic Material

All biogenic detritus is thermodynamically unstable under ambient aerobic conditions, but decomposition kinetics vary widely. Although most phytogenic material is regenerated within the near-surface waters, some seasonally variable fraction is aggregated into larger particles (by processes discussed in Chapter 6), and, together with much more refractory riverine humic substances, is transported to the sediments. Even though the more labile components are progressively eliminated during passage through the water column, "packaging" into larger particles results in increased sedimentation rates and a decrease in mean surface-to-volume ratios, so that a significant fraction of the carbon produced or introduced at the surface reaches the floor of even the deepest fjord basins. Of course, in "abiotic" fjords—such as occur in many high-latitude, glaciated areas—the organic carbon content of the bulk sediment may be very low.

Oxidation continues at the benthic interface, and at progressively slower rates during sediment burial, via a series of microbially mediated reactions: organic detritus is the sole source of energy directly or indirectly available to the benthic biota of fjord basins below the euphotic zone. In this section we are concerned primarily with aerobic respiration, which is of paramount importance, for example, in fjords where the particulate organic flux is relatively low, or where the near-surface sediments are well mixed; but the principles apply equally to anaerobic oxidation reactions discussed in Section 7.3.2.

The vertical distribution of solid-phase organic carbon (concentration C_z at a depth z downward from the interface) within near-surface fjord sediments, shown schematically in Figure 7.4, generally can be fitted closely by an exponential function:

$$C_z = C_\infty + (C_o - C_\infty) \exp[-\beta z] \quad (7.6)$$

where $C_z \rightarrow C_\infty$ as $z \rightarrow \infty$, $C_z = C_o$ at $z = 0$, and β is an empirical constant which, as will be shown, may be assigned physically meaningful values. A real value of C_∞ implies that some "background" content of organic carbon is essentially inert over the applicable sedimentation time scale. Sediment organics are known to condense into inert humic and kerogen polymers, and studies of the organic composition of fjord sediments by Brown et al. (1972) and Peake (1978) have shown increasing humic (derived both from terrestrial and autochthonous sources) and decreasing fulvic fractions with depth.

The sediment organic concentration profile of

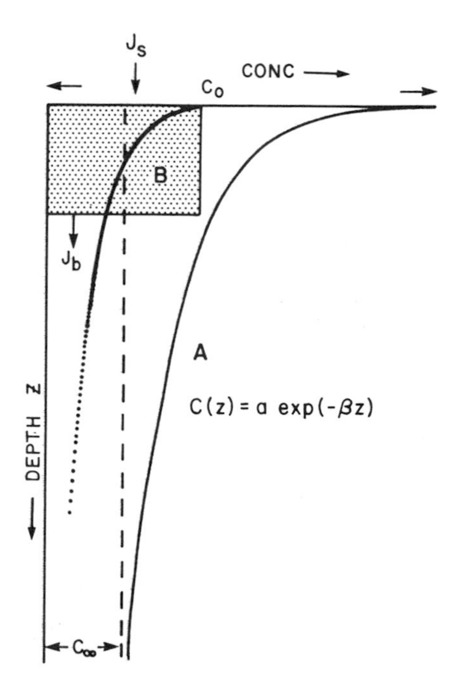

FIGURE 7.4. (A) Idealized distribution (concentration C_z) of total organic carbon (POC) with depth (z) in fjord sediments below the interface. $C_z = C_o$ at interface ($z = 0$); C_∞ is the "asymptotic" value ($z \rightarrow \infty$). (B) The mean degradation rate of organic detritus is approximated by the difference between the flux of POC arriving at the sediment surface (J_s) and the rate of burial (J_b) of refractory carbon at depth.

Figure 7.4 may be modelled assuming that degradation can be described by a simple rate law, and that sediment mixing and burial are the dominant transportation processes. A first-order carbon reaction rate term ($C_z k_d(z)$; i.e., where the rate is a function of the amount of labile material remaining at depth z) may be a reasonable simplification of the complex reaction kinetics (Aller and Yingst, 1980; Berner, 1980; Rosenfeld, 1981). The electron acceptors (the oxidants) should be present in excess, but although this condition may not be fulfilled in organic-rich sediments, Boudreau and Westrich (1984) have shown that decomposition rates are independent of oxidant concentrations in most practical situations. Values of the mean degradation rate constant (k_d) generally decrease with depth (time) as the proportion of refractory detritus increases, and as low reactivity material is synthesized in situ. Unfortunately, the range in most fjord environments is not well known. Investigations in shallow water sediments are most often concerned with anoxic environments (Section 7.3), and the oxidation rate constant is then usually assumed constant over the restricted depth range of interest. Decomposition rates of constant lability organic detritus are higher (at least double: Westrich and Berner, 1984) under oxic compared with anoxic conditions, and this may be an important factor in those fjord sediments where the redox front (Section 7.3.1) is located some distance below the interface. Degradation rates also decrease exponentially with decreasing ambient temperature. Fjord-basin sediments tend to be uniformly cold year-round, and decomposition could be up to an order of magnitude slower than in shallow water temperate environments in the summer. Conversely, Toth and Lerman (1977) have shown that the mean reaction rate constant at a given depth may increase if the organic carbon flux increases: this reflects more rapid burial and preservation of higher reactivity compounds. In the few boreal fjord sediments that have been studied in any detail, bacterial decomposition of organic material appears to be substantially confined to a relatively thin (of the order of centimeters) surface zone, as reflected by the distribution of degradation products (Section 7.2.2). In such fjords, especially where terrigenous organic detritus is a major component, a large fraction of the sedimented organic carbon may be buried and lost from the system.

Taking the sediment surface as a fixed reference horizon, sedimenting particles are progressively transported downward with burial, but are also subject to various forms of biological and physical mixing. As discussed in Chapter 6, particle mixing by the macrobenthic fauna (bioturbation) is widespread in coastal sediments. Mature communities may include organisms having a wide range of feeding and locomotion habits, but it is usual to group these processes into two categories for ease of mathematical formulation. "Conveyor-belt" organisms (Section 6.2.3.3) tend to generate a discontinuous exchange of material between nonadjacent localities. This has been modelled by Emerson et al. (1984) using a "nonlocal" exchange rate parameter. In the more general case, community bioturbation can be conceptualized as the sum effect of multiple and random disturbances over a range of time and depth scales. In an analogous fashion to that used for the modeling of soluble metabolite species (Section 7.2.2), this is commonly formulated as a vertical turbulent diffusive process. Fickian-type "biodiffusion" coefficients (D_b) have been determined for a wide variety of marine sediments, primarily using radioisotopic tracers. Values around 10^{-7} to 10^{-5} cm^2 s^{-1} have been determined for temperate coastal environments (e.g., Aller, 1982; Callender and Hammond, 1982). As yet there are no reported data for deep fjord basins, but lower values would be expected in general because of macrofaunal impoverishment.

Physical mixing of near-surface sediment is a common occurrence in many estuarine and coastal regions. Smethie et al. (1981) and Carpenter et al. (1982) have measured high rates of vertical mixing of deposits at the mouth of the Columbia River, and of shelf canyon sediments, respectively. As discussed in Chapter 5 and elsewhere throughout this volume, sediment slumping and other forms of bed-load transport are especially prevalent in steep-sided fjords. Although the diffusion analog would not be expected to describe closely a single slump event, Sugai (1985) has shown that, in one fjord, multiple physical mixing over a time scale of decades may be formulated in this fashion. Sugai and Burrell (unpublished data) have modelled slump-mixing in two adjacent fjords in southeast Alaska using a one-dimensional, two-layer model in which physical mixing is treated as a time-averaged, diffusive process confined to a uniformly mixed surface layer. Mixing coefficients in the range from 6.5 to 65 cm^2 yr^{-1} are computed for various soft-bottom localities within these fjords, with the higher values from shallower environments adjacent to major river outfalls.

A commonly used form of the time-dependent depth distribution of sedimentary organic carbon (Berner, 1980) is then:

$$\frac{\partial C}{\partial t} = \frac{\partial}{\partial z}\left(D_b \frac{\partial C}{\partial z}\right) - \omega \frac{\partial C}{\partial z} - k_d (C - C_\infty) \qquad (7.7)$$

where $\partial C/\partial t$ is the rate of change of organic carbon concentration at depth z (positive downward from the interface) and ω is the constant sedimentation rate. In this equation, constant porosity ϕ and sediment density ρ are also assumed (porosity is the ratio of pore water to total sediment volume). Although the data of Murray et al., 1978, show that this may be a reasonable simplification for fjord sediments, it is common practice to replace ω by a mass flux term r, where $r = \omega \rho (1 - \phi)$. Intuitively, bioturbation would be expected to be a function of depth, and attempts have been made to mimic decreasing infaunal activity downward from the interface by assigning an idealized functional form to D_b. For example, an exponentially decreasing function (Bruland, 1974; Nozaki, 1977), or a Gaussian distribution (Christensen, 1982) with depth have been used. However, it is more usual to assume that little error is introduced by employing both constant D_b and k_d (i.e., mean) values over the depth interval of interest. Applying these simplifications to equation (7.7), and initially assuming that the burial rate is insignificant compared with mixing, at steady state ($\partial C/\partial t = 0$):

$$D_b \frac{\partial^2 C}{\partial z^2} = k_d (C - C_\infty) \qquad (7.8)$$

which, for the boundary conditions cited previously (Fig. 7.4; eqn. 7.6), has the solution:

$$C_z = C_\infty + (C_o - C_\infty) \exp\left[-z\left(\frac{k_d}{D_b}\right)^{1/2}\right] \qquad (7.9)$$

If a value may be assigned to D_b, the first-order organic decomposition rate constant may be estimated from the slope, $-(k_d/D_b)^{1/2}$, of the linear depth plot of $\ln[(C_z - C_\infty)/(C_o - C_\infty)]$. The flux ($J_s$) of degradable carbon arriving at the benthic interface necessary to maintain the profile at steady state is:

$$J_s = (C_o - C_\infty)(k_d D_b)^{1/2} \qquad (7.10)$$

The conditions under which mixing or burial may be the dominant distributive process have been addressed by a number of people (e.g., Guinasso and Schink, 1975; Lerman, 1979; Officer, 1982). As a practical approximation, when the scale length, D_b/ω, is greater than around 1 m, the sedimentation advection term may be assumed insignificant compared with mixing. In many regions of deep fjord basins, particularly with increasing latitude, the macrofauna are sparse (Chapter 6), and bioturbation is consequently unimportant (Kristensen, 1984; Burrell and Hong, unpublished data). (Bioturbation is entirely absent in environments where anoxic conditions persist for long periods in the overlying water: see Section 7.3.) At the same time, high sedimentation rates are common. Thus sediment deposited within a fjord basin at a rate of 3 cm yr^{-1} would have a scale height of around 1 cm if the mixing coefficient were 10^{-7} cm^2 s^{-1}. Under such conditions, a more realistic reduction of equation (7.7) is:

$$\frac{\partial C}{\partial t} = -k_d(C - C_\infty) - \omega\frac{\partial C}{\partial z} \quad (7.11)$$

which, for the same boundary conditions, yields the time-invariant expression:

$$C_z = C_\infty + (C_o - C_\infty)\exp\left[-\frac{k_d}{\omega}z\right] \quad (7.12)$$

(From eqns. 7.9 and 7.12 it may be seen that vertical mixing may be ignored when $\omega^2 >> k_d D_b$.)

Berner and Westrich (1985) have noted that bioturbation generally increases the rate of subsurface degradation by enhancing the downward supply of dissolved oxygen (i.e., by deepening the oxic zone), hence preservation of organics will be favored in sediments that are not intensively mixed. It has already been noted that higher organic deposition rates result in better preservation of organic material within the sediments. However, for a constant benthic flux of labile organic detritus, although increasing the bulk sedimentation rate initially increases the fraction of organic material that is buried and preserved, at very high inorganic deposition rates (a common condition at the heads of high-latitude fjords), Aller and Mackin (1984) have shown that the fraction of organic detritus escaping degradation must decrease again. This is because the rate of burial of the oxidant progressively increases, but labile organic material is concurrently diluted with inert sediment.

In higher latitude estuaries, the assumption of a steady-state supply of organic material is likely to become increasingly less tenable. In extreme examples of high inorganic sedimentation and very low and seasonally restricted carbon input, such as characteristically occurs in glacial fjords, organic material is deposited in discrete layers within the sediment (Hoskin and Burrell, 1972; Elverhøi et al., 1983; Powell, 1983). In such cases, the non-steady-state treatment outlined by Lasaga and Holland (1976) may be more appropriately employed if the sedimentation rate is sufficiently high.

If the mean supply of degradable organic carbon to the benthic boundary is approximately constant relative to the known bulk sedimentation rate (ω or r), a simple box model approach (Fig. 7.4B) may be applied to estimate mean degradation rates. The flux of carbon arriving at the interface is given by:

$$J_s = \frac{\omega\rho C_s(1 - \phi)}{100} = \frac{C_s r}{100} \quad (7.13)$$

where C_s is the percent dry weight carbon content of the surface sediment. The carbon transport rate (the burial rate J_b) out of the "box" at the depth where the "asymptotic" concentration is approached (i.e., below which the residual carbon content, $\%C_b$, remains approximately constant) is $r \cdot C_b/100$, and the difference ($J_s - J_b$) represents degradation. It should be noted that this simple treatment is very sensitive to the assigned carbon content value at the sediment surface. Very commonly, the critical interfacial sediment is lost in core samples, and this may be a serious source of error. Kelly and Nixon (1984) have shown experimentally that freshly deposited particulate organic material may decompose very rapidly: 11 to 20% of the POC and 24 to 30% of the particulate nitrogen in less than two months. Even at high sedimentation rates, a practical surface layer sample is likely to integrate deposition over at least a one-year period. Decomposition rates (consistent with estimates of oxygen utilization and nutrient regeneration determined by independent techniques) have been obtained using carbon concentrations of sediments collected in traps moored just above the sediment surface within a deep fjord basin, in an area where sediment resuspension appears to be negligible (Hong and Burrell, unpublished data). A similar approach has been used in Norwegian fjords (Wassmann, 1983; 1984).

This discussion has emphasized time-invariant sediment profiles which, after an initiation period, would result from a persistent rain of constant-composition particles arriving at the sediment surface. Since fjords are generally relatively deep, various pelagic processes may act to smooth out irregularities in the supply of organic material to the sediments (McMahon and Patching, 1984; Sargent et al., unpublished data).

However, a characteristic feature of fjords, increasing in importance with latitude, is their "seasonality," and the spring phytoplankton bloom may lead to an enhanced flux of labile particulate organic material to the sediments through the spring-summer period (Burrell, 1983b; Wassmann, 1984). In polar fjords, the supply of organic detritus to the benthos may be almost totally confined to a few weeks of each year (Hoskin and Burrell, 1972; Elverhöi, 1984), at which time, in areas having high deposition rates, the organic component is likely to be substantially diluted with inorganic sediment. In deep sea regions, where the flux of organic detritus to the sediments is similarly very low, the sediment column may remain oxygenated if the bulk sedimentation rate remains below a certain threshold value (Müller and Mangani, 1980). Although not yet recorded, theoretically this situation could pertain to a number of polar fjords.

7.2.2 Organic Decomposition Reactions and Products

The decomposition of thermodynamically unstable organic detritus is "catalyzed" by microorganisms, which derive metabolic energy from the reactions. A number of oxidants are available in the marine environment. It appears, however, that application of the "competitive exclusion" principle, possibly with additional physiological effects (Billen, 1978), ensures that oxidation proceeds by a series of stepwise reactions of progressively decreasing free energy yield, involving different communities of bacteria. The primary electron-accepting chemical species, in order of utilization, are: O_2, NO_3^-, MnO_2, FeOOH, SO_4^{2-}, and CO_2 (Fenchel and Blackburn, 1979). Idealized decomposition reactions are shown in Table 7.2. In open ocean conditions the stoichiometry of these reactions appears to conform closely to the well-known Redfield ratios (C:N 6.5, N:P 16; Redfield, 1934), but may differ substantially in estuaries, largely because of the impact of littoral and terrestrial detritus.

In estuaries in general, the flux of reactive organic matter to the sediments is relatively high, aerobic and "suboxic" (Table 7.2) decomposition is confined to a very thin surface sediment zone, and individual reactions and their interrelationships cannot easily be distinguished. Oxygenated sediments are favored when the supply of degradable organic material is relatively low and in cold water environments where metabolic rates are reduced, conditions typical of high latitude—and especially glacial—fjords.

TABLE 7.2. Oxidation reactions of organic matter under aerobic and anaerobic conditions.

Aerobic respiration

1a. $(CH_2O)x\,(NH_3)y\,(H_3PO_4)z + x\,O_2$
$\rightarrow x\,CO_2 + y\,NH_3 + z\,H_3PO_4 + x\,H_2O$

1b. $(CH_2O)x\,(NH_3)y\,(H_3PO_4)z + (x + 2y)\,O_2$
$\rightarrow x\,CO_2 + y\,HNO_3 + z\,H_3PO_4 + (x + y)\,H_2O$

Suboxic respiration

Nitrate reduction:

2. $5\,(CH_2O)x\,(NH_3)y\,(H_3PO_4)z + 4x\,NO_3^-$
$\rightarrow x\,CO_2 + 4x\,HCO_3^- + 2x\,N_2 + 5y\,NH_3 + 5z\,H_3PO_4 + 3x\,H_2O$

Mn (IV) reduction:

3. $(CH_2O)x\,(NH_3)y\,(H_3PO_4)\,z + 2x\,MnO_2 + 3x\,CO_2 + x\,H_2O$
$\rightarrow 2x\,Mn^{2+} + 4x\,HCO_3^- + y\,NH_3 + z\,H_3PO_4$

Fe (III) reduction:

4. $(CH_2O)x\,(NH_3)y\,(H_3PO_4)z + 4x\,FeOOH + 7x\,CO_2 + x\,H_2O$
$\rightarrow 4x\,Fe^{2+} + 8x\,HCO_3^- + y\,NH_3 + z\,H_3PO_4$

Anaerobic respiration

Sulfate reduction:

5. $2\,(CH_2O)x\,(NH_3)y\,(H_3PO_4)z + x\,SO_4^{2-}$
$\rightarrow x\,H_2S + 2x\,HCO_3^- + 2y\,NH_3 + 2z\,H_3PO_4$

Methane production:

6. $2\,(CH_2O)x\,(NH_3)y\,(H_3PO_4)z$
$\rightarrow x\,CO_2 + x\,CH_4 + 2y\,NH_3 + 2z\,H_3PO_4$

Fjord-estuarine sediments, therefore, may uniquely range from being entirely anoxic (see Section 7.3) to examples having substantial depths of oxic sediment. The latter type of sedimentary environment is akin to the deep oceanic pelagic and hemipelagic substrates that have been intensively studied in recent years. Idealized depth profiles of, primarily, mobile reactants and products present in the pore water of aerated sediments are shown in Figure 7.5. It must be emphasized that, in contrast to deep ocean regions, sedimentation patterns in the shallower areas of fjords will be cyclic or episodic on various time scales, and steady-state conditions are only likely to be approached in deep basin environments.

Aerobic respiration is described by equations (1a) and (1b) of Table 7.2. Ammonia released by deamination (1a) is further oxidized by nitrifying bacteria (e.g., *Nitrosomonas* and *Nitrobacter*) to nitrate: equation (1b) represents the overall reaction. The form of the dissolved oxygen profile of Figure 7.5 indicates diffusion from the overlying water into the sediment and reaction with organic material, as described by, for example, Murray and Grundmanis (1980), and modelled by Grundmanis and Murray (1982) and Galoway and Bender (1982). Nitrate diffuses downward to serve as the next electron-acceptor favored by the bacteria. Jahnke et al. (1982) have shown that the occurrence and location of a nitrate peak in the pore water depends upon the rate of denitrification within the anoxic zone. Reduction of iron and manganese oxyhydroxides (Table 7.2; eqns. 3 and 4) is a minor source of energy to the total heterotrophic pool; but these reactions are of prime importance with regard to cycling of heavy metals within the sediment, and between the sediment and overlying water, as discussed in Sections 7.2.3 and 7.3.3.

In most fjord sediments, ammonia produced in the sediment interstitial water below the suboxic horizon migrates upward into the oxygenated zone. Ammonia (NH_4^+ in the marine environment) is thermodynamically unstable in the presence of free oxygen but exists metastably in many natural environments and is, for example, an important nutrient species in near-surface waters. The conversion of ammonia to nitrate via nitrite—nitrification—can occur only in the presence of oxygen (the catalyzing bacteria are obligate aerobes), but other factors are also important. Most importantly, Lipschultz et al. (1983) have shown that nitrification is inhibited by light in excess of a few percent of surface irradiation levels; but this will not be a factor in subeuphotic fjord basins. Whether or not ammonia is oxidized within the sediment is largely determined by the depth of the oxic zone, which (see Sections 7.2.1 and 7.2.3) depends on the rate of supply of labile carbon and dissolved oxygen, the bulk sedimentation rate and degree of sediment mixing, and the ambient temperature. Nitrification is typically undetected in fine-grained, organic-rich, temperate-zone estuarine sediments (e.g., Klump and Martens, 1981), and ammonia diffuses into the overlying water (e.g., Aller and Benninger, 1981). Elevated concen-

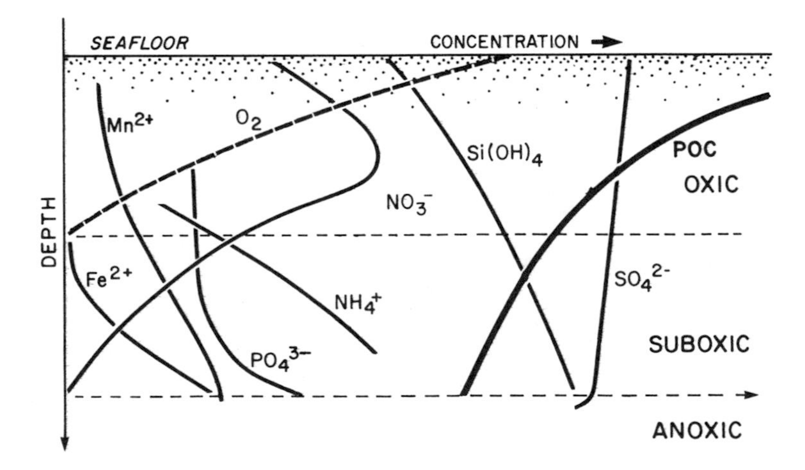

FIGURE 7.5. Idealized depth profiles of interstitial water constituents, dissolved oxygen, and total solid-phase organic carbon (see Fig. 7.4) in aerobic and suboxic zones of unmixed sediment (see Table 7.2).

trations of ammonia similarly have been observed in the bottom waters of polls and fjords where the zone of oxygenated sediment is very thin (or absent), and the dissolved oxygen content of the overlying water is low (e.g., Lännergren, 1975). Conversely, the sedimentary conversion of ammonia to nitrate has been shown to occur where organic carbon contents (Vanderborght and Billen, 1975) or temperatures (Nedwell et al., 1983) are low. The latter are common attributes of fjord sediments, which are, in addition, usually sited below euphotic zone depths.

Gradients of remineralized soluble species in sediment interstitial water, derived from the degradation of sedimented organic detritus, drive transport into the overlying water column. Figure 7.6A illustrates the accumulation of dissolved reactive silica within a subarctic fjord basin following a period of "stagnation" (influx of dense water in the summer displaces the resident water leaving a remnant at the head of the inlet: Fig. 7.6B). These processes are now generally recognized as major sources of nutrients for estuarine primary production. If the upward transport of a mobile metabolite species (con-

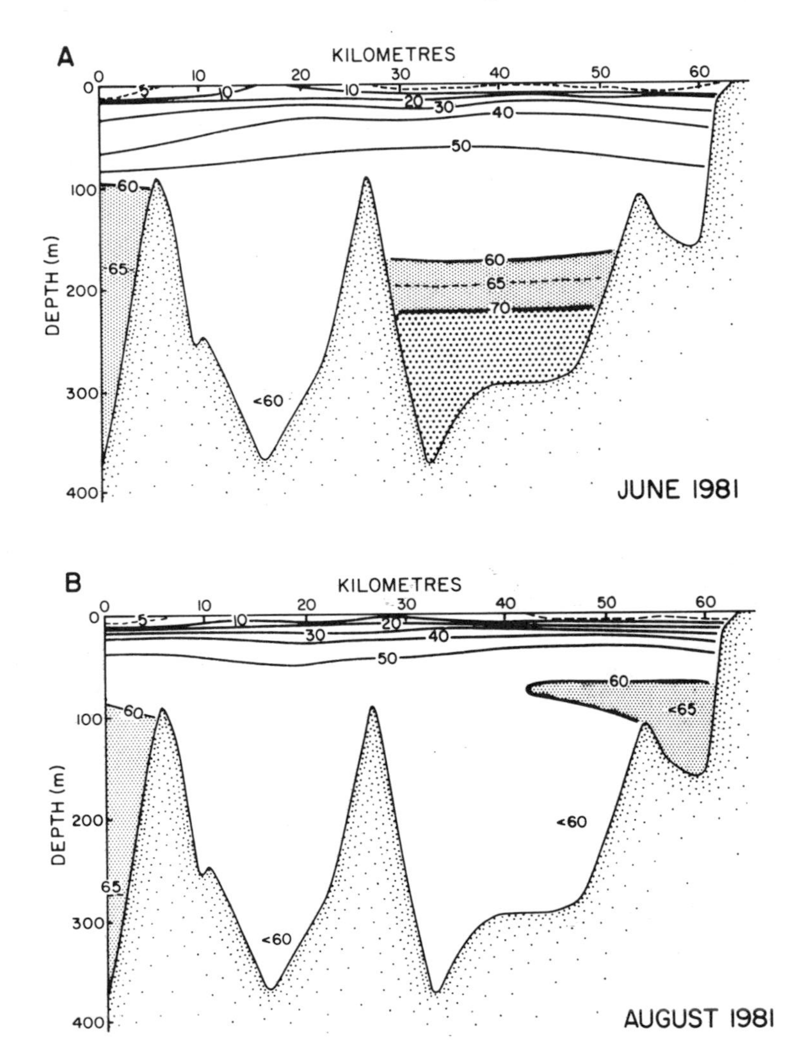

FIGURE 7.6. Longitudinal sections showing dissolved reactive silica concentrations (μM) within Boca de Quadra fjord, southeast Alaska. Benthic degradation of organic detritus generates "new" nutrients, which are eventually transported into the near-surface, euphotic zone. (A) June: Toward the end of a period when the basin water has been "isolated" for more than six months and silica has accumulated at the bottom of the basin. (B) August: Following complete flushing of the basin by water having a silica content < 60 μM. A remnant of the basin "winter water" has been displaced upward and isolated at the head of the fjord. Data from Burrell (1983b).

centration [N]) within the surface sediments may be ascribed to idealized diffusion and advection processes, the flux across the benthic interface is given by:

$$J_N = \phi D_s \frac{\partial[N]}{\partial z}\bigg|_{z=0} - \omega \phi [N]_o \quad (7.14)$$

This is the flux out of sediment of depth z positive downward from the boundary; porosity values are needed because concentrations are per unit volume of pore water. Here D_s is the effective bulk sediment diffusion coefficient, which may be approximated from known molecular diffusion values, corrected for temperature, ion pairing (Lasaga, 1979), sediment tortuosity effects (e.g., Ullman and Aller, 1982), and mixing. Values for the common metabolite species lie in the range from 10^{-7} to 10^{-5} cm^2 s^{-1} (Li and Gregory, 1974), and may hence be of the same order of magnitude as assigned "biodiffusion" coefficients. The second term of equation (7.14), which describes advection with reference to a fixed boundary horizon (the downward velocity of interstitial water is assumed equal to the sedimentation rate), is frequently omitted. However, in fjords having high sedimentation rates of relatively fine-grained sediment, this term is likely to be significant.

It is a very difficult practical exercise to measure an accurate vertical concentration gradient across the benthic boundary, but the major check to employing equation (7.14) is probably in assigning a realistic diffusion coefficient. It is possible to directly determine the interfacial flux of a number of mobile species, integrated over some time period, by serially monitoring concentration changes within a container enclosing a small area of sediment (benthic chamber or "bell-jar" experiments). As emphasized previously, the benthic interface is likely to be the critical reaction zone. Comparison of computed and measured flux values at the same locality (Aller and Benninger, 1981; Callender and Hammond, 1982), can yield valuable insights into the actual physical processes involved. In shallow water regions of temperate-boreal fjords, the flux of many soluble nutrient and heavy metal species out of the sediments varies seasonally. As noted in Section 7.3, increased deposition of labile organic detritus during the summer may cause the redox front (Section 7.3.1) to migrate upward, which, in turn, facilitates the efflux of reduced species such as Mn^{2+} and NH_4^+. Conversely, increased temperatures during the summer stimulate the bioturbating activity of the benthic fauna, which both deepens the oxygenated layer and increases the vertical transport rate of constituents solubilized in this zone. For example, van der Loeff et al. (1984) have demonstrated that the benthic flux of dissolved reactive silica at shallow sites within Gullmarsfjorden, west Sweden, is up to an order of magnitude greater in the summer than in the winter. During the latter season, when ambient temperatures may fall below 0°C, the measured transport rate of silica across the interface corresponds closely to that predicted via molecular diffusion.

The procedures referenced above can only be applied to specific point-localities. In the case of silled fjords, where the basin water may remain stagnant for periods of time that are relatively long compared with the transport rates of the nutrient species, it is possible (uniquely for estuarine environments) to estimate the total flux across the benthic interface by determining the rate of accumulation within the basin water column (Fig. 7.6A). This procedure may be simply illustrated (Fig. 7.7) by considering the mass balance at a time when advective exchange of the basin water is negligible and it may be assumed that transport across some upper, open boundary (area A_o) can be represented as a diffusional exchange process (effective coefficient K_z'), driven by a concentration gradient across the boundary. Between any two observation periods (Δt), the net integrated concentration change of the constituent within the basin (volume V) is determined by the flux (J_N) from the sediments (area A_b), and turbulent exchange across the upper boundary:

$$J_N A_b = V\frac{\Delta[N]}{\Delta t} + A_o K_z' \cdot \frac{\partial[N]}{\partial z}\bigg|_{z=o} \quad (7.15)$$

Among other simplifications, this basic model ignores supply from particulate phases within the basin column; but various levels of sophistication are possible depending on site-specific circumstances. The turbulent diffusion coefficient may be determined from the contemporaneous mass balance of a conservative parameter, such as salinity, or from the distribution

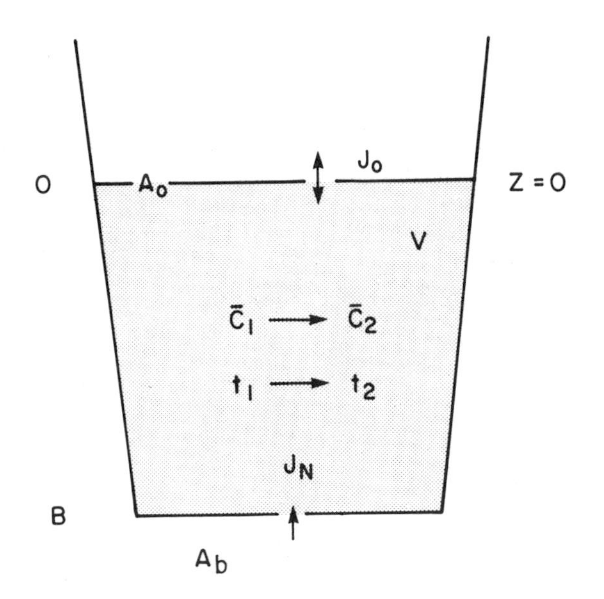

FIGURE 7.7. Box-model computation of the benthic flux of a regenerated metabolite species within an advectively isolated fjord basin. The simplistic case illustrated here assumes that organic degradation occurs only at or in the sediments. The change in mean concentration over time period Δt ($= t_2 - t_1$) is the algebraic sum of the benthic flux (J_N) and turbulent transport (conceptualized as a diffusive process) across the open upper boundary.

of natural radiotracers, as demonstrated by Smethie (1981).

7.2.3 Aerobic Reaction and Transport of Heavy Metals

In addition to the major metabolites discussed above, aerobic decomposition of organic detritus within the water column and at the benthic interface releases heavy metal constituents. This is a topic of major practical importance in all estuarine environments because of the potentiality for subsequent uptake by the benthic fauna. In recent years, a considerable body of information has been assembled from work on deep oceanic sediments. As would be expected (Section 7.1.3; Table 7.1), metals of particular interest in this environment are of the "deep regeneration" type, and there is evidence both from pore-water profiles (Klinkhammer et al., 1982) and from the study of pelagic manganese nodules (Dymond et al., 1984; Aplin and Cronan, 1985) that copper, nickel, and zinc may be released by oxic diagenetic reactions. Recent benthic chamber experiments carried out in Gullmarsfjorden, on the west coast of Sweden, have demonstrated significant effluxes of cadmium, copper, nickel, and zinc from shallow oxygenated sediments (Sundby et al., in press; Westerlund et al., in press). However, at the present time, there is more information available concerning the behavior of copper in deep fjord environments.

Since the advent in the early 1970s of reliable procedures for collecting and analyzing marine waters for a number of trace heavy metal constituents, it has become apparent that dissolved copper concentrations generally increase toward the sediment boundary. This has been demonstrated in the open ocean (e.g., Bruland, 1980; Heggie, 1982) and in deep fjord basins (Heggie and Burrell, 1975). Evidence that this distribution is at least partially the result of transport out of the sediment is provided by elevated interstitial water concentrations frequently identified at or just below the interface. This is well

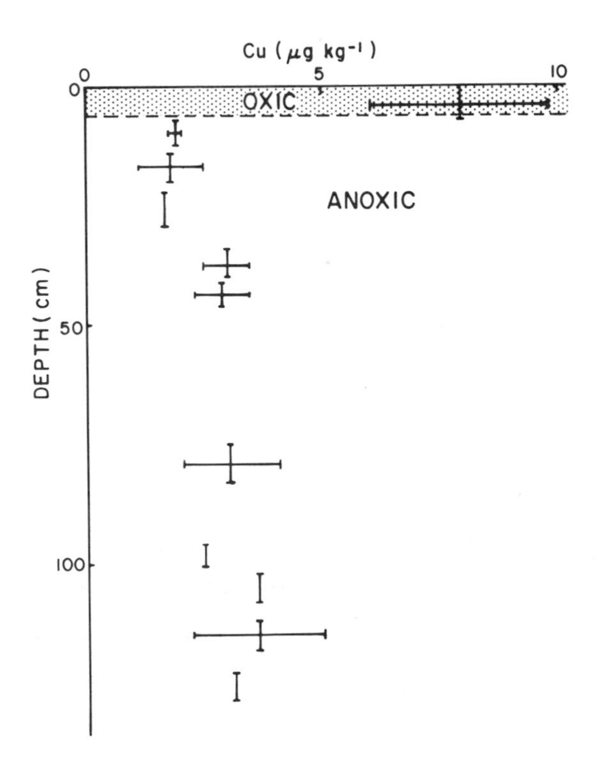

FIGURE 7.8. Vertical profile of soluble copper (μg kg^{-1}; < 0.4 μm) in interstitial water of deep (circa 290 m) basin sediment from Resurrection Bay, south central Alaska. (Composite data from 6 cores collected over a 10-mo period: the horizontal bars represent one standard deviation.) The near-surface concentration maximum occurs within the oxygenated sediment zone. Data from Heggie and Burrell (1982).

established for environments having significant depths of aerobic surface sediment, pelagic oceanic sediments (Callender and Bowser, 1980; Klinkhammer, 1980) especially, but is also found in organic-poor fjord substrates (Heggie and Burrell, 1982). Figure 7.8 illustrates a maximum in the concentration of interstitial water copper in the near-surface basin sediment of Resurrection Bay, south central Alaska. The oxic-anoxic boundary at this locality (Heggie and Burrell, 1980) was located at around 5 to 6 cm below the interface. It would be expected that sediment release and transport of copper into the overlying water generally would be of greatest quantitative importance in estuaries, which are closest to the primary terrestrial sources. (It was suggested in Section 7.1.3 that this phenomenon may partially account for reported instances of "desorption" of copper from river particulates.)

As in the case of the benthic fluxes of nutrients discussed in the previous section, fjord basins offer ideal natural laboratories for observing and understanding the cycling of important trace constituents such as copper between the sediments and the water column. Figure 7.9 (Heggie, 1983) shows the depth distribution of soluble (< 0.4 μm) copper within the basin of an Alaskan subarctic fjord in winter, which, at this locality, is a period of "stagnation" of the subsill water. Heggie (1983) has modelled this vertical distribution, between the benthic boundary and the sill, on the assumption that the primary controlling factors are vertical turbulent diffusion and removal of copper from solution via a first-order scavenging reaction. At steady state (compare with eqns. 7.8 and 7.9):

$$K_z \frac{\partial^2[\mathrm{Cu}]}{\partial z^2} = k_s\,([\mathrm{Cu}] - [\mathrm{Cu}]_\infty \qquad (7.16)$$

$$[\mathrm{Cu}]_z = [\mathrm{Cu}]_\infty + ([\mathrm{Cu}]_o - [\mathrm{Cu}]_\infty) \exp\left[-z\left(\frac{k_s}{K_z}\right)^{1/2}\right] \qquad (7.17)$$

where $[\mathrm{Cu}]_z$ is the soluble copper concentration at a height z above the interface, $[\mathrm{Cu}]_o$ and $[\mathrm{Cu}]_\infty$ are copper concentrations at the benthic boundary and at sill height, respectively, and K_z is the vertical eddy diffusion coefficient. For this cited example, the first-order removal constant (k_s) was computed to be around 6×10^{-7} cm^2 s^{-1}, so that the residence time of copper in the basin-water column is of the order of 20 days.

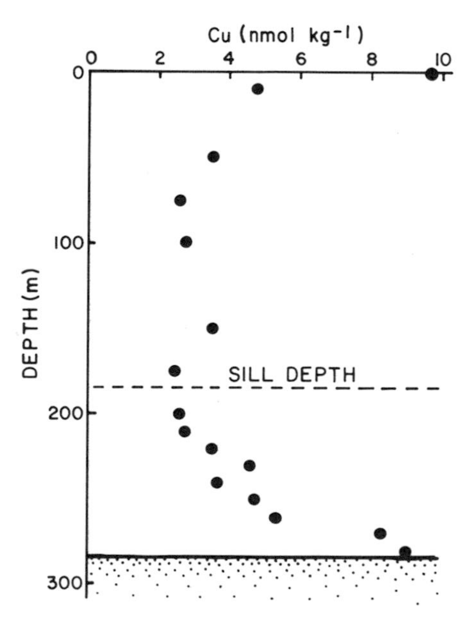

FIGURE 7.9. Vertical profile of soluble (< 0.4 μm) copper from the surface to the bottom of the deep basin of Resurrection Bay, south central Alaska, at a time (February) when the subsill water is "stagnant." Increasing concentrations toward the benthic boundary indicate a flux out of the sediments. From Heggie (1983).

At the present time, the chemical forms of soluble heavy metals transported out of the sediments are not known, but, because of the presence of natural humics and other organic substances in interstitial waters, organometallic complexation may be important. This has been advocated particularly in the case of copper because many people (e.g., Schnitzer and Khan, 1972 and Kerndorff and Schnitzer, 1980) have shown that the stability of humic-metal complexes largely parallels the well-known Irving-Williams series and that, of the common heavy metals, only mercury and iron yield organic complexes of comparable high stability. Organic complexation of interstitial water metals would also be favored by the relatively high concentrations of dissolved organic material present.

A second category of minor constituents (ammonia, considered in Section 7.3.2. is an example of a major constituent that behaves similarly) are those elements, mobilized as reduced species under anoxic conditions, that have relatively slow oxidation kinetics, and hence significant residence times in solution in the marine aerobic environment. Two such metals that have been studied in an intermittently anoxic British Columbia fjord are chromium (Emerson et al., 1979) and antimony (Bertine and Lee, 1983). However, undoubtedly the most important element within this category is manganese.

The geochemistry of manganese in estuarine and coastal waters has been extensively studied and reported in recent years. Basically, this element is mobilized as soluble Mn^{2+} under suboxic conditions (see Section 7.3.2) and is removed from solution in oxygenated waters. The removal process—precipitation or sorption onto solid surfaces—is frequently equated with oxidation. As discussed below, however, this appears to be an oversimplification. It should be noted also that, although Mn (IV) is the thermodynamically stable aerobic oxidation state, measured values of newly precipitated marine particulate manganese may be three (MnOOH) or less (Emerson et al., 1982; Grill, 1982). Regardless of the mechanics and stoichiometry, the removal of reduced, soluble manganese is characteristically slow compared with local transport rates. This results in a temporal and spatial cycling sequence of major geochemical importance: particulate manganese carried into anoxic water or sediment is reduced and may be transported in solution back into the oxygenated environment, where slow oxidation and removal as particulates completes the cycle. These processes are dramatically evident in the oxic-anoxic regions of permanently or intermittently anoxic fjord basins (Emerson et al., 1979; Skei, 1980; Skei and Melsom, 1982). But, since oxidation is kinetically controlled, soluble Mn^{2+} may penetrate oxygenated sediment (Owens et al., 1980; Balzer, 1982), and a quantitatively important flux from most estuarine and coastal sediments into the overlying water is to be expected (e.g., Aller, 1980c). In open, dynamic marine areas, this flux can only be demonstrated and quantitatively estimated from interstitial water gradients (generally possible only in deep sea areas where there is a sufficient depth of oxic sediment at the surface), or by benthic chamber enclosure experiments (e.g., Balzer, 1982) as discussed in the previous section.

Earlier in this chapter it was stressed that, during periods of bottom-water "stagnation," many physically enclosed fjord basins potentially provide excellent natural environments for studying sediment-water geochemical exchanges. This has been demonstrated for copper: the same approach may profitably be applied to describe manganese cycling. Figure 7.10 shows a gradient of particulate manganese, positive to the sediments, in the oxygenated basin water of Bunnefjord (Oslo, Norway). Skei and Melsom (1982) have demonstrated that particulate matter from the bottom of this fjord contains manganese-rich globules and aggregates and also manganese-coated diatom test frag-

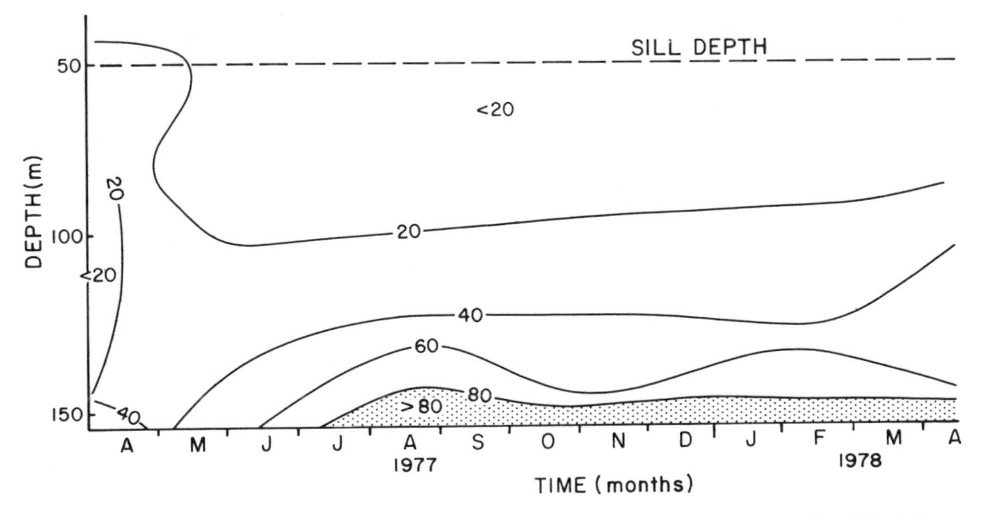

FIGURE 7.10. Time-series distribution of particulate manganese ($\mu g\ L^{-1}$) in Bunnefjord, south Norway. Sampling followed flushing of the basin in February-March (1977). Oxygen concentrations in the deep basin decreased by some 3 ml L^{-1} between March and August, concurrently with increased concentration of manganese. From Skei and Melsom (1982).

ments (Fig. 7.11), and Taylor and Price (1983) suggest that "brown flocs" recovered from deep, aerated waters of Bolstadfjord (west coast of Norway) consist of manganese oxyhydroxides. Figure 7.12 (Owens et al., 1980) illustrates covarying gradients of both soluble ($< 0.4\ \mu m$) and particulate manganese within the basin of Resurrection Bay fjord in south central Alaska.

It is commonly assumed (e.g., Wilson, 1980) that the solution-solid phase partitioning of manganese in oxygenated sea water represents oxidation of Mn^{2+} to some higher oxidation state. The rate expression for the oxidation of Mn^{2+} (Morgan, 1967a):

$$\frac{\partial[Mn^{2+}]}{\partial t} = k\,[Mn^{2+}]\,[MnO_x]\,[OH^-]^2\,[O_2] \tag{7.18}$$

shows the reaction to be autocatalytic and favored by increasing oxygen concentrations and pH. Laboratory determinations of the reaction rate constant (k) give values which are very slow (residence times, k^{-1}, of the order of 10^2 yr; see summary by Grill, 1982); even where $[Mn^{2+}]$,

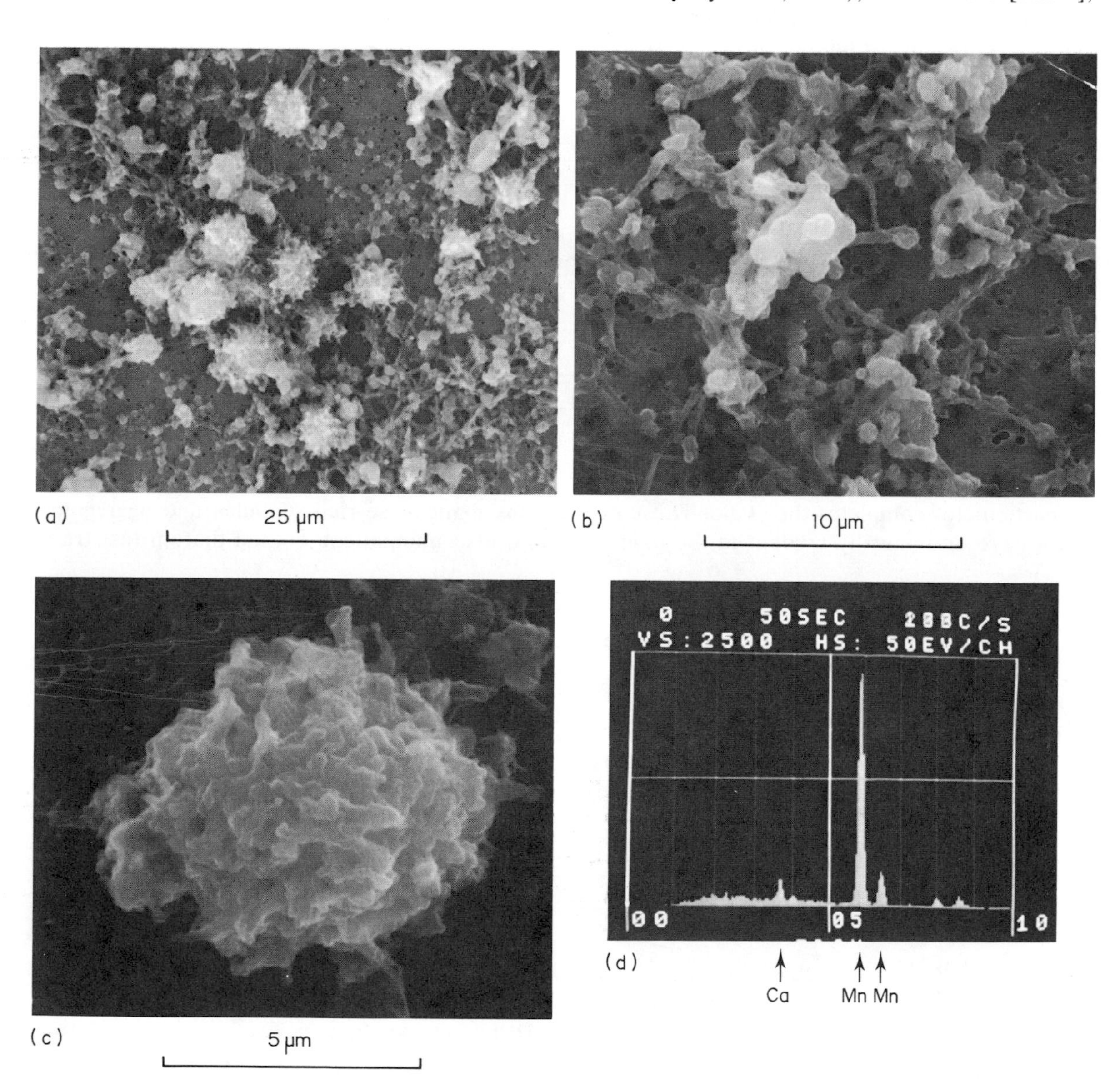

FIGURE 7.11. Scanning electron micrographs of particulate sediment collected from the bottom water of Bunnefjord, south Norway, showing the presence of manganese-rich particles, aggregates, and coatings on diatom debris. From Skei and Melsom (1982).

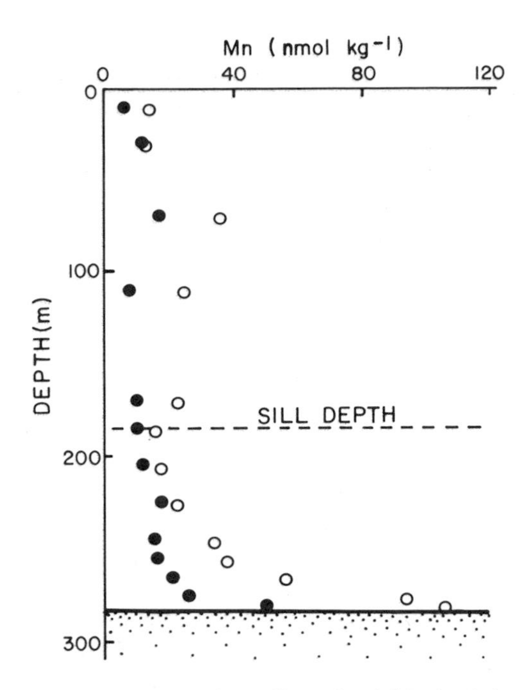

FIGURE 7.12. Vertical profiles of soluble (< 0.4 μm; closed circles) and particulate (open circles) manganese from the surface to the bottom of the deep basin of Resurrection Bay, south central Alaska, toward the close of the period when the sub sill water is "stagnant." Soluble manganese effluxed from the sediments is scavenged and resedimented. From Owens et al. (1980).

$[OH^-]$, and dissolved oxygen concentrations are maintained at much higher levels than would normally be encountered in the natural marine environment. It is apparent that removal of Mn^{2+} from solution initially need not be solely due to oxidation. Emerson et al. (1982) and Grill (1982) suggest that a two-step process occurs: partition onto solid surfaces in the reduced Mn^{2+} state, followed by oxidation. There is now also increasing evidence (Nealson and Ford, 1980; Emerson et al., 1982) for bacterial catalysis of these reactions (and for antimony oxidation under similar conditions: Bertine and Lee, 1983).

The soluble and particulate manganese profiles shown in Figure 7.12 are for a period when the subsill basin water is effectively advectively isolated. Applying the same simplifications and boundary conditions as for the treatment of copper given above, the time rate of change of soluble Mn^{2+} effluxed from the sediments (concentration [Mn]) is a function of vertical diffusive transport within the basin and removal via solid phases:

$$\frac{\partial[\mathrm{Mn}]}{\partial t} = K_z \frac{\partial^2[\mathrm{Mn}]}{\partial z^2} - R \qquad (7.19)$$

where R is the removal rate term. In this particular example the dissolved oxygen concentration varied by less than 2 ml L^{-1} through the basin so that, from equation (7.18), R is primarily a function of [Mn], and P_{Mn}, the concentration of particulate manganese. Since the residence times of both species are much less than that of the basin water, steady-state conditions are approximated:

$$K_z \frac{\partial^2[\mathrm{Mn}]}{\partial z^2} = k_2' [\mathrm{Mn}] P_{Mn} \qquad (7.20)$$

where k_2' is a pseudo second-order removal rate constant. If the particulate manganese phase is present in excess through the basin column, the solution to equation (7.20) is the same as equation (7.17). For the sub-arctic fjord cited here, the computed residence time of soluble reduced manganese in the basin is around ten days, comparable to the ranges calculated (Emerson, 1980; Emerson et al., 1982) for Saanich Inlet, a seasonally anoxic fjord in southern British Columbia.

Manganese dioxide has a $pH_{ZPC} < 3$ in the marine environment and is known to be a very efficient scavenger of cationic species (Murray, 1975). It would be expected, therefore, that the oxidative formation and sedimentation of particulate manganese within fjords would be a major control on the localized cycling and distribution of a number of other trace constituents in the near-sediment basin water. This is a likely mechanism for the scavenging of copper released from oxic surface sediments, discussed above. Berrang and Grill (1974) believe that soluble molybdenum in Saanich Inlet may be similarly removed, and Carpenter et al. (1984) show that ^{210}Pb in Puget Sound, Washington, is primarily associated with hydrous manganese oxides. Essentially the same redox reactions largely, though not entirely, explain the formation of marine manganese nodules with their incorporated minor constituents (Glasby, 1977; Dymond et al., 1984). Growth of manganese nodules has been reported in fjords, or in restricted regions within fjords, where the bulk sediment rate is very low (Calvert and Price, 1970; Grill, 1978).

Iron is also mobilized in anoxic sediments,

but, unlike manganese, Fe^{2+} oxidation kinetics are very rapid (of the order of minutes: Stumm and Lee, 1960; Murray and Gill, 1978). Soluble iron is not transported out of the sediments unless the redox front (see Section 7.3.1) occurs at or above the benthic boundary. In oxygenated fjords, therefore, there is characteristically an iron concentration peak in the sediments approximately marking the oxic-anoxic transition zone (Fig. 7.13A: here the double peak is the result of sediment slumping). Precipitation of ferric iron may co-precipitate or scavenge other ions in a fashion analogous to that described for manganese. The subsurface solid phase molybdenum peak shown in Figure 7.13B illustrates this type of internal cycling. Freshly precipitated iron oxyhydroxide theoretically should bind anionic species such as molybdate since it has a pH_{ZPC} close to the natural ambient pH range.

It has been emphasized elsewhere in this volume that fjord sediments are frequently subject to episodic bed load transport and mixing, so that buried anoxic sediment may be resuspended or mixed up into the oxygenated zone. Under these circumstances thermodynamic modeling predicts increased soluble concentrations of a number of elements (such as cadmium, copper, nickel, lead, and zinc) as a result of reactions generating new, higher solubility solid phases (Lu and Chen, 1977). However, laboratory simulations show only very slight and temporary remobilizations (Lu and Chen, 1977; Khalid et al., 1978), or no significant release (Hirst and Aston, 1983). This behavior is generally attributed to sorption onto coexisting solid phases.

7.2.4 Oxygen Utilization

Oxygen is consumed by the decomposition of organic matter within the water column and surface sediments. In most marine environments, the water column is well ventilated and dissolved oxygen is present even in the deepest parts of the oceans. A number of silled fjords, however, may periodically or episodically go anoxic, and this has a drastic impact on the indigenous biota (see Chapter 6, Section 6.2.4). Anoxia is a consequence of the rate of resupply of dissolved oxygen (primarily the supply of new, oxygenated water) failing to keep pace with the rate of degradation of organic matter. Assuming "ideal" stoichiometry (see below) aero-

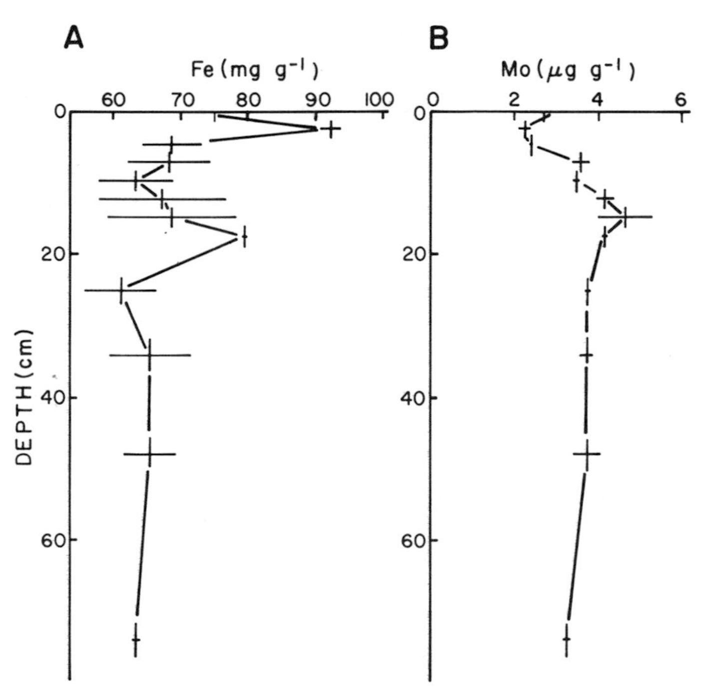

FIGURE 7.13. Depth distribution of acid-extractable metals in sediment from the deep basin (circa 385 m) of Boca de Quadra fjord, southeast Alaska. (A) Iron (mg g^{-1}). The double peak indicates sediment slumping. (B) Molybdenum (μg g^{-1}). From Hogarty (1985).

bic decomposition of 1 mg of organic carbon requires 3.4 mg of oxygen; approximately 2.4 ml, depending on the ambient physical conditions. In the simplest case, where most particulate organic material reaching the sediments is oxidized aerobically (in some deep sea environments, for example), the utilization rate of oxygen in the sediments may be estimated from the rate of loss of labile organic matter, as discussed above. This may apply also to certain polar fjords having relatively low carbon sedimentation rates. In most fjords, however, degradation of organic detritus will also occur anaerobically.

It is possible to directly measure oxygen uptake within, and at the surface of, the sediments by measuring the decrease with time in water overlying incubated (and carefully collected) core samples, and Revsbech et al. (1980) have computed oxygen uptake by direct analysis of oxygen contents in the surface sediments using microelectrodes. Under long-term, steady-state conditions, such procedures theoretically measure both aerobic respiration and also anaerobic metabolism via secondary oxidation of reduced metabolite species transported across the redox front (Section 7.3.1). However, in practice, reduced products may seasonally accumulate in the sediments (the redox front migrates with respect to the interface), while, conversely, disturbing the sample may drastically increase oxidation of reduced species—the "chemical oxidation demand"—at the surface of the core. Rate measurements from incubated samples primarily relate to the metabolism of micro-organisms and the meiofauna. Core samples containing representative macrofauna would be impossible to obtain; but this faunal fraction typically accounts for only a small part of the total benthic oxygen demand. Hall et al. (in press) have studied oxygen uptake kinetics in Gullmarsfjorden, Sweden, using benthic chambers, and show that aerobic mineralization proceeds at an approximately constant rate until dissolved oxygen at the surface is almost depleted.

It has been noted that the flux of organic detritus to the fjord bottom may vary seasonally, and that regeneration in the vicinity of the interface may be very rapid (of the order of weeks). Graf et al. (1983) and Wassmann (1984) have shown that the benthos can respond almost instantaneously to a fresh supply of reactive carbon. The rate of oxygen consumption within a fjord basin may be computed from observation of the concentration rate of change, and knowledge of transportation rates across the free upper boundary, in the same way as described (Fig. 7.6) for nutrient constitutents. By this means, Burrell (in press) has estimated a mean wintertime rate of oxygen utilization within the deep basin of a southeast Alaska fjord of around 1.5 ml L^{-1} yr^{-1}. Benthic oxygen consumption rates determined from incubated core samples retrieved from the deepest region (circa 385 m) of the same basin were some three times greater in the summer than during the winter (9.7 and 3.5 ml m^{-2} hr^{-1}, respectively; Hong and Burrell, unpublished data), hence the mean annual basin consumption rate here is probably around twice the computed winter value. Heggie and Burrell (1977, 1981) applied a more sophisticated advection-diffusion model to determine seasonal oxygen consumption within the basin of Resurrection Bay on the northern Gulf of Alaska coast. In this fjord, oxygen transport is primarily by vertical eddy diffusion in the winter, and by advection through the summer when the basin water is renewed. It was found that oxygen consumption rates in late autumn–early winter, following summer phytoplankton production, were over twice the rate in late winter–spring, and that consumption rates in the bottom 10 m water column "box" (i.e., including sediment consumption) was double that computed for the overlying (100 m deep) basin. The values determined for these two Alaskan fjords are significantly lower than the consumption range of 3.3 to 4.4 ml L^{-1} yr^{-1} determined by Barnes and Collias (1958) and Christensen and Packard (1976) for areas within Puget Sound, Washington, and the 6.6 ml L^{-1} yr^{-1} value given by Stanton (1984) for bottom waters of Stirling Basin, New Zealand, both temperate-zone fjord systems. Gade and Edwards (1980) cite a global range of 2 to 8 ml L^{-1} yr^{-1}.

7.3 Anoxic Environments

7.3.1 The Redox Front

The flux of organic detritus downward to the bottom waters and sediments of estuarine coastal waters is generally much greater than in open ocean areas, and that into physically enclosed fjord basins may be even higher. At the same time, decomposition preferentially utilizes dissolved oxygen so that, in waters where the circulation is restricted, oxygen may be removed at a faster rate than it can be supplied. The geochemical horizon—usually located within the sediment—where free oxygen no longer occurs is termed the redox discontinuity or "front."

The depth of aerobic sediment is conventionally determined from Eh profiles, which are difficult to interpret in such multicomponent, poorly poised systems, or by visual observation of a characteristic color change from an oxidized (brown) to reduced (black). However, from direct electrode measurements of interstitial dissolved oxygen concentrations, Revsbech et al. (1980) have demonstrated that the lower portion of the surface "brown layer" is devoid of oxygen and hence that the color transition convention may seriously overestimate the depth of oxygenated substrate.

Nearly all estuarine sediments are potentially anoxic a few centimeters below the surface. At any given time the location of the redox front is primarily determined by the flux of reactive carbon, and by the degree of biological and physical mixing. Bioturbation by the macro-infauna ventilates the sediment, and is hence a positive feed-back process favoring diverse habitation by a variety of aerobic organisms. Conversely, high rates of deposition of labile organic matter cause the boundary to move upward, restricting the niches available for colonization by obligate aerobic macro- and meiofauna, and hence limiting biological mixing of the near-surface deposits. In deep, cold, higher latitude fjord basins, infaunal densities and activities may be much reduced compared with shallow, temperate estuaries (see Chapter 6), and this may result in the redox front being situated at relatively shallower depths. Deep fjord substrates generally are not subject to short-term, tidal resuspension cycles, and, as long as the inorganic content is high, they tend to be fine grained and compact, factors which impede mixing and irrigation of the near-surface sediment. (Organic-rich anoxic sediments may be highly flocculant and have a very high water content.) Anaerobic organisms, being small, do not physically redistribute sediment, so that preservation of sedimentological structure (such as the annual varves described by Gross et al., 1963) is a common feature of anoxic fjords.

In shallow water, higher latitude environments, where there is a marked annual temperature cycle, a seasonal vertical migration of the redox boundary is commonly observed. Two opposing processes have been identified. In the northern Baltic, where water temperatures in winter are less than 0°C, and benthic biological activity consequently much reduced, Nedwell et al. (1983) have shown that the depth of oxygenated sediment is increased, whereas in the summer, higher temperatures and fresh supplies of reactive organic material cause the redox boundary to move upward toward the surface. However, it is more commonly observed that increased macro-faunal activity in the summer enhances the rate of bioturbation, which increases the depth of aerated sediment. This sequence has been well described, for example, in Gullmarsfjorden, western Sweden, by van der Loeff et al (1984).

In a number of fjord basins, the resupply of dissolved oxygen at times may be so curtailed that the redox front migrates out of the sediment into the overlying water column, resulting in periods when the deep basin waters are also anoxic (e.g., Skei, 1983c). It needs to be emphasized that anoxia is a function both of the local supply rate and reactivity of organic detritus, and of the rate of oxygen replenishment. Thus, water in many of the fjord basins along the contiguous coast of British Columbia and Alaska is characteristically isolated for approximately six months of every year; but anoxic waters are generated annually only in southern British Columbia (in Saanich Inlet, for example). One reason for this may be higher rates of degradation of organic detritus at the lower latitudes (see Section 7.2.4), and also perhaps a higher proportion of labile material. Basins that fluctuate between oxic and anoxic conditions at frequencies of around a year or less are unlikely to attain chemical equilibrium for all species that undergo redox transitions. However, there are a few fjords, such as Framvaren, south Norway (Skei, 1983c) where the deep water is permanently anoxic. The transition zone between oxic and anoxic waters is an especially favorable environment for the development of bacteria that derive energy from the chemical transitions. Emerson et al. (1979) determined that around 80% by weight of the particulate matter collected just above the redox front in the basin of Saanich Inlet, British Columbia, was bacterial biomass.

In Framvaren, the front is located within the euphotic zone. Skei (1983a) has shown that the mass of photosynthetic purple bacteria in this region exceeds the plankton biomass of the overlying oxygenated waters. Because of their distinctive geochemical characteristics, anoxic fjords—especially along the Norwegian coast and on Vancouver Island—are well known and have been studied for many years (Strøm, 1936; Richards, 1965; Deuser, 1975). However, it should be noted that the majority of known fjord basins around the world do not go anoxic.

7.3.2 Anoxic Organic Decomposition Reactions

In organic-rich fjord sediments, the aerobic decomposition reactions (Table 7.2; eqn. 1) described in Section 7.2.2 may be compressed into a zone a few millimeters thick at the sediment surface (or, as noted above, may be shifted into the overlying water). Decomposition of organic detritus does not, however, cease in the absence of molecular oxygen. Anaerobic micro-organisms catalyze, and derive energy from, a series of oxidative reactions utilizing other electron-accepting chemical species. The "suboxic" reactions involving nitrate and higher oxidation states of iron and manganese (Table 7.2; eqns. 2–4) play quantitatively unimportant roles in the degradation of organic material, although iron and manganese reduction reactions are essential steps in the redox cycling of these, and associated trace metal species (see Sections 7.2.3 and 7.3.2). Sulfate, a major anionic constituent of sea water, is the most important marine anaerobic oxidant, and sulfate reduction is succeeded in turn by various methane-producing reactions (Table 7.2; eqns. 5 and 6). The redox front in coastal sediments is thus essentially synonymous with the O_2–HS^- transition zone. Chemolithotrophic bacteria also derive energy from the reoxidation of upward-migrating reduced chemical species: for example, from the conversion of HS^- to S° catalyzed by *Beggiatoa* (Fenchel and Riedl, 1970; Klump and Martens, 1981), and from the oxidation of methane within the sulfate reduction zone (Devol et al., 1984). The basic sequence of chemical reactions is the same in the anoxic waters of fjord basins, and in sediments below the redox front. Geochemical differences observed (e.g., Emerson et al., 1979) in these two environments stem primarily from higher transport rates of soluble constituents and lower particulate concentrations in the water column.

Organic detritus is less efficiently oxidized anaerobically than via oxic respiration (the latter results in a greater unit energy yield), and the highest marine sediment organic contents have been recorded from fjord basins (e.g., Strøm, 1955; Skei, 1983c). Such environments are important geochemical sinks for organic carbon. In general (Toth and Lerman, 1977), a higher proportion of the organic detritus arriving at the fjord benthic boundary will be buried as bulk sedimentation rates increase. Burrell (1983b; see Fig. 6.6) has computed that less than 50% of the carbon deposited within a deep Alaskan fjord is degraded in the near-surface sediments. For comparable carbon and bulk sedimentation rates, this trend could be enhanced at higher latitudes and in deeper waters where the metabolic efficiency is reduced.

Anaerobic decomposition of organic matter (Table 7.2, eqns. 5 and 6) and inorganic redox reactions of iron and manganese, result in concentration gradients in the sediment interstitial water of the type illustrated schematically in Figure 7.14. The idealized depth distribution of a regenerated species such as ammonia, concentration [N], at depth z, may be fitted by an exponential curve of the form (cf. eqn. 7.6):

$$[N]_z = [N]_o + ([N]_\infty - [N]_o)(1 - \exp[-\beta z]) \qquad (7.21)$$

where $[N]_o$ is the concentration at the surface ($z = 0$) and $[N]_\infty$ is the "asymptotic" value at depth. Figure 7.15 shows vertical concentration profiles of soluble ammonium and phosphate in sediment from an Alaskan (55°N) fjord basin. It should be noted that the interstitial water N/P ratios are very much smaller than the predicted "ideal" Redfield stoichiometric value of 16. Suess (1976) has noted that N/P ratios of organic detritus in stagnant Baltic Sea basin sediments may be less than ten. In addition, the effective ammonium ion diffusion coefficient appears to be at least twice that of the phosphate species (Krom and Berner, 1981; Rosenfeld, 1981), and various secondary sediment-solution partition processes generally have a significant impact on the interstitial water distribution of these two constituents. Ammonia sorbs strongly, both exchangeably and irreversibly, onto solid phases, especially onto organic or clay-organic com-

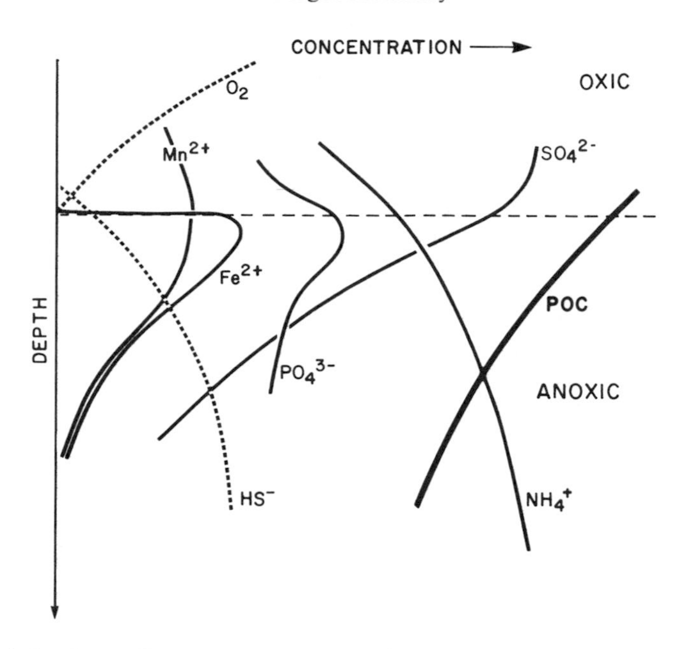

FIGURE 7.14. Idealized depth profiles of interstitial water constituents (oxidants and products; see Table 7.2), and total solid-phase organic carbon, in the anoxic zone of unmixed sediment. (The overlying suboxic and oxic zones are shown in Fig. 7.5.)

plexes (Rosenfeld, 1979; Boatman and Murray, 1982). Krom and Berner (1981) have shown that phosphate participates in the sedimentary redox cycling of iron, being specifically adsorbed (Parfitt et al., 1975) onto oxyhydroxides, and released below the redox boundary. The kinetics of phosphate adsorption have been discussed by Crosby et al. (1984) and Watanabe and Tsunogai

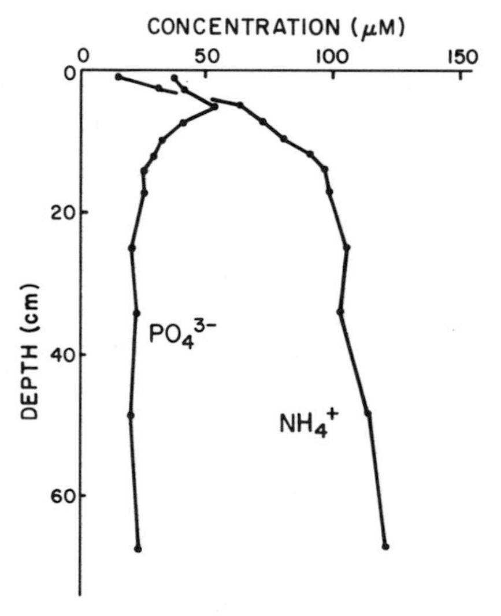

FIGURE 7.15. Depth profiles of phosphate and ammonium concentrations in the interstitial water of fjord basin sediment (Smeaton Bay, southeast Alaska).

(1984). In fjord sediments where the redox front seasonally or episodically migrates up to and beyond the benthic interface, enhanced concentrations of phosphate may be observed in the bottom waters. Paxeus et al. (in prep.) have also recorded release of humic substances (associated with solid phases of iron) under these conditions.

In addition to physical-chemical reactions, interstitial water soluble species may diffuse, or be advected vertically relative to the interface as a consequence of continuing sedimentation. The one-dimensional vertical concentration distribution of a soluble constituent may then be modeled by assuming that certain reactions and processes are dominant, using techniques that have been developed and described in detail in numerous recent publications (see especially Lerman, 1979; Berner, 1980). For the simplest case, assuming constant porosity over the depth interval of interest, and disregarding adsorption and other removal reactions, the steady-state distribution of a soluble metabolite N is given by:

$$D_s \frac{\partial^2[N]}{\partial z^2} + R_N = \omega \frac{\partial[N]}{\partial z} \qquad (7.22)$$

Here R_N is the net rate of production, and the other terms are as employed in Sections 7.2.1 and 7.2.2 (e.g., see eqn. 7.14). The time-invariant assumption ($\partial[N]/\partial t = 0$) may be more generally applicable to interstitial water conditions of deep fjord basins than for most other estuarine environments. If an oxygenated zone overlies the anaerobic sediment, bioturbation and irrigation effects will have to be taken into account (i.e., D_s will be at least of the same order of magnitude, and possibly much larger, than D_b), but where anoxic conditions extend to the interface or into the overlying water, excluding macrobenthic organisms, vertical diffusion rates may approximate molecular diffusion values. Ideally, the production term is primarily a function of the degradation (rate R_c) of organic material (i.e., from eqn. (7.6) of $C_o \exp[-\beta z]$). Thus, for example, if N were the ammonium species, then from Redfield stoichiometry, $R_N = 16/106\ R_c$. However, as described above, adsorption and other solid-liquid reaction terms are usually of comparable importance.

A common pattern of interstitial water nutrient profiles observed in subarctic Alaskan fjords is one where concentrations progressively decline, or approach a constant value, beneath maxima located at relatively shallow depths below the interface. Such patterns, in undisturbed sediment, are consistent with microbial degradation activity being predominantly confined to the near-surface sediment. However, it has been emphasized that, typically, deep fjord sediments are likely to be affected by slumping and other episodic bed-load movements, resulting in a layered sediment structure that may be reflected in the pore-water chemical profiles (Fig. 7.16). In such cases it may be presumed that the observed profiles are transient and do not approximate steady-state conditions. (Applicable mathematical treatments of non-steady-state distributions have been given by Aller, 1980c, and Sholkovitz et al., 1983.) Given the usual range of sediment interstitial water diffusion coefficients, steady state theoretically may be re-established within weeks or months following such a disturbance. However, in low temperature environments it is likely that degradation rates are limiting.

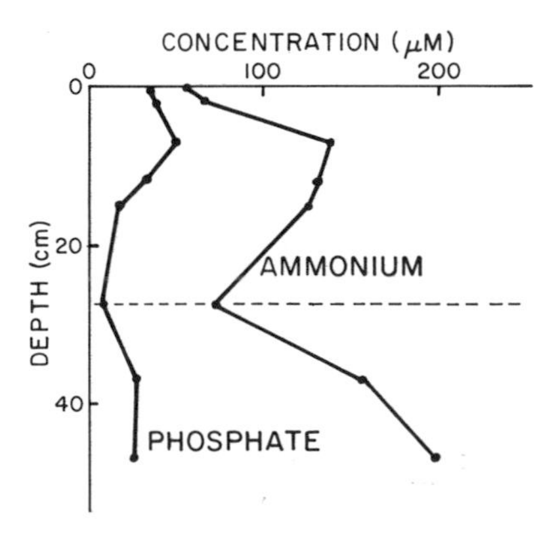

FIGURE 7.16. Depth profiles of phosphate and ammonium interstitital water concentrations in an Alaskan fjord showing effects of sediment slumping.

7.3.3 Reaction and Mobilization of Heavy Metals

Oxidized forms of sedimentary iron and manganese are chemically reduced (Table 7.2; eqns. 3 and 4) in a "suboxic" (Froelich et al., 1979) zone between regions of oxic and anaerobic sulfate respiration. It is, in fact, largely the visual distinction between the oxyhydroxide and sulfide forms of iron in vertical sections of marine sediments that conventionally designates the redox front. Reduction of iron and manganese generates concentration maxima of soluble species in the coexisting pore-water phase (Fig. 7.14), and also releases associated metals (such as nickel; Klinkhammer, 1980) into solution. (Solid phase iron is present in high concentrations in sediments, but it appears that only a small fraction is remobilized; probably primarily that present as "labile" oxide coatings.) Vertical migration may then occur upward or downward within the sediment along concentration gradients. Upward transport into the aerobic environment results in reprecipitation and recycling, both of iron and manganese as described previously, and of associated trace metals such as molybdenum (Fig. 7.13), chromium (Emerson et al., 1979), and cobalt (Kremling, 1983; Heggie et al., in press).

Within the sulfate reduction zone, it might be expected that the concentrations of trace metals in solution would be determined by the solubility of solid phase sulfides:

$$M^{2+} + HS^- \rightleftarrows H^+ + MS(s) \qquad (7.23)$$

However, many investigators (for example: Presley et al., 1972, and Lu and Chen, 1977) have shown that concentrations of the more common heavy metals in anoxic waters are far higher than would be predicted from the thermodynamic solubility products of their simple sulfides; and soluble concentrations in equilibrium with solid-solution phases in which the metal of interest constituted only a minor component would be even lower. Enhanced solubilities in interstitial waters have been ascribed to the formation of both organic and inorganic complexes, but only copper, and possibly iron and zinc, appear to form quantitatively significant organocomplexes. There is abundant field and laboratory evidence showing that copper may be substantially complexed by humic and fulvic acids in the pore waters of anoxic marine sediments (Mantoura et al., 1978; Sugai and Healy, 1978; Boulègue et al., 1982). Krom and Sholkovitz (1978) and van den Berg and Dharmvanij (1984) have also shown that much of the iron and zinc present in solution in organic-rich estuarine interstitial water may be organically bound. Concentrations of natural dissolved organic material within the water column are generally much lower than in sediment interstitial water, and humics there should be primarily complexed by the major divalent cations (Nissenbaum and Swaine, 1976). Nevertheless, Lieberman (1979) believes that 20 to 50% of the soluble copper in Lake Nitinat, a permanently anoxic fjord on Vancouver Island, is organically complexed. Unfortunately, as noted in Section 7.1.3, it is not presently feasible to accurately predict the chemical species fractionation of trace constituents such as copper in natural saline waters via thermodynamic modeling.

Gardner (1974) noted that anoxic metal solubilities would be increased in inorganic systems if bisulfide and polysulfide soluble complexes were formed:

$$M^{2+} + nHS^- \rightleftarrows M(HS)_n^{(n-2)-} \quad (7.24)$$

Emerson et al. (1983) have considered metal sulfide solubility in terms of "hard sphere" and "soft sphere" cations (Stumm and Morgan, 1981). "Soft sphere," or B-type metal ions are highly polarizable and preferentially coordinate with ligands (including sulfur-containing species) that behave similarly. Metals such as copper and cadmium (Boulègue, 1983) hence form strong complexes with bisulfide and polysulfide ligands, and total concentrations of soluble species of this category of metals increase with increasing values of [HS⁻]; i.e., initially with depth. Metals transitional between A- and B-types—iron, manganese, and cobalt, for example—do not produce strong complexes with reduced sulfur species, and concentrations of

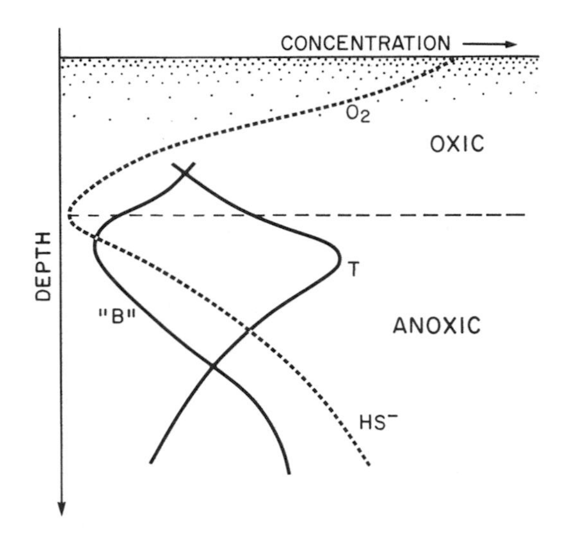

FIGURE 7.17. Depth profiles of idealized soluble transition (T) and type "B" cationic species concentrations in a fjord basin water column in the vicinity of the redox front. From Jacobs and Emerson (1982).

these metals tend to decrease with depth within the anoxic zone. These trends, shown schematically in Figure 7.17 (after Jacobs and Emerson, 1982), have been demonstrated by Kremling (1983) from anoxic regions of the fjordlike Baltic Sea.

7.3.4 Authigenic Formation of Solid Phases

Degradation reactions and migration of soluble species within the sediment interstitial water environment may lead to saturation with respect to potential solid phases, and hence, depending on reaction kinetics, to the generation of diagenetic minerals. Conversely, formation of new solid phases is usually the primary control on concentrations of soluble species in the pore waters below the surface zone. The initial solids produced may be metastable forms. For example, iron in solution in anoxic sediments generally appears to be limited by the production of amorphous or crystalline monosulfide phases (Davison, 1980). The thermodynamically stable form (pyrite), is not kinetically favored (Berner, 1967), although Skei (1983a; unpublished data) demonstrates formation of framboidal pyrite in permanently anoxic fjord waters.

Under certain favorable conditions it may be possible to directly identify the incipient formation of specific minerals (Suess, 1979), but this is usually not possible, because of low concentrations, poor crystallinity, or other practical reasons. The more usual practice is to indirectly infer the existence of likely controlling solid phases by determining saturation parameters (Ω) from analysis of the soluble concentrations of the ions participating in the reaction:

$$xM + yA \rightleftarrows M_xA_y(s)$$

$$\Omega = \frac{[M]^x\,[A]^y}{K'_{sp}} \tag{7.25}$$

where K'_{sp} is the apparent solubility product. When Ω consistently approaches unity, formation of the solid phase is assumed to occur. This approach necessitates determination of the applicable ion activity co-efficients (γ), and the free ion fraction, in order to determine K'_{sp} from known thermodynamic solubility product data ($K^\circ_{sp} = K'_{sp}\gamma_M^x\,\gamma_A^y$). Values of K°_{sp} for a large number of relatively simple solid phases are known (see examples in Table 7.3: $pK^\circ_{sp} = -\log K^\circ_{sp}$). However, in natural marine environments, amorphous and mixed solid phases (e.g., of carbonates: Suess, 1979) may be very common; particularly solid solutions incorporating trace constituents in major ion minerals.

Suess (1979) has shown that major ion carbonates and precipitates of amorphous silica are forming in sediments underlying anoxic waters of the Baltic Sea, and the precipitation of metal sulfides

$$M^{2+} + HS^- \rightleftarrows MS(s) + H^+$$

is well documented (e.g., Spencer and Brewer, 1971; Emerson et al., 1983; Skei, 1983c). Some solid phases believed to be in equilibrium with minor soluble species in various anoxic fjord environments are given in Table 7.3. Because many fjord sediments are also organic rich, early diagenesis of organic compounds may also be important (Brown et al., 1972).

7.4 Summary

As with other types of estuaries, fjords are regions where terrestrial input mixes, and where constituents may react, with sea water and with marine solid phases. For example, freshwater-soluble species may flocculate, or be adsorbed on, or incorporated in, particulate matter. Such processes can be advantageously studied in deep, steep-sided fjords since "estuarine" mixing zone reactions are here spatially segregated from reactions involving interchange between the deposited sediments and the overlying water. However, the watershed environment, and hence the character of freshwater discharge into the marine fjord, differs widely from that of shallow, temperate-zone estuaries in a number of respects. Fjords are typically located in mountainous terrain, so that the residence times of direct precipitation in the watersheds is likely to be very short. In regions of high rainfall, and where a large portion of the precipitation is initially stored as snow, concentrations of soluble constituents in the runoff, nutrients and trace metals, for example, may be very low. With in-

TABLE 7.3. Diagenic formation of mineral phases in fjord anoxic environments.

Ion	Mineral	pK°_{sp}	Environment	Reference
PO_4^{3-}	$(Mg)(NH_4)PO_4$ struvite	13.7	Saanich Inlet	Murray et al., 1978
Fe^{2+}	FeS mackinawite	17.5	Saanich Inlet	Murray et al., 1978
			Baltic Sea	Kremling, 1983
	Fe_3S_3 greigite	18.2	Saanich Inlet	Jacobs & Emerson, 1982
	FeS_2 pyrite	32.8	Framvaren	Skei, unpublished
Mn^{2+}	$MnCO_3$ rhodochrosite	10.4	Saanich Inlet	Murray et al., 1978
			Jervis Inlet	Grill, 1978
			Loch Fyne	Pederson & Price, 1982
	mixed carbonate		Baltic Sea	Kremling, 1983
	MnS	14.8	Baltic Sea	Suess, 1979
Co^{2+}	CoS	21.4	Baltic Sea	Kremling, 1983
Cd^{2+}	CdS greenockite	27.2	Saanich Inlet	Jacobs & Emerson, 1982
Zn^{2+}	ZnS	22.3	Framvaren	Skei, unpublished

creasing latitude, seasonal variations in the freshwater influx assume greater importance. In arctic fjords (except where there is a tidewater glacier) freshwater input may cease entirely in the winter. Conversely, a major fraction of the river discharge and associated particulate (and bed load) sediment may occur over a relatively short period in the summer.

The origins, transport, and degradation of particulate organic material within silled fjords has been stressed in this chapter. Fjord environments range from temperate-boreal types producing, receiving, and retaining high concentrations of organic detritus, to those in glacial regions where terrigenous carbon input may be very slight, and where the relatively low marine biogenic flux to the sediments is likely to be considerably diluted by aluminosilicate material. The relative reactivity of organic detritus incorporated in the sediments depends very much on source. Phytogenous detritus is rapidly degraded, whereas terrestrial plant debris is frequently highly inert and may be substantially buried within the sediments. Since riverine particulate influx tends to peak at certain times of the year; it is sometimes possible to temporally separate the primary fluxes of autochthonous and allochthonous carbon arriving at the benthic interface.

Major factors determining the vertical distribution of both solid phase and soluble interstitial water constituents have been briefly discussed. The sediment of "low-C input" fjords is comparable with deep sea pelagic environments inasmuch as the surface sediments may be oxygenated to a much greater depth than is usual in coastal regions. Sedimentary oxic zone reactions (such as nitrification) and transport patterns can thus be conveniently studied. Primary examples outlined here have been the regeneration and near-bottom cycling of copper, and the behavior of reduced, mobile manganese within the oxygenated sediments and bottom-water zone. Elucidation of the benthic behavior of minor constituents such as these has been facilitated because of the physical confinement of bottom waters, over certain time periods, within deep fjord basins. During these "stagnant" periods, the basins function as "geochemical buckets," and the flux of soluble chemical species out of the sediments can be directly observed and measured. In like fashion, the mean benthic flux of regenerated metabolites back into the water column may be computed.

Fjord basins in general are important sinks for organic carbon in the subarctic and temperate climatic regions. Depending primarily on the relative contribution of labile organic detritus, on the ambient temperature range (degradation rates increase exponentially with temperature), and on the rate of oxygen supply, increased carbon loading will cause the redox front to rise within the sediment. This process is accelerated with progressive elimination of the macro-infauna, which ventilate and irrigate the near-surface sediments (see Chapter 6). In fjord basins where water circulation patterns periodically prevent adequate resupply of oxygen to depth, the redox boundary may migrate into the water column.

A number of fjord basins around the world have long been regarded as ideal natural laboratories for studying anaerobic reactions and processes, both within the sediments and overlying water. In particular, recent work has afforded new insights into the mobilization and transport of heavy metals, and the authigenic formation of new solid phases.

Part 3 Implications/Applications

8

Environmental Problems: Case Histories

8.1 Introduction

As with coastal waters in most parts of the world, fjords are vulnerable to man's influences. However, because of the topographic impediments to free circulation (and hence the frequently extended flushing times) discussed in previous chapters, fjords are generally much more susceptible to pollution problems than are most other types of estuaries (e.g., Skei et al., 1972).

A large amount of data for impacted estuaries has been generated by and for government agencies and various industrial concerns. Much of this literature refers to "snapshot" baseline surveys, or reports various monitoring program parameters, without full scientific discussion or conclusions. However, during the last decade several papers on pollution of water, biota, and sediments in fjords have been presented (e.g., Loring, 1976a, b; Oftstad et al., 1978; Knutzen and Sortland, 1982; Skei, 1978, 1981a; Bjørseth et al., 1979; Skei and Paus, 1979; Mirza and Gray, 1981; Nyholm et al., 1983).

The principal categories of anthropogenic impact on fjords are: (1) effluents from pulp and paper industries, (2) urban sewage and other nonpoint source discharges, (3) fish processing wastes, (4) effluents from metal refining, and (5) rock and mill waste from onshore mining. The natural function of a number of fjords has also been drastically altered by subtracting from, or augmenting, the natural freshwater influx. In recent years, several fjords have become staging sites and major storage and trans-shipment ports for offshore oil producers. It should be noted that the physical confinement of fjords may be advantageous if the objective—as has been the case, for example, with some mining wastes—is to concentrate rather than disperse the pollutant. This characteristic, coupled with high sedimentation rates, could make some fjords preferable marine dumping sites for a number of solid wastes. Many pollutant chemicals dis-

TABLE 8.1. Pollution case histories of fjords discussed in Chapter 8 (in order of discussion).

	Environmental problem				
Fjord	Metals	Organic micropollutant	Eutrophication	Mine tailings	Organic matter
Agfardilkavsâ, Greenland	X			X	
Resurrection Bay, Alaska					X
Port Valdez, Alaska		X			X
Howe Sound, British Columbia	X	X	X	X	X
Rupert Inlet, British Columbia	X			X	
Saguenay Fjord, Quebec	X				X
Iddefjord, Norway	X	X			X
Saudafjord, Norway	X	X			
Sørfjord, Norway	X	X			
Ranafjord, Norway	X	X		X	
Lock Eil, Scotland					X
Byfjord, Sweden	X				X
Oslofjord, Norway	X		X		X
Mariager Fjord, Denmark			X		X
Alice Arm, British Columbia	X			X	
Puget Sound, Washington	X				
Kitimat Arm, British Columbia	X	X			
Frierfjord, Norway	X	X	X		X
Jössingfjord, Norway				X	
Alberni Inlet, British Columbia					X
Borgenfjorden, Norway			X		X
Bellingham Bay, Washington	X	X			

charged in solution are "particle reactive" (Chapter 7), and are similarly incorporated in sediments deposited relatively close to the input source. Remobilization of pollutant constituents from the sediment sink back into the overlying water is possible (Bothner and Carpenter, 1974; Wood, 1974), but, as discussed in Chapter 7, net flux is generally to the sediments. Uptake of exotic chemicals, especially toxic metals, concentrated in the sediments by benthic and demersal species is a major ongoing concern (e.g., Bryan, 1980), unfortunately outside the scope of this book.

Several categories of pollutants (sewage, and plant and animal processing wastes in particular) consist primarily of labile organics, and disposal of this material in fjord basins may result in anoxic conditions, or in the spread of natural anoxia. The behavior of metals in anoxic waters has been briefly outlined in Chapter 7, but a great deal more work in this field is required.

A number of specific fjord estuaries that have been polluted or impacted by man in some major way are described in this chapter. The examples chosen represent a range of pollutants in different fjord environments from various parts of the world (Table 8.1). The emphasis is on sediments, on the indigenous benthic biota, and on reactions and exchange between the sediments and the overlying water.

8.2 Agfardlikavsâ Fjord, Greenland

8.2.1 Environmental Setting

Agfardlikavsâ (or Affarlikassaa) is a fjord situated on the west Greenland coast (Fig. 8.1). The fjord is approximately 4 km long, 0.5 km wide, and with a sill depth of 23 m. The maximum depth is 76 m and the total volume of the fjord is 80×10^6 m^3 (Möller, 1984).

The freshwater runoff from the adjacent watershed to Agfardlikavsâ is dominated by meltwater from the Inland Ice at the head of the fjord. The total volume of the river discharge in one year is of about the same magnitude as the volume of the fjord (Möller, 1984). The natural sediments in the Agfardlikavsâ are grey-green silty muds, consisting predominantly of dolomitic carbonates (Bondam, 1978). The sediments

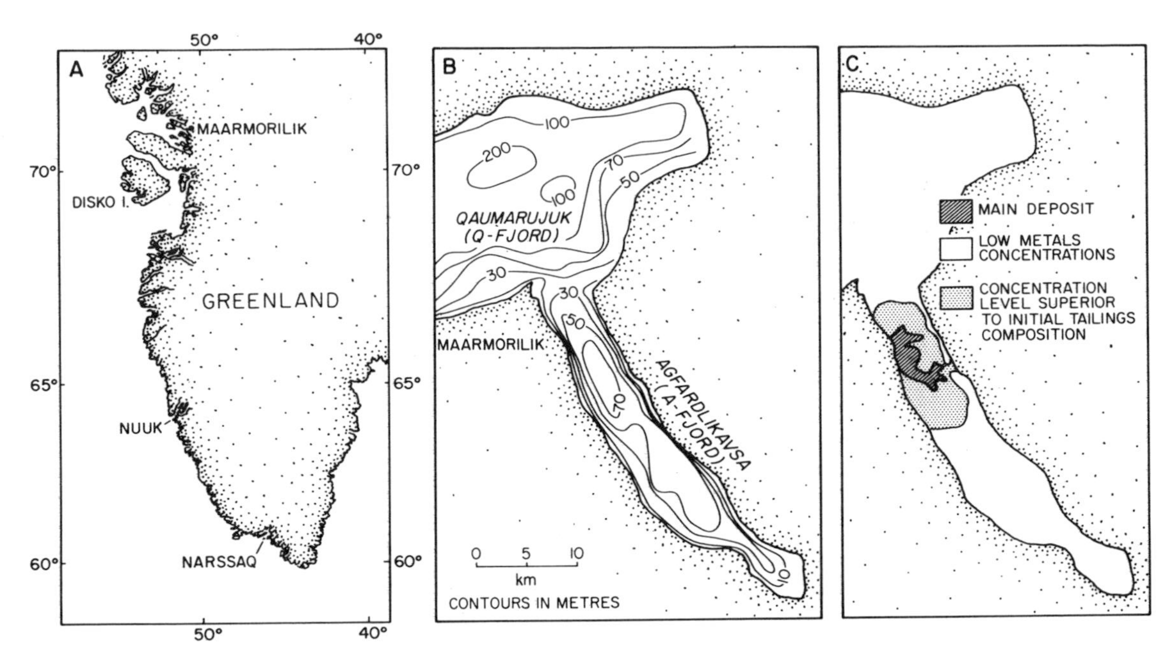

FIGURE 8.1. (A) Agfardlikavsâ, west Greenland; (B) bathymetry; and (C) distribution of tailings close to the discharge point (after Nyholm et al., 1983).

are low in organic matter. The natural sedimentation rate in the inner part of the fjord, based on foraminifera populations, has been estimated at 2 cm yr^{-1} (Willumsen, 1975).

From mid-November to mid-June, the fjord is covered by ice. During the summer, large amounts of silt are transported into the fjord, creating high turbidity in the surface water (Nyholm et al., 1983). Very little coarse material is transported into the fjord as the river water is filtered through several lakes that act as traps for coarse material. The mean particle size of uncontaminated sediments is 2.6 μm (VKI, 1979). Based on sediment trap measurements, approximately 2000 tons of suspended matter settles in Agfardlikavsâ fjord per year (Nyholm et al., 1983).

The water body below sill depth is practically stagnant during the summer. This is the period when most of the sedimentation takes places.

8.2.2 Environmental Problems

The marine disposal of tailings from the Greenex A/S lead/zinc mine and mill, situated at Marmorilik at the mouth of Agfardlikavsâ Fjord (Fig. 8.1) started in 1973. The outlet from the mill is situated at 38-m depth and the tailings are discharged as a suspension at a rate of 0.045 m^3 s^{-1} (Möller, 1984). The very dense suspension (density = 1340 kg m^{-3}) travels down the fjord slope as a turbidity current, continuously depositing its content of particles (Möller, 1984). The amount of tailings discharged into the fjord is 40×10^6 kg $month^{-1}$.

Only two years after the establishment of the mine, much of the sea bottom of Agfardlikavsâ was influenced by tailings from the flotation mill (GGU, 1977). In 1978 the tailings contained 1.1% zinc, 57 ppm cadmium, and 0.44% lead (GGU, 1980). The bottom sediments near the outlet had a lower metal content than sediments some distance away (Fig. 8.1): the finest fraction of the tailings has the highest metal content and is thus transported a longer distance owing to a slow settling rate (GGU, 1980). Sediments deposited 1 km away from the source contained twice as much zinc and lead as the initial tailings. Apparently, the grain size effect, concomitant with changes in the particle settling velocity, is very important when considering disposal of mine tailings in fjords.

Extremely high concentrations of dissolved metals have been measured in the water of Agfardlikavsâ (Asmund, 1980). Hence, the elevated levels of metals in the sediments may reflect early adsorption of dissolved metals onto river-introduced fine particles. The high dissolved

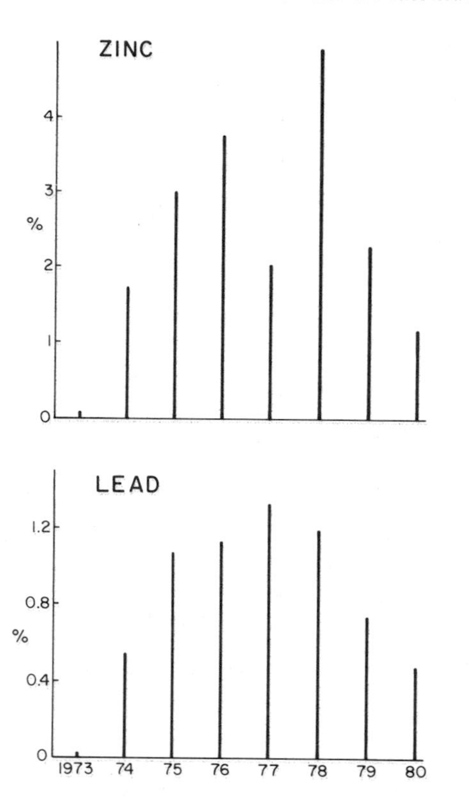

FIGURE 8.2. The abundance (by weight) of zinc and lead in surface sediments in Agfardlikavsâ, 800 m from the discharge point of tailings (after GGU, 1982).

levels of metals are not necessarily a result of dissolution of metal sulfides in sea water but may relate to the flotation technique used at the mill. The addition of lime and aluminum sulfate to the tailings (prior to discharge) has recently reduced the problem of elevated dissolved metal in the fjord waters. This has had an improved effect on the contamination of the surface sediments. Figure 8.2 illustrates the annual variations of zinc and lead in the surface sediments, 800 m from the discharge outlet. The treatment of the tailings, started in 1979, has had an important effect. Additionally, improvement of the outlet system in 1979 eliminated an earlier upward-directed spreading of tailing particles caused by air entrainment (Nyholm et al., 1983). The wide spreading of the tailings has now ceased and instead the material is deposited within a limited area of $15 \cdot 10^3$ m^2. Figure 8.1C shows the areal distribution of contaminated sediments in Agfardlikavsâ Fjord.

One important environmental question concerns the role that the contaminated sediments play as a secondary source of metals to water and biota. To make such an assessment, one has to consider the various sources and pathways of metals and their relative importance. Nyholm et al. (1983) have developed a model to study the transport of heavy metals in Agfardlikavsâ Fjord. They considered the transport of metals

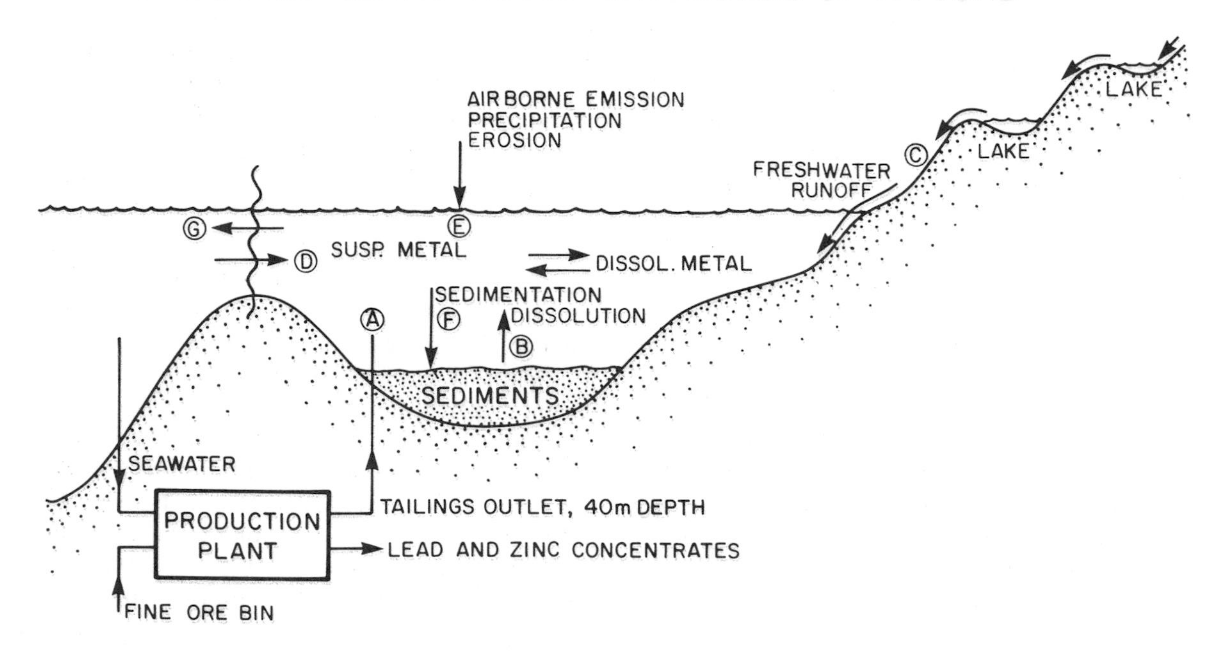

out of the fjord, natural and anthropogenic sources and the sediments as a sink or source of metals (Figs. 8.3 and 8.4). As the water below sill depth is relatively stagnant during summer, little exchange of metals to the neighboring fjord takes place. In fact, sedimentation during summer and fall is an important metal sink, as this is the period with the highest levels of silt and organic matter in the water (Nyholm et al., 1983). During winter, water exchange takes place beneath the ice, transporting dissolved metals out of the fjord. Little escape of tailing particles has occurred, however, especially after 1979 when the new discharge arrangements came into operation. A certain amount of the metals is dissolved immediately from the tailings upon entering the fjord. This is particularly significant for lead (Nyholm et al., 1983) and may also explain the exceptionally high dissolved lead concentrations in the water compared with background levels (see also Section 8.11).

Release of metals from the tailings-contaminated sediments is described in the model by simple zero order kinetics (Nyholm et al., 1983). Release rates are derived from laboratory sediment-water exchange assays (Table 8.2). Bottom areas referred to in Table 8.2 (1, 2, and 3) are those shown in Figure 8.1C. Compared with the amounts of metals in the tailings and the soluble part of those metals, the metal released from the contaminated sediments represents a minor source. However, as the primary sources are reduced (treatment of the tailings), secondary sources may become more important.

FIGURE 8.3. General system description. The following dominant transport pathways are shown in the figure (after Nyholm et al., 1983). (A) Discharge of tailings into the bottom water of the A-fjord and immediate dissolution of metals (primary source). (B) Dissolution of metals from the tailings deposit at the bottom of A-fjord (secondary source). (C) Freshwater runoff in summer carrying large amounts of glacier silt and providing a background input of metals (secondary source). (D) Inflow from Q-fjord. (E) Additional inputs of metals to the surface water from various sources, including airborne emission from the mining and milling activities, background atmospheric precipitation, and erosion of surrounding rocks (secondary source). (F) Sedimentation of metals that have become bound to particulate material (glacier silt, phytoplankton, etc.) (primary sink). (G) Outflow to Q-fjord.

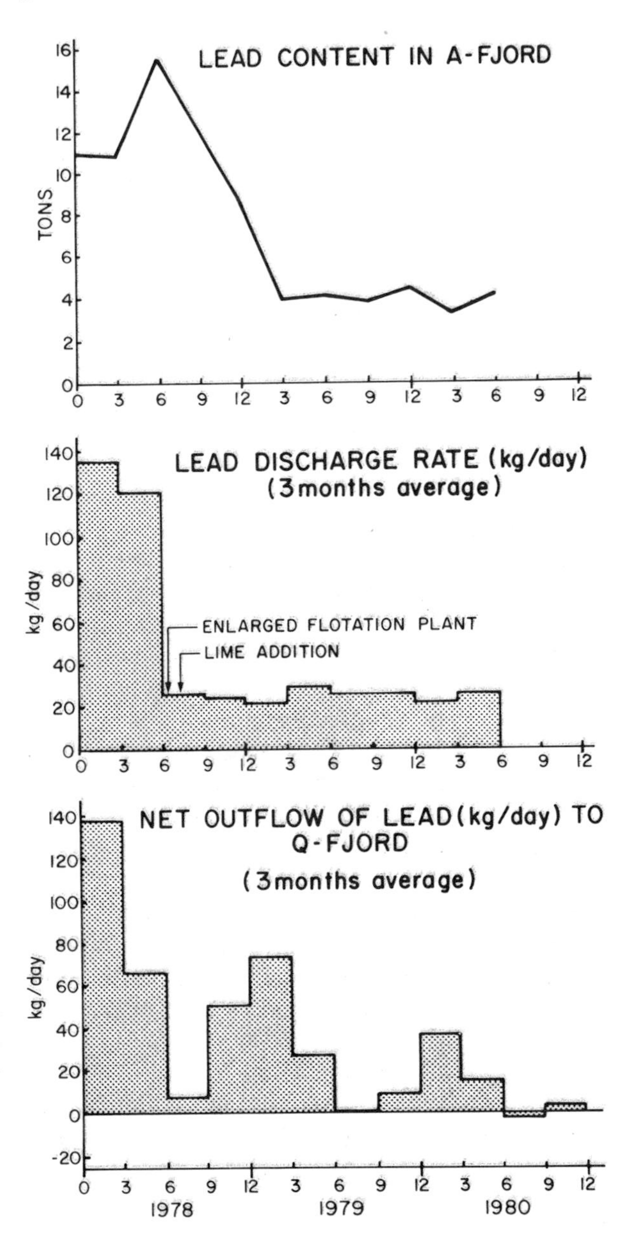

FIGURE 8.4. Graphical presentation of 3-mo mass balances for lead during the period 1978–80 (after Nyholm et al., 1983).

Sorption experiments on the capacity of silt to bind metals (zinc and lead) indicated a lack of simple sorption mechanisms (Nyholm et al., 1983). Even if arctic fjords are normally low in organic matter, it may play a significant role as a sorption substrate for metals. It has been observed that Greenland fjords may act as a sink for organic matter (Petersen, 1978), most of which originates from outer waters (see Section 2.6.1). Although the concentration of organic

TABLE 8.2. Release rate of metals from sediments.*

Section of sediments **	Area 10^3 m^2	Release rates Lead mgPb m^{-2} day^{-1}	Zinc MgZn m^{-2} day^{-1}	Cadmium mgCd m^{-2} day^{-1}
1	150	3.2	17	0.056
2	465	0.7	11	0.048
3	1325	0.5	5	0.033

* After Nyholm et al., 1983.
**Corresponds to bottom areas in Fig. 8.1C.

matter in the sediments is low, the flux may be high owing to rapid deposition. Sedimentation rates based on sediment trap measurements during the period July–October were in the order of 10 g m^{-2} day^{-1} (Nyholm et al., 1983). This corresponds to a deposition rate of zinc and lead of about 150 and 50 mg m^{-2} day^{-1}, respectively (Nyholm et al., 1983). Presumably, during the ice-covered period in winter, the deposition rate is one to two orders of magnitude less.

The model simulations made by Nyholm et al. (1983) clearly show the influence of the mill alterations, of the flotation process in 1978, and the alum flocculation used in 1979, on the levels of trace metals in the fjord environment (Fig. 8.4). The model also demonstrates a nonlinear relationship between discharge rates and future loadings of the adjacent fjord owing to sedimentation of metals in the Agfardlikavsâ (Nyholm et al., 1983).

The fate of trace metals in Agfardlikavsâ has also recently been discussed in relation to hydrodynamic conditions (Möller, 1984), considering the dissolved metals as conservative tracers. Diffusion coefficients based on salinity and temperature profiles are applied on measured metal profiles to estimate transport of metals from the bottom water to the surface water during the "stagnant" summer period. Table 8.3 indicates the relative importance of various transport mechanisms of lead. Apparently, the wind-generated entrainment in the autumn and the flushing events in winter are more important transport mechanisms than diffusion from the bottom water.

A baseline survey was carried out in 1972 prior to the start of mining and milling activities in 1973, and since then monitoring has been run twice a year by the Ministry of Greenland (Nyholm et al., 1983). The monitoring program has included determinations of metals in sea water, sediments, seaweeds, mussels, shrimps, zooplankton, fish, and seal. Elevated levels in biota have been found in the area near the Marmorilik mine but have not been considered as a health risk for the local inhabitants.

TABLE 8.3. Example of transport of lead from the bottom water of Agfardlikavsâ, west Greenland, 1981–82. The transports due to flushing are hypothetical transports.*

Month	Cause Diffusion	Entrainment	Flushing	After flushing
Apr. 81 – Aug. 81	100 kg/mo			
Sep. 81 – Nov. 81	200 kg/mo	2400 kg		
Dec. 81 – Mar. 82	100 kg/mo (?)		(2500 kg)	(700 kg/mo (?))
Total	1500 kg	2400 kg	(2500 kg)	(2100 kg)

*After Möller, 1984.

8.3 Resurrection Bay, Alaska

8.3.1 Environmental Setting

Resurrection Bay, some 30 km long and 6 to 8 km wide, is located on the Kenai Peninsula of south central Alaska (Fig. 8.5A), separated from the open Gulf of Alaska by a 185 m deep sill. The single enclosed basin has a maximum depth of around 290 m. Most of the freshwater input occurs at the head of the inlet from the Resurrection River. The annual discharge pattern is strongly unimodal (Chapter 7; Fig. 7.1): maximum precipitation and inflow in this region occur in late summer–fall when riverine discharge, which includes snow-field and glacier meltwater, is around two orders of magnitude greater than that in midwinter. However, the mean annual volume influx amounts to less than 5% of the tidal prism, and hence estuarine circulation is relatively unimportant. Niebauer (1980) suggests that axial winds may be important to the surface transport, especially in the winter season of katabatic winds and minimum upper column stability.

Royer (1975, 1983) has demonstrated that the Aleutian low pressure system is the major driving force for geostrophic flow within the subsurface mixed layer of the Gulf of Alaska through the winter. The prevailing cyclonic winter winds, directed by the coastal mountains, generate onshore convergence along the northeast Gulf coast. Conversely, this region is predominantly influenced by relatively weak high

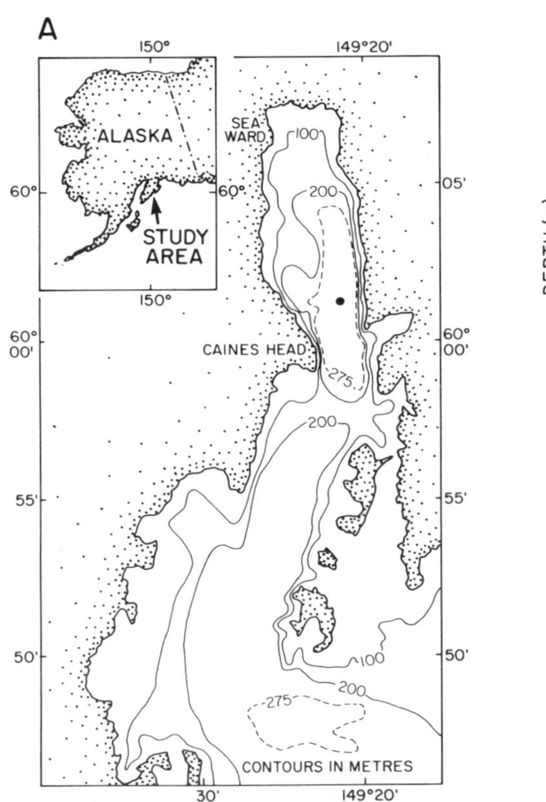

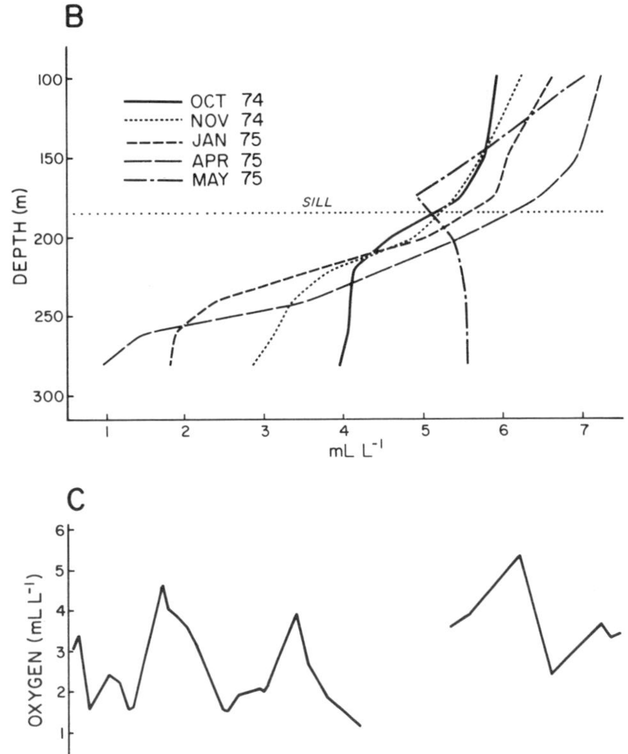

FIGURE 8.5. (A) Resurrection Bay, south central Alaska, showing bathymetry and the location of the basin sampling station. (B) Vertical profiles of oxygen concentrations (ml L^{-1}) from above the sill to the deepest part of the basin (location shown in Fig. 8.5A) through the period (October-April) when the basin water is "stagnant." The basin has been flushed by higher oxygen-content water in May. From Burrell (1983a). (C) Seasonal and yearly differences in dissolved oxygen concentrations (ml L^{-1}) at the bottom (275 m) of the Resurrection Bay basin. Seasonal fluctuations show the effect of deep water replacement during the summer, and "stagnation" through the winter. Increased mean concentrations after 1977 mark cessation of fish processing waste dumping.

pressure centers through the summer (approximately May-September). Relaxation of the intense winter onshore Ekman transport permits run-up of denser water onto the shelf and into the coastal waterways. Maximum density marine source water is then present through this season at sill height, resulting in flushing of fjords which, like Resurrection Bay, have barrier sills below the zone of minimum annual density variation (located around 150 m: Muench and Heggie, 1978). Seasonal patterns of deep water renewal in Resurrection Bay have been described by Heggie and Burrell (1977, 1981) and Burrell (in press). Influxes of water over the sill may occur as early as April-May, with complete and multiple replacements of the subsill basin water through the summer. But from September-October onward through the oceanographic winter period, deep water behind the sill is not advectively replaced, and vertical exchange within the basin column must be primarily via eddy diffusion. Since the major energy source for turbulent mixing is most likely from tidal interaction at the relatively deep sill, water within the basin is stratified and relatively stable through this season.

8.3.2 Environmental Problems

The densest replacement water that penetrates to the bottom of the Resurrection Bay basin in late summer–early fall has a relatively low dissolved oxygen content (around 4 to 5 ml L^{-1}). Consumption of oxygen in the surface sediments, and in the water column above the benthic interface, initially progressively depletes oxygen concentrations in the deep water through the winter (Fig. 8.5B), although this trend is generally reversed before the next flushing cycle. The latter feature is probably primarily a result of increased downward diffusive transport as the vertical gradient in oxygen concentrations increases, and also of a decreasing supply of labile benthic organic detritus through the winter. Prior to 1976, the major source of carbon—an annual mean of 5.6×10^8 g over the 1972–75 period—was fish processing waste discharged in the central basin region of the fjord. This material consisted predominantly of coarse fragments that are believed to have been transported largely intact into the basin—a mean flux to the sediments of around 28 gC m^{-2} yr^{-1}. From the annual oxygen consumption rates computed by Heggie and Burrell (1981), it has been estimated (Burrell, 1983b) that an approximate doubling of this anthropogenic source of labile carbon potentially could have resulted in limited periods of anoxia in the deep parts of the basin in the winter. After 1976, discharge of fish processing waste was no longer permitted in Resurrection Bay, and mean dissolved oxygen concentrations in the deep basin water increased (Fig. 8.5C).

8.4 Port Valdez, Alaska

8.4.1 Environmental Setting

Port Valdez forms a northeasterly extension of Prince William Sound, Alaska (Figs. 8.6A, 2.22). It is some 20 km long by 5 km wide, and is separated from Prince William Sound (which is itself a very large fjord) by two adjacent 110- and 128-m depth sills at the western end. The fjord floor is approximately level at around 240-m depth (247-m maximum), rising precipitously along the mountainous margins, and less steeply at the head to intertidal and deltaic regions in the vicinity of the old (pre-earthquake) and new town sites.

Mean rainfall in this region is around 160 cm yr^{-1} (only about one day in six is clear), much of which is initially stored in the surrounding Chugach Mountains as snow. A number of glaciers occur in the watershed area. Major sources of fresh water, including the Valdez Glacier stream, enter at the head, but there are numerous small peripheral streams, and a second major influx from the Shoup Glacier in the vicinity of the entrance sills. The freshwater supply is strongly seasonal, peaking in July-August. In spite of the high annual precipitation, mean freshwater volume influx is only 5 to 7% of the tidal prism (Colonell, 1980; the average tidal range is around 5 m). During the winter, freshwater inflow is very low (Carlson et al., 1969), and estuarine circulation is generally negligible at this time of year (Muench and Heggie, 1978); that is, the Port functions more as an embayment than an estuary. In the winter of 1971–72, the entire vertical column was near-homogeneous as a result of thermohaline convection (Muench and Nebert, 1973), but subsequent observations

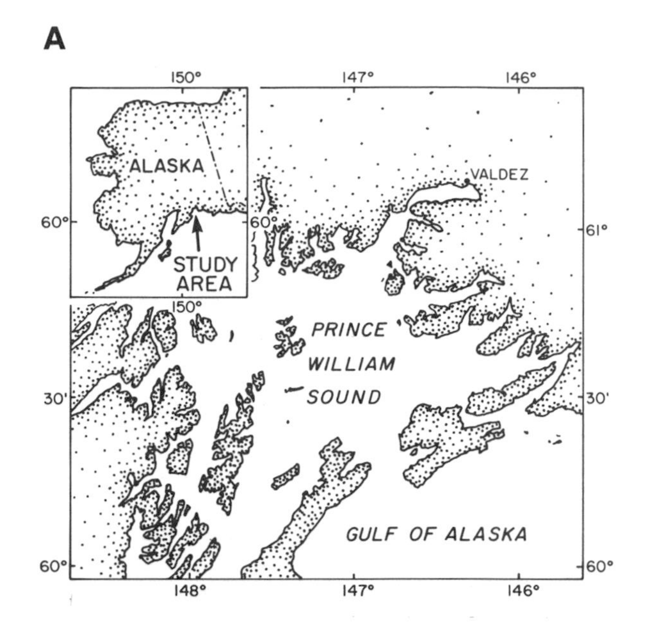

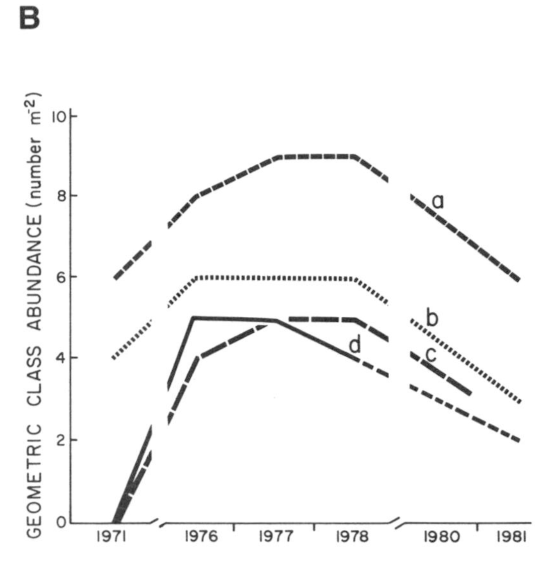

FIGURE 8.6. (A) Port Valdez is a fjord within the large fjordlike Prince William Sound, south central Alaska. The oil terminus is located near the head on the southern shore. (B) Time-series abundances (individuals m^{-2} in $\log_2$ geometric classes; i.e., Class II contains 2–3 individuals m^{-2}, Class III 4–7, etc.) of four dominant taxa at a locality (235-m depth) along the axis of the fjord opposite the oil tanker terminus. Construction activities extended over the period 1976–78. a–*Heteromastus filiformis*; b–*Lumbrineris spp.*; c–*Nephtys punctata*; d–*Eudorella emarginata*. From Feder and Jewett (in press).

suggest that this was an unusually cold winter. Near-surface winter mixing may be enhanced by easterly katabatic winds draining through gaps in the coastal mountain range. Even during the summer, estuarine circulation appears to be largely confined to the top 15 m of the water column. Muench and Heggie (1978) note that the height of the barrier sill guarding Port Valdez falls within the depth interval of minimum annual density variation within Prince William Sound. This means that there is no well-defined seasonal pattern of density maxima and minima at sill depth, and hence that deep flushing of the basin occurs irregularly and is not confined to a particular period during the year.

The surficial sediments of Port Valdez have been described by Sharma (1979). Most of the floor is blanketed with clay-sized material consisting predominantly of illite and chlorite, but there is an irregular zone of coarser grained sediment at the head (Fig. 1.6). The organic carbon content is less than 1%. Adjacent to the submarine extension of the deltas, the surficial deposits are underlain unconformably at shallow depths by graded gravel deposits. These have been attributed to slumping caused by the 1964 earthquake (Coulter and Migliaccio, 1966). Sharma and Burbank (1973) estimate that in excess of 2.5×10^{12} g of, primarily, glacial particulate sediment are carried into the fjord annually. Distinctive turbid plumes originate in the summer from the Lowe River and the Valdez Glacier stream at the head of the inlet (see Chapter 4, Fig. 4.7). Much of this material is deposited in the eastern part of the fjord, and the sedimentation flux here is roughly an order of magnitude higher than that to the western region (Feder and Matheke, 1980). The subtidal infauna of Port Valdez is dominated by deposit-feeding organisms typical of soft substrates, with polychaetous annelids dominating. Feder and Matheke (1980) suggest that the relatively low diversity, species richness, and infaunal biomass distributions indicate that the soft-bottom region is chronically disturbed by the seasonally high sedimentation loads. Cluster analysis of multi-year infaunal species abundance data has defined two major groupings of stations (Feder and

Matheke, 1980; Pearson et al., in press; see Chapter 6, Fig. 6.18), coinciding with regions of high and low sedimentation. Sessile organisms tend to be excluded from the eastern area where the sedimentation flux is greatest.

8.4.2 Environmental Problems

Construction work on the marine terminus of the trans-Alaska oil pipeline, located on the southern shore of Port Valdez, began in 1976 and was completed in mid-1978. Prior to that time the population of the entire Valdez region was around 1000 persons only, and anthropogenic impact on the fjord was insignificant. No catastrophic oil spills have occurred since oil trans-shipment began, but regulated, point-source discharge of treated ballast water discharged from oil tankers is permitted. During the summer, the organic fraction of the ballast waste (around 170 kg day^{-1}) forms a plume at about 50 to 60 m depth extending several kilometers down-fjord (Lysyj et al., 1981). However, Button et al. (1981) have shown that over 65% is microbially degraded, and most of the remainder is flushed seaward. Because of the efficient flushing characteristics of this fjord, hydrocarbon residence times are of the order of months or weeks; only a very small fraction (less than 3%) is believed to be retained in the sediments adjacent to the diffuser. Over the period since ballast water discharge was first permitted, it has been found that, adjacent to the terminus, barnacles *(Balanus balanoides)* now grow at a slower rate than formerly (Rucker, 1983); that recruitment of mussels *(Mytilus edulis)* declined in the early years of operation; and that populations of the clam *Macoma balthica* have generally declined (Colonell, 1980). There has also been a progressive enhancement of anthropogenic hydrocarbons in the sediments, and in mussels, within the near-field mixing zone. However, detailed monitoring maintained through 1983 did not reveal any major or widespread environmental impact that could be directly attributed to hydrocarbon pollution. The most pervasive impact on the benthic biota documented to date appears to be related, not directly to hydrocarbon spills or discharge, but to physical disturbance of the substrate associated with construction of the tanker terminal in the 1975–77 period. Feder and Jewett (in press) have shown that between 1976–78 there was a decrease at all stations in the proportion of rare species, and a temporary increase of moderate and high abundance taxa (Fig. 8.6B), characteristics consistent with a temporal pollution gradient, as discussed by Gray and Pearson (1982) and Pearson et al. (1983).

8.5 Howe Sound, British Columbia

There are a number of fjords located close to large urban populations where environmental concerns must address the competing interests of fisheries, recreational activities, and industrial development. Examples include Puget Sound (Washington), Howe Sound, and Oslofjord (Norway). Howe Sound, located close to metropolitan Vancouver on the southwest coast of British Columbia, may be considered typical (Fig. 2.23). Details on the environmental setting and problems, as outlined below, can be found in Hoos and Vold (1975) and reports of the Squamish Estuary Management Plan (1981).

8.5.1 Environmental Setting

Howe Sound is a geomorphically complex fjord (Fig. 8.7) with a number of U-shaped submarine valleys that have deep basins (max. depth 325 m), steep sides, and submerged sills (110 m and 70 m deep). The basins contain thick (600 m) deposits of Quaternary sediment (Syvitski and Macdonald, 1982).

Except for the winter period of low river discharge, the waters of Howe Sound are density stratified. The estuarine circulation is driven by a combination of tidal oscillations, katabatic winds, and freshwater discharge of the Squamish and Fraser rivers, which produces the major residual flow. Current velocities in the surface layer may exceed 0.8 m s^{-1}, while those of the bottom waters approach zero (0.05–0.1 m s^{-1}; Bell, 1975). Within upper Howe Sound, the residual currents are multilayered. The lower section of Howe Sound has a considerably more complex circulation pattern, a result of competition between the Fraser and Squamish river discharges, and the divergence and convergence of currents around the numerous islands. The average discharge of the Squamish and Fraser

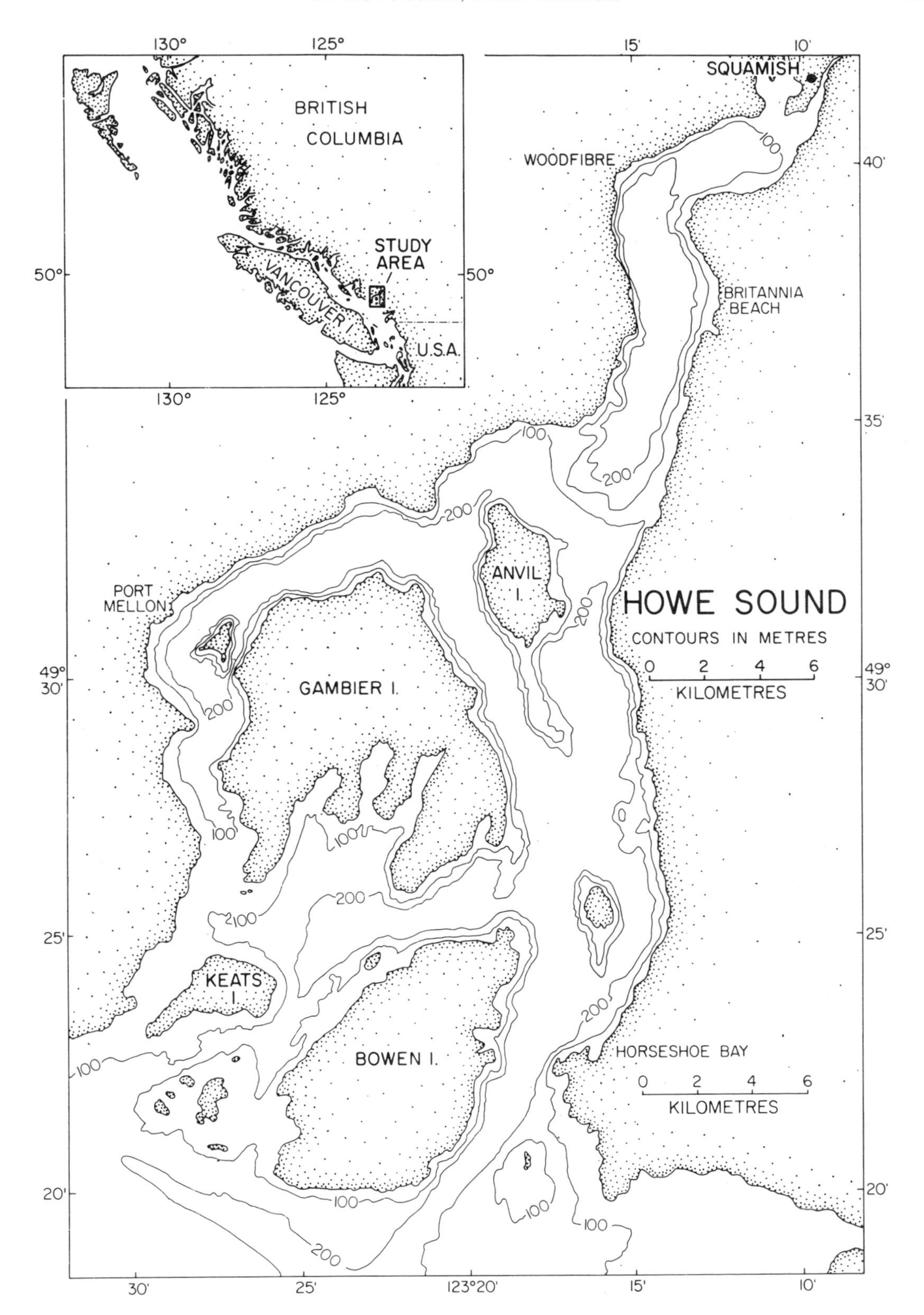

FIGURE 8.7. Location of Howe Sound, British Columbia, and its bathymetry (after Syvitski and Macdonald, 1982).

rivers is 250 and 2700 $m^3 s^{-1}$, respectively. As the Fraser River freshet is a few weeks in advance of the Squamish River freshet (i.e., snow meltwater vs. glacier meltwater), the former initially dominates the circulation pattern. Since the Squamish flows into the fjord head, it is the dominant dispersal mechanism for particulate matter within the inner basin.

Water in the basins south of the inner sill may exchange and mix to mid-depths with upper Howe Sound waters fairly frequently; complete deep water renewal occurs every 3 to 4 yr (Bell, 1975). Occasionally, the deep waters of upper Howe Sound go anoxic (Levings, 1980a).

8.5.2 Environmental Problems

The sources of pollution for Howe Sound include sewage, logging, log handling and log storage, pulp mills, mining, gravel operations, wastes from recreational activity (marinas), port facilities, and chemical industries (Fig. 8.8; McDaniels, 1973).

Discharge of sewage is not yet considered an important environmental problem, although over 3,500,000 L day^{-1} of variably treated sewage enters Howe Sound. As a result, water turbidity is locally increased, oxygen in the water is depleted, nutrients are increased, and toxic substances such as chlorine are added to the marine waters. The Squamish port facilities contribute various amounts of oil and gas to the marine waters through waste handling problems.

Much of the Squamish drainage basin is being logged. As a result of the removal of the forest cover from the margins of the rivers and streams, river water temperature has increased. This may limit suitable spawning areas, increase water turbidity, and increase the numbers of pathogenic organisms (Hoos and Vold, 1975).

Bark debris from river discharge, pulp mills, and booming grounds, is known to settle out close to local sources. Nearby, the seafloor may be littered with submerged logs and wood debris. Populations of prawns and squat lobsters may locally increase as they can use the crevices of the sunken logs for protection (McDaniels, 1973). In contrast, anemones, sea pens, bivalves, and starfish are only rarely present in these areas (ibid.). Where wood debris is especially abundant, the seafloor is anoxic and benthic life is eliminated (Levings, 1973). Local harbor spillages of wood preservative (penta- and tetrachlorophenol) are deemed responsible for a number of fish kills. As an example, in October 1973, over 1000 adult and juvenile coho and chiner were killed near the Squamish harbor (SEMP, 1981). Wood preservative and pesticide, contained within the wood debris littering the seafloor, may pose an environmental problem of yet unknown proportions.

The two pulp mills along the shores of Howe Sound (Woodfibre and Port Mellon) each produce approximately 500 t day^{-1} of Kraft pulp. Between 1912 and 1960, Woodfibre was a sulfite mill whose main waste byproduct is spent sulfite liquor composed of relatively inactive lignin ($\approx$ 50%) and the more toxic compounds of hexose and pentose sugars ($\approx$ 15%). The sugars have an extremely high biological oxygen demand. Port Mellon and, since 1961, Woodfibre, have operated as bleached Kraft mills. The main Kraft mill effluent contains a number of highly toxic constituents (hydrogen sulfide, mercaptans, resin acids, and soaps). For the mill capacity of Port Mellon or Woodfibre, such an effluent requires approximately 18,000 kg of dissolved oxygen per day (Waldichuck, 1960). Thus, Woodfibre effluent discharge may contribute to hastening the otherwise natural eutrophication of the deep waters of inner Howe Sound. Bleaching wastes, associated with these Kraft mills, includes the compound zinc hydrosulfite. Elevated levels of zinc in some B.C. shellfish (e.g., oysters) may result from such effluent discharge (SEMP, 1981).

Since 1899, a copper-zinc mine has operated at Britannia Beach. The mine was closed in 1974, but a copper precipitator continues in operation: particulate ferric hydroxide [γ-FeO(OH)] is the main byproduct. During mine operations, the tailings were dumped along the shore margins, and as a result, concentrations of Cu and Zn exceeding 1300 $\mu g\ g^{-1}$ have been found offshore in the sediments (Thompson and McComas, 1974). The shoreline tailings have slumped a number of times from the steep sidewall slopes of the fjord into the deep basins. Such debris slides may account for the two accumulations north and south of Britannia Beach (Fig. 8.9). Shore-dumped tailings have also been reworked by waves and dispersed up and down fjord from the mine site. Some of the tailings eventually exit over the inner sill, suspended in the out-

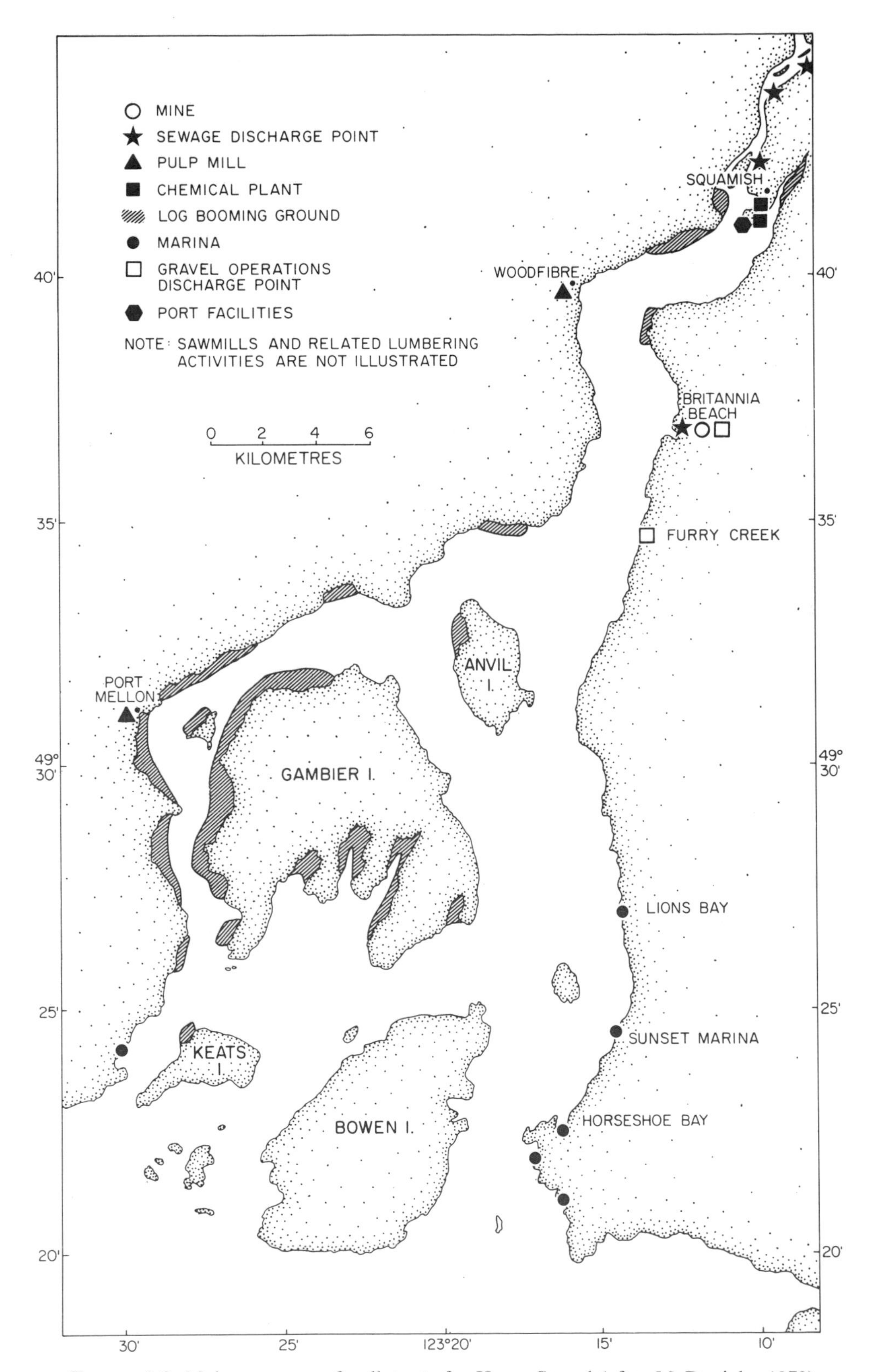

FIGURE 8.8. Major sources of pollutants for Howe Sound (after McDaniels, 1973).

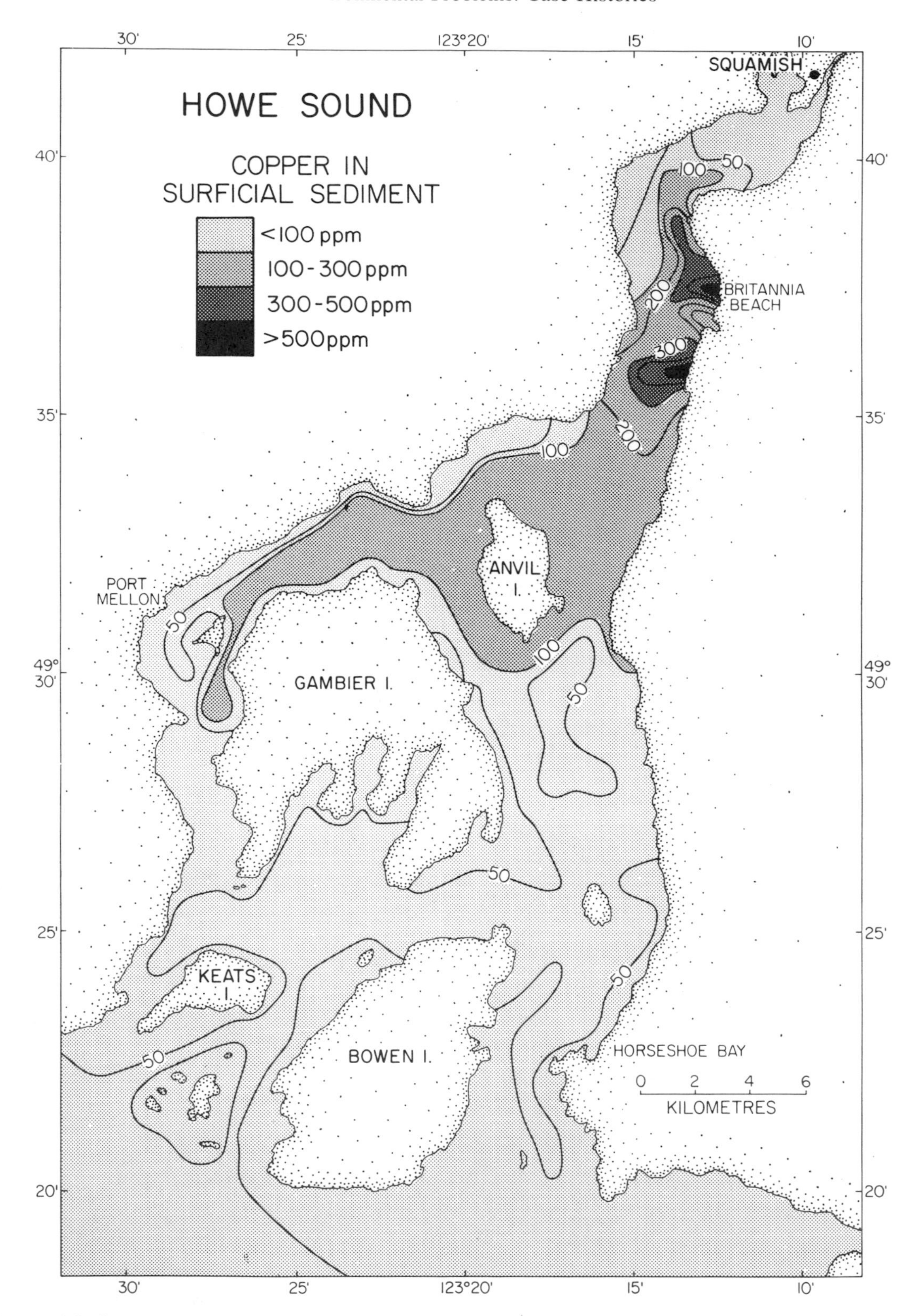

FIGURE 8.9. Copper concentrations in surficial sediments from Howe Sound (after Syvitski and Macdonald, 1982).

flowing Squamish-influenced surface current (Syvitski and Macdonald, 1982). Near the tailings outlet the seafloor is devoid of flora and fauna (McDaniels, 1973), possibly a result of critical particulate loading rather than metal toxicity. (It should be noted, however, that Rygg and Skei (1984) have correlated decreasing bottom fauna diversity with copper concentrations in fjord sediments: see Fig. 8.34.) Similarily, sand and gravel operations along the shores of Howe Sound may be responsible for nearby areas of the seafloor being devoid of a benthic community: a result of aggregate washing that causes high turbidity levels in nearby streams with suspended sediment concentrations of 15 to 36 g L^{-1}.

There are two chemical plants operating within the port of Squamish: (1) a liquid sodium chlorate plant operated by Canadian Oxy (Hooker) Chemicals Ltd.; and (2) a chlor-alkali plant operated by FMC Chemicals Ltd. The former has limited impact on the marine environment; not so with the latter. The chlor-alkali plant discharges mercury, sulfates, alkalines, acids, and chlorine, initially into a settling lagoon (a liquid residence time of 12 h). Although the pond allows for a great reduction in the toxic levels of the effluent compared with the influent, levels of mercury entering the marine environment range from 0.1 to 75 $\mu g\ L^{-1}$, at discharge rates in excess of 10^7 L day^{-1}. Mercury pollution has been the most serious environmental problem in Howe Sound. Thompson et al. (1980) have estimated that about 450 kg of mercury

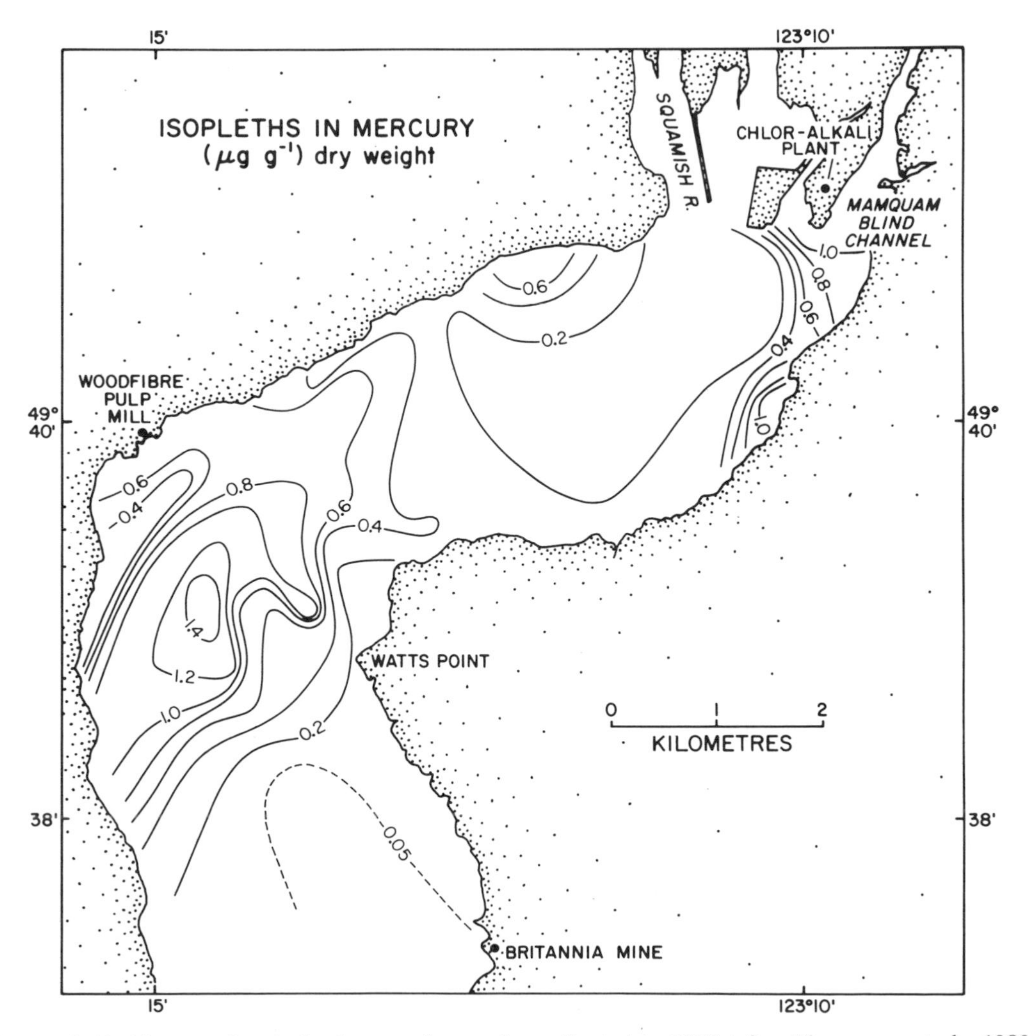

FIGURE 8.10. Mercury isopleths from grab samples collected in 1979 (after Thompson et al., 1980).

have accumulated in the sediments adjacent to the chlor-alkali plant; values as high as 33.2 $\mu g\ g^{-1}$ have been recorded in cores adjacent to the loading dock. Mercury concentrations in surface sediments remain well above background ($\approx 0.1\ \mu g\ g^{-1}$) some 5 km away from the plant; but out from the Squamish River, the levels of mercury are relatively low, as a result of dilution from high inorganic sedimentation rates (Fig. 8.10). Elevated mercury concentrations are highest in sediment with high cation exchange capacity and/or adsorption characteristics, that is, clays and organic matter.

Mercury settles through the water column attached to particles. Once on the seafloor it is eventually methylated by bacteria whereby mono- or dimethylmercury is produced. Monovalent methylmercury [CH_3Hg^+] is considered the most toxic form of mercury and readily accumulates in aquatic organisms. Methylation is known to occur under aerobic and anaerobic conditions, but is promoted in aerobic environments (SEMP, 1981). However, the content of methylmercury in surface sediments collected in 1976 was found to be less than 1% of the total mercury present (ibid.). Nevertheless, many local fauna had accumulated mercury above the permissible level (0.5 ppm) for the sale of fish and shellfish in British Columbia. Crab, ling cod, flounder, rockfish, and shrimp had values many times above this level (Hoos and Vold, 1975). Migratory species such as salmon, steelhead, and herring had unusually high levels of mercury also, but were below the permissible levels (ibid.). As a safeguard, fisheries in upper Howe Sound were closed during the 1970s.

Improvements in the settling pond have since decreased mercury discharge substantially. The highest levels of mercury are now found buried beneath the seafloor, and mercury levels in fish have decreased to acceptable levels. The one exception is the spiny dogfish *(Squalus acanthias)*. Recreational fisheries in upper Howe Sound have therefore been reopened.

The total dustfall for upper Howe Sound ranges from 1 to 9 $g\ m^{-2}\ mo^{-1}$. The FMC Chemical plant emits 540 $t\ yr^{-1}$ of mercury into the air. Thompson et al. (1980) have suggested that the high levels of mercury found in surface sediments above low water level may be related to atmospheric fallout.

8.6 Rupert Inlet, British Columbia

8.6.1 Environmental Setting

Rupert and Holberg inlets (Fig. 8.11) together form a continuous basin some 44 km long, separated from Quatsino Sound, a major indentation on the northwest coast of Vancouver Island (Fig. 2.23), by the Quatsino Narrows. The Rupert-Holberg fjord has a maximum depth of 165 m close to the Narrows entrance. The latter is a long and narrow constriction (approximately 4×0.4 km) with a minimum 18-m depth at the northern end. Approximately half of the total freshwater input into the combined inlets (Drinkwater, 1973) is via the Marble River, which enters, not at the head, but adjacent to the Narrows, so that a normal estuarine circulation pattern would not be expected. The average freshwater input is only approximately around 7% of the mean tidal prism (which, in turn, is 3% of the total fjord volume). The mean tidal range is 2.8 m, and tidal currents up to 3 $m\ s^{-1}$ have been recorded in the vicinity of the Narrows constriction.

A number of investigations of the annual hydrographic patterns in Rupert-Holberg inlets (Pickard, 1963; Drinkwater, 1973; Stucchi and Farmer, 1976) have shown that the column is well mixed vertically at all times of the year (water properties are vertically uniform and bottom-water oxygen concentrations remain high), but that there are large and rapid overall temporal changes. Drinkwater (1973) and Stucchi and Farmer (1976) have demonstrated that tidal flow through the Narrows generates intense tidal mixing within the adjacent portion of Rupert-Holberg inlets, possibly aided (Drinkwater and Osborn, 1975) by the T-junction configuration at Hankin Point, in front of the entrance. Stucchi and Farmer (1976) and Stucchi (1980) have shown that, at different seasons of the year, the turbulent tidal jet generated through the Narrows may be either negatively or positively buoyant within the fjord. As in other silled fjords bordering the northeast Pacific (e.g., see Section 8.3), denser source water is available for transport in over the shallow constriction through the summer. At this time of year, tidal exchange via a "density current" occurs at or close to the bottom, and the basin water is replaced by high-

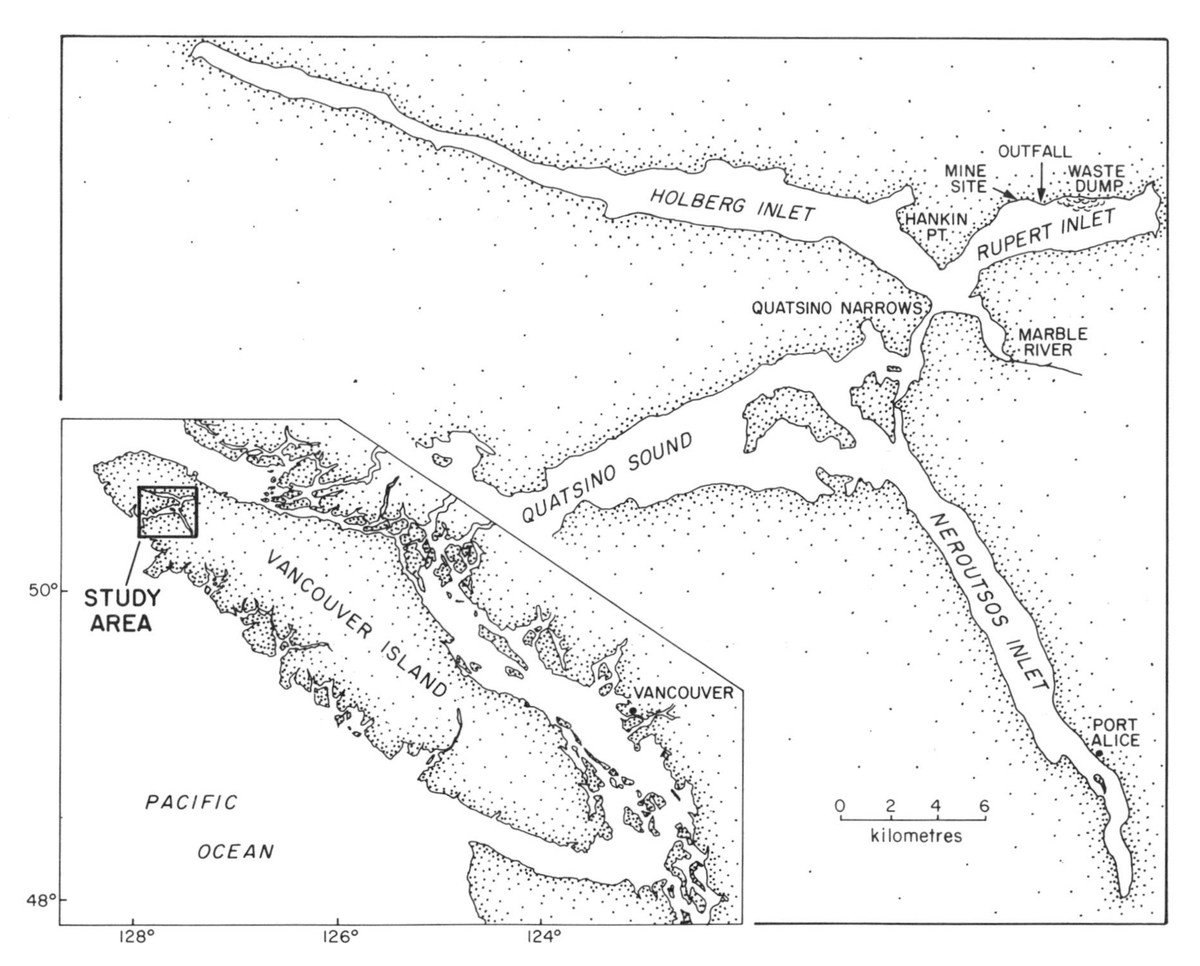

FIGURE 8.11. Location map showing Rupert and Holberg inlets, and the connection to Quatsino Sound and the Pacific Ocean via the Quatsino Narrows. The copper-molybdenum facilities and discharge localities are located on the northern shore of Rupert Inlet.

er salinity water. Conversely, mean fjord salinity (density) decreases in the fall and winter. Although through this period of the annual cycle the flood-tide jet emerges as a buoyant plume, supercritical flow is generated in the Narrows (the densimetric Froude number exceeds unity), and penetration of the denser deep water can occur. Stucchi (1980) has shown that at sufficiently high Froude numbers (> 7), incoming water can be transported to the bottom of the deep basin. This flow regime generates considerable turbulence, mixing denser water upward to the near-surface zone, where it may be transported out into Quatsino Sound on the ebb tide (see Chapter 4, Fig. 4.19E).

Johnson (1974) showed that, especially at times of deep water renewal, sediment was eroded from the deep region of the fjord between the Narrows and Hankin Point and deposited further up-inlet. Stephens and Sibert (1976) have observed that phytoplankton production is relatively high in Rupert-Holberg, and have ascribed this to decreased vertical stability (i.e., an enhanced supply of "new" nutrients into the euphotic zone as discussed in Chapter 6).

8.6.2 Environmental Problems

Marine dumping of waste products from a copper-molybdenum mine located on the northern shore of Rupert Inlet (Fig. 8.11) was first authorized in late 1971. Waste discharge is primarily of two types: mill tailings and overburden. The finely ground tailings (50% less than 23 μm or medium silt size) are discharged into Rupert Inlet from a single pipeline at a depth of 50 m. The discharge rate of this material is of the order of 450 kg s^{-1} over the expected 25-yr life of the mine. Early submarine tailings distribution and sedimentation patterns were described by Johnson (1974). One major feature has been slumping away from the discharge zone and turbidity flow down channels that extend along the axis of Rupert Inlet (Hay, 1982; Hay et al., 1983a). By this means, the mine tailings are transported away from the near-field discharge zone into the dynamically active region of maximum scour at the junction, and on into Holberg Inlet. Figure 8.12 illustrates the spread of tailings along the fjord floor between November 1971 and June 1974. The quantity of non-ore waste rock produced (1350 kg s^{-1}) is around three times the mass of tailings. This material is discarded in the littoral zone immediately adjacent to the mine (see Fig. 8.12).

Evidence of greatly enhanced turbidity at the surface of the fjords appeared in the spring of 1972, some 6 months after tailings disposal commenced. Goyette and Nelson (1977) show

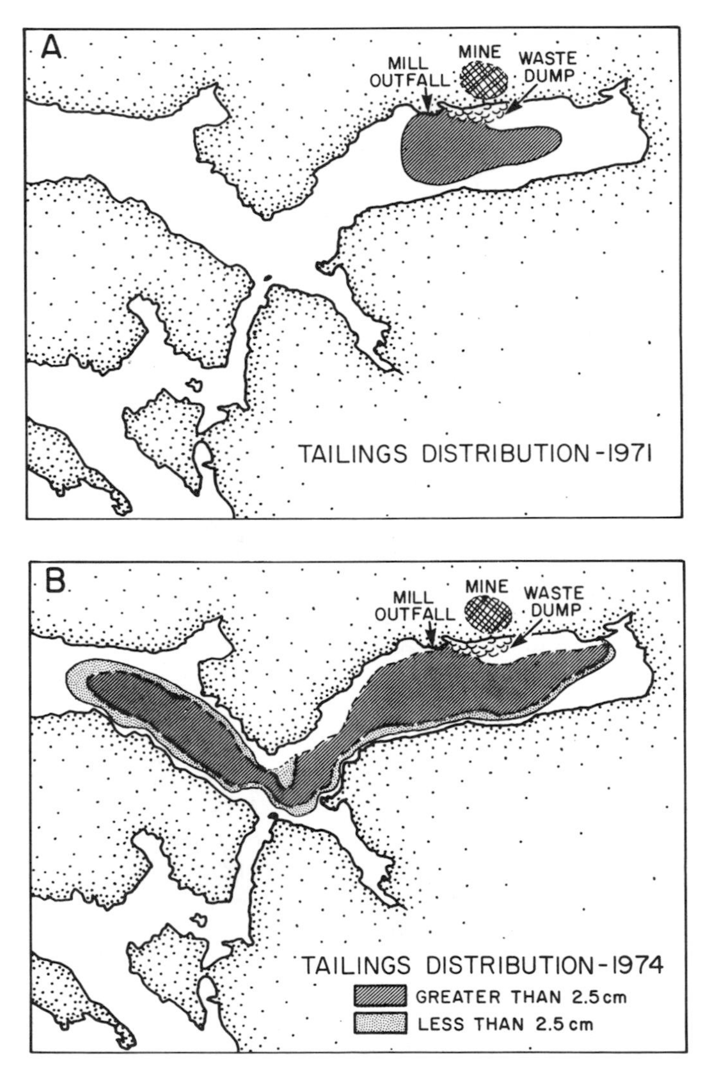

FIGURE 8.12. The spread of mill-tailings away from the discharge point in Rupert Inlet between November 1971 (A) and June 1974 (B) as determined from surficial grab samples. From Goyette and Nelson (1977).

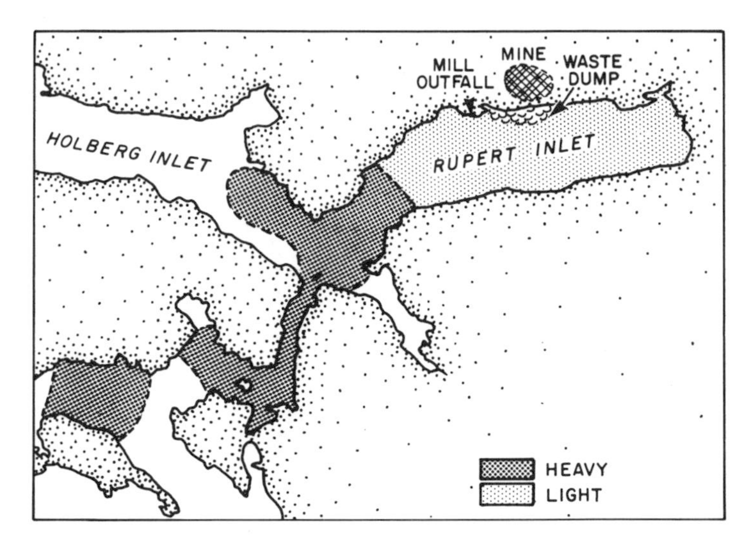

FIGURE 8.13. Distribution of surface turbidity within Rupert-Holberg Inlets, from aerial photographs, at the termination of a flood tide, August 1973. From Goyette and Nelson (1977).

that turbidity is greatly increased over background below around 40-m depth throughout the basin, but that increases at the surface are particularly associated with vigorous tidal mixing in the vicinity of the Quatsino Narrows. Figure 8.13 illustrates surface turbidity around Hankin Point, and within and outside the Narrows, at the termination of a flood tide in August 1973.

The mill tailings contain 700 ppm copper and 40 ppm molybdenum (Poling, 1982), approximately 16 and 25 times greater concentrations, respectively, than in the natural sediment. Consequently, copper concentrations in the surficial sediments through much of the Rupert-Holberg fjord have progressively increased since discharge began (Fig. 8.14), and copper contents serve as a useful assay for tailings contamination of the natural sediment (Hay et al., 1983b). Goyette and Nelson (1977) suggest that elevated concentrations of copper occur in some bivalve species within Rupert Inlet, especially close to the mineral concentrate loading dock. Pedersen (unpublished data) has computed the transport rate of soluble copper, remobilized from the tailings as a result of oxidation of residual sulfides, into the overlying water column. Pedersen (1984) also notes that the usual sedimentary regeneration and benthic flux of soluble inorganic phosphate (see Chapter 7) appear to be suppressed in these sediments, probably because of the addition of lime during the milling treatment.

8.7 Saguenay Fjord, Quebec

8.7.1 Environmental Setting

The Saguenay Fjord is a 90 km long, 1 to 6 km wide basin separated from the St. Lawrence Estuary by a shallow (20 m deep) sill. A second, deeper (80 m) sill some 20 km up-fjord separates the smaller, outermost basin from the main (275 m deep) basin (Fig. 8.15). High (366 m) rock walls surround the fjord (Loring, 1976a). In excess of 90% of the freshwater input into the fjord is from the Saguenay River at the head which drains the extensive (78,000 km^2: Fig. 2.6) Lac St. Jean watershed. The mean monthly flow rate is in excess of 1500 m^3 s^{-1}, but, although discharge is now regulated by a series of hydroelectric dams (Loring, 1976a), discharge rates around four times the mean flow have been recorded during spring freshet periods. Surface salinities increase seaward from around 0.5‰ at the river mouth to 28‰ near the entrance sill.

The circulation of the upper Saguenay waters is as a classical two-layer estuarine circulation. Renewals below sill depth occur on a semi-continuous basis. The salinity of the bottom water may reach 31‰ (Reid et al., 1976); the temperature may be as low as 1°C (Sundby and Loring,

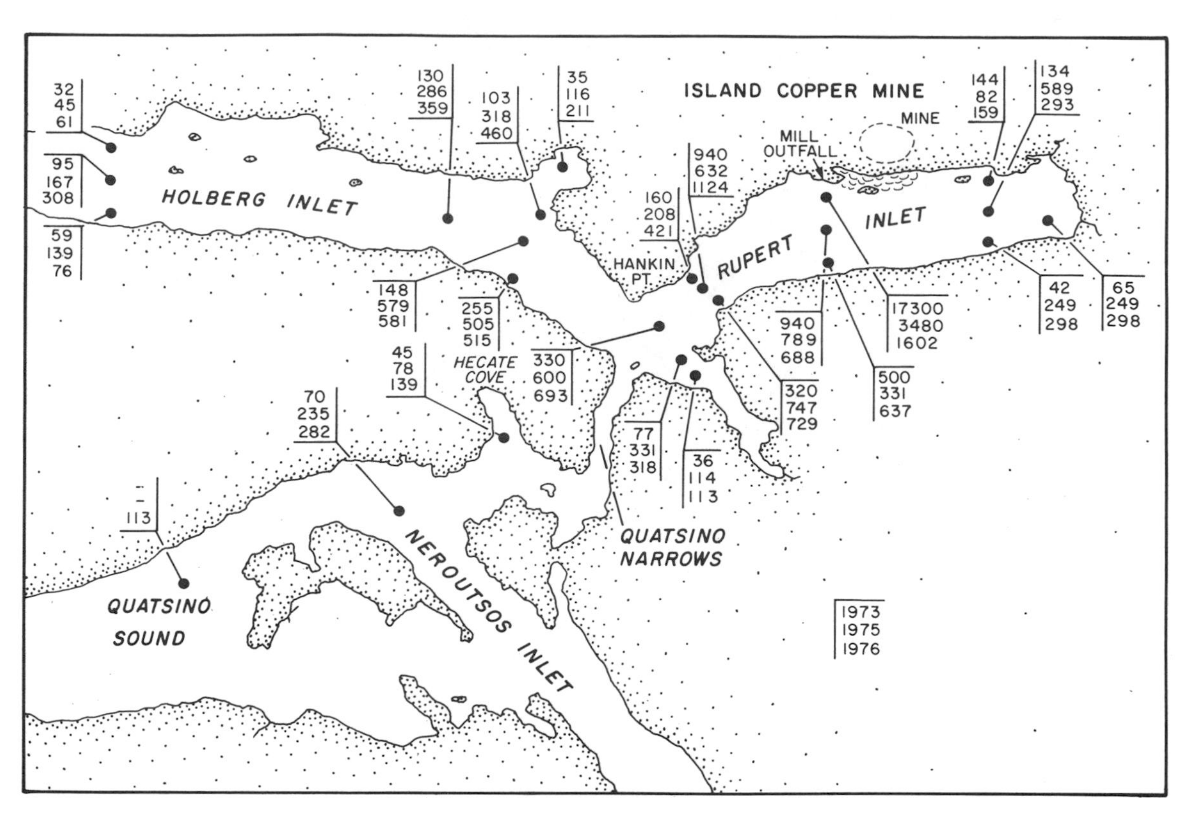

FIGURE 8.14. Copper contents (mg kg^{-1} dry weight) of surficial sediments in Rupert-Holberg Inlets, British Columbia in 1973, 1974, and 1975. From Goyette and Nelson (1977).

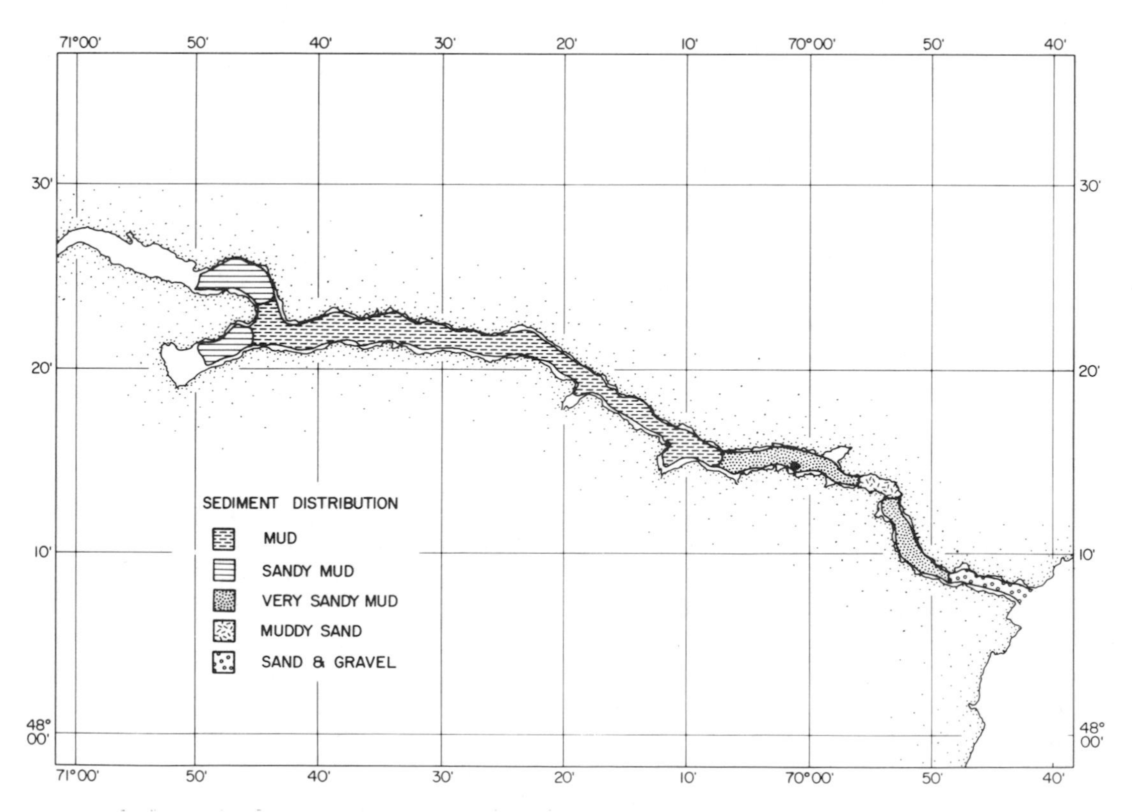

FIGURE 8.15. Distribution of sedimentary facies in the Saguenay Fjord, Quebec (after Loring, 1976a).

1978). The frequency of bottom renewals is sufficient to maintain oxygen levels above 4 ml L^{-1} in the deep water (Taylor, 1975). However, several months of isolation of the water below sill depth have been recorded (Sundby and Loring, 1978).

A number of mineral processing and pulp and paper mills are sited on the Saguenay River and its tributaries: the latter contribute heavily to the particulate organic input into the fjord. Total particulate sediment concentrations range from 10 to 20 mg L^{-1} at the head to less than 0.3 mg L^{-1} in the deep basin (Sundby and Loring, 1978). Inorganic particulate sediment within the fjord is mineralogically similar to the mud-sized fraction of the bottom sediments, consisting of minerals—primarily quartz, feldspar, and amphiboles—characteristic of the surrounding crystalline rocks of the Canadian Shield (Loring, 1976a). The distribution of particulate matter in the Saguenay Fjord reflects two sources: (1) the dispersal of particulate matter in the upper waters induced by the Saguenay River, and (2) erosion and resuspension of bottom sediments (Fig. 8.16). Major ion:aluminum ratio distributions remain relatively constant in time and space (Sundby and Loring, 1978). The large variations in the distribution of particulate Fe and Mn basically appear to follow the expected behavior of these elements outlined in Chapter 7. Thus, near-bottom particulate sediment is enriched with Mn, which has probably been transported across the benthic interface. This is similar to findings in deep Norwegian fjords (Price and Skei, 1975). Yeats and Bewers (1976) and Bewers and Yeats (1979) have demonstrated that Fe behaves nonconservatively and is largely sedimented within the fresh-marine mixing zone at the head of the fjord. In this region, Smith and Ellis (1982) record a preponderance of poorly sorted clay- and silt-sized sediment believed to have been deposited as flocculated aggregates. Sundby and Loring (1978) also believe that quantities of Mn-enriched particulate material are carried into the fjord from the St. Lawrence Estuary.

Loring (1976a) and Smith and Walton (1980) have divided the fjord sediments into three broad lithotopes (Fig. 8.16). In the arms at the head of the fjord, the sediment is characterized by alternating layers of high organic content muds, and coarser, quartz-rich material (Smith et al.,

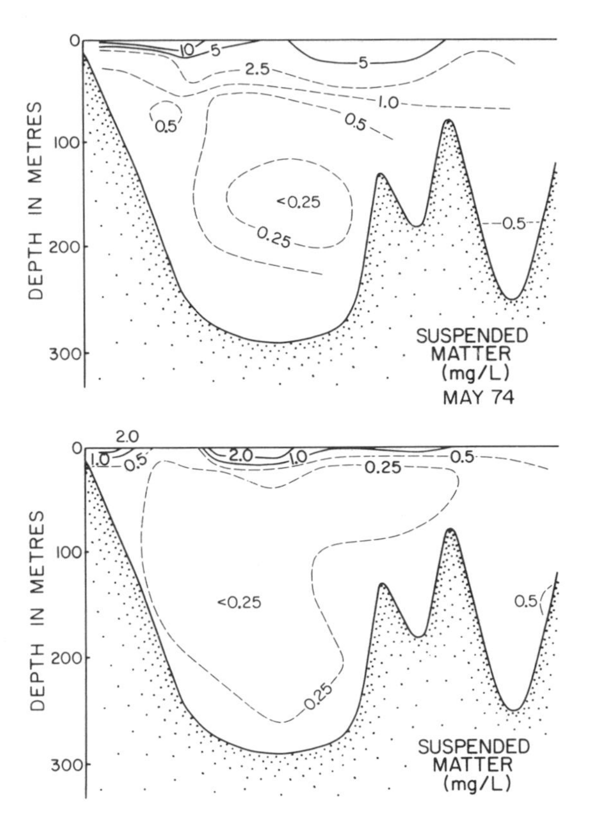

FIGURE 8.16. The distribution of suspended particulate matter in May and September (after Sundby and Loring, 1978).

1980). Basin sediments consist of black anoxic sandy (5–10%) muds, merging seaward into grey muds within the deep parts of the principal basin. Coarse-grained sandy (> 30%) muds occur within the outermost basin, and there are sandy-gravels at the mouth. Sedimentation rates are highest (around 25 cm yr^{-1}; J.N. Smith, unpublished B.I.O. data) near the head of the fjord where the river carrying competence abruptly decreases. In this region, Pocklington (1976), Schafer et al. (1980) and others have shown that two independent sedimentation mechanisms operate. The sandy mud layers result from particle-by-particle deposition of riverine sediment that has been heavily supplemented with anthropogenic carbon waste. The clay layers appear to have been "pulsed" into the fjord as episodic events, the most prominent of which was caused by a massive landslide (25 million tons) into the river that occurred in May 1971 (Fig. 4.17), the effects of which are also evident in sediment profiles several kilometers seaward (Smith and Walton, 1980). The depauperate

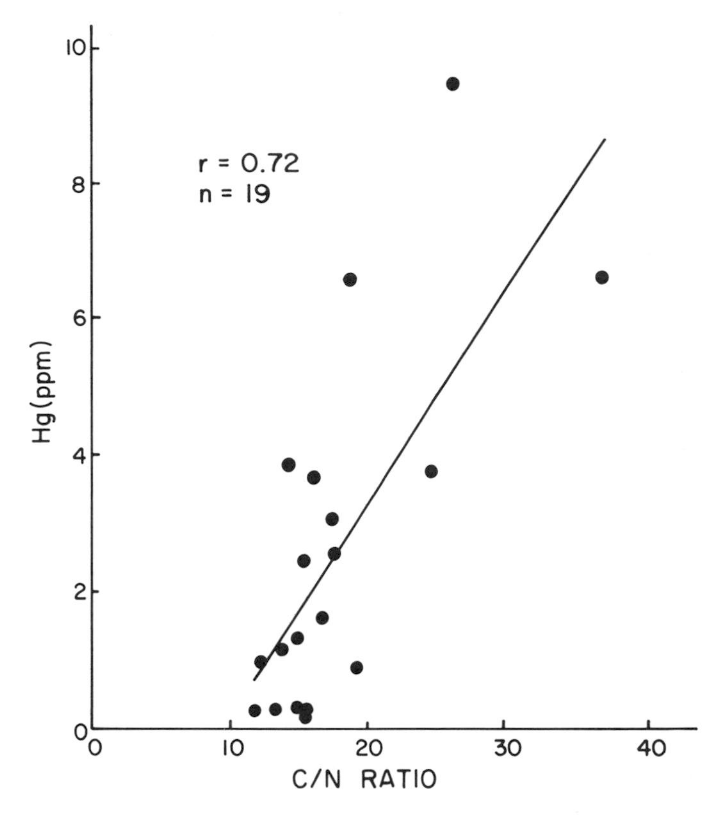

FIGURE 8.17. Relationship between total mercury and C/N ratio (by weight) in the Saguenay Fjord sediments (after Loring, 1975).

sediments in this area have not been mixed by bioturbation, and Smith and Ellis (1982) have been able to construct a very detailed sedimentation chronology from the inverse correlation of unsupported ^{210}Pb activity and the mean monthly river discharge record (Fig. 4.17A). The lead is associated primarily with the finest grained, Pb-poor material.

Pocklington and Leonard (1979) have demonstrated distinctive differences between sedimentary organic matter deposited in the outermost and the main (western) basins. The finer grained inner basin sediments contain 2 to 3% organic matter, including lignin, having a mean C/N ratio of around 20. This material is thought primarily to originate as waste from the numerous wood and pulp processing plants located along the Saguenay River, supplemented with natural terrigenous detritus. A terrestrial source has been confirmed by Tan and Strain (1979) who show that $\delta^{13}C$(PDB) values throughout the main basin span a very narrow and low range (-26.4 to $-25.8‰$) corresponding to that expected for terrestrial organic matter. It appears, therefore, that the contribution of autochthonous carbon to the sediments in the principal fjord basin is negligible. Largely because of high turbidity, in-situ phytoplankton production is low (Therriault and Lacroix, 1975). However, it has been shown that relatively dense water from the St. Lawrence Estuary may be tidally pumped into the fjord (Drainville, 1968; Seibert et al., 1979), and that this flow transports large quantities of particulate organic carbon to deep regions within the fjord (Therriault et al., 1980, 1984; see also Chapter 6). Particulate organic carbon influxed from the St. Lawrence Estuary is believed to enhance secondary pelagic production, but it produces no identifiable signal within the sediments, except possibly within the outermost basin where C/N ratios are notably lower than in the main body of the fjord.

8.7.2 Environmental Problems

As described above, the influx of organic detritus from numerous wood processing plants sited along the Saguenay River far exceeds natural rates and gives rise to anoxic sediment within most parts of the main basin. However, frequent influxes of external water into the well-mixed basin appear to prevent the development of anoxic conditions within the water column. Since 1947, a chlor-alkali plant, situated some 25 km up-stream from the head of the fjord has introduced large quantities of mercury into the marine fjord environment (Loring, 1975; Loring and Bewers, 1978; Smith and Loring, 1981). Prior to the implementation of discharge regulations in 1971, it has been estimated that the maximum annual rate of Hg discharge was of the order of 8×10^3 kg yr^{-1}. From 1948 to 1971, 136 tons of mercury were discharged from the chlor-alkali plant (Loring et al., 1983). About half of the fraction not retained within the river is believed to have been rapidly deposited (residence time in the water column of around 2 mo) at the head of the fjord. Loring and co-workers have shown that the anthropogenic Hg enters the fjord primarily in association with lignin detritus (Pocklington and Leonard, 1979) released by the pulp and paper mills. Hence, there is a strong correlation between mercury and C/N ratios in the sediments (Fig. 8.17). There is also a strong correlation between the mercury loading in the sediment and mercury discharge rates, except in the surface sediments (Fig. 8.18). The latter may relate to uncertainties in the losses of mercury in recent years (Loring et al., 1983) and influence of mercury-deficient landslide sediments (Schafer et al., 1980).

The extent of mercury methylation in the sediments has not been investigated in the Saguenay, but according to Loring and Bewers (1978) about 2 kg km^{-2} yr^{-1} is likely to be methylated and released into the water above (i.e., approximately 10% of the mass balance of mercury). Deposition of unpolluted or less polluted sediments on top of the contaminated sediments, in addition to the methylation process, will gradually decontaminate the seafloor. Loring et al. (1983) have estimated a response time of mercury in the sediments of 2 years. This implies that if anthropogenic mercury discharges were curtailed, the mercury content in the water masses would decrease to natural levels within 2 yr.

Other metals, notably lead, cadmium, and zinc, also appear to be elevated in Saguenay Fjord sediments compared with adjacent St. Lawrence Estuary sediments of similar type

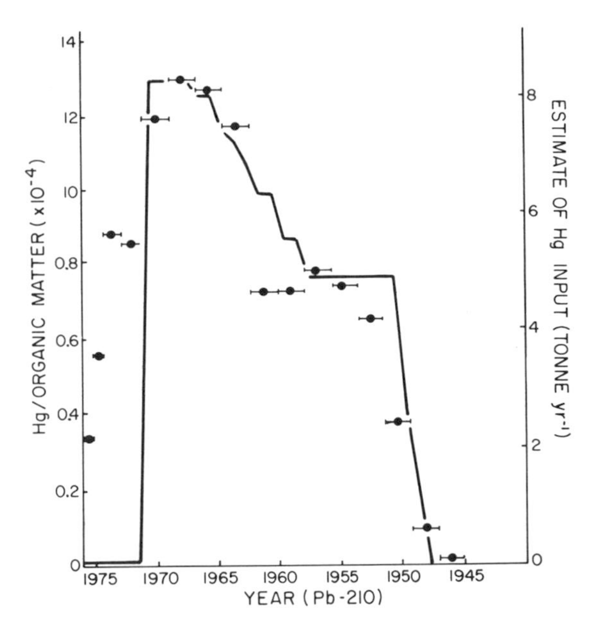

FIGURE 8.18. Calculated mercury discharges from the chlor-alkali plant (based on production data and strict compliance with regulations) vs. mercury/organic matter ratios, down the core (after Loring et al., 1983).

(Loring, 1976a). As is generally the case, the heavy metals are predominantly concentrated in the nondetrital fraction, positively correlated with particle surface area (fineness of grain: Loring, 1976b). Nondetrital Zn, Pb, and Cu contents also correlated with the total organic carbon content, and with the type of organic material as characterized by the C/N ratio. The elevated concentrations of Hg, Pb, and Zn released from industrial sites along the river primarily appear to be transported into the marine fjord complexed with, or sorbed onto, organic detritus. The intimate association with organic waste is a good example of the potential synergistic effects that may result from mixed industrial effluents. High concentrations of particulate lead and cadmium in the surface layer of the fjord suggest that these metals are introduced by the Saguenay River (Cossa and Poulet, 1979).

8.8 Iddefjord, Norway/Sweden

8.8.1 Environmental Setting

Iddefjord is situated at 59°6′N and 11°20′E, on the border between Sweden and Norway. The fjord is 25 km long. The total water volume is 400×10^6 m^3. The fjord receives fresh water

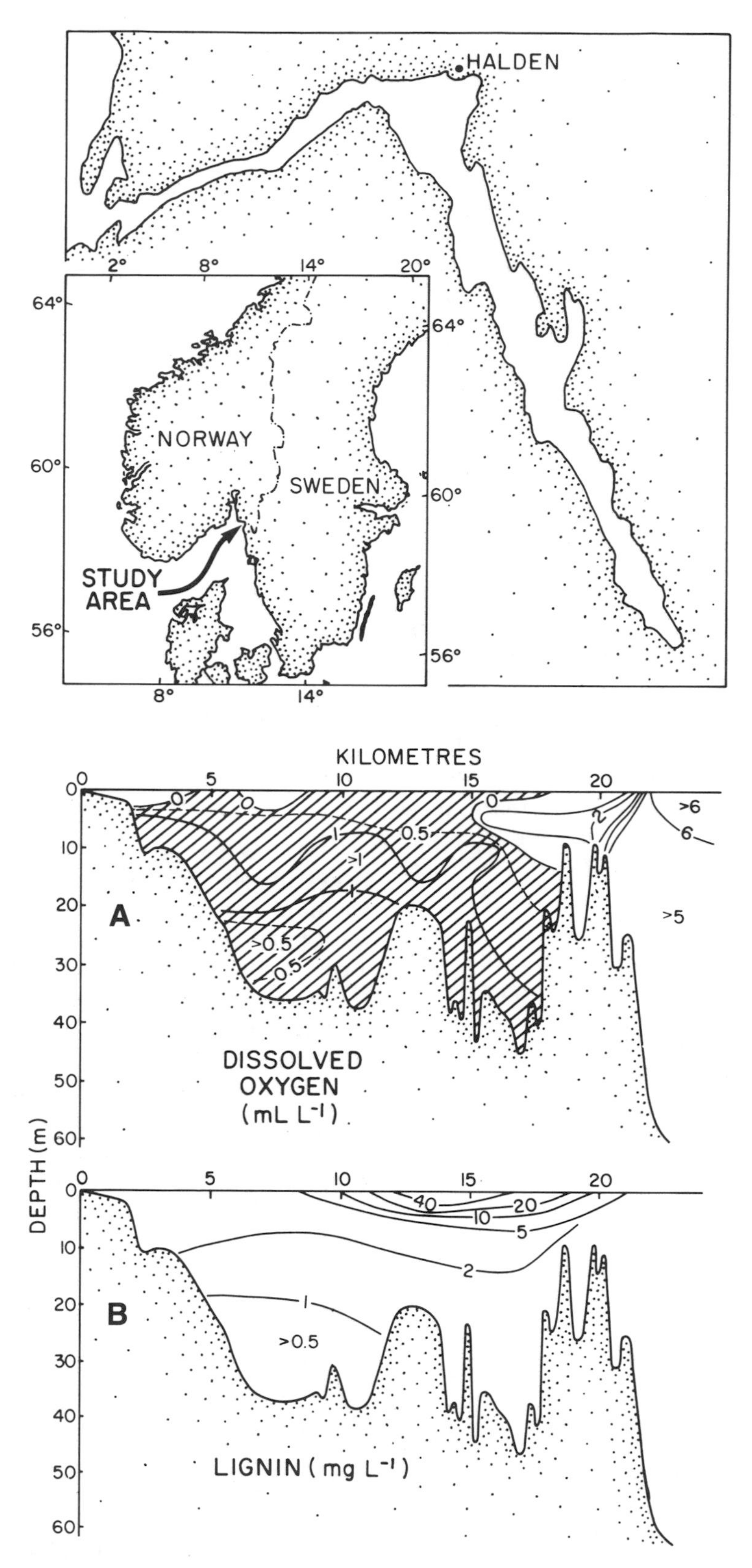

FIGURE 8.19. Map of Iddefjord with depth profile from the head (right) to the mouth (left) of the fjord. Source of pulp and paper pollution is located near Halden. (A) Isopleths of oxygen and hydrogen sulfide (shaded area) in Iddefjord in August 1974 (ml L^{-1}) (after Engström, 1975). (B) Isopleths of lignin sulfonates in the water of Iddefjord in April 1975 (after Josefsson and Nyquist, 1976).

mainly through two rivers, one at the head and one in the middle part of the fjord. The total average discharge is ≈ 40 $m^3 s^{-1}$. The fjord is separated from the adjoining fjord by two sills (8–9 m depth). The maximum basin depth is ≈ 45 m.

The surface salinity varies between 10 and 20‰, while the bottom-water salinity may exceed 30‰. Because of a shallow sill, the bottom water (below 20-m depth) is near stagnant for long periods. The fjord is separated into two major basins (Fig. 8.19), and even if deep water exchange takes place in the outermost basin, the inner basin may still be anoxic. A typical situation following a deep water exchange is exhibited in Figure 8.20, showing the difference in oxygen levels in the two basins.The oxygen-depleted water at intermediate depths (5–15 m) is vertically displaced anoxic water. Periodically, hydrogen sulfide reached the surface water as a result of deep water flushing (Fig. 8.19B).

The riverine sediment input to Iddefjorden is not known. However, the rivers are likely to dump much of the sediment load in lakes upstream, and the sediment discharge to the fjord is thought to be small. The visibility in the surface water, based on Secchi-depth measurements, is very low. This is not due to large quantities of mineral particles but to discharge of fibers and colored ligno-sulfonates from a pulp and paper factory.

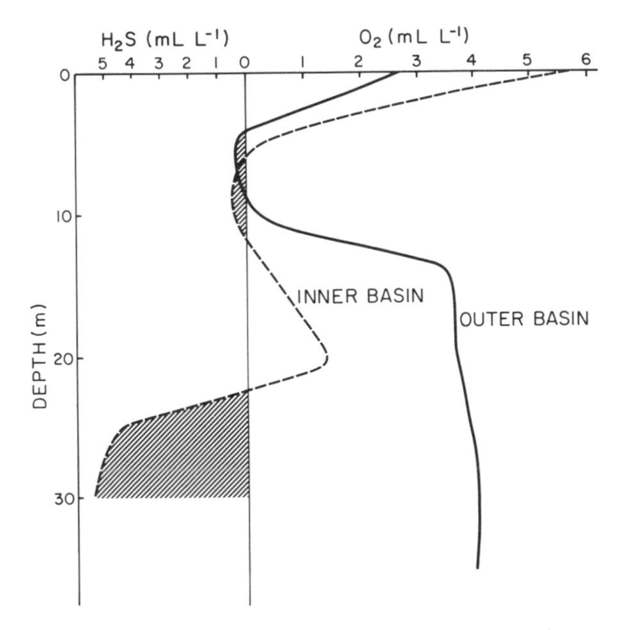

FIGURE 8.20. Depth profiles of oxygen/hydrogen sulfide in the outer and inner basin of Iddefjord, after a flushing event (October 1975).

8.8.2 Environmental Problems

The fjord has a long history of pollution. Planning mills were built as early as 1866 and followed two years later by a wood pulp factory. In 1908 a cellulose factory was establish, and in 1915 a paper factory. This implies that Iddefjord has received waste from the pulp and paper industry for more than 70 years. Investigations of the fjord in 1920 did not indicate any signs of pollution, but in 1947 symptoms of oxygen deficiency were evident. The organic loading to a fjord with such a shallow sill and infrequent deep water exchange inevitably results in oxygen depletion and sulfate reduction in the bottom water. For short periods, hydrogen sulfide has even reached the surface of Iddefjord (Dybern, 1972; Engstrøm, 1975).

The organic effluents are discharged into Iddefjord in the middle reaches of the fjord (Fig. 8.19). Measurements of ligno-sulfonates in the water indicate transport both toward the head as well as the mouth of the fjord (Fig. 8.19B). This distribution pattern is presumably caused by tidal activity, transporting surface water back and forth. During periods of large freshwater discharge, the transport of effluents is mainly directed seaward.

Significant amounts of metals have also been discharged into the fjord. Mercury was used as a slimicide from 1964 to 1968. Heavy metals like copper, zinc, lead, and cadmium are present in sludges from the roasting of sulfide ores. These metals were released into the fjord from 1908 to 1980. The sludge discharge rate during this period is shown in Figure 8.21.

The history of pollution has been recorded in the bottom sediments of Iddefjord. The sediments of the deep basins and close to the source of pollution are highly anoxic, with a strong smell of hydrogen sulfide. Formation of gas bubbles (methane) occurs frequently, and large patches of bottom sediments have been observed rising to the surface, owing to gas release. This occurs in the area of large fibre deposits accumulating on the bottom.

The distribution of metals in the surface sediments of Iddefjord collected in June 1977 is displayed in Figure 8.22. Generally, the highest metal concentrations occur in the middle of the

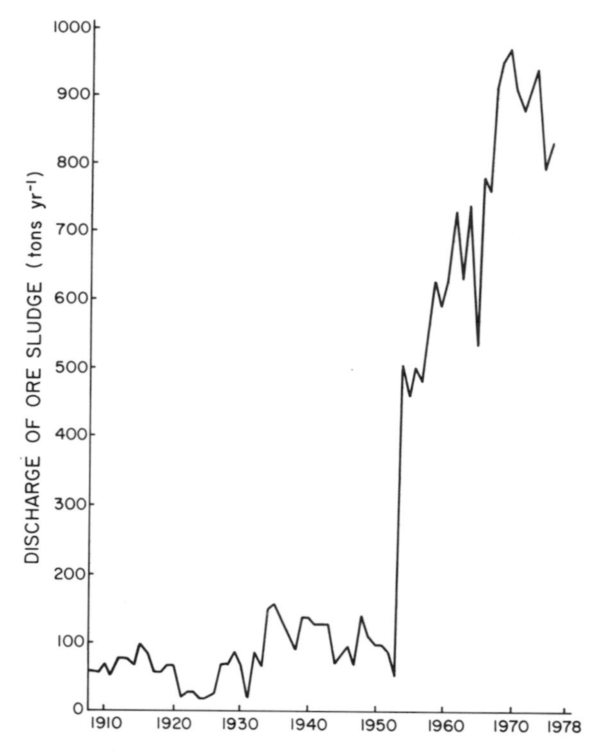

FIGURE 8.21. Annual discharge rates of ore sludge from SO_2-production.

fjord where the pollutant discharge occurs. Decreasing concentrations both seaward and toward the head of the fjord suggest that, similar to ligno-sulfonates (Fig. 8.19B), the metals are transported by tidal water. Based on the concentration gradients and the levels of metals at the mouth of Iddefjord, it may be concluded that most of the metals discharged into the fjord are trapped in the sediments, presumably as little-soluble metal sulfides. The surface sediments may be exposed to oxygenated bottom water in some periods and anoxic water during other periods. To what extent this will affect the mobility of the metals is not known. Metals associated with redox-sensitive elements like iron and manganese would fluctuate according to redox conditions. During oxic periods these metals would co-precipitate with iron and manganese, to be released at the onset of anoxicity (cf. Förstner and Wittmann, 1979; Salomons and Förstner, 1984).

Lead-210 datings of one sediment core, some 3 km away from the pulp and paper discharge (Station 1 Fig. 8.22), indicated a deposition rate of 2.7 mm yr^{-1}. Close to the factory, the sediment accumulation rates are expected to vary with the rate of fibre deposition.

The vertical distributions of mercury in one proximal and one distal sediment core indicate differences in sedimentation rates (Fig. 8.23). The maxima of mercury in the two cores may represent the period of use of mercury as a slimicide (1964–68). The general increase of mercury in the proximal core (Station 1) from the beginning of this century may be explained by the discharge of metal-enriched sludges from the SO_2 production. If we assume that the mer-

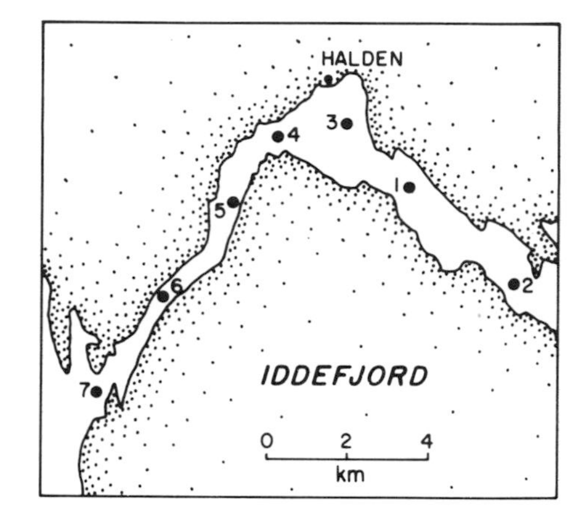

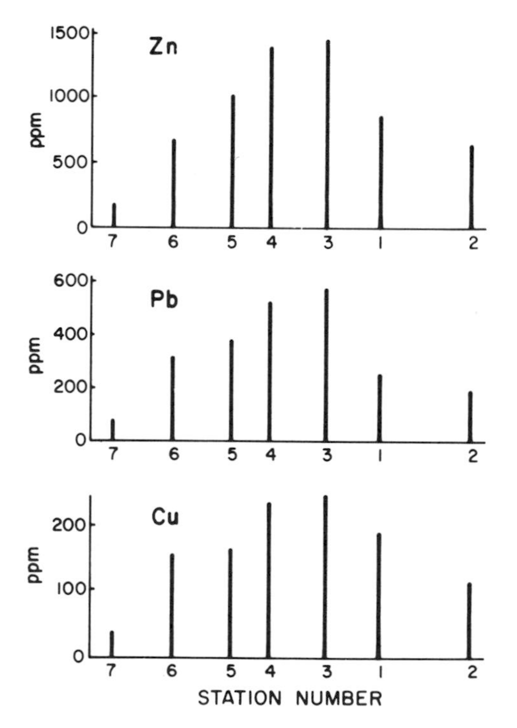

FIGURE 8.22. Concentrations of zinc, lead, and copper in surface sediments (0–4 cm) of Iddefjord.

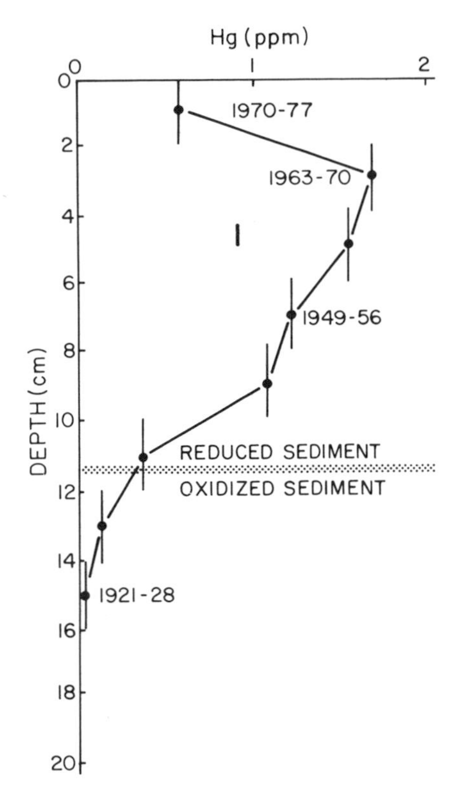

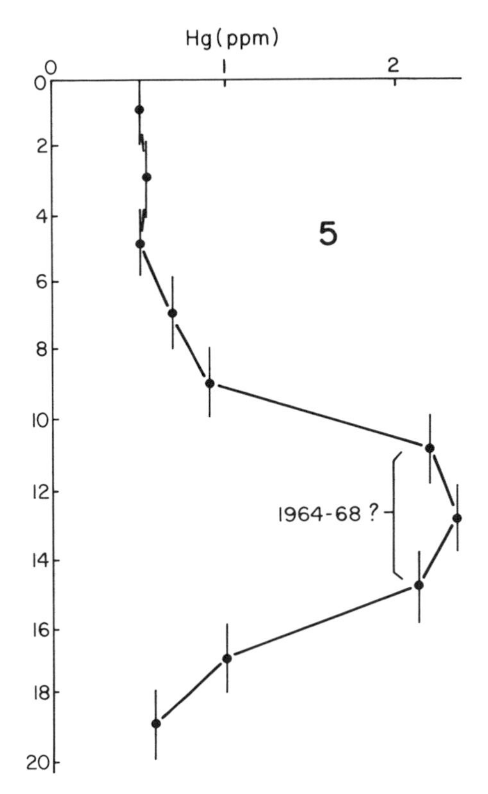

FIGURE 8.23. Depth profiles of total mercury in sediment cores from the inner basin (1) and outer basin (5) in Iddefjord. (The interface between sediments deposited during oxic and anoxic conditions is marked.)

cury maximum observed at 10–16-cm depth in distal core (5) represents sediments deposited from 1964 to 1968, the rate of sedimentation at this locality exceeds 1 cm yr^{-1}. The difference in accumulation rates in the landward and the seaward basin suggest that the dominant sediment transport direction is seaward, generated by the freshwater flow.

Based on the thickness of anoxic mud overlaying the clayey silt in core (1), a geochemical change occurred in the early 1940s (Fig. 8.23). This was the time when the pulp and paper production expanded and the organic load on the fjord increased. It also correlates with observed oxygen depletion in the bottom water in 1947. A close similarity between sedimentation rates and effluent discharge values from pulp mills has also been shown elsewhere (Stanley et al., 1981).

The down-core vertical change in concentrations of heavy metals, originating from the discharge of ore processing, is different from that of mercury (Fig. 8.24). It reflects the steady increase in metal loading from the beginning of

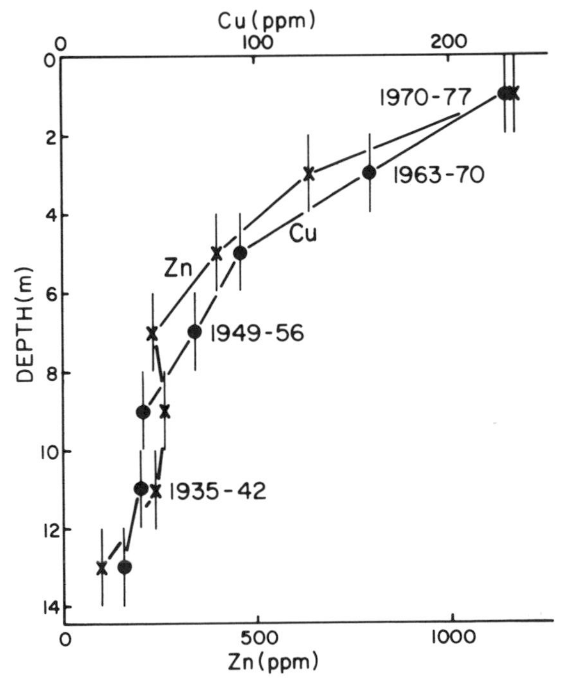

FIGURE 8.24. Vertical distribution of copper and zinc in the dated sediment core from the inner basin in Iddefjord.

the 1950s to the middle of the 1970s (Fig. 8.21), when the sediment investigations were done. The good agreement between the history of pollution and the readings in the sediments is a result of the lack of bioturbation or other mechanical disturbances of the sediments.

8.9 Saudafjord, Southwest Norway

8.9.1 Environmental Setting

Saudafjord, on the southwest coast of Norway, is situated at 59°37′N and 6°19′E (Fig. 8.25). The fjord is 13 km long with a maximum basin depth of 390 m and a deep sill ($\approx$ 200 m). The bottom water is frequently flushed, and high oxygen concentrations are maintained.

The average annual freshwater discharge is 40 $m^3\ s^{-1}$, of which 75% passes through a hydroelectric power station. Hence, the sediment supply to the fjord is low. The overall rate of sedimentation over the past 50 years is approximately 1.4 mm yr^{-1} in the middle of Saudafjord (Skei, 1981a).

The sediments are silty and sandy and contain 0.8 to 16.6% organic matter (as loss on ignition). The highest concentrations were recorded in the harbor basin of Sauda and are mainly due to input of terrestrial organic matter. Because of a high deposition rate in this area, anoxic conditions develop in the subsurface sediments.

8.9.2 Environmental Problems

The main contributor of pollutants to Saudafjord is a ferroalloy smelter (Fig. 8.25), established in 1923. The smelter produces ferromanganese and silicomanganese. Water from the scrubber (2300 $m^3\ h^{-1}$ in 1976) was initially discharged at 5-m depth, mixed into the brackish surface layer and transported seaward. In recent years, treatment plants have come into operation and the effluent discharge has been reduced.

The main pollutants in the effluent from the smelter are polycyclic aromatic hydrocarbons (PAH), cadmium, lead, and zinc. This has

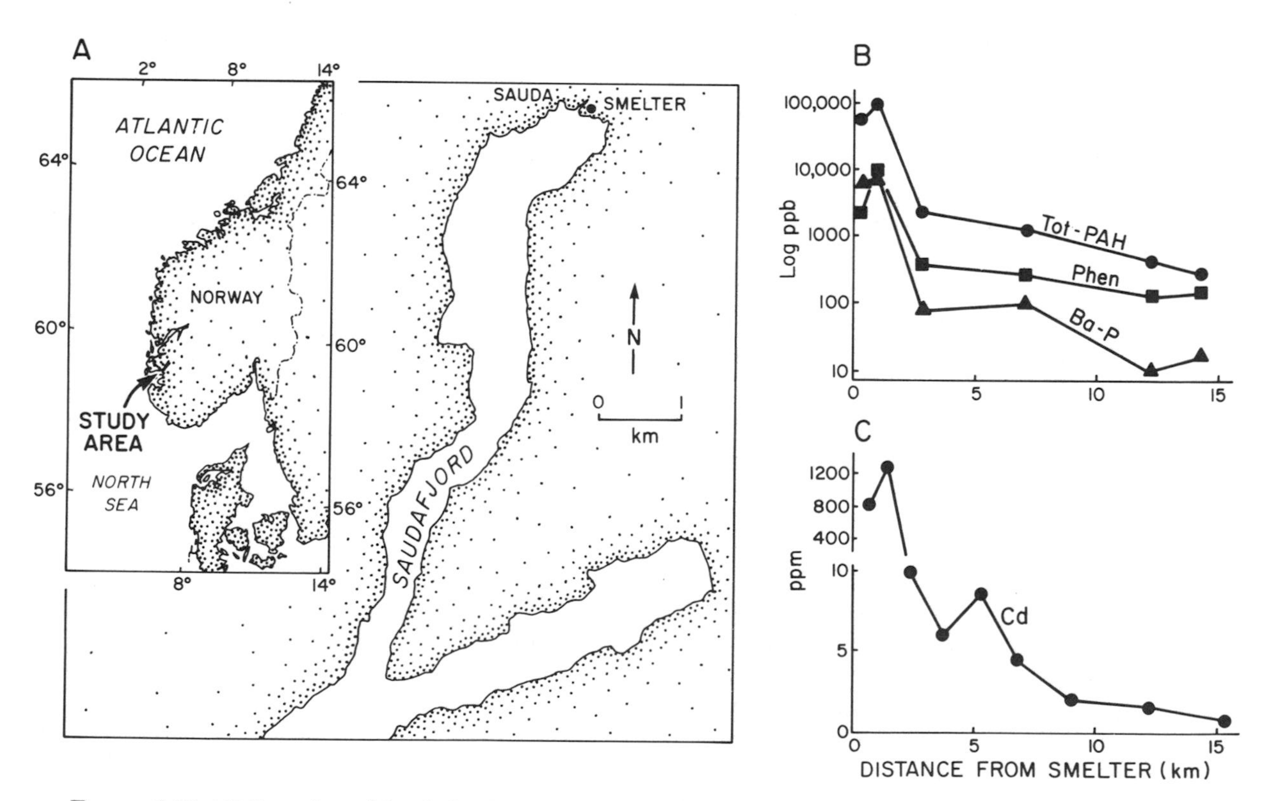

FIGURE 8.25. (A) Location of Saudafjord and position of smelter at the head of the fjord. Also shown is the distribution down-fjord from the smelter of: (B) total polycyclic aromatic hydrocarbons (PAH), phenanthrene (Phen), and benzo (a) pyrene (Ba-P); and (C) cadmium (Cd), in the surface sediments of the fjord.

caused enrichment of these pollutants in the sediments and the biota (mussels) (Bjørseth et al., 1979; Skei, 1981a). The lateral distribution of total PAH, phenanthrene, and benzo (a) pyrene in the surface sediments is shown in Figure 8.25B. It is evident that the gradient of PAH is very steep within 2 to 3 km from the source of pollution, although elevated levels are observed along the total length of the fjord. The steep gradient of PAH in sediments away from a source has been observed in several Norwegian fjords (NIVA, unpublished reports). Apparently, PAH is discharged in a particulate form with a grain size that allows immediate sedimentation (Bjørseth et al., 1979). The sediments near the discharge outlet are black colored and fragments of coke and cinders have been identified. In addition to the immediate sedimentation of coarse waste containing PAH, a long-distance transport of PAH occurs. Sediments collected 15 km away from the source are clearly contaminated. There are two likely modes of transport: (1) PAH attached to fine dust particles is transported seaward by the surface brackish water flow; and (2) PAH is transported atmospherically by wind blowing from the smelter and down the fjord. Mussels show elevated concentrations of PAH more than 15 km from the source.

The relative abundance of PAH components, as compared with total abundance of PAH, may or may not vary with distance from the source. Benzo (a) pyrene shows little relative change and is normally $\approx$ 10% of total PAH. The relative abundance of phenanthrene increases away from the smelter. Either there is preferential settling of other PAH components close to the source or dissolution or biodegradation of other PAH compounds in the sediments away from the source (Bjørseth et al., 1979). A contributing factor may also be the existence of two sources of PAH: the smelter and naturally occurring PAH (cf. Sporstöl et al., 1983). Natural PAH from forest fires, geochemical processes, and biosynthesis is known to contain a high percentage of phenanthrene. With increasing distance from the smelter in Saudafjord, the contribution from natural sources becomes more important.

The effluents from the smelter also contain heavy metals like cadmium, zinc, and lead. The concentrations of cadmium in the sediments are extremely high, with levels above 1200 ppm locally (Fig. 8.25C). The falloff of concentration seaward follows that of PAH. Within 2 km from the source, the concentration of cadmium is below 10 ppm and at the mouth of Saudafjord the cadmium level is down to 0.8 ppm in the surface sediments. It is apparent that cadmium and PAH have the same origin and that they are associated in the sediments (cf. Müller et al., 1977). Coal ash contains high concentrations of heavy metals (Erlenkeuser et al., 1978), including cadmium. Despite the high levels of cadmium in the sediments, only moderate levels of cadmium in water and biota have been observed (NIVA, unpublished reports). This suggests that cadmium present in coal ash and cinders shows little sign of mobilization. This is very different from the general behavior of cadmium in sediments (cf. Förstner and Wittmann, 1979).

8.10 Sørfjord, West Norway

8.10.1 Environmental Setting

Sørfjord is a north-south trending extension of Hardangerfjord, 1 to 2 km wide and 40 km long (Fig. 8.26A). The fjord has a steplike seafloor mostly with a flat bottom. The maximum depth is 390 m (Fig. 8.26B). Most of the runoff enters the fjord at its head ($\approx$ 50 m^3 s^{-1}), partly as glacial runoff. The annual amount of sediment being transported into Sørfjord is approximately 8000 tons (Skei, 1975). The majority of the silt is deposited within Sørfjord while the finest clay fraction is transported further into Hardangerfjord. In the central parts of Sørfjord an average sediment accumulation rate of 2 mm yr^{-1} or 1.28 kg m^{-2} yr^{-1} has been estimated from excess ^{210}Pb profiles (Skei, 1981c).

Circulation within the fjord is predominantly two-layer estuarine. The depth of the halocline varies with season and location in the fjord (Fig. 8.27). The sill at the mouth of Sørfjord is deep ($\approx$ 225 m) and frequent deep water exchange maintains oxygen concentrations in the bottom water above critical levels ($>$ 4 ml L^{-1}).

The sediments of Sørfjord are silty and contain 6 to 7% organic material except in the innermost part of the fjord where the organic content doubles (NIVA, unpublished). The sediments are generally oxic and do not contain hydrogen sulfide, except locally in the harbor basin of Odda.

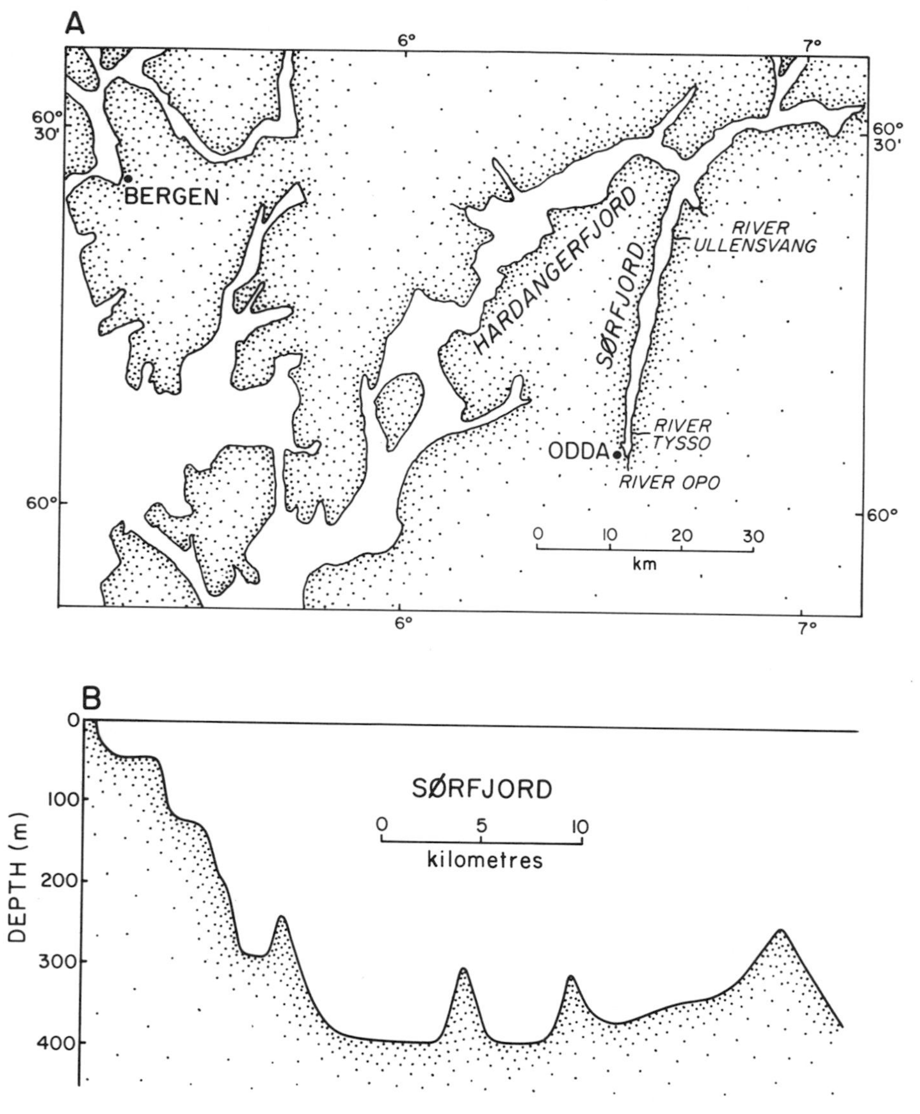

FIGURE 8.26. (A) Map of Sørfjord area, western Norway, with (B) longitudinal profile along Sørfjord.

8.10.2 Environmental Problems

Sørfjord has been considered the most metal-polluted fjord in the world, with several percent of heavy metals in the sediments (Skei et al., 1972) and several thousands of $\mu g\ L^{-1}$ of metals in the water (Skei, 1981c).The main source of the metals is a zinc plant, situated at the head of the fjord at the industrial town of Odda (Fig. 8.26A). Additionally, Sørfjord has received industrial effluents from an aluminium smelter (shut down in 1981) and a carbide-dicyanamide smelter. The history of smelting at the town of Odda extends some 60 to 70 years back, which implies that Sørfjord has received industrial waste for a long time. All compartments of the fjord environment are therefore severely polluted.

Environmental investigations started in Sørfjord in about 1970. Since then the environment has been monitored regularly, except for the years 1975 to 1978. The investigations have included water, biota, and sediments, and have focused on the heavy metals (zinc, lead, cadmium, and mercury) discharged from the zinc plant. The daily discharge of metals to Sørfjord in 1972 is shown in Table 8.4. The most important source today is a jarosite residue with high metal content that is discharged at 20 to 30 m depth as a slurry with density 1090 kg m^{-3}. As

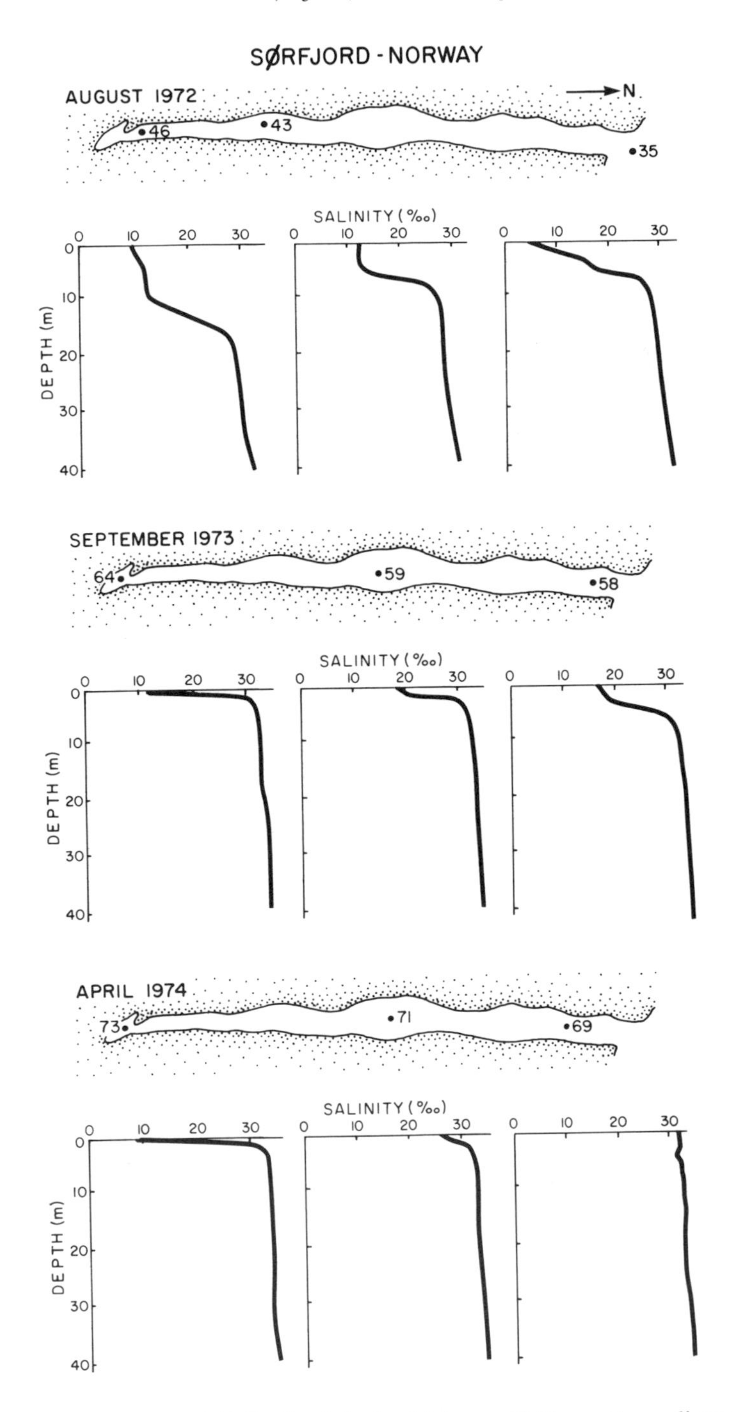

FIGURE 8.27. Salinity profiles collected in Sørfjord during high runoff (August), medium runoff (September), and low runoff (April) (after Skei, 1981c).

the density of this suspension is higher than the receiving sea water, it travels as a density current seaward. This is very similar to the behavior of mine tailings at Agfardlikavsâ in Greenland (see Section 8.2). The jarosite consists of particles varying in size between 1 and 20 μm and also contains soluble metal salts. Upon entering the fjord, the soluble fraction of the metals is released immediately and with the finest solid fraction, transported long distances. The coarser grained part of the solid waste is deposited close to the discharge point. Jarosite residue deposited

TABLE 8.4. Metals in industrial waste discharged into Sørfjord (1972).*

Metal	Compound in waste	Discharged weight per day (in tons)
Zn	$ZnO \cdot Fe_2O_3$, $ZnSO_4$	6
Cu	$CuSO_4$, CuO_2, Cu-jarosite	0.3
Cd	$CdSO_4$, CdS, $CdO \cdot Fe_2O_3$	0.03
Pb	$PbSO_4$	4.5
As	$FeAsO_4$	0.09
Sb	$FeSbO_4$	0.06
Hg	HgS, HgSe, Hg_2Cl_2, Hg, Hg^{2+}	0.003
Fe	$(NH_4)_2Fe_6(SO_4)_4(OH)_{12}$, $ZnO \cdot Fe_2O_3$, Fe_2O_3	23
Mn	$MnSO_4$, MnO_2	0.12

*After Environmental Coimmittee Report, 1973.

in the shallow part of the fjord is also subject to resuspension by the eroding effect of wind-induced waves, tidal currents, and sporadic deep water flushing events.

The lateral distribution for dissolved zinc (Fig. 8.28A) illustrates a horizontal plume situated below the compensating current (25–100 m depth). During the 10 years of monitoring, the same general picture of metal distribution in Sørfjord has been evident, that is, high levels at the surface, low, or moderate concentrations in the compensatory underflow and a second maximum underneath, indicating that a multilayered flow is present (Fig. 8.28A; Skei, 1981c). When considering a particulate pollutant, such as iron present in jarosite, a different distribution pattern is evident. Figure 8.28B shows that the plume initially sinks because of the gravity flow, gradually mixing into the surrounding water, eventually to lose its identity. The distribution of both soluble and particulate components in the jarosite residue (Fig. 8.28) clearly demonstrates that the water mass in the entire fjord is influenced by this discharge. It is also evident from the levels of heavy metals (zinc and others) at the mouth of Sørfjord that Hardangerfjord is influenced. Typical levels of metals in the surface water in 1984, at the mouth of Sørfjord, were $\approx 100\ \mu g\ L^{-1}$ zinc, $\approx 1.5\ \mu g\ L^{-1}$ lead, and $\approx 1\ \mu g\ L^{-1}$ cadmium. The severe pollution of the surface water, despite the fact that the regular discharges from the zinc smelter today occur at 20-m depth or deeper, has lead to intensified investigations of other potential sources. Attention has focused recently on land dump sites near the smelter, which are flooded during high tide. These sites are acidic residue dumps from the earlier days of production, containing 10 to 15% zinc and lead, as well as high concentrations of associated metals. During ebb tide, sea water with low pH (3–5) and extremely high metal content (830 mg L^{-1} zinc, 22 mg L^{-1} cadmium, 2.4 mg L^{-1} lead, and 5.2 mg L^{-1} copper, all maximum values) enters the surface water of the fjord. This may contribute significantly to the surface pollution of Sørfjord. In fact, elevated levels of cadmium and zinc (in mussels and seaweeds) in Hardangerfjord, some 100 km from the source, suggest a large-scale pollution problem with impact on natural resources.

The majority of metals released from the Zn plant are present as solids. The extent to which metals present in the solid fraction of the jarosite residue may solubilize, is not yet quantified. Particularly, the kinetics of the reaction between jarosite and sea water are not known. However, vertical profiles of metals in the water above the sediments strongly influenced by jarosite residue indicate an increase toward the seabed, suggesting a certain amount of release (Fig. 8.29).

Metals released in a dissolved state may become absorbed onto waste particles, inorganic silicates (clay and silt), or natural organic matter (humus, plankton, etc.). The amount of natural particles suspended in the water is seasonally dependent. The natural sediment transport is largest during the snow melting season in early summer (May-June), creating turbid water above the halocline (Fig. 8.30). Similar to the Greenland case (8.2), the sediment-laden melt-water is first lake-filtered, where a considerable

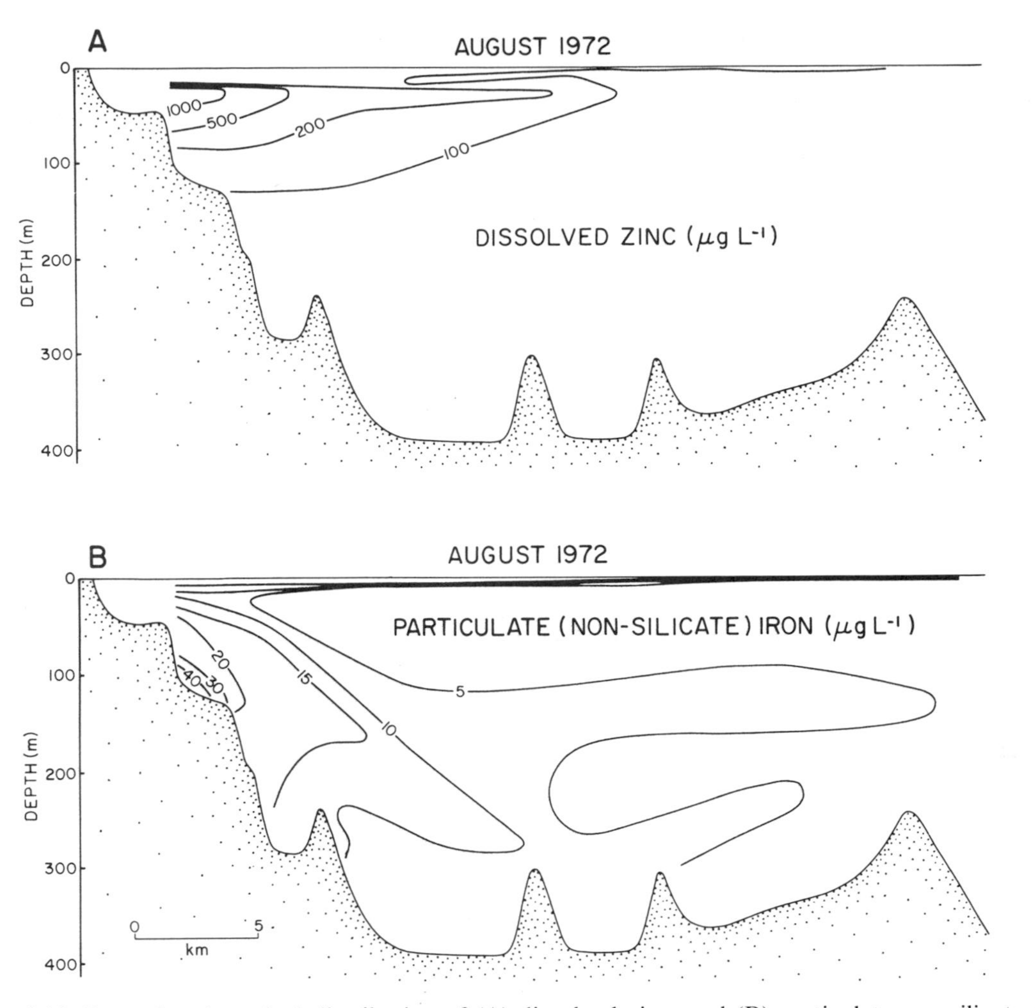

FIGURE 8.28. Lateral and vertical distribution of (A) dissolved zinc, and (B) particulate, nonsilicate iron, during August 1972 in Sørfjord.

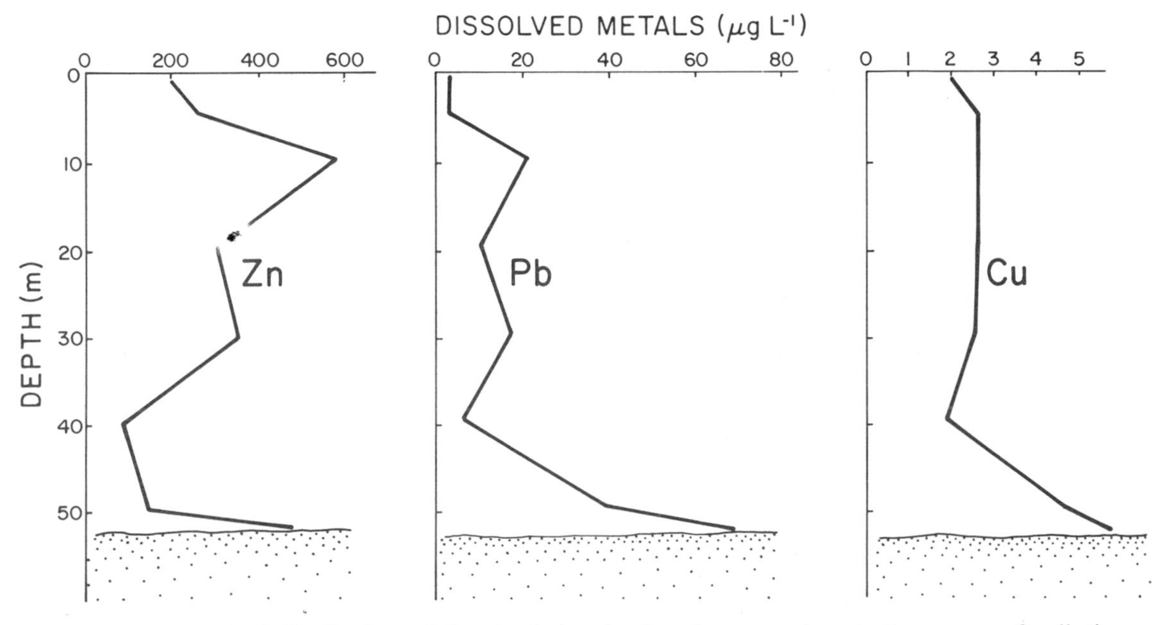

FIGURE 8.29. Vertical distribution of dissolved zinc, lead, and copper close to the source of pollution.

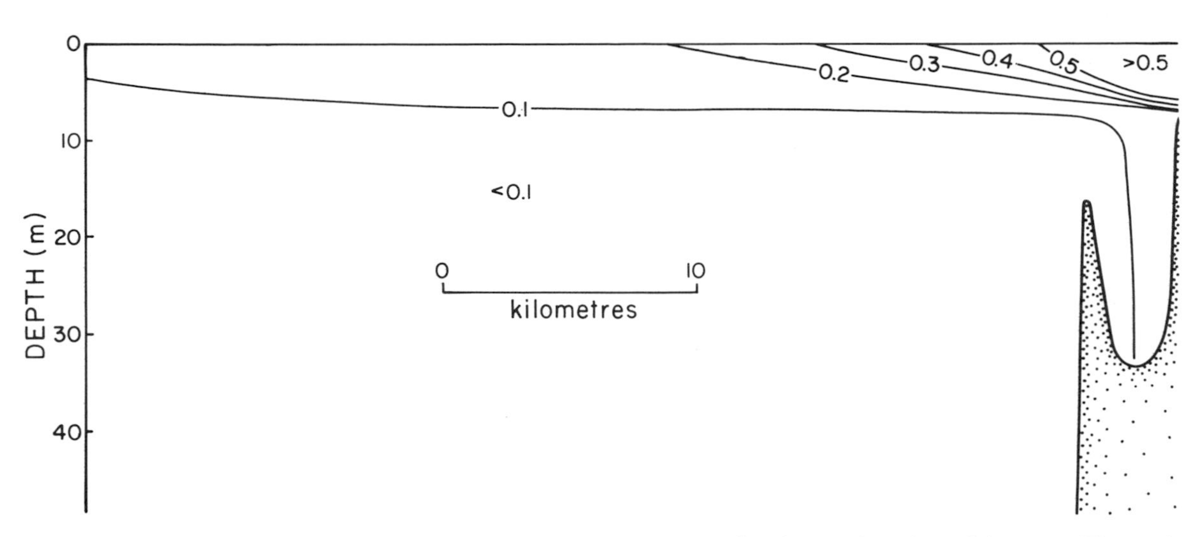

FIGURE 8.30. Distribution of the coefficient of irradiance attenuation from mineral particles ($\lambda \approx 470$ nm, in values of m^{-1}) for Hardangerfjord in July 1967 (after Aas, 1971).

amount of fluvial sediment is lost. Hence, detrital silicates may not be a very important carrier of pollutants in Sørfjord. Plankton, however, may act as an important adsorption surface for heavy metals (Knauer and Martin, 1973). Analyses of mercury in both zooplankton and phytoplankton from Sørfjord (Skei et al., 1976) clearly indicated enrichment, particularly in the phytoplankton, with increasing concentrations toward the fjord head.

The transfer of metals from the water column to the seafloor may occur by the following mechanisms: (1) settling of solid waste particles; (2) adsorption of metals onto natural inorganic particles; and (3) adsorption or uptake of metals on plankton and other organisms. The relative importance of these transfer mechanisms varies with distance from the source of pollution and with season. Close to the waste outlet ($\approx$ 1 km^2), the sediments are red-colored by industrial waste. This is a direct effect of discharge of solid jarosite. With increasing distance from the source, the other transfer mechanisms become relatively more important. The lateral gradients of metal concentrations in the surface sediments of Sørfjord show an exponential increase within 5 km from the discharge pipe (Fig. 8.31). In this area, sedimentation is completely dominated by industrial waste. As a result, the sediments are devoid of macroscopic life, except for a few polychaetes (Rygg and Skei, 1984). Here, concentrations of 1 to 1.50% zinc have been measured in the surface sediments. At the mouth of Sørfjord the surface sediments contain 400 to 500 ppm zinc (Fig. 8.32A). The lateral distribution of the other heavy metals analyzed (lead, copper, cadmium, and mercury) is very similar to zinc, except that cadmium concentrations reach background levels in the central part of the fjord (Figs. 8.32B, 8.35). Cadmium shows the widest areal distribution, based on accumulation in mussels. Apparently, the sediments are not a permanent sink for cadmium. It is known to be a mobile element (Salomons and Eysink, 1981), which is easily complexed (chloro-complexes). Considering the detrimental effects of cadmium (Wittmann, 1979), this causes environmental concern.

The vertical distribution of solid metals in a sediment core illustrates the historical development of pollution in Sørfjord (Fig. 8.33). The core was taken from the deepest basin of the fjord, some 20 km from the smelter. An increase in metals started to occur at about 1930, corresponding to the establishment of the zinc smelter in 1929. Concentrations 20 cm below the seafloor are constant for all metals, representing pre-industrial background values. The linear increase of metal concentrations in the surficial 10 cm of the seafloor suggests a steady increase of pollutants since the early 1940s. At the core location, the sediments are only moderately polluted and bioturbation is still expected to occur. Investigations of the soft-bottom fauna in Sørfjord have shown a drastic decrease in diversity and number of species from the

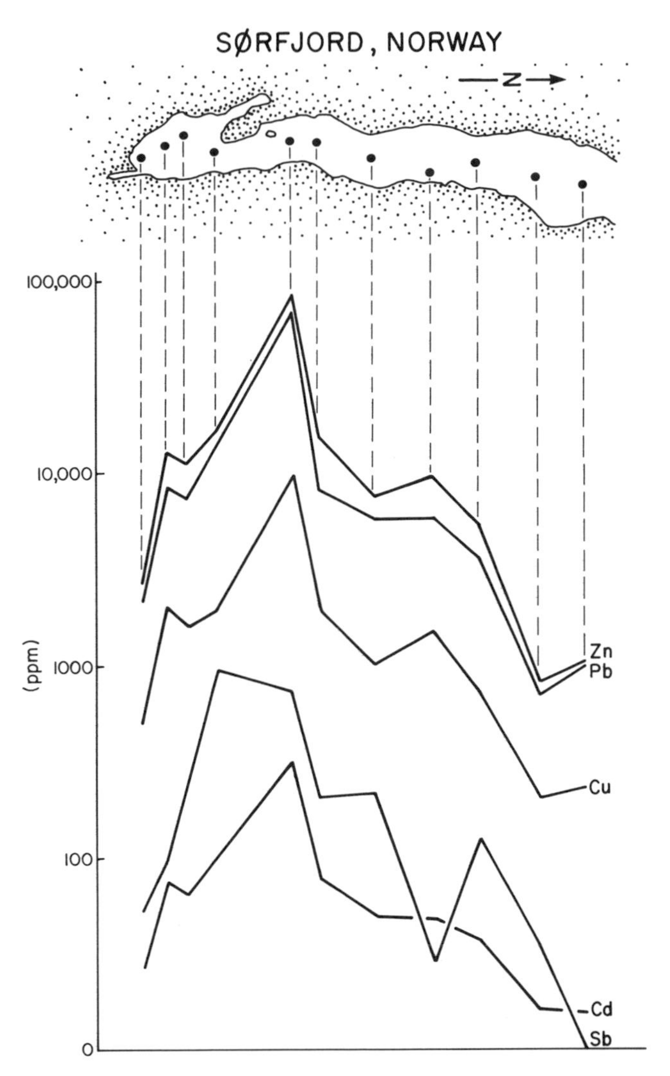

FIGURE 8.31. Distribution of metals in the surface sediments on a 5-km longitudinal profile at the head of Sørfjord (after Skei et al., 1972).

mouth to the head of the fjord (Rygg and Skei, 1984).

The faunal diversity is apparently strongly influenced by the concentrations of copper in the surface sediments (Fig. 8.34). This suggests that copper is highly toxic to bottom-living organisms (Bryan and Hummerstone, 1971). A correlation coefficient of −0.79 (59 samples) between copper and faunal diversity was observed when considering 9 polluted fjords (Rygg and Skei, 1984). The corresponding correlation coefficients for lead and cadmium were −0.50 and −0.35. The impact of elevated pollutant concentrations in sediments on the diversity of benthic organisms has also been documented elsewhere (Pearson and Rosenberg, 1978).

Mass-balance calculations of metals in fjords are subject to large uncertainties, but are important in the assessment of the fate of pollutants. For instance, it is important to know the exact discharge rate of metals from industry. However, data are often based on a limited number of analyses and more average discharge conditions when irregular discharge conditions may be more important. A similar sparse data set exists for the transport calculation on the quantity of a pollutant advected out of the fjord. The quantity of a metal accumulating within bottom sediments may be based on data extrapolation from a number of basin cores (assuming that polluted sediments occur only within the basins and not on the steep sides).

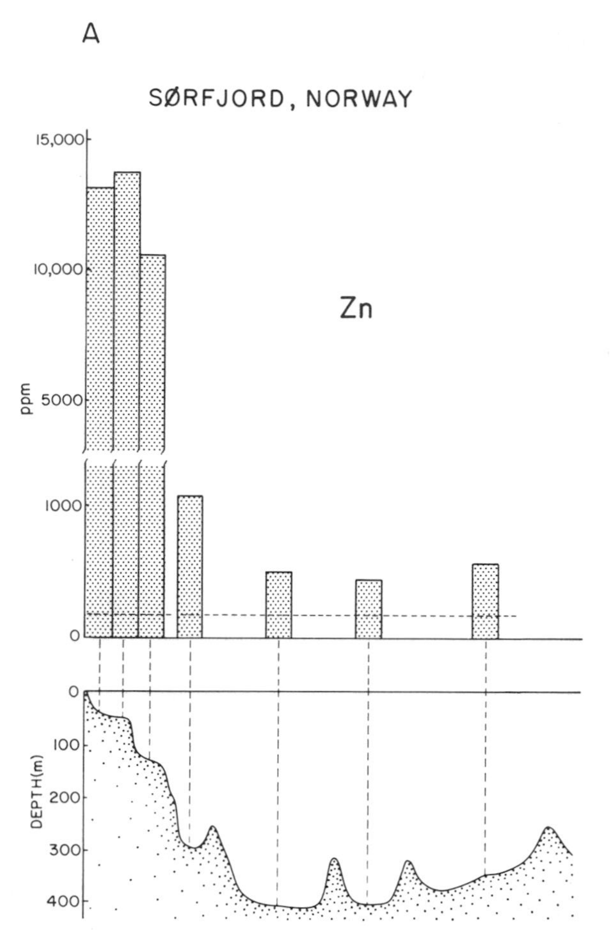

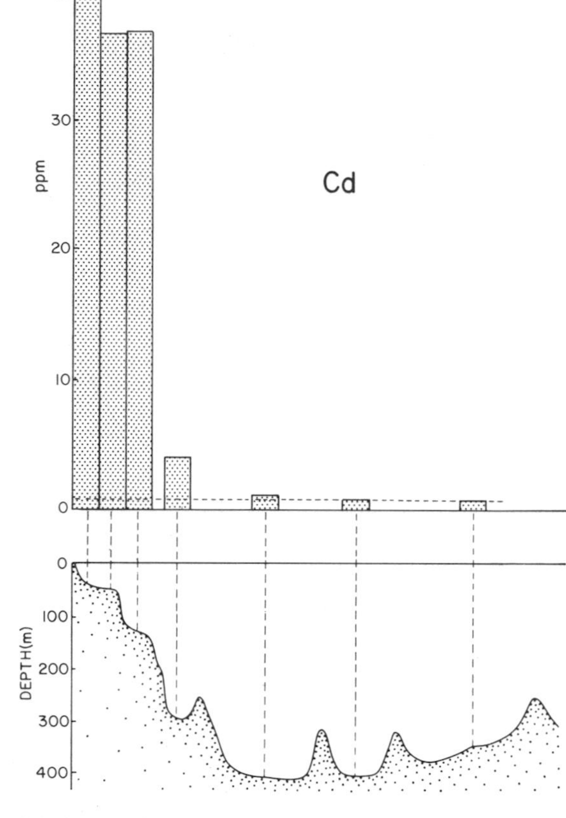

FIGURE 8.32. The abundance of zinc (A) and cadmium (B) in the surface sediments of Sørfjord, 1978. The top stippled line indicates an upper background value of 150 ppm for zinc and lower lines indicates an upper background value of 0.6 ppm for cadmium.

Such a calculation may be compared with estimates from the smelter discharge to make an assessment on the percentage of pollutant metal deposited within the fjord. Additional assessment may be based on lead-210 core determinations to calculate the annual metal fluxes to the sediments. For zinc, lead, and mercury this gave retention percentages of 20 to 25, 35, and 80, respectively (Skei, 1981c). This implies that the majority of zinc and lead · transported out of Sørfjord and into Hardangerfjord, in agreement with observations of elevated levels of these metal in biota in Hardangerfjord. Mercury, however, appears to be incorporated to a larger extent in the sediments of Sørfjord. This is in agreement with biota observations and the general assessment that mercury shows a strong particle affinity (Lindqvist et al., 1984). The mass-balance estimates also agree with the measurements of metals in the water, that show high concentrations of zinc at the fjord entrance while the concentrations of mercury are more normal.

Another important issue of sediment pollution in Sørfjord is the extent to which the polluted sediments will continue to contribute heavy metals to the water and biota. At the present discharge rates from the smelter, the primary outlet source far exceeds the release from the sediments. However, if the primary discharge is reduced, the secondary sources will gradually become more important. Experiments performed on the shaking of highly polluted sediments with sea water revealed a considerable release of dissolved zinc, lead, and cadmium. This suggests that vigorous resuspension of the sediments in the innermost part of the fjord may be associated with massive release of heavy

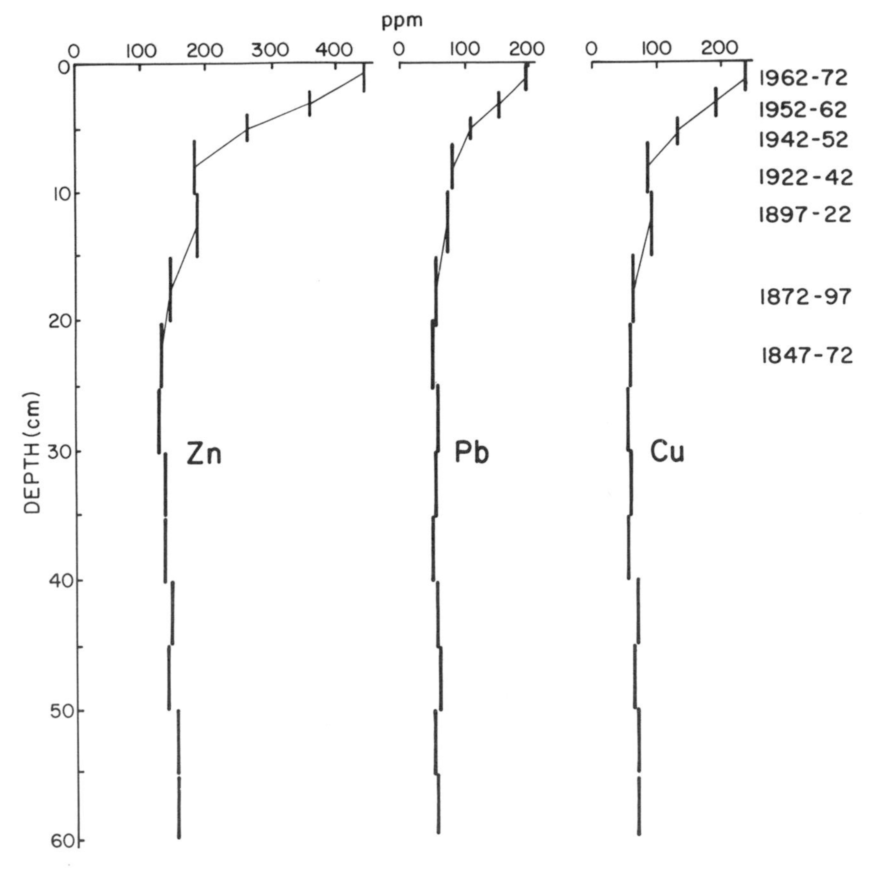

FIGURE 8.33. The vertical distribution of total zinc, lead, and copper in a ^{210}Pb dated core from the central part of Sørfjord.

metals. Sea water is quite aggressive with respect to mobilizing heavy metals from solids. By using zinc concentrate (pure sphalerite) and performing shaking experiments with sea water, a substantial release of zinc and associated metals was observed.

8.11 Ranafjord, Northern Norway

8.11.1 Environmental Setting

Ranafjord, situated at 66°15′N in northern Norway (Fig. 8.36), is 540 m deep, fully oxygenated, and receives on average 340 $m^3\ s^{-1}$ of fresh water. The sill depth is approximately 150 m. A two-layer estuarine circulation is observed in the upper waters, and the bottom water appears to be frequently exchanged. The fjord is surrounded by Cambro-Silurian sedimentary rocks and Caledonian intrusives.

The natural sediment supply to Ranafjord, from the river situated at the fjord head, has been estimated at 27,000 tons yr^{-1} (NIVA, unpubl.). This is based on measurements of particulate matter and discharge during 1980–81.

Vertical profiles of particulate aluminum in the deep basin of Ranafjord demonstrate that the transport of aluminosilicates is directly related to level of runoff (NIVA, unpublished report). Increase of particulate aluminum in the bottom water (collected in December) may be from bottom resuspension during winter flushing events.

8.11.2 Environmental Problems

Since the beginning of this century, zinc, lead, and copper have been mined in the area and tailings from a flotation plant have been discharged into the fjord near its head (Fig. 8.36). Additionally, a coke plant and an iron works are situated near the river mouth. A total of 2.0 million tons yr^{-1} of solid industrial waste is discharged into Ranafjord, close to 100 times the natural sediment supply. This implies that the

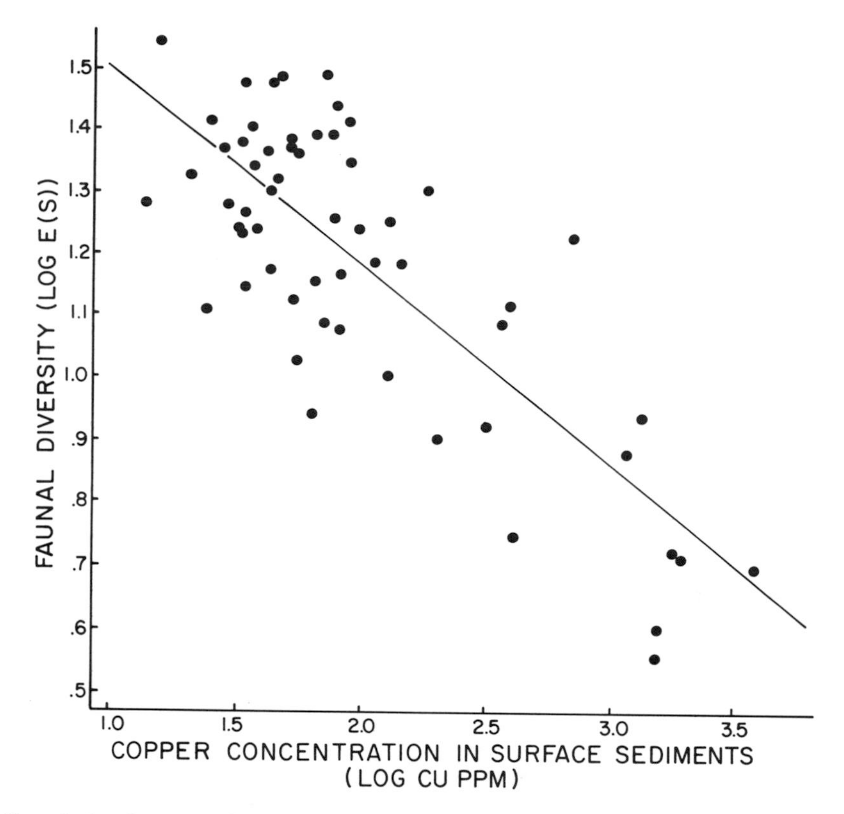

FIGURE 8.34. Correlation between fauna diversity and copper concentration in surface sediments in polluted fjords in Norway, Sørfjord included (after Rygg and Skei, 1984).

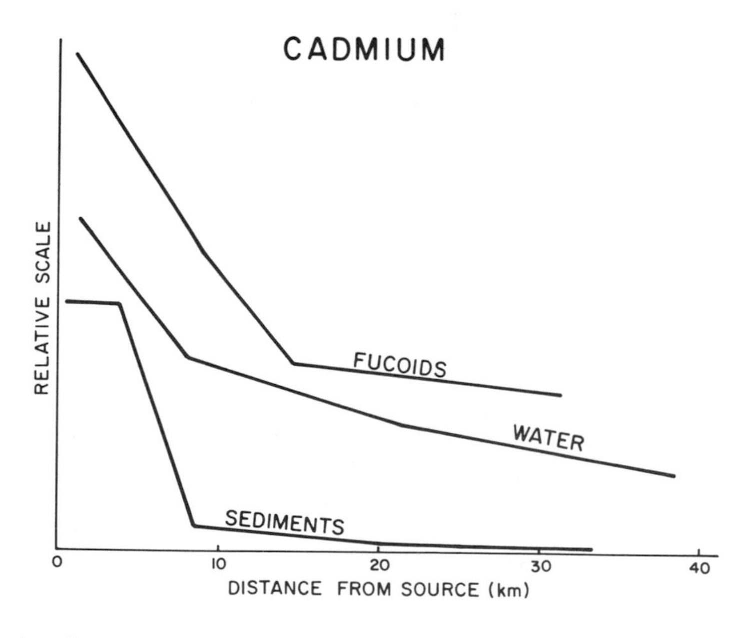

FIGURE 8.35. Cadmium in fucoids, water, and surface sediments with increasing distance from the source of pollution in Sørfjord.

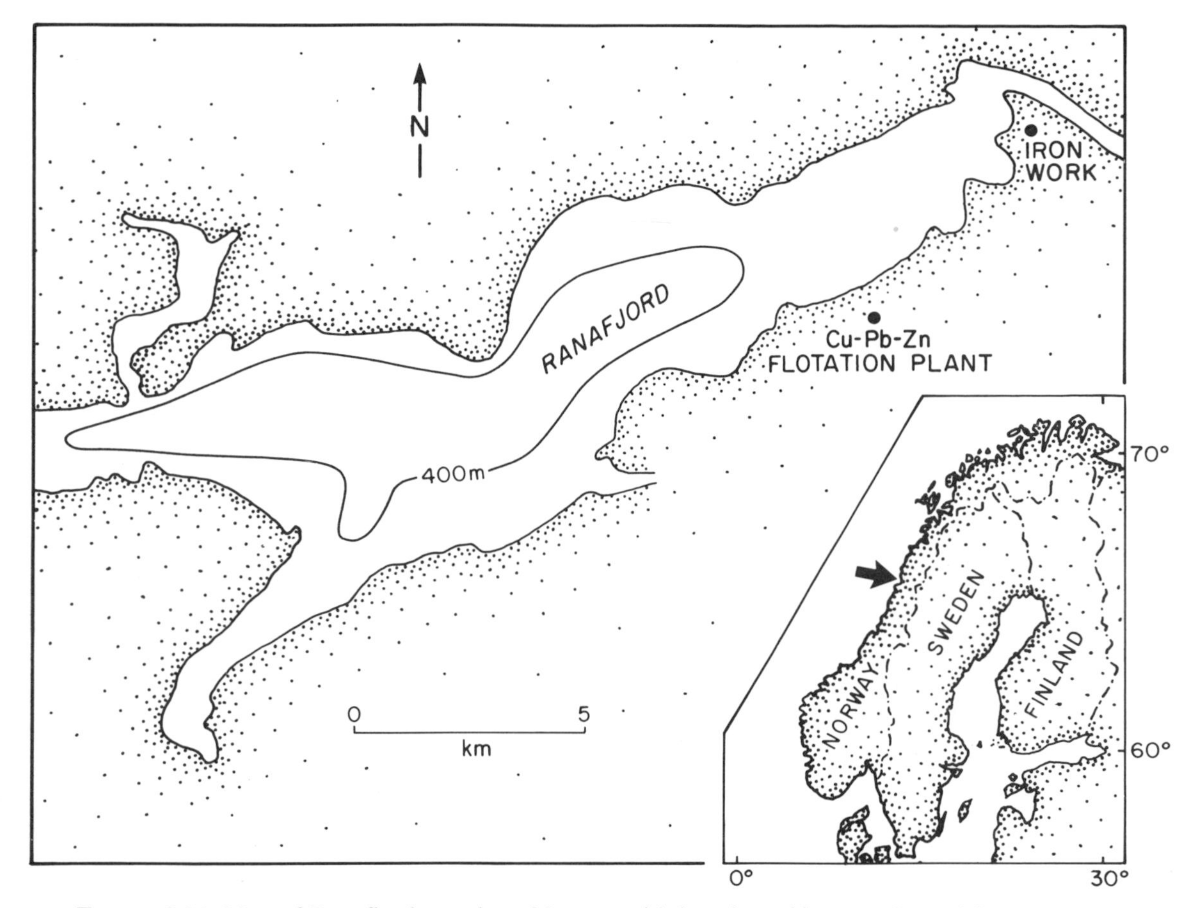

FIGURE 8.36. Map of Ranafjord, northern Norway with location of iron works and flotation mill.

bottom sediments are extensively influenced by tailings in the inner parts of the fjord. The iron works is the main contributor of solids (since 1964), contributing nearly 90% of the total tailings input. This waste has low metal content, except for iron, and is released in a suspension at 30-m depth at a rate of 40 kg s^{-1} (Tesaker, 1978). Transmission measurements at the discharge site demonstrate the impact of the suspension on transmission (Fig. 8.37A). Measurements of particulate manganese, phosphorus, and iron, which are all major components in the waste from the iron works, show maximum concentrations at the depth of minimum transmission (Fig. 8.37B). Apparently, the suspension plume sinks fairly rapidly. Some 2 to 3 km away from the discharge point, the iron concentration at the 50-m water depth had decreased by 97% (NIVA, unpublished report). Instead, the iron maximum is observed at the 200-m depth, close to the seafloor.

Because of a steep slope near the tailings discharge point and the large quantities of tailings, the material accumulates under unstable conditions (Tesaker, 1978), very similar to those observed in Rupert Inlet in British Columbia (Hay, 1982). A 50 m deep canyon was eroded in loose deposits until the bare rock was exposed. The eroded volume during 7 yr was 10^7 m^3 (Tesaker, 1978). A suspension current in the bottom of Ranafjord showed a velocity of 25 cm s^{-1}, measured 1 m above the bottom (Fig. 8.38; Tesaker, 1978). Calculations of sediment transport by the suspension current suggest that $\approx$ 18% of the discharged tailings is transported in suspension, the remainder is carried as bed load (Tesaker, 1978). Light scatterance measurements made in March 1971 in Ranafjord indicate turbid water above the halocline and an underwater plume in the innermost 5 km of the fjord (Fig. 8.39).

The discharge from the lead-zinc flotation plant causes extremely elevated levels of metals in the sediments in an area of $\approx$ 5 km^2 (Fig.

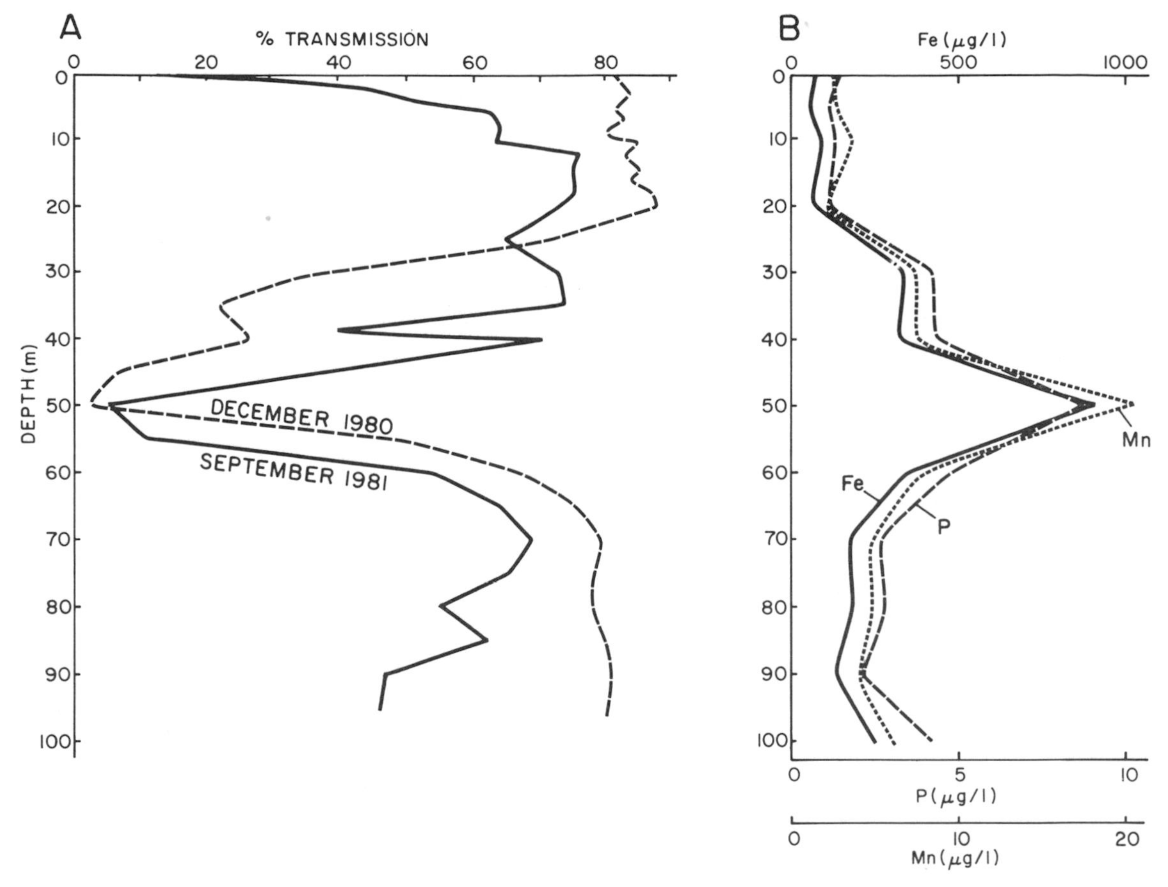

FIGURE 8.37. The uppermost 100 m of the water column, close to the discharge outlet from the iron works (Ranafjord) in terms of (A) light transmission, and (B) profiles of particulate manganese, phosphorous, and iron.

8.40B). Compared with the Greenex lead-zinc mine in Greenland (Section 8.2), the discharge from the mine in Ranafjord is small (15,000 tons of tailings per month, compared to ≈ 40,000 tons of tailings per month in Greenland). As a result, the metal concentrations in the sediment are lower in Ranafjord. Elevated levels of dissolved lead in the water were observed within 1 km from the source (Fig. 8.40C). Lead appears to be mobilized in sea water to a larger extent than zinc, cadmium, and copper. This has been confirmed by laboratory experiments on mine tailings and sea water (NIVA, unpublished report).

The distribution of excess heavy metals in the sediments of Ranafjord extends some 25 km away from the industrial sources. Two transport mechanisms are in operation: (1) slumping and sediment gravity flow, and (2) buoyant surface water transport. The gravity flow occurs intermittently, following buildup of tailings on the steep slopes and results in slides. The gravity flow may erode the basin sediments, leaving coarse, sandy sediments behind. As a result, the basin sediments in the inner part of the fjord have a lower metal content than sediments in the central part of the fjord (Fig. 8.40A).

The surface transport of pollutants is less important. However, spherical iron particles are observed in the surface water as far as 12 km away from the iron works. These particles are less than 1 μm, and energy-dispersive scanning electron microscopy analysis suggests they are pure iron oxides (hematite or magnetite), indicating an iron works origin.

A sediment core taken about 15 km distant from the industrial area was analyzed using different extraction techniques (Skei and Paus, 1979), to study the partitioning of metals. The vertical profiles of total lead and zinc clearly demonstrate increased concentrations in the upper 10 cm of the core, corresponding to sediments deposited since the turn of the century (based on ^{210}Pb datings) (Fig. 8.41A). This may be explained by the establishment of the zinc-

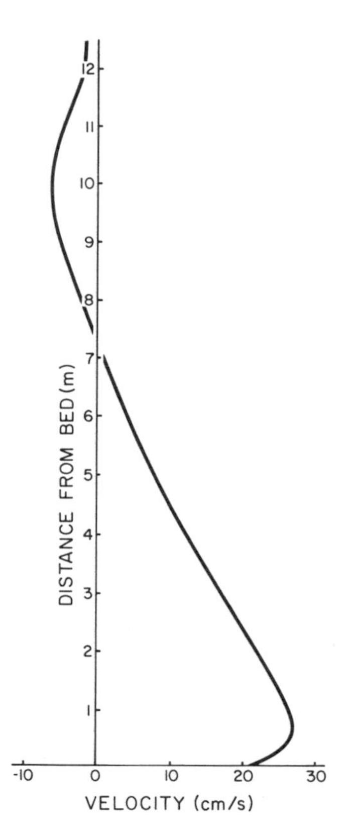

FIGURE 8.38. Velocity profile in the eroded canyon in Ranafjord, 1971 (after Tesaker, 1978).

lead flotation plant at the same time. The lead and zinc depth profiles are very different from the vertical distribution of mercury and manganese (Figs. 8.41B, C). The general increase in mercury toward the sediment surface may be interpreted as a result of increased input of mercury to the environment on a global scale. There are no local inputs of mercury in the area, and the levels even in the surface sediment (0.18 ppm) must be considered low. Similarly, the profile of manganese is not attributed to an anthropogenic effect: the abrupt increase in the top 4 to 6 cm of the core is thought to be the result of diagenetic processes (Lynn and Bonatti, 1965) in which manganese in the reducing part of the core migrates upward and is reprecipitated in the oxidized zone (Fig. 8.41C). Leaching of the sediments with acetic acid shows that more than 90% of the manganese in the upper 2 cm of the core may be diagenetic (Fig. 8.41D). As the distributions of the other metals analyzed are totally different, manganese is not likely to be important as a scavanger in Ranafjord. Apparently, the anthropogenic inputs govern their distribution rather than natural scavanging processes.

There is very little evidence of impact of metals on organisms in Ranafjord. This may indicate

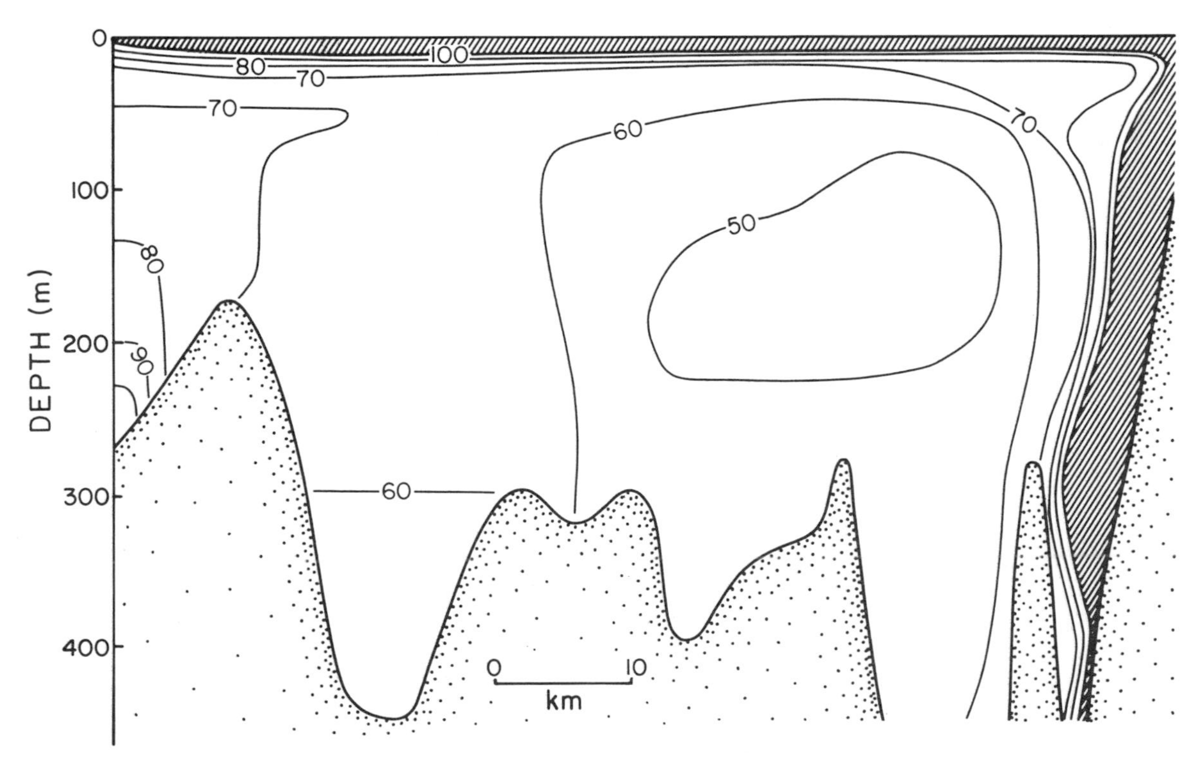

FIGURE 8.39 Distribution of monochromatic light scatterance (λ = 406 nm, relative units) in Ranafjord, April 1971. Hatched areas are for values that exceed 100 (after Aas, 1976).

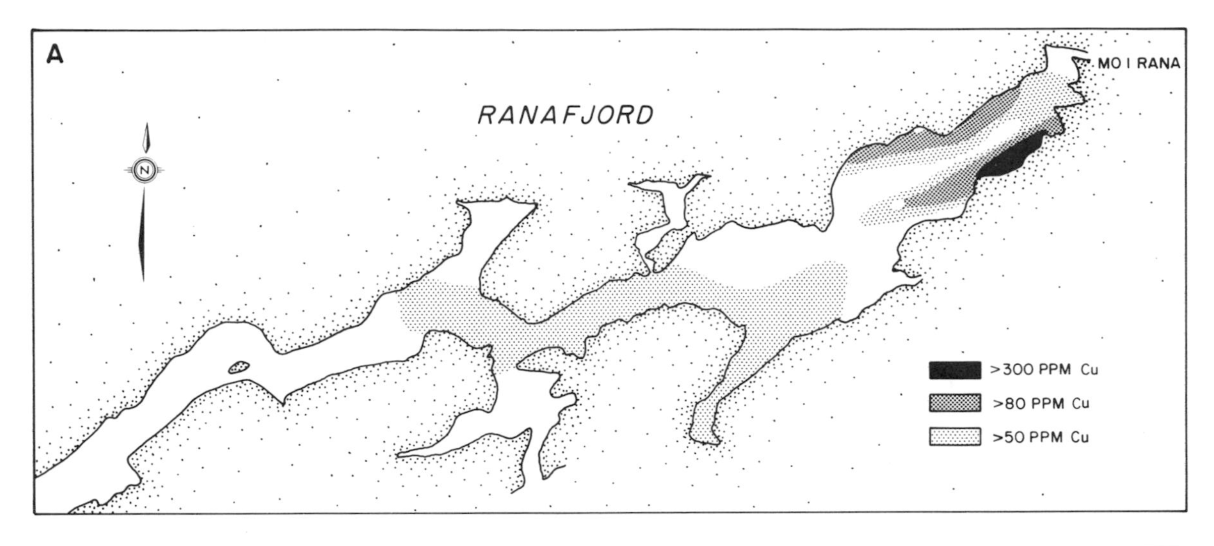

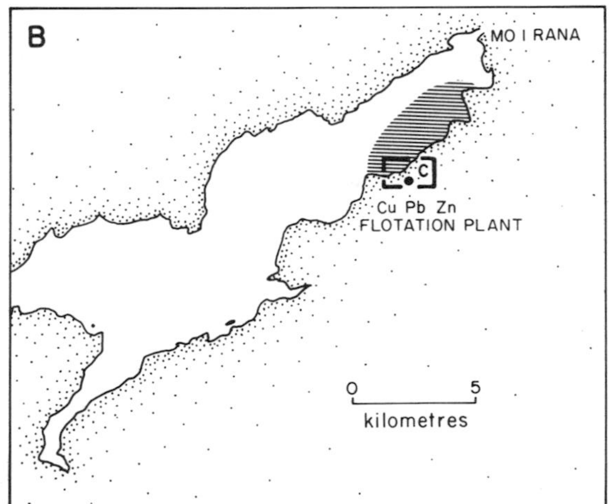

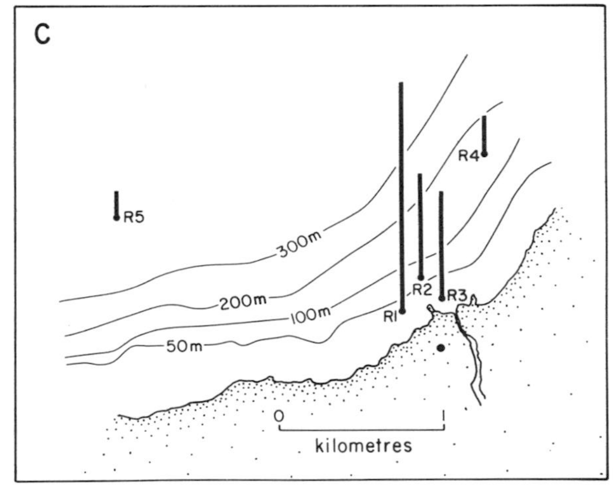

FIGURE 8.40. (A) The distribution of copper in the surface sediments; (B) general distribution of high metal surface sediments; and (C) dissolved lead in the upper waters, near the flotation mill in Ranafjord, Norway.

that particulate metals in the tailings are not generally bio-available (cf. Ellis, 1982).

8.12 Loch Eil, Scotland

8.12.1 Environmental Setting

Loch Eil (maximum depth 79 m) is a fjord at the head of Loch Linnhe on the west coast of Scotland (Fig. 8.42). Together they constitute the submarine western end of the Great Glen rift fault. Lochs Eil and Linnhe are divided at the Annat narrows by a 5 to 9 m deep sill, and Loch Linnhe is separated from the open water of the Firth of Lorne by a second (18 m deep) sill at the Corran Narrows. Bottom temperatures and salinities within Loch Eil are in the range 6.8 to 12.6°C and 25 to 30‰ (Pearson, 1970; Milne, 1972). Freshwater input is relatively high, and the water column is well stratified above the sill (Fig. 2.10). In spite of this, and the shallow sill barrier, low oxygen concentrations do not naturally occur in the bottom waters. Edwards et al. (1980) have observed that the density of the deep Loch Eil water lies in the range of the Loch Linnhe source water at sill height, from which it may be concluded that subsill water inside the Annat Narrows is regularly flushed (period of months or less). These authors show that, although the basin water may stagnate during neap-tide periods, renewal occurs at spring tides (Fig. 8.43). Pearson (1970) has recorded tidal

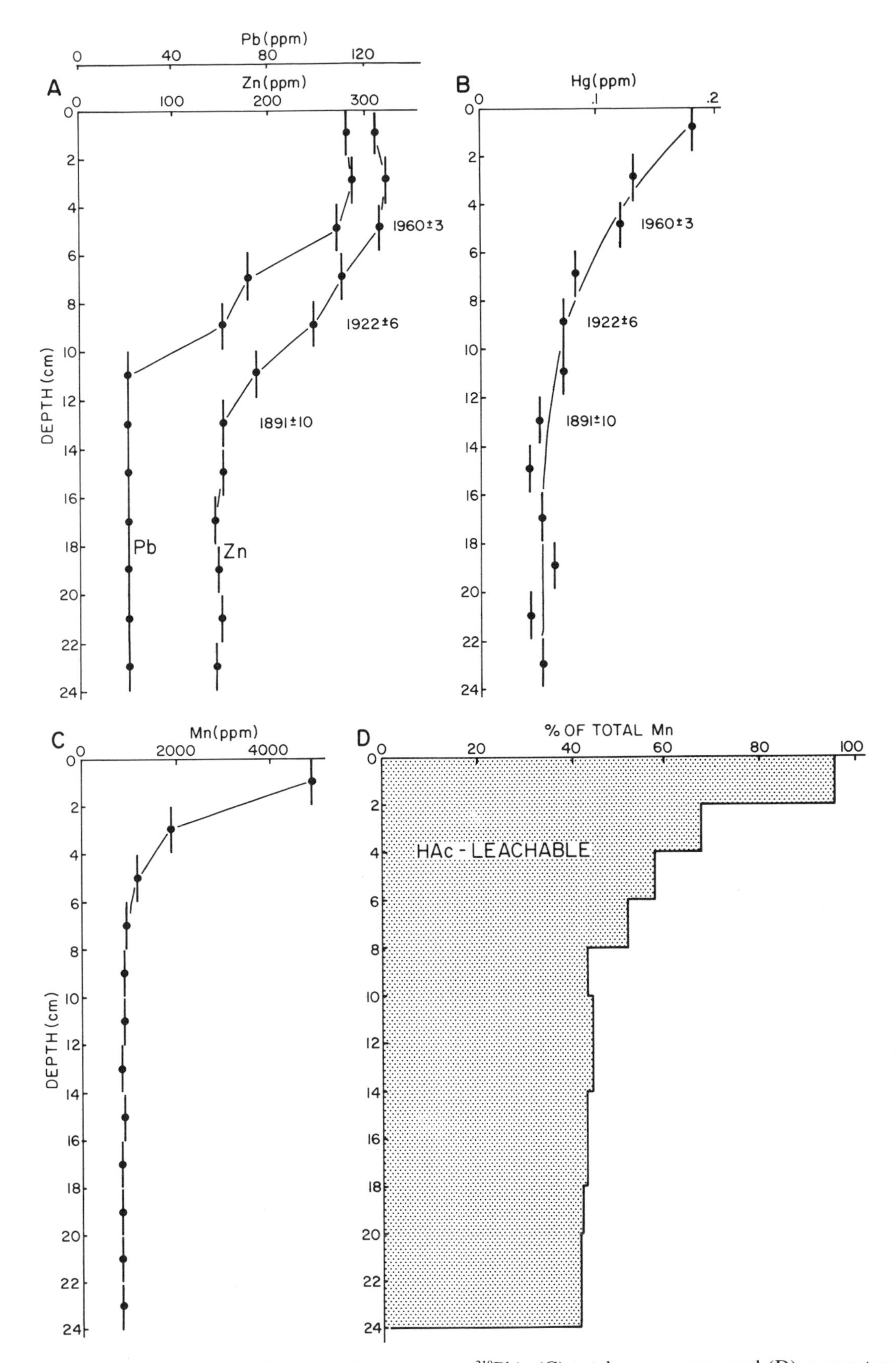

FIGURE 8.41. Vertical change down a sediment core collected from Ranafjord, in: (A) total lead and zinc; (B) total mercury (age determinations based on excess ^{210}Pb); (C) total manganese; and (D) percentage of HAc-leachable manganese (after Skei and Paus, 1979).

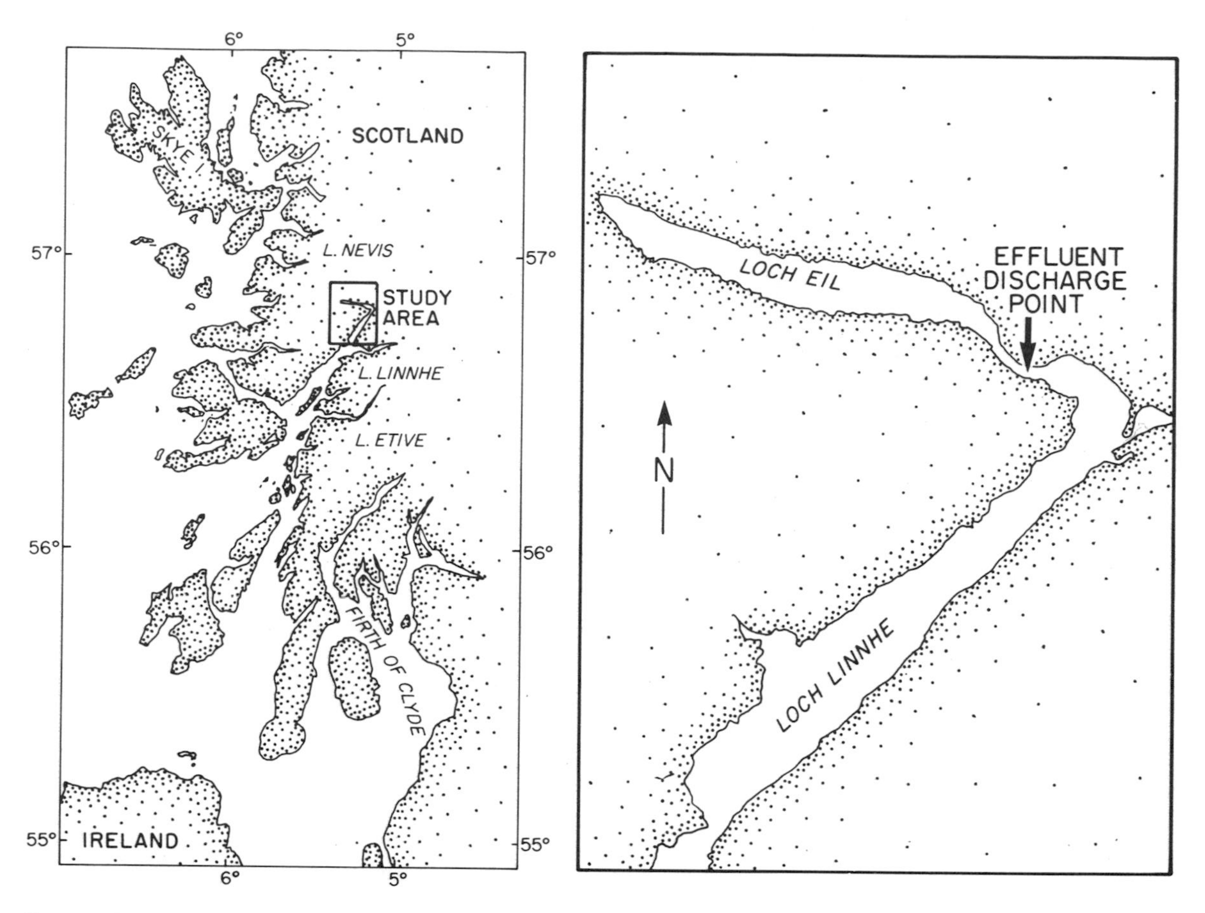

FIGURE 8.42. Lochs Eil and Linnhe on the west coast of Scotland. Pulp mill effluent was discharged at the Annat sill region, which separates the two sea lochs. Loch Linnhe is further partially separated from the open waters of the Firth of Lorne at the Corran Narrows.

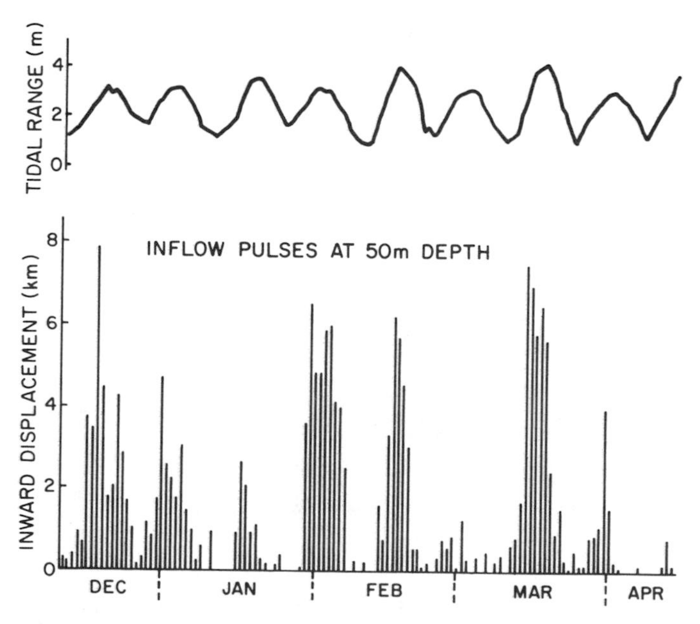

FIGURE 8.43. Tidal range showing principal neap-spring tide pattern December 1975–April 1976 in Loch Eil, and corresponding daily net up-fjord displacements measured at a locality adjacent to the Annat sill. From Edwards et al. (1980).

currents in excess of 250 cm s^{-1}, although velocities at the bottom of the basins are generally very low. As in most fjord areas, most of the riverine particulate sediment discharge follows, with only a short lag time, periods of maximum rainfall. The mean annual influx of terrigenous particulate carbon (POC) is relatively low, however (10 g m^{-2} yr^{-1} at the floor of the loch): around 30% of the autochthonous carbon is sedimented (Pearson, 1982). Natural POC contents of the sediments are less than 3% in the basins and around 1 to 2% in the vicinity of the narrows. Sedimentation rates were noted to be an order of magnitude higher at the time of spring tides (Stanley et al., 1981; see Fig. 4.20), and this is attributed to enhanced resuspension of bottom sediments during these periods. Pearson (1970) has recorded predominantly mud-sized material blanketing the basin floors, with coarser grained material in shallow water regions.

8.12.2 Environmental Problems

In 1966, effluent from a pulp and paper mill was first discharged in the Eil-Linnhe system in the vicinity of the Annat Narrows (Fig. 8.42). The industrial process used here (the Stora process) eliminates release of digestion and bleaching liquors (Pearson, 1980b), and waste discharge consisted primarily of nontoxic cellulose fiber. Waste input doubled in 1970–71 (but fluctuated monthly), and the subsequent mean flux of anthropogenic carbon to the sediments has been estimated at 245 g m^{-2} yr^{-1} (Pearson, 1982); or around 85% of the total carbon input. Mean carbon contents of the basin sediments initially increased to around 5%, but remained approximately constant thereafter. Rapid increases in populations of marine cellulytic bacteria accompanied the discharge of fibre wastes (Pearson, 1980b), and Vance et al. (1982) demonstrated that decomposition rates rapidly balanced input rates. Duff (1981) estimated the supply of nutrients refluxed from the sediments as a result of the greatly enhanced sedimentation of relatively labile carbon (Hofsten and Edberg, 1972) to be at least an order of magnitude greater than the level necessary to sustain phytoplankton production in the overlying euphotic zone. As discussed in Chapter 7, much of the natural terrestrial organic matter transported into temperate and boreal fjords is refractory and tends to be buried with the sediments.

Pearson (1971, 1975, 1982) and Pearson et al. (1982) have discussed in detail the response of soft-bottom infaunal communities to increasing temporal and spatial gradients of organic enrichment in Loch Eil. The increased flux of labile carbon caused the redox front to migrate upward (see Chapter 7); organisms were increasingly confined to the near-surface zone; and suspension feeders were replaced by detritivores. With increasing carbon load, initially there was an increase in the abundance and biomass of rapidly maturing species responding to the enhanced energy resource (Fig. 8.44), followed by a systematic decline in the density and number of all species. Rare species tended to be eliminated first, leading to diagnostic changes in community dominance (Pearson et al., 1983).

At about the time that anthropogenic carbon fluxes were increasing in Loch Eil, the reverse sequence was taking place in the Saltkällefjord on the west coast of Sweden. Organic pollution in this estuary was drastically reduced in 1966, and the subsequent pattern of changes to the macrobenthic community structure (Fig. 8.45: Rosenberg, 1972, 1976) was generally the reverse of that documented for Loch Eil (Pearson

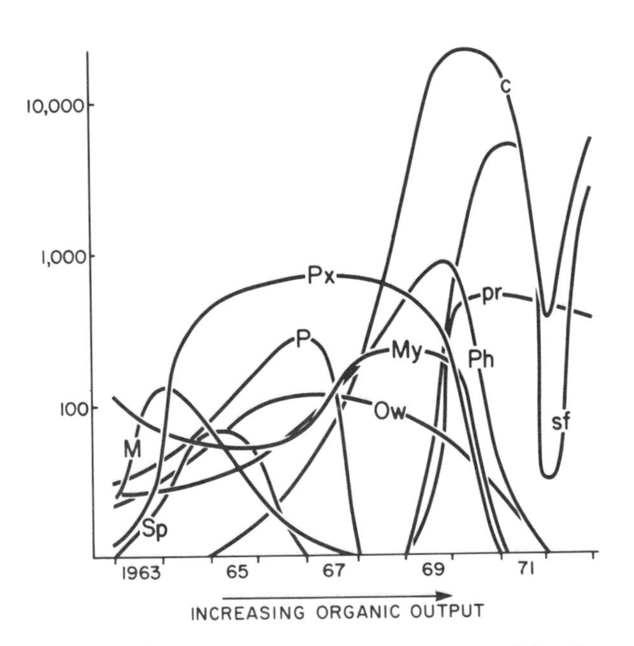

FIGURE 8.44. Species succession in an area of Loch Eil subject to increasing inputs of anthropogenic carbon. c, *Capitella capitata*; sf, *Scolelepis fuliginosa*; pr, *Protodorvillea kefersteini*; Ph, *Pholoe minuta*; My, *Myrtea spinifera*; Ow, *Owenia fusiformis*; P, *Prionospio cirrifera*; M, *Myriochele heeri*; Sp, *Spiophanes kroyeri*; Px, *Praxillura longissiima*. From Pearson and Rosenberg (1978).

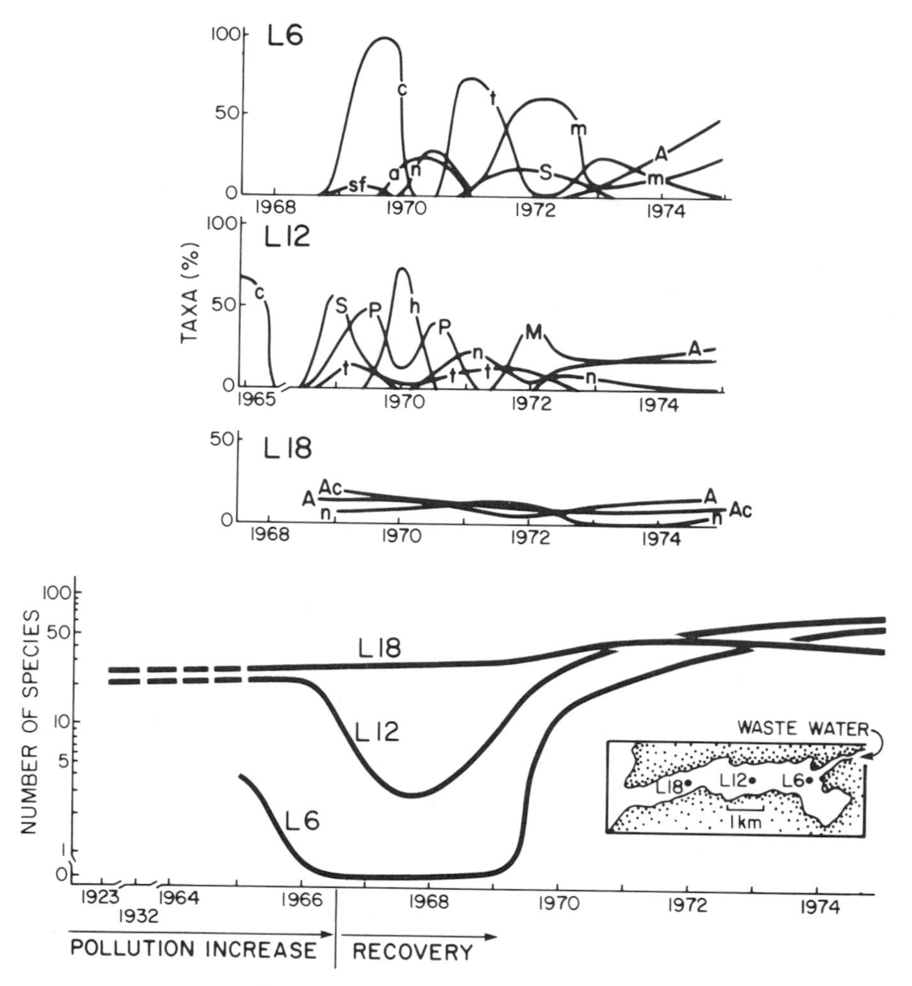

FIGURE 8.45. Recovery of the benthic macrofauna at three localities within the Saltkällefjord, western Sweden, following organic pollution abatement in 1966. After approximately 6 years, community recovery appeared to be complete at the innermost and most heavily polluted station. c, *Capitella capitata*; sf, *Scolelepis fuliginosa*; a, *Abra alba*; n, *A. nitida*; t, *Thyasira spp.*; S, *Scalibregma inflatum*; m, *Mysella bidentata*; A, *Amphiura filiformis*; h, *Heteromastus filiformis*; M, *Myriochele oculata*; P, *Polyphysia crassa*; Ac, *Amaphiura chiajei*. From Rosenberg (1976) and Pearson and Rosenberg (1978).

and Rosenberg, 1976). It took several years for the major infaunal taxa to become re-established in the innermost parts of the Saltkällefjord (Cato et al., 1980), and about 8 yr for complete community recovery. Pearson and Rosenberg (1978) have characterized the responses of soft-bottom, macrobenthic communities to organic enrichment gradients (see also Fig. 6.13) in terms of the generalized SAB (species-abundance-biomass) plots shown in Figure 8.46. Starting from a heavily enriched, defaunated condition, with decreasing carbon input, density and biomass curves reflect the initial peak of opportunistic species. (Pearson, 1981, has noted that the suite

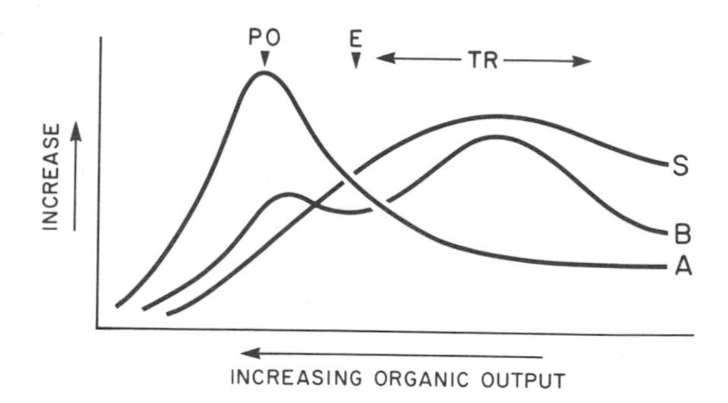

FIGURE 8.46. Generalized species-abundance-biomass (SAB) curves along a gradient of decreasing organic enrichment (partially based on data of Figs. 8.44 and 8.45). S–species numbers; A–total abundance; B–total

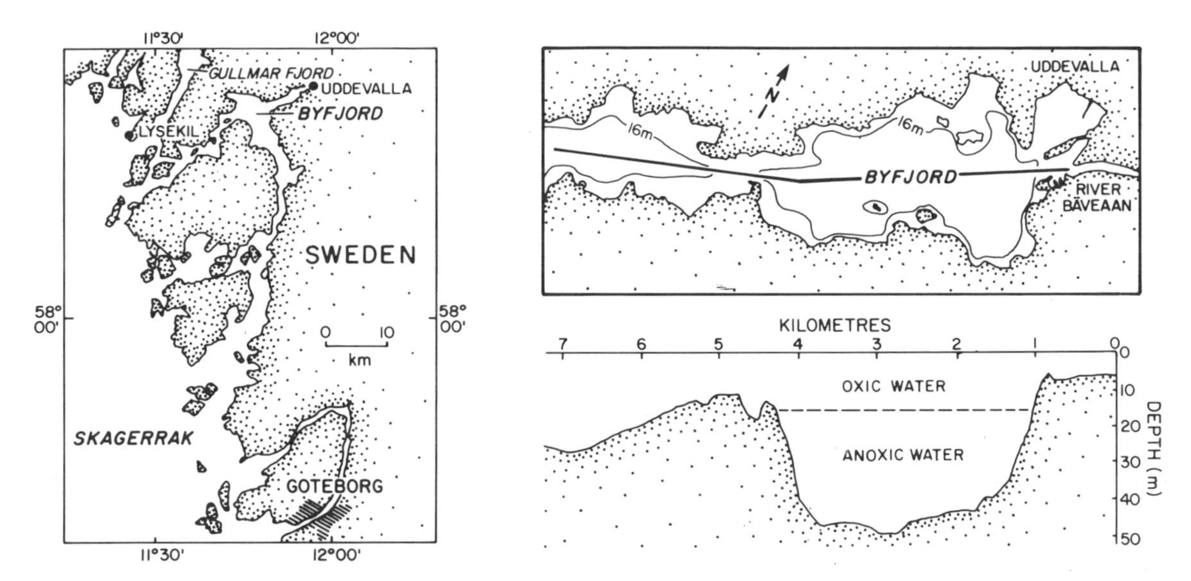

FIGURE 8.47. Location map of Byford, Sweden, also showing plan and longitudinal section of fjord (after Rosenberg et al., 1977).

of opportunists that may appear following severe defaunation is probably unpredictable, depending on chance circumstances.) Beyond this there is an "ecotone" or transition phase which, with time, may stabilize into a normal "equilibrium" community. These characteristics illustrate some of the expected responses of the infauna to pervasive, nontoxic stress discussed in Chapter 6.

biomass; PO–peak of opportunists; E–ecotone point; TR–transition zone. From Pearson and Rosenberg (1978).

8.13 Byfjord, Sweden

8.13.1 Environmental Setting

Byfjord (Fig. 8.47) is an almost nontidal estuary located on the west coast of Sweden. It is some 4 km long, with a maximum depth of approximately 50 m, and is separated from the adjacent open Havstensfjord coastal region by a 12 m deep sill. The volume of fresh water entering at the head is relatively slight, and appears not to markedly affect the hydrography (Gøransson and Svensson, 1975), but, because coastal waters in this region are comparatively brackish, surface salinities range between 22 and 30‰, and ice forms in the winter. The absence of strong tidal currents and the sill constriction prevent regular flushing of the basin. Bottom water is replaced only sporadically, at frequencies in excess of a year, and the basin is virtually permanently anoxic below about the 18-m depth.

8.13.2 Environmental Problems

Byfjord receives large quantities of sewage and industrial waste from the town of Uddevalla (population circa 50,000) situated at the head, and, in this respect, is typical of a number of such inlets around the northwest European coast. Danielsson et al. (1975) have estimated the riverine influx of organic carbon to be of the order of 5×10^5 kg yr^{-1}, but an approximately equal quantity is exported into Havstensfjord. From an estimated mean annual primary production of 175 gC m^{-2} yr^{-1}, Rosenberg et al. (1977) have constructed an energy flow model that provides for an annual flux of carbon to the sediments of approximately 3×10^5 kg; although, based on mean sedimentation rates, Olausson (1973) had previously advocated a value nearly an order of magnitude greater than this.

Approximately 45% of the fjord sediment surface area is in contact with oxygenated water for most of the year. Rosenberg (1977) has shown that the soft-bottom macrofauna consist predominantly of small and short-lived species, probably as a consequence of the seasonally variable salinity and dissolved oxygen regime. There is a very significant correlation between the biomass of both the macro- and meiofauna components and ambient dissolved oxygen concentrations. The redox front has been shown to migrate upward in the summer, although periods of anoxia of the order of days need not adversely affect the macro-infauna. Rosenberg (1977) notes that, in this fjord, polychaetes appear to be more tolerant of very low oxygen tensions than are mollusc species, and that the most successful polychaetes are not the same opportunistic species that may be dominant under carbon-enriched stress conditions (Pearson and Rosenberg, 1978; see also Section 8.12). As in other euryhaline environments, the meiofaunal component is relatively well developed compared with the macrofaunal (Rosenberg et al., 1977, give a meio:macro abundance ratio of 2000:1).

8.14 Summary of Impacts in Other Fjords

The most common anthropogenic impact in fjords is enhanced organic loading: wood and fish processing waste, sewage and oil (see Sections 8.3, 8.5, 8.7, and 8.8). This type of discharge is of particular concern because of the characteristically restricted circulation patterns within fjords, and hence the propensity for anoxia. Long in-fjord residence times are especially a feature of Scandinavian fjords where tidal flushing is relatively ineffective. In addition to the Swedish west coast estuaries described previously, pollution and the spread of anoxic conditions within Oslofjord has been recognized for at least 50 yr (Beyer, 1968). For instance, Mirza and Gray (1981) note the decrease of bottom-fauna diversity toward the innermost part of Oslofjord, a result of organic loading and periodic anoxic conditions. [The lowest diversity, however, is in the central (western) part of Oslofjord where the seafloor is affected by the outfall of a nearby cement factory.] Prior to 1900, there was little evidence of anoxia in the innermost part of Oslofjord, although low oxygen in the bottom water was recorded (Hjort and Gran, 1900). By 1915, periods of anoxic conditions began to occur (Petersen, 1915), although relatively permanent anoxic conditions at the sediment-water interface had developed in local embayments as a result of an increased loading of sewage from the city of Oslo. The innermost basin (Bunnefjord) now has a 15-cm surface layer of black, H_2S-smelling mud, situated above a light-colored silty clay: indicative of the onset of anoxic conditions some 50 years ago (Skei and Melsom, 1982).

Wood and fish waste detritus is perhaps the major source of organic pollution in boreal regions where the human population density is low. Such natural products are basically nontoxic. Problems are primarily caused by the BOD demand: a cultural eutrophication effect. In this regard it should be noted that the bulk of these organic wastes, including pulped cellulose fiber, are relatively labile. However, wood processing usually encompasses pulp and paper-making industries, which generate additional adverse impacts and toxic byproducts (Pearson, 1980b). The two most common pulping techniques for producing wood fibre are the Kraft and sulfite processes. The former, in particular, generates toxic effluents, and also discoloration of the receiving waters which adversely affects phytoplankton production (see Chapter 6).

In more temperate environments, anthropogenic organic loading in fjords may be predominantly from urban and industrial waste. One of the major documented estuarine discharges—drainage of nutrients from agricultural land—generally has an insignificant impact on fjords at present, although there are notable exceptions. Beyer (1981) has shown that anoxic conditions in the shallow Mariager Fjord on the east coast of Jutland (Denmark) is primarily attributable to agricultural runoff. With increased demand for agricultural development in more northern climates (in Alaska, for example), this source of eutrophication is expected to increase.

It has long been the practice to discharge untreated sewage into coastal waters (Topping, 1976), especially in the more sparsely populated parts of the world. Sewage per se is a relatively benign addition and, so long as dispersion and dilution within the recipient body of water is sufficient, the cost of secondary treatment may not be justified (e.g., Officer and Ryther, 1977). Even in patently unsuitable localities, such as in a number of the poorly flushed, partially enclosed fjords and inlets along the west coast of Sweden, massive discharge of raw urban waste was a common occurrence until recent times (Cato, 1977; Cato et al., 1980). The effects of gradients of sewage sludge on the resident benthos (Tulhki, 1968; Olsson et al., 1973) are essentially similar to those induced by organic enrichments from lumber processing waste disposal discussed above.

Urban point-source discharge generally integrates a variety of industrial and household wastes which, together with storm drainage, are likely to introduce a wide range of toxic substances into the marine environment. The influx of potentially harmful substances may be even greater from nonpoint-source urban surface runoff. Elevated concentrations of heavy metals are of particular interest because of potential complexation reactions with the coexisting organic material. This has been noted previously in the case of mercury and wood processing wastes within the Saguenay Fjord. The interaction of sewage and other organic wastes with industrial heavy metals has also been described by Cato et al. (1980). In one respect, compared with most other estuarine environments, fjords are superior receptacles for solid wastes (including scavenged heavy metals). Fjord sedimentation rates tend to be relatively high, and resuspension in basins is likely to be slight, so that, on cessation of pollution, the contaminated sediments may be rapidly buried and isolated from the active interfacial zone. However, even here, bioturbation and episodic physical mixing may delay this process.

Urban runoff also discharges large amounts of hydrocarbons into adjacent estuarine and coastal waters (NAS, 1975), and, to date, this has undoubtedly been a more important source than oil industry activities. However, because of increasing exploration and production in more northerly waters, it is to be expected that contamination of coastal waters and fjords from oil spills will increase in the future. Low molecular weight aromatics are some of the more toxic constituents of crude oil (Johnston, 1984). Most of this fraction is normally lost via evaporation, but, at low temperatures, such as would be commonplace in boreal and arctic fjords, the residence times of the more acutely toxic constituents would be extended (Elmgren et al., 1980).

Goldberg (1976) has emphasized that the popular conception that petroleum in the marine environment is rapidly or entirely degraded is fallacious. Many factors appear to limit the rate of biodegradation, and the reaction will be significantly retarded in cold water. In Port Valdez, Alaska (Section 8.4), it was noted that some 65% of the ballast waste was degraded. The water within this particular fjord is flushed relatively rapidly, but, in general, dispersion, and hence the assimilative capacity for pollutants (Goldberg, 1979), is poor in fjords compared with most other types of estuaries. Slow dispersion of spilled oil was also observed in Swedish coastal waters (Elmgren et al., 1980), and extended residence times would undoubtedly pose problems in a number of fjords potentially susceptible to oil pollution (e.g., in Sullom Voe, Shetland Islands: Pearson and Eleftheriou, 1981).

Several specific examples of the disposal of mine and rock waste into fjords have been given above, and it seems inevitable that this type of impact will greatly increase in the future. For example, it is proposed to dispose of the mill waste from a very large molybdenum mining operation (some 10% of the known world's reserves) into one or more fjords in southeast Alaska (Burrell, 1982b). This type of solid waste appears to be largely nontoxic, but extensive smothering of the resident benthos occurs, and there is a need to determine rates and processes of recolonization once waste dumping is terminated. Mine waste was discharged into Alice Arm, a subsidiary fjord at the head of Observatory Inlet in northern British Columbia over two time periods: between 1967 and 1972 and, following a 9 year hiatus, for an 18 month period between April 1981 and November 1982. Two years after termination of the initial impact, Ellis (1977) reported that infauna had been rehabilitated on the tailings within the fjord basin. Similarly, Kathman et al. (1983, 1984) surveyed the basin benthic community structure during, and some 11 months following cessation of the second discharge period, by which time infaunal densities within a region of maximum impact had increased fourfold.

Discharge of drill cuttings and fluids may potentially elevate the concentrations of petroleum hydrocarbons and several toxic heavy metals in the surrounding sediments, but Cook Inlet, Alaska, is the only fjord that has been the direct recipient of such material to date. Elevated concentrations of barium have been identified immediately adjacent to one point-source discharge in this inlet, but salmon fry in live-box experiments were shown to be unaffected (Lees and Houghton, 1980). However, Cook Inlet has 4- to 5-m tides, and tidal currents in certain areas in excess of 100 cm s^{-1}, and so is a very atypical fjord environment.

Smelters (aluminum, ferromanganese, zinc, copper, etc.) are important sources of metals and polycylic aromatic hydrocarbons (PAHs) in fjords (see Sections 8.9 and 8.10). Their location at the coastline is often a result of the demand for electricity from nearby power plants and shipping access. The smelters create two problems: (1) emission of pollutants via smoke stacks to the atmosphere; and (2) waste water discharge directly to the fjord. Fjords surrounded by high mountains have conditions of little air exchange and the aeolian pollutants tend to fall close to the source. Investigations of the sediments of Puget Sound (Washington) linked the elevated heavy metal concentrations to the discharges from the Tacoma copper smelter (Crecèlius et al., 1975). The elevated concentrations of arsenic in the sediments are, to a large extent, related to wind-borne particles. Atmospherically transported PAHs have also been considered as a source for many fjord sediments (Sporstöl et al., 1983). In the case of Kitimat Arm (British Columbia) sediments, however, PAH contamination is associated with discharges from a nearby aluminum smelter and atmospheric transport only plays a secondary role (Cretney et al., 1983).

The chlor-alkali industry has, with earlier discharges from the pulp and paper industry, been responsible for the largest mercury threat to fjords (see Section 8.7). Similar to waste generated by smelters, the pollutant is introduced into both the air and the water. A chlor-alkali plant in Bellingham Bay, Puget Sound, caused elevated mercury concentrations in surface sediments within 7 km of the plant (Crecelius et al., 1975). In Frierfjord, SE Norway, a chlor-alkali plant contaminated the sediments within 20 km from the source. A semi-enclosed bay close to the plant contained sediments with a mercury content varying between 90 and 350 ppm (Skei, 1978). The upper 5 cm of the sediments contained approximately 10 tons of mercury within an area of 1 km^2 as a result of 25 years of mercury discharge (Skei, 1978).

Finally, because they are typically located in mountainous terrain, fjords are particularly susceptible to manipulation for hydroelectric power generation. This may involve either augmentation or depletion of the natural freshwater discharge, which, in turn, is reflected in changes in the seasonal hydrography and water circulation patterns within the fjord (Kaartvedt, 1984). Edwards and Edelsten (1976) have shown how regulation of the freshwater input into Loch Etive on the west coast of Scotland has decreased the frequency of deep water renewal. In Norway, the most obvious physical effects of regulation include changes in salinity, modified current systems, and changes in temperature caused by altered mixing depth (Kaartvedt, 1984). A practical problem induced by the latter is the formation of ice during winter (a result of increased freshwater flow) and winter cooling restricted to the surface layer. Fjords in western Norway, influenced by regulated discharge, have displayed temperatures 5°C above normal in the marine water below the brackish layer (Kaartvedt, 1984). Further, a better developed winter stratification may prevent nutrients from being mixed into the euphotic zone, influencing the production conditions. Regulation will also influence sediment transport, which may have both a positive and a negative effect. With the elimination of peak discharges, the light conditions in a fjord will improve as a result of lowered particle transport: primary production may increase. The associated reduced rate of sedimentation, which will occur as a consequence of lowered fluvial sediment transport, implies that in polluted fjords the rehabilitation time will be extended, as it takes longer to cover up the polluted sediment with natural sediment. Biological consequences to river regulation include: (1) less freshwater discharge in summer will reduce the ecological stress on organisms (Kaartvedt, 1984); (2) the associated decreased rate of sedimentation will reduce the ecological stress on benthic organisms (see Farrow et al., 1983); (3) the exchange rates of plankton between the coastal waters and fjords will alter as the circulation pattern is influenced; and (4) the change in sea-water temperature may influence the spawning of fish and the metabolism in other organisms.

9

Future Fjord Research

Most of the concepts and data described within this volume have survived years of scrutiny and provide the necessary foundation for scientific advancement in both the fields of theory and application. Fjord science is relatively new, however, and some topics presented remain hotly contested, awaiting further testing in new localities or by new approaches. Fjord oceanographers must now undertake a critical evaluation and systematic synthesis of processes from the ever-increasing descriptive data. We need to know the net effect or end product of a particular process, or set of processes, in a fjord system. We have identified the principal variables and end-member conditions in fjords; now we must increase our ability to predict these products, both in time and space, especially where man tampers with natural conditions. On a global basis, we must understand the role of fjord processes as an interface and buffer between land and sea. To elucidate many of the concepts outlined within this book, a universal model must be developed to intercompare fjords and fjord systems, such as through mass-balance and residence time calculations, and the cycling of inputs and outputs of sedimentary material.

Provided below are topics that we believe should be addressed individually or as part of large, integrated experiments. We list them by discipline groupings, but realize that the solution of one problem may impact on the advancement in other fields of fjord research and that more than one discipline will search for its solution using different techniques. We also realize that to solve some of these problems, advancements in laboratory or field instrumentation and methodology must be forthcoming.

9.1 Oceanographic Problems and Projects

1. The development of a generalized circulation model for tidewater-glacier-dominated fjords (see Section 4.5.4). In the past, danger and cost have prohibited both the long-term

measurements of the vertical component of current velocities at the ice front and a means for estimating the kinetics associated with iceberg calving and subsequent melting.

2. The effects of wind mixing on the overall circulation of a fjord are enormous, although our knowledge base is built on relatively few experiments carried out over a relatively narrow range of surface layer density stratification. We have almost no information on the effects of very strong cross-fjord winds (i.e., from tributary valleys) on general circulation. We do not reasonably understand the effect of the timing of these events in relation to sea-ice formation or breakup.
3. Models of fjord circulation need a more adequate representation of the scales and intensities of turbulent flows that may account for the effects of relatively brief periods of intense mixing events (Farmer and Freeland, 1983). Further, these dynamic (stochastic) circulation models must begin to account for the time-dependent nature of real flows, especially in regard to rapid changes in wind intensity, the pulsing nature of river discharge (order of magnitude changes over a few days) and intermittent renewal events. Such models are necessary if we are truly to represent the pathways of pollutants discharged into fjords.
4. There has been an excellent series of experiments conducted on the dynamics of mixing and turbulence associated with sills (see Section 4.4). However, sill morphology is highly variable and complex, and experiments (both lab and field) must be undertaken to include a more realistic spectrum of sill morphologies and associated bottom roughness elements. We may then begin to appreciate the natural range in the contribution of sill energetics to deep water turbulence of the inner basin(s). Studies could focus on the points of flow separation, on whether lee waves are formed or a hydraulic jump takes place, and whether a train of internal waves or an internal bore develops. The bending and breaking of internal waves along sinuous fjord walls should be examined. These experiments would also provide useful information on gravity flow dynamics, which in turn could help in our modeling of sediment gravity flows (e.g., turbidity currents).
5. We are just beginning to appreciate the effect of brine rejection, associated with sea-ice formation, on deep water renewal events and on its contribution to deep water turbulence (see Section 4.5.1). Future field experiments should examine the destabilizing effect of brine rejection in regions where sea ice forms while estuarine circulation is still well established: possible sites include Lake Melville, Labrador, or the Saguenay Fjord, Quebec.
6. We have identified a variety of shear flow instabilities near the river mouth (see Section 4.1.2), that is, vorticity shedding along the plume margins, breaking internal waves, bottom friction along side walls, and isobathic-opposing tongues of up-fjord/down-fjord currents. We now need to incorporate the dynamics associated with frontal mixing in time-dependent three-dimensional circulation models.
7. Specifics on the redistribution of energy in deep fjord basins remain to be detailed. For instance, we know little about the contribution of double-diffusive instabilities (see Section 4.6.1) and intermittent sediment gravity flows (see Section 5.3) to deep water turbulence.
8. Farmer and Freeland (1983) note that new observational approaches are needed, especially in the area of remote sensing application. For instance, high frequency echo sounding and range-gated acoustic Doppler velocity sensing have recently been used to monitor small- to medium-scale processes associated with tidal mixing and advection, sediment gravity flows, and submarine springs: the observation of hyperpycnal flows associated with through-ice glacier discharges or turbid river plume events seems inevitable. Farmer and Freeland (1983) further note that a new generation of microstructure instruments for the detection and measurement of turbulence is needed, as well as a new class of chemical tracers (such as exotic flurocarbons).

9.2 Biogeochemical Problems and Projects

1. Past research and experiments have examined the separate components involved in the cycling of nutrients in fjords (see Chap-

ters 6 and 7). We now need to develop a more comprehensive nutrient cycling model that accounts for seasonal changes in both coupled and decoupled euphotic and benthic reactions.

2. Better documentation of the biogeochemical reactions involving particulate organic carbon during settling through the water column. Similar experiments should be conducted in a spectrum of deep fjords with different vertical diffusive and thermal energy regimes. These results would allow for a more realistic appraisal of the effects of the ever-increasing disposal of anthropogenic waste.
3. Fjords that have recently begun the process of rapid deglaciation offer scientists the exciting possibility of documenting the chemical evolution of a marine basin dominated by siliclastic sediment. Possible field sites include Muir Inlet, Alaska (deglaciation began 100 years ago), and near the Columbia Glacier, Alaska (delgaciation began within the last few years).
4. The modelling of diagenetic reactions associated with non-bioturbated sediments and a seasonally stable ambient thermal environment but under a seasonally forced cycle. Such models would require more information on the diagenesis of natural organic compounds in both oxic and anoxic environments.
5. A better understanding of the kinetics of boundary reactions at: (a) the redoxcline within the sediments; (b) the water-sediment interface; (c) the redoxcline between water masses; and (d) the interface between fresh water and marine water. Of special importance are the mineral-solution transitions, the associated bioclines (e.g., rapid changes with depth of a particular type of bacteria), and the effect of boundary mobility in the establishment of new steady-state conditions.
6. What is the input effect of a seasonally variable fluvial dissolved load, including humic and fulvic acids, on mixing-zone biogeochemical reactions near the river mouth and under highly stratified conditions?
7. What is the effect of seafloor disturbances (i.e., slides and other mass flow processes) on the geochemical environment through the exposure of older marine sediments or remoulded but more recent sediments? Of interest is the rate of re-establishment of steady-state conditions.
8. More detailed measurements of chemical fluxes from the sediment into near-stagnant bottom waters under both oxic and anoxic conditions. Of particular interest is the effect of density convection through the sediment-water interface just after a deep water exchange. This is important information for mass balance and residence time calculations of chemical species in fjord waters (see Sections 7.2.3 and 7.3.3).
9. Further observations on the relationship between sediment trap material and seafloor sediments, in an attempt to clarify early diagenetic reactions of inorganic and organic substances and minerals.
10. The use of state-of-the-art technology to quantify the geochemical environments, including: (a) in-situ (manned submersible or remote probes) pore-water extractions and ion-sensitive probe profiling; (b) multibarrel corers for spatial variability resolution experiments; (c) streamlined and programmable, isopycnal micro-water samplers; and (d) in-situ reaction chambers.

9.3 Biological Problems and Projects

1. Ecological studies of fjord epilithics in sill regions, on rocky exposures of fjord wall slopes, and on ice-rafted or otherwise transported bouldery debris isolated on the muddy basin floors (see Section 6.2.2). The role of turbulence as a limiting factor is of particular interest in sill regions (see Section 4.6.1).
2. Further field and lab experiments on the stress tolerance of different organisms (see Section 6.2.4), including: (a) adaptations to reduced oxygen; (b) adaptations to high particulate loads and or high accumulation rates; (c) the effect of short-lived high-energy events; (d) the effect of sea-ice conditions; and (e) anthropogenic discharges. Re-colonization patterns and community structures should be emphasized. Similarly, we need more data on the influence of fresh water, suspended particles, light, and temperature as stress factors on the pelagic ecosystem.

3. The degree of coupling of primary production-pelagic phytoplankton to benthic production, with special emphasis on feeding habits. We are only beginning to evaluate the seasonal variations in the type and vertical flux of organic detritus. We therefore need to know what causal relationship exists between the vertical flux of organic and inorganic particulate matter and benthos composition.
4. In what form and under what influences does the transport of organisms from the coast and shelf waters into fjords occur? Here we are interested in the recruitment of organisms along a gradient.
5. Longer term (i.e., 10-yr) observations of the benthic community structure, including in-situ camera or submersible experiments, for the development of actualistic ecosystem models. Such observational experiments are ideally suited for studies on the long-term detrimental effects of polluted sediments on the bottom fauna, or the successional evolution of a recently deglaciated fjord.
6. The documentation of the changing energy supply to subeuphotic benthos—with depth and season. Another similar project might be the effect of foreign (distant-generated) turbulence, such as that associated with sills, on the supply of nutrients to depleted euphotic zones.
7. Observations on the vertical zonation and niche structure in soft-bottom sediments (macro-, meio-, microbenthos associations). Of special interest are the early successional changes associated with rapid deglaciation and the adaptation to new seafloor environments associated with slides and mass flow deposits (e.g., overconsolidated substrates).

9.4 Geological-Related Problems and Projects

1. The effects of weathering of fresh glacial sediments (subaerial versus subaqueous), including the process of burial diagenesis. In particular, is there a style of clay formation unique to fjord environments (i.e., highly charged primary minerals, short marine water column residence time, extremes in accumulation rates)?
2. On the processes of flocculation and agglomeration, we need more information on the in-situ settling velocity and structural changes in the fabric of suspended particles with depth, and the interaction of mineral floccules with planktonic grazing. Also of interest is the effect of fine structure with associated pycnoclinal surfaces and shear zones on the characteristics of suspended particulate matter.
3. The development of deterministic two-layer estuarine models to predict the concentration, vertical flux and grain size distribution of suspended particulate matter in fjords based on sediment influx as a function of meteorological parameters (temperature, snowfall, rainfall, glacier melt discharges, etc.).
4. How does the distribution of permafrost affect the development of arctic sandurs (see Section 3.4.1)? In particular, what is the level of permafrost-controlled freshwater discharge into fjords via submarine springs (see Section 4.5.7)?
5. On the dynamics of delta progradation, especially over the foreset beds, we need: (a) more detailed current and SPM measurements near the seafloor to detect short-lived underflows, grain flows, or turbidity currents; (b) experiments to determine the effect of internal waves breaking on the prodelta and possibly generating sediment failure; and (c) experiments to simulate earthquake accelerations and large-scale delta front failure.
6. There is a great need for high-resolution seismic surveys tied in with well-positioned long cores for more rigorous bio-, seismo-, and lithostratigraphic facies models of: (a) glaciomarine deposits; (b) modern deposits associated with delta progradation, sidewall slope, and sill environments; and (c) coupled shelf-fjord sedimentation events. These models must be related to the subaerial Quaternary history and must be time-progressing to further our understanding of the evolution and maturing of young estuaries.
7. The development of semiquantitative models that can relate fjord delta and basin processes to long-term climatic variations (e.g., discharge dynamics) and the effects asso-

ciated with relative sea level fluctuations (e.g., fluvial cannibalism, shoreline tilting, changes in tidal or wave conditions).

8. What is the rate of sediment delivery to the fjord basin by: (a) aeolian transport; (b) sea-ice rafting; (c) iceberg rafting; and (d) mountain slope colluvium?
9. What is the relative importance of the various styles of slope failure (Table 5.2) in the sedimentary record? More quantitative information should be obtained on the dynamics of slope failure release mechanisms and transport processes through: (a) realistic back-calculations; and (b) real-time measurements of these processes.
10. We need more information on: (a) the mass-sediment properties associated with rapid sedimentation; (b) resuspension characteristics of cohesive organic-rich versus organic-poor sediments; and (c) the real-time variations in excess pore pressure.
11. On ice-proximal sedimentation processes, we need more information on: (a) basal shear stresses of tidewater glaciers and the development of push moraines; (b) ice-tunnel discharge processes; (c) the effects of "tagsaq" (iceberg-generated) waves on the seafloor and shoreline; (d) jökolhlaup dynamics and deposits; and (e) variations in the when and where of ice-front advances and retreats.
12. What is the effect of deep water renewals on the transport of suspended particulate matter in and out of fjords (see Section 4.4)?

9.5 Approaches

In the past, the most profitable approach has been the small, single-objective study often undertaken by the graduate or post-doctoral research student. Historically, estuaries of all kinds have been training grounds for developing oceanographic students. These projects have taken advantage of suitable fjords close to population centers and accessible by smaller research vessels. Such university-led projects should continue to emphasize seasonally synoptic sampling over a number of years and through a number of students. Anchorable research platforms (e.g., barges) could augment ship-based surveys, and possibly a number of universities could join forces to administer the student exchange and platform logistics.

With faster paced integrated programs, there has been less involvement with supposedly independent student topics, especially if the "original idea" concept is of concern. These larger, interdisciplinary programs have recently become very successful; a number are mentioned in Chapter 1. Many of these projects have succeeded in merging both logistics and the expertise of university scientists (and occasionally students) with those from government and industry. Such an approach is needed where hostile and generally inaccessible fjords are being studied and where there is little cost difference between a simple survey and a more complex temporally focused investigation.

Integrated projects are also warranted where contemporaneous sampling is prerequisite. In the Sedimentology of Arctic Fjords Experiment (Syvitski and Schafer, 1985), a mother ship with ice-breaking capabilities was used in conjunction with two well-instrumented launches, two smaller Boston whalers, and a helicopter, in addition to land-based parties. For instance, a large seiche event was detected by the mother ship through CTD measurements. The mother ship was able to radio the oceanographic-equipped launch, which then began high-frequency acoustic monitoring and CTD profiling to record the onset of an internal bore breaking on a prodelta slope. Another approach was the deployment by helicopter of current meter-thermistor chain moorings just after ice breakup in the fjord, at a time that would prevent ship-based operations. The moorings were retrieved later by ship, during the more open water season.

Future approaches will increasingly use moored instrument packages; many of these will be programmable "event-triggered" instruments. For example, during a deep water renewal event, a programmed microprocessor would increase the sampling rate after some current velocity threshold had been exceeded. Similarly, moored attenuance or acoustic meter arrays may be set-up to begin recording at the passing of a turbidity current. If these packages were linked to communication buoys, they could conceivably communicate via satellite to base stations. This near state-of-the-art technology may allow the short-notice diversion of a research ship to augment the sampling by the moored instruments.

In all approaches, field observations should be designed to provide the needed quantitative input to mathematical models and scaled-down laboratory simulation experiments. The models must be then used for prediction and these, in turn, compared with new observations and the model subsequently refined. This second data-gathering stage has historically been a weak link in past models and must be undertaken in a more systematic fashion.

In summary, fjords are of scientific interest for their environmental uniqueness and their contribution to the earth and oceanographic knowledge base. Future advances must be understood and applied by scientists who monitor the impact of humanity on its environment.

References

Aarseth, I. 1980. Glaciomarine sedimentation in a fjord environment: Example from Sognefjord. In: O. Orheim (ed.) Glaciation and Deglaciation in Central Norway. Field guide to excursion, Symposium on processes of glacier erosion and sedimentation, Geilo, Norway, 1980. Norsk Polar Instit., Oslo 1985.

Aarseth, I., Bjerkli, K., Björklund, R., Böe, D., Holm, J.P., Lorentzen-Styr, T.J., Myhre, L.A., Ugland, E.S., and Thiede, J. 1975. Late Quaternary sediments from Korsfjorden, west Norway. Sarsia 58: 43–66.

Aarthun, K.E. 1961. The natural history of the Hardangerfjord. 2. Submarine daylight in a glacier-fed Norwegian fjord. Sarsia 1: 7–20.

Aas, E. 1971. The natural history of the Hardangerfjord. 9. Irradiance in Hardangerfjord 1967. Sarsia 46: 59–78.

———. 1976. The influence of freshwater on light conditions in fjords. pp. 129–138. In: S. Skreslet, R. Leinebö, J.B.L. Mathews, and E. Sakshaug (eds.) Freshwater on the Sea. Publ. by Assoc. Norweg. Oceanogr., Oslo.

Adams, C.J. 1981. Uplift rates and thermal structure in the Alpine Fault zone and Alpine Schists, southern Alps, New Zealand. Thrust and Nappe Tectonics. The Geological Society of London. No. 9.

Adams, J. 1980. Contemporary uplift and erosion of the Southern Alps, New Zealand. Geol. Soc. of Am. Bull., Part II, 91: 1–114.

Ahlmann, H.W. 1941. The main morphological features of North-East Greenland. Geogr. Ann. 23: 148–183.

Aitken, A.E., and Gilbert, R. 1981. Biophysical processes on intertidal flats at Pangnirtung fiord, Baffin Island, N.W.T. Final research report, Dept. of Geol., Queens University, Kingston, Ont., 92 p.

Albright, L.J. 1983. Influence of river-ocean plumes on bacterio plankton production of the Strait of Georgia, British Columbia. Mar. Ecol. Prog. Ser. 12: 107–113.

Alexander, V., and Horner, R. 1976. Primary productivity of sea-ice algae. pp. 289–304. In: D.W. Hood, and D.C. Burrell (eds.) Assessment of the Arctic Marine Environment. Occas. Publ. No. 4, Inst. Mar. Sci., Univ. Alaska, Fairbanks.

Allen, C.R., O'Brien, M.G., and Sheppard, S.M.F. 1976. The chemical and isotopic characteristics of some northeast Greenland surface and pingo waters. Arctic Alpine Res. 8: 297–317.

Allen, G.P., Castaing, P., and Klingebiel, A. 1971. Preliminary investigations of the surficial sediments in the Cap-Breton Canyon (southwest France) and the surrounding continental shelf. Mar. Geol. 10: M27–M32.

Aller, R.C. 1978. Experimental studies of changes produced by deposit feeders on pore water, sediment, and overlying water chemistry. Am. J. Sci. 278: 1185–1234.

———. 1980a. Quantifying solute distributions in the bioturbated zone of marine sediments by defining an average microenvironment. Geochim. Cosmochim. Acta 44: 1955–1965.

———. 1980b. Relationships of tube-dwelling benthos with sediments and overlying water chemistry. pp. 285–308. In: K.R. Tenore, and B.C. Coull (eds.) Marine Benthic Dynamics. Univ. South Carolina Press, Columbia.

———. 1980c. Diagenetic processes near the sediment-water interface of Long Island Sound. I. Decomposition and nutrient element geochemistry (S, N, P). pp. 237–350. In: B. Saltzman (ed.) Estuarine Physics and Chemistry: Studies in Geophysics, Vol. 22. Academic Press, New York.

———. 1980d. Diagenetic processes near the sediment-water interface of Long Island Sound. II. Fe and Mn. pp. 351–415. In: B. Saltzman (ed.) Estuarine Physics and Chemistry: Studies in Long Island Sound. Advances in Geophysics, Vol. 22. Academic Press, New York.

———. 1982. The effects of macrobenthos on chemical properties of marine sediment and overlying water. pp. 53–101. In: P.L. McCall, and M.J.S. Tevesz (eds.) Animal-Sediment Relations. Plenum Press, New York.

Aller, R.C., and Benninger, L.K. 1981. Spatial and temporal patterns of dissolved ammonium, manganese, and silica fluxes from bottom sediments of Long Island Sound, U.S.A. J. Mar. Res. 39: 295–314.

Aller, R.C., and Mackin, J.E. 1984. Preservation of reactive organic matter in marine sediments. Earth Planet. Sci. Lett. 70: 260–266.

Aller, R.C., and Yingst, J.Y. 1980. Relationships between microbial distributions and the anaerobic decomposition of organic matter in surface sediments of Long Island Sound, U.S.A. Mar. Biol. 56: 29–42.

Almagor, G., and Wiseman, G. 1982. Submarine slumping and mass movements on the continental slope of Israel. pp. 95–128. In: S. Saxon and J.K. Nieuwenhuis (eds.) Marine Slides and Other Mass Movements. Plenum Press, New York.

Ambler, D.C. 1974. Runoff from a small Arctic watershed. pp. 45–49. In: Permafrost Hydrology, Proceedings of a Workshop Seminar, Canadian National Committee of the International Hydrological Decade.

Amos, C.L., and Alfoldi, T.T. 1979. The determination of suspended sediment concentration in a macrotidal system using Landsat Data. J. Sed. Petrol. 49: 159–173.

Amos, C.L., and King, E.L. 1984. Bedforms of the Canadian eastern seaboard: A comparison with global occurrences. Mar. Geol. 57: 167–208.

Anderson, J.J., and Devol, A.H. 1973. Deep water renewal in Saanich Inlet an intermittently anoxic basin. Est. Coast. Mar. Sci. 1: 1–10.

Anderson, J.S., 1977. Primary production associated with sea ice at Godhaven Disko, West Greenland. Ophelia 16: 205–220.

Anderson, L.W. 1978. Cirque glacial erosion rates and characteristics of Neoglacial tills, Pangnirtung Fjord area, Baffin Island, N.W.T. Arctic Alp. Res. 10: 749–760.

Andreae, M.O. 1978. Distribution and speciation of arsenic in natural waters and some marine algae. Deep-Sea Res. 25: 391–402.

Andresen, A., and Bjerrum, L. 1967. Slides in subaqueous slopes in loose sand and silt. pp. 221-239. In: A.F. Richards (ed.) Marine Geotechnique. University Illinois Press, Urbana.

Andrews, J.T. 1975. Support for a stable Late Wisconsin ice margin (14,000 to ≈ 9,000 B.P.): A test based on glacial rebound. Geology 3: 617–619.

Andrews, J.T., Jull, A.J.T., Donahue, D.J., Short, S.K., and Osterman, L.E. 1985. Sedimentation rates in Baffin Island fiord cores from comparative radiocarbon dates. Can. J. Earth Sci. 22: 1827–1834.

Andrews J.T., and Matsch, C.L. 1983. Glacial marine sediments and sedimentation: An annotated bibliography. GEO Abstracts No. 1, Short Run Press, Exeter.

Andrews, J.T., Basory, R.G., and Drapier, L. 1970. An inventory of the present and past glaciation of Home Bay and Okoa Bay, East Baffin Island, N.W.T. and paleoclimatic considerations. J. Glaciology 9: 337–362.

Andrews, J.T., Osterman, L., Kranitz, J., Jennings, A., Stravers, J., Williams, K., and Mothersill, J. 1983. Quaternary piston cores. pp. 14–1 to 14–70. In: J.P.M. Syvitski and C.P. Blakeney (comp.) Sedimentology of Arctic Fjord Experiment: HU 82-031 data report, Vol. 1. Can. Data Rep. Hydrogr. Oc. Sci. 12.

Andru, P., and Allen, W.G. 1979. River training for a deep sea harbour on the Squamish Delta. pp. 316–327. In: River Basin Management, Fourth Nat. Hydrotechnical Conf.

Ankar, S. 1977. The soft bottom ecosystem of the northern Baltic proper with special reference to the macrofauna. Contributions from the Askö Laboratory, Univ. Stockholm, Sweden 19: 1–62.

Aoki, S., and Oinuma, K. 1981. Distribution of clay minerals in surface sediments of Suruga Bay, central Japan. J. Geol. Soc. Japan 87: 429–438.

Aplin, A.C., and Cronan, D.S. 1985. Ferromanganese oxide deposits from the central Pacific Ocean. II. Nodules and associated sediments. Geochim. Cosmochim. Acta 49: 437–451.

Armstrong, J.E. 1966. Glaciation along a major fjord valley in the Coast Mountains of British Columbia Canada. G.S.A. special paper, 101: 7.

Arntz, W.E. 1981. Zonation and dynamics of macrobenthos biomass in an area stressed by oxygen deficiency. pp. 215–225. In: G.W. Barrett and R. Rosenberg (eds.) Stress Effects on Natural Ecosystems. John Wiley, Chichester.

Arntz, W.E., and Rumohr, H. 1982. An experimental study of macrobenthic colonization and succession, and the importance of seasonal variation in temperate latitudes. J. Exp. Mar. Biol. Ecol. 64: 17.

Ashley, G.M. 1978. Bedforms in the Pitt River, British Columbia. In: A.D. Miall (ed.) Fluvial Sedimentology. Can. Soc. Petrol. Geol. Mem. 5: 89–104.

Ashley, G.M. 1979. Sedimentology of a tidal lake, Pitt Lake, British Columbia, Canada. pp. 327–345. In: C. Schlucter (ed.) Moraines and Varves. INQUA Proc. A.A. Balkema Publ.

Ashley, G.M. 1980. Channel morphology and sediment movement in a tidal river, Pitt River, British Columbia. Earth Surface Proc. 5: 347–368.

Asmund, G. 1980. Water movements traced by metals dissolved from mine tailings in a fjord in North-West Greenland. pp. 347–353. In: H.J. Freeland, D.M. Farmer, and C.D. Levings (eds.) Fjord Oceanography. Plenum Press, New York.

Asvall, R.P. 1976. Effects of regulation of freshwater runoff. pp. 15-20. In: S. Skreslet, R. Leinebö, J.B.L. Matthews, and E. Sakshaug (eds.) Fresh Water on the Sea. Assoc. Norw. Oceanogr., Oslo.

Bagnold, R.A. 1966. An approach to the sediment transport problem from general physics. U.S. Geol. Survey Prof. Paper 422-I, 37 p.

———. 1973. The nature of saltation and of 'bedload' transport in water. Proc. Royal Society (London) 322A: 473–504.

Baker, E.T., Canon, G.A., and Curl, H.C. 1983. Particle transport processes in a small marine bay. J. Geophys. Res. 88: 9661–9669.

Balistrieri, L.S., and Murray, J.W. 1979. Surface of goethite (α FeOOH) in seawater. pp. 275–298. In: E.A. Jenne (ed.) Chemical Modeling in Aqueous Systems. American Chemical Society, Washington, D.C.

———. 1983. Metal-solid interactions in the marine environment: estimating apparent equilibrium binding constants. Geochim. Cosmochim. Acta 47: 1091–1098.

Balistrieri, L.S., Brewer, P.G., and Murray, J.W. 1981. Scavenging residence time of trace metals and surface chemistry of sinking particles in the deep ocean. Deep-Sea Res. 28A: 101–121.

Balzer, W. 1982. On the distribution of iron and manganese at the sediment-water interface: Thermodynamic versus kinetic control. Geochim. Cosmochim. Acta 46: 1153–1161.

Barnes, C.A., and Collias, E.E. 1958. Some consideration of oxygen utilization rates in Puget Sound. J. Mar. Res. 17: 68–80.

Barnes, N.E., and Piper, D.J.W. 1978. Late Quarternary history of Mahone Bay, Nova Scotia. Can. J. Earth Sci. 15: 586–593.

Barrett, P.J. 1975. Textural characteristics of Cenozoic preglacial and glacial sediments at site 270, Ross Sea, Antarctica. In: D.E. Hayes, L.A. Frakes et al. (eds.) Initial Reports of the Deep Sea Drilling Project 28: 757–768.

Barrie, C.Q., and Piper, D.J.W. 1982. Late Quaternary geology of Makkovik Bay, Labrador. Geol. Sur. Canada Paper 81-17: 1–37.

Bartsch-Winkler, S., Ovenshine, A.T., and Kachadoorian, R. 1983. Holocene history of the estuarine area surrounding Portage, Alaska as recorded in a 93 m core. Can. J. Earth Sci. 20: 802–820.

Bayne, B.L. 1985. Responses to environmental stress: tolerance, resistance and adaptation. pp. 331–349. In: J.S. Gray and M.E. Christiansen (eds.) Marine Biology of Polar Regions and Effects of Stress on Marine Organisms. John Wiley and Sons, Ltd., Chichester.

Beck, K.C., Reuter, J.H., and Perdue, E.M. 1974. Organic and inorganic geochemistry of some coastal plain rivers of the southeastern United States. Geochim. Cosmochim. Acta 38: 341–364.

Bell, L.M. 1975. Factors influencing the sedimentary environment of the Squamish River delta in southwestern British Columbia. Unpubl. M.A.Sc. thesis, Geol. Sci., University of British Columbia, Vancouver, 145 p.

Bell, L.M., and Kallman, R.J. 1976. The Kitimat River Estuary. Special Estuary Series 6, Status of Environment of Knowledge to 1976. Environ. Can.

Bell, W.H. 1973. The exchange of deep water in Howe Sound. Pac. Mar. Sci. Dep. Environ. Rep. 73–13: 1–111.

———. 1975. The Howe Sound current metering program. Pac. Mar. Sci.Dep. Environ. Rep. 75-7, 1300 p.

Bennett, L.C., Jr., and Savin, S.M. 1963. Studies of the sediments of parts Ytre Samlafjord with continuous seismic profiler. The natural history of the Hardangerfjord, Sarsia 14: 79–94.

Bergsma, B.M., Svoboda, J., and Freedman, B. 1983. Retreating glacier at Ellesmere Island N.W.T. releases on intact pre-Little Ice Age plant community. Abst. 12th Arctic Workshop. Univ. Massachusetts at Amherst. Contribution 44: 14.

Berner, R.A. 1967. Thermodynamic stability of sedimentary iron sulfides. Amer. J. Sci. 265: 773–785.

———. 1971. Principles of Chemical Sedimentology. McGraw-Hill, 240 p.

———. 1980. Early Diagenesis—A Theoretical Approach. Princeton Univ. Press, Princeton, 241 p.

Berner, R.A., and Westrich, J.T. 1985. Bioturbation and the early diagenesis of carbon and sulfur. Amer. J. Sci. 285: 193–206.

Berrang, P.G., and Grill, E.V. 1974. The effect of manganese oxide scavenging on molybdenum in Saanich Inlet, British Columbia. Mar. Chem. 2: 125–148.

Berthois, L. 1969. Contribution a L'étude sedimentologique du Kangerdlugssuag, côte ouest du Grønland. Meddelelser om Grønland, 187 p.

Bertine, K.K., and Lee, D.S. 1983. Antimony content and speciation in the water column and interstitital waters of Saanich Inlet. pp. 21–38.In: E.S. Wong, E. Boyle, K.W. Bruland, J.D. Burton, and E.D. Goldberg (eds.) Trace Metals in Seawater. Plenum Press, New York.

———. 1979. The behavior of trace metals in estuaries of the St. Lawrence basin. Nat. Can. 106: 149–161.

Beyer, F. 1968. Zooplankton, zoobenthos, and bottom sediments as related to pollution and water exchange in the Oslofjord. Helgolander wiss. Meeresunters. 17: 496–509.

———. 1976. Influence of freshwater outflow on the distribution and production of plankton in the Dramsfjord. pp. 165–171. In: S. Skreslet, R. Leinebo, J.B.L. Matthews, E. Sakshaug (eds.). Freshwater on the Sea. Assoc. Norw. Oceanogr., Oslo.

Beyer, J.E. 1981. Aquatic Ecosystems. Univ. Washington Press, Seattle, 315 p.

Bienfang, P.K., Harrison, P.J., and Quarmby, L.M. 1982. Sinking rate response to depletion of nitrate, phosphate and silicate in four marine diatoms. Mar. Biol. 67: 295–302.

Billen, G. 1978. The dependence of the various kinds of microbial metabolism on the redox state of the medium. In: Biogeochemistry of Estuarine Sediments. UNESCO, Paris.

Billett, D.S.M., Lampitt, R.S., Rice, A.L., and Mantoura, R.F.C. 1983. Seasonal sedimentation of phytoplankton to the deep-sea benthos. Nature 302: 520–522.

Bjerrum, L. 1971. Subaqueous slope failures in Norwegian fjords. Norweg. Geotech. Inst. Publ. 88, 8 p.

Björseth, A., Knutzen, J., and Skei, J.M. 1979. Determination of polycyclic aromatic hydrocarbons in sediments and mussels from Saudafjord, W. Norway, by glass capillary gas chromatography. The Science of the Total Environment 13: 71–86.

Blake, W., Jr. 1975. Pattern of post glacial emergence Cape Storm and South Cape Fjord, Southern Ellesmere Island, N.W.T. Geol. Surv. Canada Paper 75-1C: 69–177.

Blake, W. 1977. Iceberg concentrations as an indicator of submarine moraines, eastern Queen Elizabeth Islands, District of Franklin. Geol. Surv. Canada Paper 77-B: 281–286.

Blenkarn, K.A. 1970. Measurement and analysis of ice forces on Cook Inlet structures. pp. 365–378. Offshore Technology Conf., Houston, Texas.

Boatman, C.D., and Murray, J.W. 1982. Modeling exchangeable NH_4^+ adsorption in marine sediments: Process and controls of adsorption. Limnol. Oceanogr. 27: 99–110.

Bodungen, B., Bröckel, K., Smetacek, V., and Zeitzschel, B. 1981. Growth and sedimentation of the phytoplankton spring bloom in the Bornholm Sea (Baltic Sea). Kieler Meeresforsch. 5: 49.

Boesch, D.F., and Rosenberg, R. 1981. Response to stress in marine benthic communities. pp. 179–200. In: G.W. Barrett, and R. Rosenberg (eds.) Stress Effects on Natural Ecosystems. John Wiley, Chichester.

Boesch, D.F., Wass, M.L., and Virnstein, R.W. 1976. The dynamics of estuarine benthic communities. pp. 177-196. In: M. Wiley (ed.) Estuarine Processes, Vol. 1. Academic Press, New York.

Bogen, J. 1983. Morphology and sedimentology of deltas in fjord and fjord-valley lakes. Sed. Geol. 36: 245–267.

Bokuniewicz, H.J., and Gordon, R.B. 1980. Sediment transport and deposition in Long Island Sound. pp. 69–106. In: B. Saltzman (ed.) Estuarine Physics and Chemistry: Studies in Long Island Sound. Academic Press, New York.

Bondam, J. 1978. Recent bottom sediments in Agfardlikavsâ and Qaumarujuk fjords in Marmorilik, West Greenland. Bull. Geol. Soc. Denmark 27, Special Issue: 39–45.

Bornhold, B.D. 1983a. Fiords. Geos 1: 1–4.

———. 1983b. Sedimentation in Douglas Channel and Kitimat Fjord System. In: R.W. Macdonald (ed.) Proceedings of a Workshop on the Kitimat Marine Environment. Can. Tech. Dept. of Hydrography and Ocean Science, No. 18: 88–114.

Bothner, M.H., and Carpenter, R. 1974. Sorption-desorption reactions of mercury with suspended matter in the Columbia River. IAEA-SM-15815: 73–87.

Boudreau, B.P., and Westrich, J.T. 1984. The dependence of bacterial sulfate reduction on sulfate concentration in marine sediments. Geochim. Cosmochim. Acta 48: 2503–2516.

Boulègue, J. 1983. Trace metals (Fe, Cu, Zn, Cd) in anoxic environments. pp. 563–577. In: E.S. Wong, E. Boyle, K.W. Bruland, J.D. Burton, and E.D. Goldberg (eds.) Trace Metals in Seawater. Plenum Press, New York.

Boulègue, J., Lord, C.J., and Church, T.M. 1982. Sulfur speciation and associated trace metals (Fe, Cu) in the pure waters of Great Marsh, Delaware. Geochim. Cosmochim. Acta 46: 453–464.

Boulton, G.S. 1975. Processes and patterns of subglacial sedimentation: a theoretical approach. In: A.E. Wright and F. Moseley (eds.) Ice Ages: Ancient and Modern. Geol. J. Spec. Issue 6: 7–42.

Boulton, G.S., and Eyles, N. 1979. Sedimentation by valley glaciers: A model and genetic classification. pp. 11–25. In: C. Schlüchter (ed.) Moraines and Varves. Proceedings of an INQUA symposium on Genesis and Lithology of Quaternary Deposits, 1978.

Boulton, G.S., Dickson, J.H., Nichols, H., Nichols, M., and Short, S.K. 1976. Late Holocene glacier fluctuations and vegetation changes at Maktak Fjord Baffin Island, N.W.T. Artic and Alpine Research 8: 343–356.

Boulton, G.S., Chroston, P.N., and Jarvis, J. 1981. A marine seismic study of the late Quaternary sedimentation and inferred glacier fluctuations along western Inverness-shire, Scotland. Boreas 10: 39–51.

Bouma, A.H., Hampton, M.A., and Orlando, R.C. 1977. Sand waves and other bedforms in lower Cook Inlet, Alaska. Mar. Geotechnology 2: 291–308.

Bouma, A.H., Hampton, M.A., Rappeport, M.L., Whitney, J.W., Teleki, P.G., Orlando, R.C., and Torreson, M.C. 1978. Movement of sand waves in lower Cook Inlet, Alaska. Offshore Technology Conf. Paper 3311: 2271.

Bourg, A.C.M. 1983. Role of freshwater/seawater mixing on trace metal adsorption phenomena. pp. 195–208. In: C.S. Wong, E. Boyle, K.W. Bruland, J.D. Burton, and E.D. Goldberg (eds.) Trace Metals in Seawater. Plenum Press, New York.

Bowden, K.F. 1967. Circulation and diffusion. In: S.H. Lauff (ed.) Estuaries. Am. Ass. Advanc. Sci. Publ. 83: 15–36.

Bowen, A.J., Normark, W.R., and Piper, D.J.W. 1984. Modelling of turbidity currents on Navy Submarine Fan, California Continental Borderland. Sedimentology 31: 169–185.

Boyd, L.A. 1935. The fjord region of east Greenland. Am. Geogr. Soc. Spec. Publ. 18, 369 p.

———. 1948. The coast of northeast Greenland. Am. Geogr. Soc. Spec. Publ. 30, 339 p.

Boyer, D.J., and Pheasant, R. 1974. Delimination of weathering zones in the Narpaing Fjord area of eastern Baffin Island, Canada. Geol. Soc. Am. Bull. 85: 805–810.

Boyle, E.A., Edmond, J.M., and Sholkovitz, E.R. 1977. The mechanism of iron removal in estuaries. Geochim. Cosmochim. Acta 41: 1313–1324.

Boyle, E., and Huested, S. 1983. Aspects of the surface distribution of copper, nickel, cadmium and lead in the north Atlantic and north Pacific. pp. 379–394. In: C.S. Wong, E. Boyle, K.W. Bruland, J.D. Burton, and E.D. Goldberg (eds.) Trace Metals in Seawater. Plenum Press, New York.

Bøyum, A. 1973. Salsvatn, a lake with old sea water. Swiss J. Hydrol. 35: 262–277.

Braarud, T. 1961. The natural history of Hardangerfjord. Sarsia 1: 3–6.

———. 1974. The natural history of the Hardangerfjord. 11. The fjord effect upon the phytoplankton in late autumn to early spring 1955-56. Sarsia 55: 99–114.

———. 1975. The natural history of the Hardangerfjord. 12. The late summer water exchange in 1956, its effect upon phytoplankton and phosphate distribution, and the introduction of an offshore population into the fjord in June 1956. Sarsia 58: 9–30.

———. 1976. The natural history of the Hardangerfjord. 13. The ecology of taxonomic groups and species of phytoplankton related to their distribution patterns in a fjord area. Sarsia 60: 41–62.

Braarud, T., Hofsvang, B.F., Hjelmfoss, P., and Overland, A.K. 1974. The natural history of the Hardangerfjord, 10. The phytoplankton cycle in the fjord waters and in the offshore coastal waters. Sarsia 55: 63–98.

Brattegard, T. 1966. The natural history of the Hardangerfjord. 7. Horizontal distribution of the fauna of rocky shores. Sarsia 22: 1–54.

———. 1980. Why biologists are interested in fjords. In: H.J.Freeland, D.M. Farmer, and C.D. Levings (eds.) Fjord Oceanography, Nato. Conf. Ser. 4: 53–66.

Bremmeng, G.S. 1974. Strandvatn, northern Norway, a lake with old seawater. Swiss J. Hydrol. 36: 351–356.

Brenchly, G.A. 1981. Disturbance and community structure: an experimental study of bioturbation in marine soft-bottom environments. J. Mar. Res. 39: 767–790.

Bretschneider, C.L. 1952. The generation and decay of wind waves in deep water. Trans. Am. Geophys. Union 33: 381–389.

Brochu, M. 1971. Le processus de deglacement du Fjord de Marcourt au nouveau-Zuebec: description et interpretation. Rev. Geogr. Montr. 25: 43–52.

Brodie, J.W. 1962. The Fiordland Shelf and Milford

Sound. New Zealand Oceanographic Institute, Dept. of Scientific and Industrial Research, Wellington.

Brodie, J.W. 1964. The Fiordland Shelf and Milford Sound. In: T.M. Skerman (ed.) Studies of a southern fiord. New Zealand Oceanographic Inst. Mem. No. 17: 15–23.

Brogersma-Sanders, M. 1957. Mass mortality in the sea. In: J.W. Hedgepath (ed.) Treatise on Marine Ecology and Paleoecology, Geol. Soc. Am. Mem. 67: 941–1010.

Brown, F.S., Baedecker, M.J., Nissenbaum, A., and Kaplan, I.R. 1972. Early diagenesis in a reducing fjord, Saanich Inlet, British Columbia. III. Changes in organic constituents of sediments. Geochim. Cosmochim. Acta 36: 1185–1203.

Bruland, K.W. 1974. Pb-210 geochronology in the coastal marine environment. Unpublished Ph.D. thesis, Univ. California, San Diego.

———. 1980. Oceanographic distributions of cadmium, zinc, nickel and copper in the North Pacific. Earth. Planet. Sci. Lett. 47: 59–68.

Brunswig, D., Arntz, W.E., and Rumohr, H. 1976. A tentative field experiment on population dynamics of macrobenthos in the western Baltic. Kieler Meeresforsch. 3: 49–59.

Bruun, A.F., Brodie, J.W., and Fleming, C.A. 1955. Submarine geology of Milford Sound, New Zealand, N.Z. J. Sci. Tech., Sect. B36: 397–410.

Bryan, G.W. 1980. Recent trends in research on heavy-metal contamination in the sea. Helgoländer Meers. 33: 6–25.

Bryan, G.W., and Hummerstone, L.G. 1971. Adaption of the polychaete *Nereis diversicolor* to estuarine sediments containing high concentrations of heavy metals. I. General observations and adaption to copper. J. Mar. Biol. Assoc. U.K. 52: 845–863.

Buckley, J.R. 1977. The currents, winds and tides of northern Howe Sound. Unpublished Ph.D. thesis, University of British Columbia, Vancouver, B.C., 238 p.

Buckley, J.R., and Pond, S. 1976. Wind and the surface circulation of a fjord. J. Fish. Res. Bd. Can. 33: 2265–2271.

Burd, B.J., and Brinkhurst, R.O. 1985. The effect of oxygen depletion on the galatheid crab *Munida quadrispina* in Saanich Inlet, British Columbia. pp. 435–443. In: J.S. Gray and E. Christiansen (eds.) Marine Biology of Polar Regions and Effects of Stress on Marine Organisms. John Wiley and Sons, Ltd., Chichester.

Burrell, D.C. 1972. Suspended sediment distribution patterns within an active turbid-outwash fjord. pp. 227–245. In: S.S. Wetteland and P. Bruun (eds.) Proceed. Int. Conf. Port. Ocean Engin., Tech. Univ.Norway, Trondheim.

———. 1980. Captain Cook: The political, social and scientific background. pp. 29–40. In: A. Shalkop (ed.) Exploration in Alaska. Cook Inlet Historical Society, Anchorage, Alaska.

———. 1982a. Marine environmental studies in the Wilson Arm-Smeaton Bay system: Chemical and physical 1981. Unpublished report, Inst. Mar. Sci., Univ. Alaska, Fairbanks, 285 p.

———. 1982b. The pre-impact biogeochemical environment of Boca de Quadra and Smeaton Bay fjords, southeast Alaska. pp. 311–341. In: D.V. Ellis (ed.) Marine Tailings Disposal. Ann Arbor Science, Ann Arbor.

———. 1983a. Patterns of carbon supply and distribution and oxygen renewal in two Alaskan fjords. Sed. Geol. 36: 93–115.

———. 1983b. The biogeochemistry of Boca de Quadra and Smeaton Bay, southeast Alaska: A summary report on investigations 1980–1983. Unpublished Report, Inst. Mar. Sci., Univ. Alaska, Fairbanks, 630 p.

———. Interactions between silled fjords and coastal regions of the Gulf of Alaska. In: D.W. Hood and S. Zimmerman (eds.) The Gulf of Alaska: Physical and Biological Environment. NOAA, U.S. Gov. Print Office, Washington D.C. (in press).

Burrell, D.C., and Matthews, J.B. 1974. Glacial and turbid outwash fjords. pp. 1–19. In: H.T. Odum, B.J. Copeland, and E.A. McMahon (eds.) Coastal Ecological Systems of the United States, Vol. 3, Chapter D-1. Estuarie Res. Fed., Washington D.C.

Button, D.K., Robertson, B.R., and Craig, K.S. 1981. Dissolved hydrocarbons and related microflora in a fjordal sea port: Sources, sinks, concentrations and kinetics. Appl. Environ. Microbiol. 42: 708–719.

Callender, E., and Bowser, C.J. 1980. Manganese and copper geochemistry of interstitial fluids from manganese nodule-rich pelagic sediments of the northeastern equatorial Pacific Ocean. Amer. J. Sci. 280: 1063–1096.

Callender, E., and Hammond, D.E. 1982. Nutrient exchange across the sediment-water interface in the Potomac River estuary. Est. Coast. Shelf Sci. 15: 395–413.

Calvert, S.E., and Price, N.B. 1970. Composition of manganese nodules and manganese carbonates from Loch Fyne, Scotland. Contr. Mineral. Pet. 29: 215–233.

Cannon, G.A. 1975. Observations of bottom-water flushing in a fjord-like estuary. Est. Coast. Mar. Sci. 3: 95–102.

Cannon, G.A., and Laird, N.P. 1980. Characteristics of flow over a sill during deep-water renewal. pp. 544–557 In: H.J. Freeland, D.M. Farmer, and C.D. Levings (eds.) Fjord Oceanography. Plenum Press, New York.

Carbonnel, M., and Bauer, A. 1968. Exploitation des

couvertures photographiques aériennes répétèes du front des glaciers velant dans Disko Bugt et Umanak Fjord, juin-juillet 1964. Meddr. Grønland Bd. 173.

Carlson, P.R., Wheeler, M.C., and Molnia, B. 1979. Neoglacial sedimentation in west arm of Glacier Bay, Alaska. U.S.G.S. Prog. and Abstr. 72.

Carlson, R.F., Wagner, J., Hartman, C.W., and Murphy, R.S. 1969. Freshwater studies. pp. 7–41. In: Baseline Data Survey for Valdez Pipeline Terminal. Inst. Mar. Sci. Rept. R69-17, Univ. Alaska, Fairbanks.

Carpenter, R., Peterson, M.L., and Bennett, J.T. 1982. ^{210}Pb-derived sediment accumulation and mixing rates for the Washington continental slope. Mar. Geol. 48: 135–164.

Carpenter, R., Peterson, M.L., Bennett, J.T., and Somayajulu, B.L.K. 1984. Mixing and cycling of uranium, thorium and ^{210}Pb in Puget Sound sediments. Geochim. Cosmochim. Acta 48: 1949–1963.

Carstens, T. 1970. Turbulent diffusion and entrainment in two-layer flow. Proc. Am. Soc. Civil. Eng. WW: 97–104.

Carstens, T., and Rye, H. 1980. Controlling impact of changes in fjord hydrology. pp. 341–347. In: H.J. Freeland, D.M. Farmer, and C.D. Levings, (eds.) Fjord Oceanography. Plenum Press, New York.

Carstens, T., and Tesaker, E. 1972. Erosion by artificial suspension current. Paper presented at the 13th Coastal Engineering Conference, Vancouver. 4 p.

Carter, L. 1973. Surficial sediments of Barkley Sound and the adjacent continental shelf, west coast Vancouver Island. Can. J. Earth Sci.10: 441–459.

———. 1976 Seston transport and deposition in Pelorus Sound, South Island, New Zealand. N.Z. Journ. Mar. Freshwater Res. 10: 263–282.

Carter, L., Carter, R.M., and Griggs, G.B. 1982. Sedimentation in the Conway Trough, a deep nearshore marine basin at the junction of the Alpine transform and Hikurangi subduction plate boundary, New Zealand. Int'l. Assoc. Sed. Sp. Publ. 8: 475–497.

Casagrande, A. 1936. The determination of the preconsolidation load and its practical significance. Proceedings of the First Int. Conf. Soil Mechanics, Cambridge, Mass. 1: 60–64.

Cato, I. 1977. Recent sedimentological and geochemical conditions and pollution problems in two marine areas in southwestern Sweden. Straiae No. 6, 158 p.

Cato, I., Olsson, I., and Rosenberg, R. 1980. Recovery and decontamination of estuaries. pp. 403-440. In: E. Olausson and I. Cato (eds.) Chemistry and Biogeochemistry of Estuaries. John Wiley and Sons, Chichester.

Chapman, C.J., and Rice, A.L. 1971. Some direct observations on the ecology and behaviour of the Norway lobster *Nephrops norwegicus*. Mar. Biol. 10: 321–329.

Chien, N. 1954. Meyer-Peter Formula for bed-load transport and Einstein bed-load function. Univ. Col. Inst. Eng. Res. No. 7.

Christensen, E.R. 1982. A model for radionuclides in sediments influenced by mixing and compaction. J. Geophys. Res. 87: 566–572.

Christensen, J.P., and Packard, T.T. 1976. Oxygen utilization and plankton metabolism in a Washington fjord. Est. Coast. Mar. Sci. 4: 339–347.

Christian, R., and Thompson, S.M. 1978. Loop rating and grading of suspended sediment in the Maranoa. New Zeal. J. Hydrol. 17: 50–53.

Christie, H., and Green, N.W. 1982. Changes in the sublittoral hard bottom benthos after a large reduction in pulpmill waste to Iddefjorden, Norway, Sweden. Neth. J. Sea Res. 16: 474–482.

Church, M. 1972. Baffin Island Sandurs: A study of arctic fluvial processes. Geol. Surv. Canada Bull. 216: 208 p.

———. 1974. Hydrology and permafrost with reference to northern North America. pp. 7–20. Permafrost Hydrology, Proceedings of Workshop Seminar. Can. National Comm. of the Inter. Hydrological Decade.

———. 1978. Paleohydrological reconstructions from a Holocene valley fill. In: A.D. Miall (ed.) Fluvial Sedimentology. Can. Soc. Petrol. Geol. Mem. 5: 743–772.

Church, M., and Gilbert, R. 1975. Proglacial fluvial and lacustrine environments. pp. 22–100. In: A.V. Jopling and B.C. McDonald (eds.) Glaciofluvial and Glaciolacustrine Sedimentation. Soc. Econ. Paleon. Mineral. Spec. Publ. No. 23.

Church, M., and Ryder, R.M. 1972. Paraglacial sedimentation: A consideration of fluvial processes conditioned by glaciation. Geol. Soc. Am. 83: 3059–3072.

Church, M., Stock, R.F., and Ryder, J.M. 1979. Contemporary sedimentary environments on Baffin Island, N.W.T. Canada: Debris slope accumulations. Arct. Alp. Res. 11: 371–402.

Church, M.A., and Russell, S.O. 1977. The characteristics and management of the Bella Coola Indian Reserve. Report by Sigma Engineering Ltd. for Canada DINA, 39 p.

Cimberg, R.L. 1982. Comparative benthic ecology of two southeast Alaskan fjords. pp. 343–356. In: D.V. Ellis (ed.) Marine Tailings Disposal. Ann Arbor Science, Ann Arbor, Michigan.

Clague, J.J., and Bornhold, B.D. 1980. Morphology and littoral processes of the Pacific coast of Canada. pp. 329–349. In: S.B. McCann (ed.) The Coastline of Canada. Littoral processes and shore morphology. Geol. Surv. Pap. 80–10.

Clague, J.J., and Mathews, W.H. 1973. The magnitude of jokulhlaups. J. Glaciology 12: 501-504.

Clague, J.J., Luternauer, J.L., and Hebda, R.J. 1983. Sedimentary environments and post- glacial history of the Fraser Delta and Lower Fraser Valley, British Columbia. Can. J. Earth Sci. 20: 1314–1326.

Clapperton, O.M. 1971. Geomorphology of the Stromness Bay-Cumberland Bay Area, South Georgia. British Antarctic Survey Scientific Reports 70, 25 p.

Clark, J.A. 1980. A numerical model of worldwide sea level changes on a viscoelastic earth. pp. 525–534. In: N-A. Morner (ed.) Earth Rheology, Isostasy and Eustasy. John Wiley & Sons, London.

Clark, J.A. Farrell, W.E. Peltier, W.R. 1978. Global changes in post-glacial sea level: A numerical calculation. Quatern. Res. 9: 265–287.

Clasby, R.C., Horner, R., and Alexander, V. 1973. An *in situ* method for measuring primary productivity of arctic sea ice algae. J. Fish. Res. Bd. Can. 30: 835–838.

Cline, J.D., and Richards, F.A. 1969. Oxygenation of hydrogen sulfide in seawater at constant salinity, temperature and pH. Environ. Sci. and Technol. 3: 838–843.

Cline, J.D., Katz, C., and Young, A.W. 1979. Distribution of hydrocarbons in Cook Inlet, Alaska: Annual report to Outer Continental Shelf Environment Assessment Program. Environ. Res. Labs. Nat. Oceanic and Atmospheric Adm., Boulder, Colorado, 57 p.

Clukey, E., Cacchione, D.A., and Nelson, C.H. 1980. Liquefaction potential of the Yukon prodelta, Bering Sea. Offshore Tech. Conf. 12. Houston. Proc. 2: 315–325.

Collier, R.W., and Edmond, J.M. 1983. Plankton compositions and trace element fluxes from the surface ocean. pp. 789–809. In: C.S. Wong, E. Boyle, K.W. Bruland, J.D. Burton, and E.D. Goldberg (eds.) Trace Metals in Seawater. Plenum Press, New York.

Collier, R.W., and Edmond, J.M. 1984. The trace element geochemistry of marine biogenic particulate matter. Prog. Oceanogr. 13: 113–199.

Collinson, J.D. 1969. The sedimentology of the Grindslow Shales and the Kinderscout Grit: A deltaic complex in the Namurian of northern England. J. Sed. Pet. 39: 194–221.

Colonell, J.M. (ed.). 1980. Port Valdez, Alaska: Environmental Studies 1976–1979. Inst. Mar. Sci. Occas. Publ. No. 5, Univ. Alaska, Fairbanks, 373 p.

Cone, R.A., Neidell, N.S., and Kenyon, K.E. 1963. Studies of the deep-water sediments with the continuous seismic profiler. The natural history of Hardangerfjord. Sarsia 14: 61–78.

Conover, R.J. 1974. Production in marine plankton communities. pp. 159-163. In: Proceedings of the First International Congress of Ecology. Centre for Agricultural Publishing and Documentation, Wageningen, Netherlands.

Cook, J., and King, J. 1784. A Voyage to the Pacific Ocean. 3 vols. T. Cadell and W. Davies, London.

Cooney, R.T. 1986. The seasonal occurrence of *Neocalanus cristatus, Neocalanus plumchrus* and *Eucalanus bungii* over the shelf of the northern Gulf of Alaska. Contin. Shelf. Res. *5:* 541–553.

Cooper, W.S. 1937. The problem of Glacier Bay. Alaska Geol. Rev. 37.

Cossa, D., and Poulet, S.A. 1979. Survey of trace metal contents of suspended matter in the St. Lawrence Estuary and Saguenay Fjord. J. Fish. Res. Bd. Can. 35: 338–345.

Côté, R., and Lacroix, G. 1978. Tidal variability of physical, chemical and biological characteristics in the Saguenay Fjord. J. Fish. Res. Bd. Can. 35: 338–345.

Coulter, H.W., and Migliaccio, R.R. 1966. Effects of the earthquake of March 27, 1974 at Valdez, Alaska. U.S. Geol. Survey Prof. Paper No. 542C, 35 p.

Craig, P.C., Griffiths, W.B., Johnson, S.R., and Schell, D.M. 1984. Trophic dynamics in an arctic lagoon. pp. 347–380. In: P.W. Barnes, D.M. Schell, and E. Reimnitz (eds.) The Alaskan Beaufort Sea: Ecosystems and Environments. Academic Press Inc., Orlando.

Craig, R.E. 1954. A first study of the detailed hydrography of some Scottish west highland sea lochs (Loch Inchard, Kanaird and the Cairnbaan group). Int. Council for the Explor. of the Sea, Annales Biol. 10: 16–19.

Cranston, R.E. 1980. Cr species in Saanich Inlet and Jarvis Inlets. pp. 689–693. In: H.J. Freeland, D.M. Farmer, and C.D. Levings (eds.) Fjord Oceanography. Plenum Press, New York.

Cranston, R.E. 1983. Chromium in Cascadia Basin, Northeast Pacific Ocean. Mar. Chem. 13: 109–125.

Crean, P.B. 1967. Physical Oceanography of Dixon Entrance, British Columbia. Bulletin 156, Fish. Res. Bd., Canada, Ottawa.

Crecèlius, E., Bothner, M.H., and Carpenter, R. 1975. Geochemistries of arsenic, antimony, mercury and related elements in sediments of Puget Sound. Environ. Sci. Technol. 9: 325–333.

Cretney, W.J., Wong, C.S., Macdonald, R.W., Erichson, P.E., and Fowler, B.R.1983. Polycyclic aromatic hydrocarbons in surface sediments and age-related cores from Kitimat Area, Douglas Channel and adjoining waterways. pp. 162–194. In: R.W. Macdonald (ed.) Proceedings of a Workshop on the Kitimat Marine Environment. Can. Tech. Dept. of Hydrography and Ocean Sciences No. 18.

Crisp, D. 1974. Factors influencing settlement of marine invertebrate larvae. pp. 177–265. In: P.T. Grant and A.M. Mackie (eds.) Chemoreception in Marine Organisms. Academic Press, New York.

Crosby, S.A., Millward, G.E., Butler, E.I., Turner, D.R., and Whitfield, M. 1984. Kinetics of phosphate adsorption by iron oxyhydroxides in aqueous systems. Est. Coast. Shelf. Sci. 19: 257–270.

Cullingford, R.A. 1979. Late glacial raised shorelines and deglaciation in the Earn-Tay area. pp. 15–32. In: J.M. Gray and J.J. Lowe (eds.) Studies in the Scottish Late glacial Environment. Pergamon Press, Oxford.

d'Anglejan, B.F., and Ingram, R.G. 1976. Time-depth variations in tidal flux of suspended matter in the Saint Lawrence Estuary. Estuar. Coast. Mar. Sci. 4: 401–416.

d'Anglejan, B.F., and Smith, E.C. 1973. Distribution, transport and composition of suspended matter in the St. Lawrence Estuary. Can. J. Earth Sci. 10: 1380–1396.

Danielsson, L.G., Dyrssen, D., Johansson, T., and Nyquist, G. 1975. The Byfjord: Chemical investigations. Statens Naturv. PM 609: 1–85.

Darnell, R.M. 1979. The pass as a physically-dominated, open ecological system. pp. 383–393. In: R.J. Livingstone (ed.) Ecological Processes in Coastal and Marine Systems. Plenum Press, New York.

Darwin, C. 1846. Geological Observations on South America Being the Third of the Geology of the Voyage of the "Beagle" under the Command of Capt. Fitzroy, R.N. during the years 1832 to 1836, pp. VIII + 279, pls 5, map.

Das, B.M. 1983. Advanced Soil Mechanics. McGraw-Hill Book Co., New York, 520 p.

Davies, J.M. 1975. Energy flow through the benthos in a Scottish sea loch. Mar. Biol. 31: 353–362.

Davies, J.M., and Payne, R. 1984. Supply of organic matter to the sediment in the northern North Sea during a spring phytoplankton bloom. Mar. Biol. 78: 315–324.

Davis, J.C. 1975. Minimal dissolved oxygen requirements of aquatic life with emphasis on Canadian species: a review. J. Fish. Res. Bd. Can. 32:2295–2332.

Davison, W. 1980. A critical comparison of the measured solubilities of ferrous sulfides in natural waters. Geochim. Cosmochim. Acta 44:803–808.

Deegan, C.E., Kirby, R., and Rae, I. 1973. The superficial deposits of the Firth of Clyde and its sea lochs. Inst. of Geol. Sci. Rep. 73/9, 135 p.

de Geer, G. 1910. Kontinentale Niveauveränderungen im Norden Europas. C.R. II Congres Geol. Int. Stockholm.

Delure, A.M. 1983. The effect of storms on sediments in Halifax Harbour, Nova Scotia. M.Sc. thesis, Dalhousie Univ., Halifax.

Dethlefsen, V., and von Westernhagen, H. 1983. Oxygen deficiency and effects on bottom fauna in the eastern German Bight 1982. Meeresforschung 30: 42–53.

Deuser, W.G. 1975. Reducing environments. pp. 1–37. In: J.P. Riley and G. Skirrow (eds.) Chemical Oceanography, Vol. 3, 2nd ed. Academic Press, London.

Devol, A.H., Anderson, J.J., Kuivila, K., and Murray, J.W. 1984. A model for coupled sulfate reduction and methane oxidation in the sediments of Saanich Inlet. Geochim. Cosmochim. Acta 48: 993–1004.

de Wilde, P.A.W. 1976. The benthic boundary layer from the point of view of a biologist. pp. 81–94. In: I.N. McCave (ed.) The Benthic Boundary Layer. Plenum Press. New York.

Donner, J. 1977. Investigations of Quaternary geology of the west coast of Disko, central west Greenland. Rapp. Grønlands Geol. Unders 85: 45–46.

Dowdeswell, E.K., and Andrews, J.T. 1985. The fiords of Baffin Island: Description and classification. pp. 93–123. In: J.T. Andrews (ed.) Quaternary Environments: Eastern Canadian Arctic, Baffin Bay and West Greenland. Allen and Unwin.

Drainville, G. 1968. Le fjord du Saguenay: Contributions à l'océanographie. Nat. Can. 95: 809–855.

Drake, D.E., and Gorsline, D.S. 1973. Distribution and transport of suspended particulate matter in Hueneme, Redondo, Newport and La Jolla submarine canyons, California. Geol. Soc. Am. Bull. 84: 3949–3968.

Drewry, D.J. 1976. Deep-sea drilling from Glomar Challenger in the Southern Ocean. Polar Rec. 18: 47–77.

Drinkwater, K.F. 1973. The role of tidal mixing in Rupert and Holberg Inlets. M.Sc. Thesis (unpubl.) Univ. of British Columbia, Vancouver, 58 p.

Drinkwater, K.F., and Osborn, T.R. 1975. The role of tidal mixing in Rupert and Holberg Inlets, Vancouver Island. Limnol. Oceanogr. 20: 518–529.

DuBoys, M.P. 1879. Le Rhône et les Rivières a Lit affouillable. Me. Doc., Ann. Pont et Chausses, series 5, vol. XVIII.

Duff, L. 1981. The Loch Eil project: Effect of organic matter input on interstitial water chemistry of Loch Eil sediments. J. Exp. Mar. Biol. Ecol. 55: 315–328.

Dugdale, R.C., and Goering, J.J. 1967. Uptake of new and regenerated forms of nitrogen in primary production. Limnol. Oceanogr. 12: 196–206.

Dunbar, M.J. 1951. Eastern Arctic waters. Fish. Res. Bd. Bull. 88: 1–31.

———. 1973. Glaciers and nutrients in arctic fiords. Science 182: 398.

Dybern, B. 1972. Idefjorden—en forstörd marin miljö. Fauna och Flora 67: 90.

Dyer, K.R. 1979. Estuarine hydrography and sedimentation. A handbook. Cambridge Univ. Press, 230 p.

Dyke, A.S. 1979. Glacial and sea-level history of south-western Cumberland Peninsula, Baffin Island, N.W.T. Canada. Arct. Alp. Res. 11: 179–202.

Dymond, J., Lyle, M., Finney, B., Piper, D.Z., Murphy, K., Conard, R., and Pisiar, N. 1984. Ferromanganese nodules from MANOP sites H, S and R: Control of mineralogical and chemical composition by multiple accretionary processes. Geochim. Cosmochim. Acta 48: 931–949.

Dyrssen, D. 1980. Sediment surface reactions in fjord basins. pp. 645–658. In: H.J. Freeland, D.M. Farmer, and C.D. Levings (eds.) Fjord Oceanography. Plenum Press, New York.

Eagle, R.A. 1975. Natural fluctuations in a soft bottom benthic community. J. Mar. Biol. Ass. U.K. 55: 865–878.

Ebbesmeyer, C.C., and Barnes, C.A. 1980. Control of a fjord basin's dynamics by tidal mixing in embracing sill zones. Estuar. Coast. Mar. Sci. 11: 311–330.

Ebbesmeyer, C.C., Barnes, C.A., and Langby, C. 1975. Application of an advective-diffuse equation to a water parcel observed in a fjord. Est. Coast. Mar. Sci. 3: 249–268.

Eckert, J.M., and Sholkovitz, E.R. 1976. The flocculation of iron, aluminum and humates from river water by electrolytes. Geochim. Cosmochim. Acta 40: 847–848.

Eckman, J.E., Nowell, A.R.M., and Jumars, P.A. 1981. Sediment destabilization by animal tubes. J. Mar. Res. 39: 361–374.

Edgers, L., and Karlsrud, K. 1982. Soil flows generated by submarine slides—Case studies and consequences. Norweg. Geotech. Inst. Publ. 143: 10 p.

Edwards, A., and Edelsten, D.J. 1976. Control of fjordic deep water renewal by runoff modifications. Hydrol. Sci. Bull. 21: 445–450.

———. 1977. Deep water renewal of Loch Etive: A three basin Scottish fjord. Est. Coast. Mar. Sci. 5: 575–595.

Edwards, A., Edelsten, D.J., Saunders, M.A., and Stanley, S.O. 1980. Renewal and entrainment in Loch Eil, a periodically ventilated Scottish fjord. pp. 523–530. In: H.J. Freeland, D.M. Farmer, and C.D. Levings (eds.) Fjord Oceanography. Plenum Press, New York.

Edzwald, J.K., and O'Melia, C.R. 1975. Clay distributions in recent estuarine sediments. Clays Clay Miner. 29: 39–44.

Edzwald, J.K., Upchurch, J.B., and O'Melia, C.R. 1974. Coagulation in estuaries. Environ. Sci. Tech. 8: 58.

Eilertsen, H.C., and Taasen, J.P. 1984. Investigations on the plankton community of Balsfjorden, northern Norway: The phytoplankton 1976–1978, environmental factors, dynamics of growth, and primary production. Sarsia 69: 1–15.

Eilertsen, H.C., Schei, B., and Taasen, J.P. 1981. Investigations on the plankton community of Balsfjorden, Northern Norway: The phytoplankton 1976–1978. Abundance, species composition, and succession. Sarsia 66: 129–141.

Einstein, H.A. 1942. Formulas for the transportation of bed-load. Trans. Am. Soc. Civil Engrs. Vol. 107.

———. 1950. The bed-load function for sediment transportation in open channel flows. U.S. Dept. Agric., Soil Conserv. Ser., T.B. No. 1026.

Eisbacher, G.H., and Clague, J.J. 1981. Urban landslides in the vicinity of Vancouver, British Columbia, with special reference to the December 1979 rainstorm. Can. Geotech. J. 18: 205–216.

Ekman, F.L. 1875. Om de strømninger som uppstar i närheten of flodmynninger. Ofvers. to Kungl Vetensk. Akad. Forhandlingar Nr. 7.

Elderfield, H., McCaffrey, R.J., Luedtke, N., Bender, M., and Truesdale, V.W. 1981. Chemical diagenesis in Narragansett Bay sediments. Am. J. Sci. 281: 1021–1055.

Ellis, D.V. 1977. Effectiveness of existing pollution controls as shown by four case-histories, and recommendations for future control. Submission to the Public Enquiry into Pollutions Control Objectives for the Mining, Mine-milling and Smelting Industries. Vancouver, B.C.

———. (ed). 1982. Marine Tailings Disposal. Ann Arbor Science, Ann Arbor. 368 p.

Elmgren, R. 1976. Baltic benthos communities and the role of the meiofauna. Contributions from the Askö Lab., Univ. Stockholm, Sweden, 14, 31 p.

———. 1980. Structures and dynamics of Baltic benthic communities with particular references to the relationships between macro- and meiofauna. Kieler Meeresforsch. 4: 324.

Elmgren, R., Weshin, L., and Linden, O. 1980. Baltic perspective. pp. 29-32. In: J.J. Kineman, R. Elmgren, and S. Hansson (eds.) The Tsesis Oil Spill. U.S. Dept. Commerce, Boulder.

El-Sabh, M.I. 1979. The lower St. Lawrence estuary as a physical oceanographic system. Naturaliste Can. 106: 55–73.

Elverhöi, A. 1984. Glaciogenic and associated marine sediments in the Weddell Sea, fjords of Spitzbergen and the Barents Sea: A review. Mar. Geol. 57: 53–88.

Elverhöi, A., and Roaldset, E. 1983. Glaciomarine sediments and suspended particulate matter, Weddell Sea Shelf, Antarctica. Polar Res. 1: 1–21.

Elverhöi, A., Liestöl, O., and Nagy, J. 1980. Glacial erosion, sedimentation and microfauna in the inner part of Kongsfjorden, Spitsbergen. Norsk. Polarinstitutt Skr. 172: 33–58.

Elverhöi, A., Lönne, O., and Seland, R. 1983. Glaciomarine sedimentation in a modern fjord environment, Spitzbergen. Polar Res. 1: 127–149.

Emerson, S. Trace metal cycling in anoxic fjords. Deep-Sea Res. In press.

———. 1980. Redox species in a reducing fjord: The oxidation rate of manganese II. pp. 681–687. In: H.J. Freeland, D.M. Farmer, and C.D. Levings (eds.) Fjord Oceanography. Plenum Press, New York.

Emerson, S., Cranston, R.E., and Liss, P.S. 1979. Redox species in a reducing fjord: Equilibrium and kinetic considerations. Deep-Sea Res. 26A: 859–878.

Emerson, S., Kalhorn, S., Jacobs, L., Tebo, B.M., Nealson, K.N., and Rosson, R.A. 1982. Environmental oxidation rate of manganese (II): Bacterial catalysis. Geochim. Cosmochim. Acta 46: 1073–1079.

Emerson, S., Jacobs, L., and Tebo, B. 1983. The behavior of trace metals in marine waters: Solubilities at the oxygen-hydrogen sulfide interface. pp. 579–608. In: C.S. Wong, J.D. Burton, E. Boyle, K.K. Bruland, and E.D. Goldberg (eds.) Trace Metals in Seawater. Plenum Press, New York.

Emerson, S., Jahnke, R.L., and Heggie, D. 1984. Sediment-water exchange in shallow water estuarine sediments. J. Mar. Res. 42: 709–730.

England, J. 1983. Isostatic adjustments in a full glacial sea. Can. J. Earth Sci. 20: 895–917.

Engström, S. 1975. Idefjord Report I-II, Report No. 192 and 193. Havfiskelaboratoriet, Lyskil, Sverige.

Environmental Committee Report. 1973. Resipientundersøkelser: Sörfjorden 1972 (unpublished).

Eppley, R.W., Holmes, R.W., and Strickland, J.D.H. 1967. Sinking rates of marine phytoplankton measured with a fluorometer. J. Exp. Mar. Biol. Ecol. 1: 191.

Erga, S.R., and Heimdal, B.R. 1984. Ecological studies on the phytoplankton of Korsfjorden, western Norway. The dynamics of a spring bloom seen in relation to hydrographical conditions and light regime. J. Plank. Res. 6: 67–90.

Erga, S.R., and Sørensen, K. 1982. Petrokjemianlegg på Kårstö. Primaerproduksjon februar-november. Planteplanktonets biomasse, hydrography, lys og naeringsalter. NIVA-Rapport, 0-80070-02, I–284.

Erickson, P., and Stukas, V. 1983. Trace metals in Boca de Quadra using the vacuum intercept pumping system. Unpublished report to U.S. Borax and Chemical Corporation, Seakem Oceanography Ltd., Victoria, Canada, 20 p.

Erlenkeuser, H., Suess, E., and Willhorn, H. 1974. Industrialization affects heavy metal and carbon isotope concentrations in recent Baltic Sea sediments. Geochim. Cosmochim. Acta 39: 823–842.

Esmark, J.E. 1824. Bidrag til vor jordklodes historie. Mag. for Naturvitenskap III.

Evans, R.A., Gulliksen, B., and Sandnes, O.K. 1980. The effect of sedimentation on rocky bottom organisms in Balsfjord, northern Norway. pp. 603-607. In: H.J. Freeland, D.M. Farmer, and C.D. Levings, (eds.) Fjord Oceanography. Plenum Press, New York.

Evansen, D. 1974. The benthic algae of Borgenfjorden, North-Trøndelag, Norway. K. norske Vidensk. Selsk., Museet, Miscellanea 16: 1–18.

Exon, N. 1972. Sedimentation in the outer Flensburg Fjord area, Baltic Sea, since the last glaciation. Meyniana 22: 5–62.

Eyles, N., Eyles, C.N., and Miall, A.D. 1983. Lithofacies types and vertical profile models: An alternative approach to the description and environmental interpretation of glacial diamict and diamictite sequences. Sedimentology 30: 393–410.

Fairbridge, R.W. (ed.). 1968. The Encyclopedia of Geomorphology. Reinhold, New York, 1295 p.

———. 1980. The estuary: Its definition and geodynamic cycle. pp. 1–35. In: E. Olausson and I. Cato (eds.) Chemistry and Biogeochemistry of Estuaries. John Wiley & Sons, Chichester.

Farmer, D.M., and Freeland, H.J. 1983. The physical oceanography of fjords. Prog. in Oceanogr. 12: 147–220.

Farmer, D.M., and Osborne, T.R. 1976. The influence of wind on the surface layer of a stratified inlet. Part I. Observations. J. Phys. Oceanogr. 6: 931–940.

Farmer, D.M., and Smith, J.D. 1978. Nonlinear internal waves in a fjord. In: J.C.J. Nihoul (ed.) Hydrodynamics of Estuaries and Fjords. Elsevier Oceanogr. Ser. 23, Amsterdam: 465–493.

———. 1980a. Tidal interaction of stratified flow with a sill in Knight Inlet. Deep- Sea Res. 27A: 239–254.

———. 1980b. Generation of lee waves on the sill in Knight Inlet. pp. 259–269. In: H.J. Freeland, D.M. Farmer, and C.D. Levings (eds.) Fjord Oceanography. Plenum Press, New York.

Farrell, W.E., and Clark, J.A. 1976. On postglacial sea level. Geophys. Jour. Astr. Soc. (1976) 4b: 647–667.

Farrow, G.E. 1983. Bottom fauna and bioturbation. In: J.P.M. Syvitski and C.P. Blakeney (comp.) Sedimentology of Arctic Fjords Experiment: HU 82–031 data report, Vol. 1. Can. Data Rep. Hydrogr. Ocean Sci. 12: 9–1 to 9–26.

Farrow, G.E., Syvitski, J.P.M., and Tunnicliffe, V. 1983. Suspended particulate loading on the macrobenthos in a highly turbid fjord: Knight Inlet, British Columbia. Can. J. Fish. Aquat. Sci. 40: 273–288.

Farrow, G.E., Syvitski, J.P.M., Atkinson, R.J.A., Moore, P.G., and Andrews, J.T. 1985. Baffin Island fjord macrobenthos: Bottom communities and environmental significance. Arctic (in press).

Feder, H.M., and Jewett, S.L. The subtidal benthos. In: D.W. Hood and S. Zimmerman (eds.) The Gulf

of Alaska. NOAA. U.S. gov. Print Office, Washington D.C. (in press).
Feder, H.M., and Keiser, G.E. 1980. Intertidal biology. pp. 143–233. In: J.M. Colonell (ed.) Port Valdez, Alaska: Environmental Studies 1976-1979. Occas. Publ. No. 5, Inst. Mar. Sci., Univ. Alaska, Fairbanks.
Feder, H.M., and Matheke, G.E.M. 1980. Subtidal benthos. pp. 235–324. In: J.M. Colonell (ed.) Port Valdez, Alaska: Environmental Studies 1976–1979. Occas. Publ. No. 5, Inst. Mar. Sci., Univ. Alaska, Fairbanks.
Feely, R.A., and Massoth, G.T. 1982. Sources, composition, and transport of suspended particulate matter in Lower Cook Inlet and Northwestern Shelikof Strait, Alaska. NOAA Tech. Rept. ERL. 415 PMEL 34, 28 p.
Fenchel, T., and Blackburn, T.H. 1979. Bacteria and Mineral Cycling. Academic Press, London, 225 p.
Fenchel, T.M., and Riedl, R.J. 1970. The sulfide system: A new biotic community underneath the oxidized layer of marine sand bottoms. Mar. Biol. 7: 255–268.
Field, W.O. 1947. Glacier recession in Muir Inlet, Glacier Bay. Alaska Geogr. Rev. 37.
Field, W.O. 1969. Current observations on three surges in Glacier Bay, Alaska, 1964–1968. Can. J. Earth Sci. 6: 831–839.
Fisher, D.A., and Koerner, R.M. 1980. Some aspects of climate change in the High Arctic during the Holocene as deduced from ice cores. pp. 249-271. In: W.C. Mahaney and C. William (eds.) Quaternary Paleoclimate. Geoabstracts, Norwich.
Fisher, J.B., Lick, W., McCall, P.L., and Robbins, J.A. 1980. Vertical mixing of lake sediments by tubificid oligochaetes. J. Geophys. Res. 85: 3997–4006.
Fjeldskaar, W., and Kaneström, R. 1980. Younger Dryas geoid-deformation caused by deglaciation in Fennoscandia. pp. 569–574. In: N-A. Morner (ed.) Earth Rheology, Isostasy and Eustasy. John Wiley & Sons, London.
Flaate, K., and Janbu, N. 1975. Soil Exploration in a 500 m deep fjord, western Norway. Mar. Geotechnol. 1: 117–139.
Folger, D.W., Meade, R.H., Jones, B.F., and Cory, R.L. 1972. Sediments and waters of Somes Sound, a fjordlike estuary in Maine. Limnol. Oceanogr. 17: 394–402.
Folving, S. 1979. New observations on meltwater and landscape conditions in the Søndre Strømfjord region from Landsat 1 images. Geogr. Tidsskr. 78/79: 28–40.
Forbes, D.L. 1984. Coastal geomorphology and sediments of Newfoundland. In: Current Res. part B, Geol. Surv. Canada 84-1B: 11–24.
Fosshagen, A. 1980. How the zooplankton community may vary within a single fjord system. pp. 399–405. In: H.J. Freeland, D.M. Farmer, and C.D. Levings (eds.) Fjord Oceanography. Plenum Press, New York.
Foster, R.H., and Heiberg, S. 1971. Erosion studies in a marine clay deposit at Romerike, Norway. Norges Geotekniske Institutt Publ. 88: 9 p.
Förstner, U., and Wittmann, G.T.W. 1979. Metal pollution in the aquatic environment. Springer-Verlag, New York. 486 p.
Fournier, J.A., and Levings, C.D. 1982. Polychaetes recorded near two pulp mills on the coast of northern British Columbia: A preliminary taxonomic and ecological survey. Syllogens Series No. 40, National Museums of Canada, Ottawa, 91 p.
Fox, L.E. 1983. The removal of dissolved humic acid during estuarine mixing. Est. Coast. Shelf Sci. 16: 431–440.
Frankel, L., and Mead, D.J. 1973. Mucilaginous matrix of some estuarine sands in Connecticut. J. Sed. Pet. 43: 1090–1095.
Freeland, H.J. 1980. The hydrography of Knight Inlet, B.C. in the light of Long's model of fjord circulation. pp. 271–277. In: H.J. Freeland, D.M. Farmer, and C.D. Levings (eds.) Fjord Oceanography. Plenum Press, New York.
Freeland, H.J., and Farmer, D.M. 1980. Circulation and energetics of a deep strongly stratified inlet. Can. J. Fish. Aquatic. Sci. 37: 1398–1410.
Freeland, H.J., Farmer, D.M., and Levings, C.D. 1980. Fjord Oceanography. Plenum Press, New York, 715 p.
Frey, R.W. 1973. Concepts in the study of biogenic sedimentary structures. J. Sed. Pet. 43: 6–19.
———. 1975. The Study of Trace Fossils. Springer-Verlag, New York. 562 p.
Froelich, P.N., and Andreae, M.O. 1981. The marine geochemistry of germanium. Science 213: 205–207.
Froelich, P.N., Klinkhammer, G.P., Bender, M.L., Luedtke, N.A., Heath, G.R., Cullen, D., Dauphin, P., Hammond, D., Hartman, B., and Maynard, V. 1979. Early oxidation of organic matter in pelagic sediments of the eastern equatorial Atlantic: Suboxic diagenesis. Geochim. Cosmochim. Acta 43: 1075–1090.
Funder, S. 1972. Deglaciation of the Scoresby Sund fjord region, northeast Greenland. Polar Geomorphology, Inst. of British Geographers Spec. Publ. 4: 33–42.
Gaarder, T. 1916. De vestlandske fjorders hydrografi. I. Surstoffet i fjordene (in norwegian). Bergen Mus. Aarb. 1915–1916. Naturvidensk. Rekke 2: 1–200.
———. 1932. Untersuchungen über Produktions- und Lebensbedingungen in norwegischen Autern-Pollen. Bergen. Mus. Arb. Naturvidensk Rekke 3: 1–64.

Gaddis, B.L. 1974. Suspended-sediment transport relationships for four Alaskan glacier streams. M.Sc. thesis, Univ. of Alaska, Fairbanks, 102 p.

Gade, H.G. 1968. Horizontal and vertical exchanges and diffusion in the water masses of the Oslo Fjord. Helgol. Meer. 17: 462–475.

———. 1970. Hydrographic investigations in the Oslofjord: A study of water circulation and exchange processes. Geophysical Inst. Rept. 24, Univ. of Bergen, Norway, 193 p.

———. 1976. Transport mechanism in fjords. pp. 51–56. In: S. Skreslet, R. Leinebø, J.B.L. Matthews, and E. Sakshaug (eds.) Freshwater on the Sea, Publ. by Assoc. Norweg. Oceanogr., Oslo.

———. 1983. Estuaries and fjords. pp. 141–169. In: G. Kullenberg (ed.) Pollutant Transfer and Transport in the Sea, Vol. 2. CRC Press, Boca Raton.

Gade, H.G., and Edwards, A. 1980. Deep-water renewal in fjords. pp.453–489. In: H.J. Freeland, D.M. Farmer, and C.D. Levings (eds.) Fjord Oceanography. Plenum Press, New York.

Gade, H.G., Lake, R.A., Lewis, E.L., and Walker, E.R. 1974. Oceanography of an Arctic Bay. Deep-Sea Res. 21: 547–571.

Gage, J. 1972a. Community structure of the benthos in Scottish sea-lochs. I. Introduction and species diversity. Mar. Biol. 14: 281–297.

———. 1972b. A preliminary survey of the benthic macrofauna and sediments in Lochs Etive and Creran, sea lochs along the west coast of Scotland. J. Mar. Biol. Ass. U.K. 52: 237–276.

Gamble, J.C., Davies, J.M., and Steele, J.H. 1977. Loch Ewe bag experiment, 1974. Bull. Mar. Sci. 27: 146–175.

Gardner, L.B. 1974. Organic versus inorganic trace metal complexes in sulfidic marine waters—Some speculative calculations based on available stability constants. Geochim. Cosmochim. Acta 38: 1297–1302.

Gardner, W. 1980. Sediment trap dynamics and calibration: A laboratory evaluation. J. Mar. Res. 38: 17–39.

Gargett, A. 1980. Turbulence measurements through a train of breaking internal waves in Knight Inlet, B.C. pp. 277–283. In: H.J. Freeland, D.M. Farmer, and C.D. Levings (eds.) Fjord Oceanography. Plenum Press, New York.

Garner, D.M. 1964. The hydrology of Milford Sound. In: T.M. Skerman (ed.) Studies of a Southern Fjord. New Zealand Oceanogr. Inst. Mem. 17: 25–33.

Gatto, L.W. 1976. Circulation and sediment distribution in Cook Inlet, Alaska. pp. 205–227. In: D.W. Hood and D.C. Burrell (eds.) Assessment of Arctic Marine Environment, Selected Topics. Occasional Publ. No. 4, Inst. Mar. Sci., Univ. of Alaska, Fairbanks.

Geyer, W.R., and Cannon, G.A. 1982. Sill processes related to deep-water renewal in a fjord. J. Geophys. Res. 87: 7985–7996.

GGU. 1977. Recipient undersøgelse 1976-77. Agfardlikavsa. Qaumarujuk. Rapport til Ministeriet for Grönland, 135 p.

———. 1980. Recipient undersøgelse ved. Marmoroilik 1979. Rapport til Røstofforvaltningen for Grönland, 123 p.

———. 1982. Recipient undersøgelse ved Marmoroilik 1979–80 Grönlands Fiskeri Undersøgelse og Grönlands Geologiske Undersøgelse, 89 p.

Gibbs, R.J. 1977. Clay mineral segregation in the marine environment. J. Sed. Pet. 47: 237–243.

Gibbs, R.J., Matthews, M.D., and Link, D.A. 1971. The relationship between sphere size and settling velocity. J. Sed. Pet. 41: 7–18.

Gilbert, G.K. 1890. Lake Bonneville. U.S. Geol. Surv. Mon. 1: 438 p.

Gilbert, R. 1975. Sedimentation in Lilloet Lake, British Columbia. Can. J. Earth Sci. 12: 1697–1711.

———. 1978. Observations on oceanography and sedimentation at Pangnirtung fiord, Baffin Island. Maritime Sediments 14: 1–9.

———. 1980a. Environmental studies in Maktak, Coronation and North Pangnirtung Fiords, Baffin Island, N.W.T. unpubl. Petro-Canada and NSERC supported Ms. 96 p.

———. 1980b. Observations on the sedimentary environments of fjords on Cumberland Peninsula, Baffin Island. pp. 633–638. In: H.J. Freeland, D.M. Farmer, and C.D. Levings (eds.) Fjord Oceanography. Plenum Press, New York.

———. 1982a. Contemporary sedimentary environments on Baffin Island, N.W.T. Canada: Glaciomarine processes in fjords of eastern Cumberland Peninsula. Arc. Alp. Res. 14: 1–12.

———. 1982b. The Broughton Trough on the continental shelf of eastern Baffin Island, Northwest Territories. Can. J. Earth Sci. 19: 1599–1607.

———. 1983. Sedimentary processes of Canadian arctic fjords. In: J.P.M. Syvitski and J.M. Skei (eds.) Sedimentology of Fjords. Sed. Geol. 36: 147–175.

———. 1984. Observations at Pangnirtung Fiord. pp. 3-1 to 3-6. In: J.P.M. Syvitski (comp.) Sedimentology of Arctic Fjords Experiment: HU 83–028 Data Report, Vol. 2. Can. Data Rep. Hydrogr. Oc. Sci. 28.

———. 1985. Glaciomarine sedimentation interpreted from seismic surveys of fiords on Baffin Island, N.W.T. Arctic 38: 271–280.

Gilbert, R., and MacLean, B. 1983. Geophysical studies based on conventional shallow and Huntec high resolution seismic surveys of fiords on Baffin Island. pp. 15-1 to 15-90. In: J.P.M. Syvitski and C.P. Blakeney (comp.) Sedimentology of Arctic

Fjords Experiment: HU 82-031 data report, Vol. 1. Can. Data Rep. Hydrogr. Ocean Sci. 12, 935 p.
Gilbert R., and Shaw, J. 1981. Sedimentation in proglacial Sunwapta Lake, Alberta. Can. J. Earth Sci. 18: 81–93.
Gilbert, R., Syvitski, J.P.M., and Taylor, R.B. 1985. Reconnaissance study of proglacial Stewart Lakes, Baffin Island, District of Franklin. In: Current Research, part A, Geol. Surv. Canada, paper 85-1A: 505–510.
Gilmartin, M. 1962. Annual cyclic changes in the physical oceanography of a British Columbia fjord. J. Fish. Res. Bd. Can. 19: 921–974.
Glasby, G.P. (ed.). 1977. Marine Manganese Deposits. Elsevier, Amsterdam, 523 p.
———. 1978a. Historical Note In: Fiord Studies: Caswell and Nancy Sounds, New Zealand. N.Z., Ocean. Inst. Mem. 79: 94 p.
———. 1978b. Sedimentation and sediment geochemistry of Caswell, Nancy and Milford Sounds. pp. 19–38. In: G.P. Glasby (ed.) Fiord Studies: Caswell and Nancy Sounds, New Zealand. N.Z. Ocean. Inst. Mem. 79.
Glenne, B., and Simensen, T. 1963. Tidal current choking in the landlocked fjord of Nordasvannet. Sarsia 11: 43–73.
Goering, J.J., Shiels, W.E., and Patton, C.J. 1975. Primary production. pp. 251–278. In: D.W. Hood, W.E. Shiels, and E.J. Kelley (eds.) Environmental Studies of Port Valdez. Occas. Publ. No. 3, Inst. Mar. Sci. Univ. Alaska, Fairbanks.
Goldberg, E.D. 1976. The Health of the Oceans. The Unesco Press, Paris.
———. (ed.). 1979. Assimilative capacity of U.S. coastal waters for pollutants. Working Paper No. 1, Federal Plan for Ocean Pollution Research Development and Monitoring. U.S. Dept. Commerce, Boulder, Colorado, 284 p.
Goldthwait, R.P. 1968. Hydrologic hazards from earthquake. The Great Alaska Earthquake of 1964. Inst. of National Acad. of Sci. 3: 405–414.
Goldthwait, R.P., Loewe, F., Ugolin, F.C., Decker, H.F., DeLong, D.M., Trantman, M.B., Good, E.E., Merrell, T.R., and Rudolph, E.D. 1966. Soil Development and Ecological Succession in a Deglaciated Area of Muir Inlet; Southeast Alaska. Inst. of Polar Studies, Rept. 20, 18 p.
Goloway, F., and Bender, M. 1982. Diagenetic models of interstitial nitrate profiles in deep sea suboxic sediments. Limnol. Oceanogr. 27: 624–638.
Gøransson, C.G., and Svensson, T. 1975. The Byfjord: Studies of water exchange and mixing processes. Statens Naturvardsverk (Swedish Environmental Protection Board) PM No. 594, Stockholm, Sweden, 152 p.
Goyette, D., and Nelson, H. 1977. Marine environmental assessment of mine waste disposal into Rupert Inlet, British Columbia. Report PR-77-11, Environmental Protection Service, W. Vancouver, 93 p.
Graf, G., Bengtsson, W., Diesner, V., Schutz, R., and Theede, H. 1982. Benthic reponse to sedimentation of a spring phytoplankton bloom: Process and budget. Mar. Biol. 67: 201–208.
Graf, G., Schulz, R., Peinert, R., and Meyer-Reil, L.A. 1983. Benthic response to sedimentation events during autumn to spring at a shallow-water station in the western Kiel Bight. Mar. Biol. 77: 235–246.
Gran, H.H. 1900. Hydrographical-biological studies of the North Atlantic Ocean and the coast of Norland. Rep. Norw. Fish. and Mar. Inv. 1: 92 p.
Grant, A.C. 1975. Seismic reconnaissances of Lake Melville, Labrador. Can. J. Earth Sci. 12: 2103–2110.
Grant, W.D., and Madsen, O.S. 1979. Combined wave and current interaction with a rough bottom. J. Geophys. Res. 84: 1797–1808.
Grassle, J.F., and Sanders, H.L. 1973. Life histories and the role of disturbance. Deep-Sea Res. 20: 643–659.
Gray, J.S. 1974. Animal-sediment relationships. Oceanogr. Mar. Biol. Ann. Rev. 12: 223–261.
———. 1979. Pollution-induced changes in populations. Phil. Trans. Roy. Soc., Series B 286: 545–561.
———. 1981. The Ecology of Marine Sediments. Cambridge University Press, Cambridge, 185 p.
Gray, J.S., and Christie, H. 1983. Predicting long-term changes in marine benthic communities. Mar. Ecol. Prog. Ser. 13: 87–94.
Gray, J.S., and Pearson, T.H. 1982. Objective selection of sensitive species indicative of pollution-induced change in benthic communities. I. Comparative methodology. Mar. Ecol. Prog. Ser. 9: 111–119.
Gregory, J.W. 1913. The Nature and Origin of Fjords. John Murray, London.
Greisman, P. 1979. An upwelling driven by the melt of ice shelves and tidewater glaciers. Deep-Sea Res. 26: 1051–1065.
Grill, E.B. 1978. The effect of sediment-water exchange on manganese deposition and nodule growth in Jervis Inlet, British Columbia. Geochim. Cosmochim. Acta 42: 485–494.
———. 1982. Kinetic and thermodynamic factors controlling manganese concentrations in oceanic waters. Geochim. Cosmochim. Acta 46: 2435–2446.
Gross, M.G., Gucluer, S.M., Creager, J.S., and Dawson, W.A. 1963. Varved marine sediments in a stagnant fjord. Science 141: 918–919.
Grøntved, J. 1960. On the productivity of microbenthos and phytoplankton in some Danish fjords. Medd. Dan. Fisk. Havunders 3: 55–92.

Grund, A. 1909. Strömungsbeobachtungen im Byfjord bei Bergen, und in anderen Norwegischen Fjorden. Intern. Revued. ges. Hydrobiologie u. Hydrographie 11: 31–61.

Grundmanis, V., and Murray, J.W. 1982. Aerobic respiration in pelagic marine sediments. Geochim. Cosmochim. Acta 46: 1101–1120.

Gucluer, S.M., and Gross, M.G. 1964. Recent marine sediments in Saanich Inlet, a stagnant marine basin. Limnol. Oceanogr. 9: 359–376.

Guinasso, N.L., and Schink, D.R. 1975. Quantitative estimates of biological mixing rates in abyssal sediments. J. Geophys. Res. 80: 3032–3043.

Gulliksen, B. 1978. Rocky bottom fauna in a submarine gulley at Loppkalven, Finnmark, northern Norway. Est. Coast. Mar. Sci. 7: 361–372.

———. 1980. The macrobenthic rocky-bottom fauna of Borgenfjorden, North-Trøndelag, Norway. Sarsia 65: 115–138.

———. 1982. Sedimentation close to a near vertical rocky wall in Balsfjorden, northern Norway. Sarsia 67:21–27.

Gulliksen, B., Holte, B., and Jakola, K.J. 1985. The soft bottom fauna in Van Mijenfjord and Randfjord, Svalbard. pp. 199–211. In: J.S. Gray and M.E. Christiansen (eds.) Marine Biology of Polar Regions and Effects of Stress on Marine organisms. John Wiley & Sons, Chichester.

Gustavson, T.C. 1975. Sedimentation and physical limnology in proglacial Malaspina Lake, southeastern Alaska. pp. 249–263. In: A.V. Jopling and B.C. McDonald (eds.) Glaciofluvial and Glaciolacustrine Sedimentation. Soc. Econ. Paleont. Mineral. Spec. Pub. 23.

Hahn, H.J., and Stumm, W. 1970. The role of coagulation in natural waters. Amer. J. Sci. 268: 354–368.

Hånkanson, L., and Jansson, M. 1983. Principles of Lake Sedimentology. Springer-Verlag, New York 316 p.

Hall, P.O.J., Anderson, L.G., Iverfeldt, Å, van der Loeff, M.M.G., Sundby, B., and Westerlund, S.F.G. Oxygen uptake kinetics across the benthic boundary layer (in prep.).

Hamblin, P.F., and Carmack, E.C. 1978. River-induced currents in a fjord lake. J. Geophys. Res. 83: 885–899.

Hambrey, M.J., and Harland, W.B. 1979. Analysis of pre-Pleistocene glacigenic rocks: Aims and problems. pp. 271–276. In: C.H. Schluchter (ed.) Moraines and Varves: Origin, Genesis, Classification. A.A. Salkema.

Hamilton, P., Gunn, J.T., and Cannon, G.A. 1985. A box model of Puget Sound. Est. Coast. Shelf Sci. 20: 673–692.

Hampton, M.A., Bouma, A.H., Torresan, M.E., and Colburn, I.P. 1978a. Analysis of microtextures on quartz sand grains from lower Cook Inlet, Alaska. Geology (Boulder, USA) 6: 105–110.

Hampton, M.A., Bouma, A.H., Carlson, P.R., Molnia, B.F., Clukey, E.C., and Sangrey, D.A. 1978b. Quantitative study of slope instability in the Gulf of Alaska. Offshore Technol. Conf. 2307–2312.

Hansen, D.V., and Rattray, M. 1966. New dimensions in estuary classification. Limnol. Oceanogr. 11: 319–326.

Hansom, J.D. 1983. Ice-formed intertidal boulder pavements in the sub-antarctic. J. Sed. Pet. 53: 135–145.

Hargrave, B.T. 1969. Similarity of oxygen uptake by benthic communities. Limnol. Oceanogr. 14:801–805.

———. 1973. Coupling carbon flow through some pelagic and benthic communities. J. Fish. Res. Bd. Can. 30: 1317.

———. 1980. Factors affecting the flux of organic matter to sediments in a marine bay. pp. 243–263. In: K.R. Tenore and B.C. Coull (eds.) Marine Benthic Dynamics. Univ. S. Carolina Press, Columbia.

Hargrave, B.T., and Burns, W. 1979. Assessment of sediment trap collection efficiency. Limnol. Oceanogr. 24: 1124–1136.

Hargrave, B.T., and Taguchi, S. 1978. Origin of deposited material sedimented in a marine bay. J. Fish. Res. Bd. Can. 35: 1604–1613.

Harleman, D.R.F. 1961. Stratified flow. pp. 26-1–26-21. In: V.L. Streeter (ed.) Handbook of Fluid Dynamics. McGraw-Hill, New York.

Harris, G.P. 1984. Phytoplankton productivity and growth measurements: past, present and future. J. Plankt. Res. 6: 219–237.

Harris, P.W. 1976. Sedimentation in a proglacial lake: S.E. Jokalsolan, Iceland. Unpublished Ph.D. thesis, University of East Anglia, Norwich.

Harrison, P.J., Fulton, J.D., Taylor, F.J.R., and Parsons, T.R. 1983. Review of the biological oceanography of the Strait of Georgia: Pelagic environment. Can. J. Fish. Aquat. Sci. 40: 1064–1094.

Harrison, W.G., Platt, T., and Irwin, B. 1982. Primary production and nutrient assimilation by natural phytoplankton populations of the eastern Canadian Arctic. Can. J. Fish. Aquat. Sci. 39: 335–345.

Hartley, C.H., and Dunbar, M.J. 1938. On the hydrographic mechanism of the so-called brown zones associated with tidal glaciers. J. Mar. Res.1: 305–311.

Haselton, G.M. 1965. Glacial geology of Muir Inlet, southeastern Alaska. Inst. Polar Studies Rep. 18: 1–34.

Hattersley-Smith, G. 1964. Rapid advance of a glacier in northern Ellesmere Island. Nature 201: 176.

———. 1969. Glacial features of Tanguary Fiord and adjoining areas of northern Ellesmere Island, N.W.T. J. Glaciology 8: 23–50.

Hattersley-Smith, G., and Serson, H. 1964. Stratified water of a glacial lake in northern Ellesmere Island. Arctic 17: 109–110.

———. 1966. Reconnaissance oceanography over the ice of the Nansen sound fiord system. Operation Tanguary Defense. Res. Bd. Dept. Nat 1. Defense Hazen 28: 1–13.

Hattersley-Smith, G., Fuzesy, A., and Evans, S. 1969. Glacier depths in northern Ellesmere Island: Airborne radio echo sounding in 1966. Defence Res. Bd., Ottawa, Rep. Geophys., Hazen 36.

Hattersley-Smith, G., Keys, J.E., Serson, H., and Mielke, J.E. 1970. Density stratified lakes in northern Ellersmere Island. Nature 225: 55–56.

Hay, A.E. 1981. Submarine channel formation and acoustic remote sensing of suspended sediment and turbidity currents in Rupert Inlet, B.C. Ph.D. Thesis, Dept. Oceanography, Univ. British Columbia, Vancouver.

———. 1982. The effects of submarine channels on mine tailings disposal in Rupert Inlet, B.C. pp. 139–181. In: D.V. Ellis (ed.) Marine Tailings Disposal. Ann Arbor Science, Ann Arbor, Michigan.

———. 1983a. Acoustic remote sensing. pp. 17-1 to 17-3. In: J.P.M. Syvitski and C.P. Blakeney (Comp.) Sedimentology of Arctic Fjords Experiment: HU 82-031 Data Report, Vol. 1. Can. Data Rep. Hydrogr. Oc. Sci. 12.

———. 1983b. Cambridge Fiord polynya experiment: The buoyant jet. pp. 7-1 to 7-4. In: J.P.M. Syvitski and C.P. Blakeney (Comp.) Sedimentology of Arctic Fjords Experiment: HU 82-031 Data Report, Vol. 1. Can. Data Rep. Hydrogr. Oc. Sci. 12.

Hay, A.E., Burling, R.W., and Murray, J.W. 1982. Remote acoustic detection of a turbidity current surge. Science 217: 833–835.

Hay, A.E., Murray, J.W., and Burling, R.W. 1983a. Submarine channels in Rupert Inlet, British Columbia: I. Morphology. Sed. Geol. 36: 289–315.

Hay, A.E., Murray, J.W., and Burling, R.W. 1983b. Submarine channels in Rupert Inlet, British Columbia: II. Sediments and sedimentary structures. Sed. Geol. 36: 317–339.

Heezen, B.C., Menzies, R.J., Schneider, E.D., Ewing, W.M., and Granelli, N.C.L. 1964. Congo submarine canyon. Bull. Am. Assoc. Petrol. Geol. 48: 1126–1149.

Heggie, D.T. 1982. Copper in surface waters of the Bering Sea. Geochim. Cosmochim. Acta 46: 1301–1306.

———. 1983. Copper in the Resurrection fjord, Alaska. Est. Coast. Shelf Sci. 17: 613–635.

Heggie, D.T., and Burrell, D.C. 1975. Distributions of copper in interstitial waters and the water column of an Alaskan fjord. Amer. Geophys. Union. Trans. 56: 1006.

———. 1977. Hydrography, nutrient chemistry and primary productivity of Resurrection Bay, Alaska. Report R77-2, Inst. Mar. Sci., Univ. Alaska, Fairbanks, 111 p.

———. 1980. Sediment-seawater exchanges of nutrients and transition metals in an Alaskan fjord. pp. 675–680. In: H.J. Freeland, D.M. Farmer, and C.D. Levings (eds.) Fjord Oceanography. Plenum Press, New York.

———. 1981. Deep water renewals and oxygen consumption in an Alaskan fjord. Est. Coast. Shelf. Sci. 13: 83–99.

———. 1982. Depth distributions of copper in the water column and interstitial waters of an Alaskan fjord. pp. 317–334. In: K.A. Fanning and F.T. Manheim (eds.) The Dynamic Environment of the Ocean Floor. D.C. Heath & Co., Lexington, Mass.

Heggie, D.T., Lewis, T., and Saravo, E. Cobalt in pore waters of marine sediments. Nature (in press).

Hegseth, E.N. 1982. Chemical and species composition of the phytoplankton during the first spring bloom in Trondheimsfjorden, 1975. Sarsia 67: 131–141.

Hein, F.J., and Longstaffe, F.J. 1983. Geotechnical, sedimentological and mineralogical investigations in arctic fjords. pp. 11-1 to 11-158. In: J.P. Syvitski and C.P. Blakeney (Comp.) Sedimentology of Arctic Fjords Experiment: HU 82-031 data report, Vol. 1. Can. Data Rep. Hydrogr. Oc. Sci. 12.

———. 1985. Sedimentologic, mineralogic, and geotechnical descriptions of fine-grained slope and basin deposits, Baffin Island Fiords. Geo-Marine Lett. 5: 11–16.

Hein, J.R., Bouma, A.H., Hampton, M.A., and Ross, C.S. 1979. Clay mineralogy, suspended sediment dispersal and inferred current patterns, lower Cook Inlet and Kodiak Shelf, Alaska. Sed. Geol. 24: 291–306.

Heling, D. 1974. Investigations on recent sediments in the Fiskenaesset region, Southern west Greenland. Rapp. Grønlands Geol. Unders. 65: 65–68.

———. 1977. Mineralogy of recent pelitic sediments from the Fiskenaesset region, southern west Greenland. Rapp. Grønlands Geol. Unders. 73: 100–108.

Helland-Hansen, B. 1906. Current measurements in Norwegian fjords, the Norwegian Sea and the North Sea in 1906. Berg. Mus. Årbok 1907 15: 1–61.

Helle, H.B. 1978. Summer replacement of deep water in Byfjord, Western Norway: Mass exchange across the sill induced by coastal upwelling. pp. 441–464. In: J.C.J. Nihoul (ed.) Hydrodynamics of Estuaries and Fjords. Elsevier, Amsterdam.

Henkel, D.J. 1972. The role of waves in causing submarine landslides. Geotechnique 20: 75–80.

Herman, Y., O'Neil, J.R., and Drake, C.L. 1977. Micropaleontology and paleotemperatures of postglacial SW Greenland fjord cores. pp. 357–381. In: Y. Vasari, H. Hyvarinen, and S. Hicks (eds.) Climatic Changes in Arctic Areas during the last 10,000 years. Acta Univ. Ouluensis Ser. A.

Hill, P.R., Moran, K.M., and Blasco, S.M. 1982. Creep deformation of slope sediments in the Canadian Beaufort Sea. Geo-Marine Lett. 2: 163–170.

Hines, M.E., Orem, W.H., Lyons, W.B., and Jones, G.E. 1982. Microbial activity and bioturbation-induced oscillations in pore water chemistry of estuarine sediments in spring. Nature 299: 433–435.

Hirst, J.M., and Aston, S.R. 1983. Behavior of copper, zinc, iron and manganese during experimental resuspension and reoxidation of polluted anoxic sediment. Est. Coast. Shelf Sci. 16: 549–558.

Hjort, J., and Gran, H.H. 1900. Hydrographic-biological investigations of the Skagerrak and the Christiania fjord. Rep. Norw. Fishery Mar. Invest. 1: 1–40.

Hjulström, F. 1935. Studies of the morphological activities of rivers as illustrated by the River Fyris. Univ. of Uppsala, Geol. Inst. Bull. 25: 221–527.

Hobbie, J.E., and Lee, G. 1980. Microbial production of extracellular material: Importance in benthic ecology. pp. 341–346. In: K.R. Tenore and B.C. Coull (eds.) Marine Benthic Dynamics. Univ. South Carolina Press, Columbia.

Hodgson, D.A. 1973. Landscape and late-glacial history, head of Vendom Fiord, Ellesmere Island. Geol. Surv. Canada Paper 73-1, part B: 129–136.

Hofsten, B.V., and Edberg, N. 1972. Estimating the rate of degradation of cellulose fibres in water. Oikos 23: 29–34.

Hogarty, B.J. 1985. The fate of molybdenum and other heavy metals in two southeast Alaska fjords. Unpublished M.S. thesis, Univ. Alaska, Fairbanks, 151 p.

Holland, H.D. 1984. The Chemical Evolution of the Atmosphere and Oceans. Princeton University Press, Princeton, 582 p.

Holliday, L.M., and Liss, P.S. 1976. The behavior of dissolved iron, manganese and zinc in the Beaulieu estuary, S. England. Est. Coast. Mar. Sci. 4: 349–353.

Holtedahl, H. 1955. On the Norwegian continental terrace, primarily outside Möre - Romsdal: Its geomorphology and sediments. Univ. Bergen Arb. Natur. 14: 1–209.

———. 1960. Mountain, fjord, strandflat geomorphology and general geology of parts of western Norway. Int. Geol. Congr. 21, Guide to excursion A_6, C_3. Oslo.

———. 1965. Recent turbidites in the Hardangerfjord, Norway. Colston Res. Soc. Proc. 17, Bristol: 107–141.

———. 1967. Notes on the formation of fjord and fjord valleys. Georgr. Ann. 49, Ser. A: 188–203.

———. 1975. The Geology of the Hardangerfjord, West Norway. Norges Geol. Unders., Bull 36 (323): 1–87.

Honjo, S. 1980. Material fluxes and modes of sedimentation in the mesopelagic and bathypelagic zones. J. Mar. Res. 38: 53–97.

Honjo, S., and Roman, M.R. 1978. Marine copepod fecal pellets: Production, preservation and sedimentation. J. Mar. Res. 36: 45–57.

Hoos, L.M., and Vold, C.L. 1975. Special Estuaries Series 2, The Squamish River Estuary. Status of Environmental Knowledge to 1974. Environment Canada.

Hopkins, C.C.E. 1981. Ecological investigations on the zooplankton community of Balsfjorden, Northern Norway: Changes in zooplankton abundance and biomass in relation to phytoplankton and hydrography, March 1976–February 1977. Kieler Meeresforsch. Sonderh. 5: 124–139.

Höpner, T., and C. Orliczek. 1978. Humic matter as a component of sediments in estuaries, pp. 70–74. In: Biogeochemistry of Estuarine Sediments. UNESCO, Paris.

Horne, E.P.W. 1985. Ice-induced vertical circulation in an Arctic fiord. J. Geophys. Res. 90: 1078–1086.

Horner, R.A. 1976. Sea ice organisms. Oceanogr. Mar. Biol. Ann. Rev. 14: 167–182.

Hoskin, C.M. 1977. Macrobenthos from three fjords in western Prince William Sound, Alaska. Report R77-1, Inst. Mar. Sci., Univ. Alaska, Fairbanks 26 p.

Hoskin, C.M., and Burrell, D.C. 1972. Sediment transport and accumulation in a fjord basin, Glacier Bay, Alaska. J. Geol. 80: 539–551.

Hoskin, C.M., Burrell, D.C. and Freitag, G.R. 1976. Suspended sediment dynamics in Queen Inlet, Glacier Bay, Alaska. Mar. Sci. Commun. 2: 95–108.

———. 1978. Suspended sediment dynamics in Blue Fjord, Western Prince William Sound, Alaska. Est. Coast Mar. Sci. 7: 1–16.

Huggett, W.S., and Wigen, S.O. 1983. Surface currents in the approaches to Kitimat. pp. 34–65. In: R.W. Macdonald (ed.) Proceedings of a Workshop on the Kitimat Marine Environment. Can. Tech. Rep. Hydrogr. Oc. Sci. 18.

Hughes, R.G. 1984. A model of the structure and dynamics of benthic marine invertebrate communities. Mar. Ecol. Prog. Ser. 15: 1–11.

Huizenga, D.L., and Kester, D.R. 1983. The distribution of total and electrochemically available copper in the northwestern Atlantic Ocean. Mar. Chem. 13: 281–291.

Hulings, N.C., and Gray, J.S. 1976. Physical factors controlling abundance of meiofauna on tidal and atidal beaches. Mar. Biol. 34: 77–83.

Hunter, J.A., and Godfrey, R.J. 1975. A shallow marine refraction survey, Cunningham Inlet, Somerset Island, N.W.T. Geol. Surv. Canada Paper 75-1, Part B: 19–22.

Hunter, K.A. 1983. On the estuarine mixing of dissolved substances in relation to colloid stability and surface properties. Geochim. Cosmochim. Acta 47: 467–473.

Hunter, K.A., and Liss, P.S. 1982. Organic matter and the surface change of suspended particles in estuarine waters. Limnol. Oceanogr. 27: 322–335.

Huppert, H.E. 1980. Topographic effects in stratified fluids. pp.117–141. In: H.J. Freeland, D.M. Farmer, and C.D. Levings (eds.) Fjord Oceanography. Plenum Press, New York.

Inderbitzen, A.L. 1970. Empirical relationships between mass physical properties for recent marine sediments off southern California. Mar. Geol. 9: 311–329.

Irwin, J. 1978. Bathymetry of Caswell and Nancy Sounds. pp. 11–18. In: G.P. Glasby (ed.) Fiord Studies: Caswell and Nancy Sounds, New Zealand, N.Z. Oceanogr. Inst. Mem. 79.

Jacobs, L., and Emerson, S. 1982. Trace metal solubility in an anoxic fjord. Earth Planet. Sci. Lett. 60: 237–252.

Jacobs, L., Emerson, S., and Skei, J. 1985. Partitioning and transport of metals across the O_2/H_2S interface in a permanently anoxic basin: Framvaren Fjord, Norway. Geochim. Cosmochim. Acta. 49: 1433–1444.

Jahn, A. 1967. Some features of mass movement of Spitsbergen slopes. Geografiska Annaler, 49A: 213–225.

Jahnke, R.A., Emerson, S.R., and Murray, J.W. 1982. A model of oxygen reduction, denitrification, and organic matter mineralization in marine sediments. Limnol.Oceanogr. 27: 610–623.

Jenne, E.A. 1968. Controls of Mn, Fe, Co, Ni, Cu and Zn concentrations in water: The significant role of hydrous Mn and Fe oxides. pp. 337–387. In: Trace Inorganics in Water. American Chemical Society, Washington, D.C.

Jerlov, N.G. 1953. Influence of suspended and dissolved matter on the transparency of sea water. Tellus 5: 59–65.

Jerneløv, A., and Rosenberg, R. 1976. Stress tolerance of ecosystems. Environ. Conserv. 3: 43–46.

Johansson, S. 1980. Impact of oil on the pelagic ecosystem. pp. 61–80. In: J.J. Kineman, R. Elmgren, and S. Hansson (eds.) The Tsesis Oil Spill. U.S. Dept. Commerce, Boulder, Colorado.

Johns, R.E. 1962. A study of the density structure and water flow in the upper 10 m in a selected region of Bute Inlet. M.Sc. Thesis (unpubl.) Univ. British Columbia, Vancouver, B.C. 74 p.

Johnson, D.W. 1915. The Nature and Origin of Fjords. John Murray, London.

Johnson, R.D. 1974. Dispersal of recent sediments and mine tailings in a shallow-silled fjord, Rupert Inlet, British Columbia. Unpublished Ph.D. thesis, Univ. British Columbia, Vancouver, 181 p.

Johnson, R.G. 1970. Variations in diversity within benthic marine communities. Am. Nat. 104: 285–292.

———. 1974. Particulate matter at the sediment-water interface in coastal environments. J. Mar. Res. 32: 313–330.

Johnston, R. 1984. Oil pollution and its management. pp. 1433–1582. In: O. Kinne (ed.) Marine Ecology, Vol. 5. John Wiley & Sons, Chichester.

Johnstone, C.S., Jones, R.G., and Hunt, R.D. 1977. A seasonal carbon budget for a laminarian population in a Scottish sea-loch. Helgol. wiss Meeresunters. 30: 527–545.

Jordan, G.F. 1962. Redistribution of sediments in Alaskan bays and inlets. Geograph. Review 52: 548–558.

Jörestad, F. 1968. Leirskred i Norge. Norsk. geogr. Tidsskr. 22: 214–219.

Josefson, A.B. 1981. Persistence and structure of two deep macrobenthic communities in the Skagerrak. J. Exp. Mar. Biol. Ecol. 50: 63–97.

Josefsson, B., and Nyquist, G. 1976. Fluorescence tracing of the flow and dispersion of sulfite wastes in a fjord system. Ambio 5: 183–187.

Joyce, J.R.F. 1950. Notes on ice foot development, Nerig Fjord, Graham Land, Antarctica. J. Geol. 58: 646–649.

Jumars, P.A., and Fauchald, K. 1977. Between-community contrast in successful polychaete feeding strategies. pp. 1–20. In: B.C. Coull (ed.) Ecology of Marine Benthos. Univ. S. Carolina Press, Columbia, S.C.

Kaartvedt, S. 1984. Effects on fjords of modified freshwater discharge due to hydroelectric power production (in Norwegian). Fisken Hav. 3: 1–104.

Kalinske, A.A. 1947. Movement of sediment as bedload in rivers. Trans. Am. Geophys. Union Vol. 28, No. 4.

Kane, J.E. 1967. Organic aggregates in surface waters of the Ligarian Sea. Limnol. Oceanogr. 12: 287–294.

Kaplin, P.A. 1962. Fiord-indented shores of the Soviet Union, Moscow, 260 p.

Karlsrud, K. 1982. Stability evaluations for submarine slopes—Summary report. Norweg. Geotech. Inst. Rep. 52207–10, 30 p.

Karlsrud, K., and By, T. 1981. Stability evaluations for submarine slopes.Results of new investigations of the Orkdalsfjord slide. Norweg. Geotech. Inst. Rep. 52207-09, 32 p.

Karlsrud, K., and Edgers, L. 1982. Some aspects of submarine slope stability. Norweg. Geotech. Inst. Rep. 52207-05, 21 p.

Kathman, R.D., Brinkhurst, R.O., Woods, R.E., and Jeffries, D.C. 1983. Benthic studies in Alice Arm and Hastings Arm, B.C., in relation to mine tailing dispersal. Can. Tech. Rep. Hydrogr. Ocean. Sci. No. 22, Inst. Ocean Sci., Sidney, B.C., 30 p.

Kathman, R.D., Brinkhurst, R.O., Woods, R.E., and Cross, S.F. 1984. Benthic studies in Alice Arms, B.C., following cessation of mine tailings disposal. Can. Tech. Rep. Hydrogr. Ocean. Sci. No. 37, Inst. Ocean Sci., Sidney, B.C., 57 p.

Kattner, G., Hammer, K.D., Eberlein, K., Brockmann, W.H., Jahnke, J., and Krause, M. 1983. Nutrient and plankton development in Rosfjorden and enclosed ecosystems captured from changing water bodies during POSER. Mar. Ecol. Prog. Ser. 14: 29–43.

Kaula, W.M. 1980. Problems in understanding vertical movements and earth rheology. pp. 577–588. In: N-A. Morner (ed.). Earth Rheology, Isostasy and Eustasy. John Wiley & Sons, London.

Keen, M.J., and Piper, D.J.W. 1976. Kelp, methane, and an impenetrable reflector in a temperate bay. Can. J. Earth Sci. 13: 312–318.

Kelly, J.R., and Nixon, S.W. 1984. Experimental studies of the effect of organic deposition on the metabolism of a coastal marine bottom community. Mar. Ecol. Prog. 17: 157–169

Kelly, M. 1973. Radiocarbon dated shell samples from Nordre Strømfjord, West Greenland, with comments on models of glacio-isostatic uplift. Grønlands Geol. Unders. Rap. 59, 20 p.

Kerndorff, H., and Schnitzer, M. 1980. Sorption of metals on humic acid. Geochim. Cosmochim. Acta 44: 1701–1708.

Ketchum, B.H. 1954. Relation between circulation and planktonic populations in estuaries. Ecology 35: 191–200.

Keys, J.E. 1978. Water regime of Disraeli Fjord, Ellesmere Island. Defence Research Establishment Ottawa, Rep. 792: 1–54.

Keys, J., Johannesson, O.M., and Long, A. 1969. The oceanography of Disraeli Fjord northern Ellesmere Island. Canada Defence Res. Bd. Geophysics. Hazen 34.

Khalid, R.A., Patrick, W.H., and Gambrell, R.P. 1978. Effect of dissolved oxygen on chemical transformations of heavy metals, phosphorus and nitrogen in an estuarine sediment. Est. Coast. Mar. Sci. 6: 21–35.

Kharkar, D.P., Turekian, K.K., and Bertine, K.K. 1968. Stream supply of dissolved silver, molybdenum, antimony, selenium, chromium, cobalt, rubidium and cesium to the oceans. Geochim. Cosmochim. Acta 32: 285–298.

Kiørboe, T. 1979. The distribution of benthic invertebrates in Holbaek Fjord (Denmark) in relation to environmental factors. Ophelia 18: 61–81.

Kitching, J.A., Ebling, F.J., Gamble, J.C., Hoave, R., McLeod, A.A., and Norton, T.A. 1976. The ecology of Lough Ine. XIX. Seasonal changes in the western trough. J. Anim. Ecol. 45: 731–758.

Kjemperud, A. 1981. A shoreline displacement investigation from Frosta in Trondheimsfjorden, Nord-Trøndelag, Norway. Norsk Geologisk Tidsskrift 61: 1–15.

Kjerfve, B. (ed.) 1978. Estuarine Transport Processes. Univ. South Carolina Press, 300 p.

Klinck, J.M., O'Brien, J., and Svendsen, H. 1981. A simple model of fjord and coastal circulation interaction. J. Phys. Oceanogr. II: 1612–1626.

Klinkhammer, G.P. 1980. Early diagenesis in sediments from the eastern equatorial Pacific. II. Pore water metal results. Earth Planet. Sci. Lett. 49: 81–101.

Klinkhammer, G., Heggie, D.T., and Graham, D.W. 1982. Metal diagenesis in oxic marine sediments. Earth Planet. Sci. Lett. 61: 211–219.

Klump, J.V., and Martens, C.S. 1981. Biogeochemical cycling in an organic rich coastal marine basin. II. Nutrient sediment-water exchange processes. Geochim. Cosmochim. Acta 45: 101–121.

Knauer, G.A., and Martin, J.U. 1973. Seasonal variations of cadmium, copper, manganese, lead and zinc in water and phytoplankton in Monterey Bay, California. Limnol. Oceanogr. 18: 597–605.

Knight, R.J. 1971. Distributional trends in the recent marine sediments of Tasiujaq Cove of Ekalugad Fiord, Baffin Island, N.W.T. Marine Sediments 7: 1–18.

Knutzen, J., and Sortland, B. 1982. Polycyclic aromatic hydrocarbons (PAH) in some algae and invertebrates from moderately polluted parts of the coast of Norway. Water Res. 16: 421–428.

Kostaschuk, R.A., and McCann, S.B. 1983. Observations on delta-forming processes in a fjord-head delta, British Columbia. In: J. Syvitski and J. Skei (eds.) Sedimentology of Fjords. Sed. Geol. 36: 269–288.

Kranck, K. 1973. Flocculation of suspended sediment in the sea. Nature 246: 348–350.

———. 1975. Sediment deposition from flocculated suspension. Sedimentology 22: 111–123.

———. 1981. Particulate matter grain-size characteristics and flocculation in a partially mixed estuary. Sedimentology 28: 107–114.

Krause, E.P., and Lewis, A.G. 1979. Ontogenetic migration and the distribution of *Eucalanus bungii* in British Columbia inlets. Can. J. Zool. 57: 2211–2222.

Krauskopf, K.B. 1956. Factors controlling the concentrations of thirteen rare metals in seawater. Geochim. Cosmochim. Acta 9: 1–32B.

Kremling, K. 1983. The behavior of Zn, Cd, Cu, Ni, Co, Fe and Mn in anoxic Baltic waters. Mar. Chem. 13: 87–108.

Kremling, K., Wench, A., and Osterroht, C. 1983. Variations of dissolved organic copper in marine waters. pp. 609–620. In: C.S. Wong, E. Boyle, K.W. Bruland, J.D. Burton, and E.D. Goldberg (eds.) Trace Metals in Seawater. Plenum Press, New York.

Kristensen, E. 1984. Effect of natural concentrations on nutrient exchange between a polychaete burrow in estuarine sediment and the overlying water. J. Exp. Mar. Biol. Ecol. 75: 171–190.

Krom, M.D., and Berner, R.A. 1981. The diagenesis of phosphorus in a nearshore marine sediment. Geochim. Cosmochim. Acta 45: 207–216.

Krom, M.D., and Sholkovitz, E.R. 1978. On the association of iron and manganese with organic matter in anoxic marine pore waters. Geochim. Cosmochim. Acta 42: 607–611.

Krone, R.B. 1978. Aggregation of suspended particles in estuaries. pp. 177–190. In: B. Kjerfve (ed.) Estuarine Transport Processes. University South Carolina Press, Columbia.

Kudusov, E.A. 1973. Integrated value and rate of abrasion of the fjord coast of Kamchatka during the Holocene. Geomorfologiya 3: 73–75.

Lafond, C., and Pickard, G.L. 1975. Deep water exchanges in Bute Inlet, British Columbia. J. Fish. Res. Bd. Can. 32: 2075–2089.

Lake, R.A., and Walker, E.R. 1975. Notes on the oceanography of d'Ilberville Fiord. Arctic 26: 222–229.

———. 1976. A Canadian arctic fjord with some comparisons to fjords of the Western Americas. J. Fish. Res. Bd. Can. 33: 2277–2285.

Lännergren, C. 1975. Phosphate, silicate, nitrate and ammonia in Lindåspollene, a Norwegian land-locked fjord. Sarsia 59: 53–66.

———. 1979. Buoyancy of natural populations of marine phytoplankton. Mar. Biol. 54: 1–10.

Larsen, B. 1977. Investigations of the sediment thickness in Söndre Strömfjord, West Greenland. Grönlands Geologiske Undersøgelse 85: 47–98.

Larsen, T. 1980. A way to measure the residual flow in the Limfjord, Denmark. pp. 333-341. In: H.J. Freeland, D.M. Farmer, and C.D. Levings (eds.) Fjord Oceanography. Plenum Press, New York.

Lasaga, A.C. 1979. The treatment of multi-component diffusion and ion pairs in diagenetic fluxes. Amer. J. Sci. 279: 324–346.

Lasaga, A.C., and Holland, H.D. 1976. Mathematical aspects of non-steady state diagenesis. Geochim. Cosmochim. Acta 40: 257–266.

Lavrushin, Y.A. 1968. Features of deposition and structure of the glacialmarine deposits under conditions of a fiord-coast. Translations from Lithology and Economic Minerals 3: 63–79.

Lees, D.C., and Houghton, J.P. 1980. Effects of drilling fluids on benthic communities at the lower Cook Inlet C.O.S.T. well. pp. 209–350. In: Symposium on research on environmental fate and effects of drilling fluids and cuttings. Contemporary Associates, Washington, D.C.

Leppakoski, E. 1969. Transitory return of the benthic fauna of the Bornholm Basin after extermination by oxygen insufficiency. Cah. Biol. Mar. 10: 163–172.

Lerman, A. (ed.). 1978. Lakes: Chemistry, Geology, Physics. Springer-Verlag, New York. 363 p.

———. 1979. Geochemical Processes: Water and Sediment Environments. John Wiley, & Sons, New York. 481 pp.

Letson, J.R.J. 1981. Sedimentology of southwestern Mahone Bay, Nova Scotia. Unpubl. M.Sc. thesis, Dalhousie Univ., Halifax, N.S., 199 p.

Levings, C.D. 1973. Intertidal benthos of the Squamish estuary. Fish. Res. Bd. Can. MS. Rep: 1218: 60 p.

———. 1980a. Benthic biology of a dissolved oxygen deficiency event in Howe Sound, B.C. pp. 515–522. In: H.J. Freeland, D.M. Farmer, and C.D. Levings (eds.) Fjord Oceanography. Plenum Press, New York.

———. 1980b. Demersal and benthic communities in Howe Sound basin and their responses to dissolved oxygen deficiency. Can. Tech. Rep. Fish. Aquat. Sci. 951, 127 p.

Levings, C.D., Foreman, R.E., and Tunnicliffe, V.J. 1983. Review of the benthos of the Strait of Georgia and contiguous fjords. Can. J. Fish. Aquat. Sci. 40: 1120–1141.

Levington, J.S. 1979. Deposit-feeders, their resources, and the study of resource limitation. pp. 117–141. In: R.J. Levingston (ed.) Ecologial Processes in Coastal and Marine Systems. Plenum Press, New York.

Levington, J.S., and Bambach, B.K. 1970. Some ecological aspects of bivalve mortality patterns. Amer. J. Sci. 268: 97–112.

Lewin, J.C., and Mackas, D. 1972. Blooms of surf-zone diatoms along the coast of the Olympic Peninsula, Washington I. Physiological investigations of *Chaetoceras armatum* and *Asteronella sociales* in laboratory cultures. Mar. Biol. 16: 171–181.

Lewis, A.G., and Syvitski, J.P.M. 1983. The interaction of plankton and suspended sediment in fjords. Sed. Geol. 36: 81–92.

Lewis, E.L., and Perkin, R.G. 1982. Seasonal mixing processes in an Arctic fjord system. J. Phys. Oceanogr. 12: 74–83.

Lewis, E.L., and Walker, E.R. 1970. The water structure under a growing sea ice sheet. J.Geophys. Res. 75: 6836–6845.

Lewis, K.B. 1971. Slumping on a continental slope inclined at 1° to 4°. Sedimentology 16: 97–110.

Lewis, M.R., and Platt, T. 1982. Scales of variability in estuarine ecosystems. pp. 3–20. In: V.S. Kennedy (ed.) Estuarine Comparisons. Academic Press, New York.
Lewis, T. 1983. Bottom water temperature variations as observed, and as recorded in the bottom sediments, Alice Arm and Douglas Channel, British Columbia. Can. Tech. Rep. Hydrogr. Ocean Sci. 18: 138–161.
Lewis, C.F.M., Blasco, S.M., Bornhold, B.D., Hunter, J.A.M., Judge, A.S., Keer, J.W., McLaren, P., and Pelletier, B.R. 1977. Marine geological and geophysical activities in Lancaster Sound and adjacent fiords. Geol. Surv. Can. Pap. 77-1A: 495–506.
Li, Y.H. 1981. Ultimate removal mechanisms of elements from the ocean. Geochim. Cosmochim. Acta 45: 1659–1664.
Li, Y.H., and Gregory, S. 1974. Diffusion of ions in sea water and in deep-sea sediments. Geochim. Cosmochim. Acta 38: 703–714.
Li, Y.H., Burkhardt, L., Buchholtz, M., O'Hara, P., and Santschi, P.H. 1984. Partition of radiotracers between suspended particles and seawater. Geochim. Cosmochim. Acta 48: 2011–2019.
Lie, W., and Evans, R.A. 1973. Long-term variability in the structure of subtidal benthic communities in Puget Sound, Washington, USA. Mar. Biol. 21: 122–126.
Lieberman, S. 1979. Stability of copper complexes with seawater humic substances. Ph.D. thesis, University of Washington, Seattle.
Liestøl, O. 1969. Glacial surges in West Spitsbergen. Can. J. Earth Sci. 6: 895–897.
Lindqvist, O., Jernelöv, A., Johansson, K., and Rhode, H. 1984. Mercury in the Swedish Environment. Global and Local Sources. National Swedish Environment Protection Board Report 1816, 105 p.
Lipschultz, F., Fox, L.E., and Wofsy, S.C. 1983. Transformations of nitrogen in the tidal Delaware river. EOS 64: 1074.
Lliboutry, L. 1965. Glaciers variations du climat sols gelés. In: Masson and Cie (eds.) Traite de Glaciologie TomeII: 429–504, 517–520, 680–682, 763–764.
Løken, O.H., and Hodgson, D.A. 1971. On the submarine geomorphology along the east coast of Baffin Island. Can. J. Earth Sci. 8: 185–195.
Løken, T., and Torrance, J.K. 1971. The geochemistry of leached Drammen marine clay. Geol. Foreningens Stockholm Förhandlingar 93: 171–175.
Long, R.R. 1975. Circulations and density distribution in a deep strongly stratified, two-layer estuary. J. Fluid Mech. 71: 529–540.
———. 1980. The fluid mechanical problem of fjord circulations. pp.67–116. In: H.J. Freeland, D.M. Farmer, and C.D. Levings (eds.) Plenum Press, New York.
Longhurst, A., Sameoto, D., and Herman, A. 1984. Vertical distribution of arctic zooplankton in summer: Eastern Canadian archipelago. J. Plank. Res. 6: 137–168.
Loring, D.H. 1975. Mercury in the sediments of the Gulf of St. Lawrence. Can. J. Earth Sci. 12: 1219–1237.
———. 1976a. The distribution and partitions of zinc, copper and lead in the sediments of the Saguenay fjord. Can. J. Earth Sci. 13: 960–971.
———. 1976b. Distribution and partition of cobalt, nickel, chrome, and vanadium in the sediments of the Saguenay fjord. Can. J. Earth Sci. 13: 1706–1718.
———. 1984. Trace metal geochemistry of sediments from Baffin Bay. Can. J. Earth Sci. 21: 1368–1378.
Loring, D.H., and Bewers, J.M. 1978. Geochemical mass balances for mercury in a Canadian fjord. Chem. Geol. 22: 309–330.
Loring, D.H., Rantala, R.T.T., and Smith, J.N. 1983. Response time of Saguenay Fjord sediments to metal contamination. In: Hallberg, R. (ed.) Environmental Biogeochemistry. Ecol. Bull. (Stockholm) 35: 59–72.
Lu, J.C.S., and Chen, K.Y. 1977. Migration of trace metals in interfaces of seawater and polluted surficial sediments. Environ. Sci. Tech. 11: 174–182.
Luternauer, F.L., and Swan, D. 1978. Kitimat submarine slump deposit(s) a preliminary report. Current Research Part A, Geol. Surv. Can. Paper 78-1A: 327–332.
Lynn, D.C., and Bonatti, E. 1965. Mobility of manganese in diagenesis of deep-sea sediments. Mar. Geol. 3: 457–474.
Lyons, J.B., Savin, S.M., and Tamburi, A.J. 1971. Basement ice, Ward Hunt ice shelf, Ellesmere Island, Canada. J. Glaciology 10: 93–100.
Lysgard, L. 1969. Forelöbig oversigt over Grönlands Klima. Meddr. dansk Met. Inst. 21, 35 p. (in danish).
Lysyj, I., Perkins, G., Farlow, J.S., and Morris, R.W. 1981. Distribution of hydrocarbons in Port Valdez, Alaska. pp. 47–53. In: Proceedings 1981 Oil Spill Conference, Publ. No. 4334, Amer. Pet. Inst., Washington D.C.
Macdonald, R.W. 1983a. Proceedings of a workshop in the Kitimat marine environment. Can. Tech. Rep. Hydrogr. Ocean Sci., No. 18, 218 p.
———. 1983b. The distribution and dynamics of suspended particles in the Kitimat Fjord system. Can. Tech. Rep. Hydrog. Ocean Sci. 18: 116–137.
Macdonald, R.W., and Murray, J.W. 1973. Sedimentation and manganese concretions in a British Columbia fjord (Jervis inlet). Geol. Surv. Can. paper 73–23, 67 p.

Macdonald, R.W., Cretney, W.J., Wong, C.S., and Erickson, P. 1983. Chemical characteristics of water in the Kitimat fjord system. pp. 67–87. In: R.W. Macdonald (ed.) Proceedings of a Workshop on the Kitimat Marine Environment. Can. Tech. Rep. Hydrogr. Ocean Sci., No. 18.

Mackie, G.O., and Mills, C.E. 1983. Use of the Pisces IV submersible for zooplankton studies in coastal waters of British Columbia. Can. J. Fish. Aquat. Sci. 40: 763–776.

Mackiewicz, N.E., Powell, R.D., Carlson, P.R., and Molnia, B.F. 1984. Interlaminated ice-proximal glacimarine sediments in Muir Inlet, Alaska. Mar. Geol. 57: 113–147.

Madsen, O.S., and Grant, W.D. 1976. Quantitative description of sediment transport by waves. In: Proc. 15th Conf. Coastal Engineering, Honolulu, A.S.C.E. New York 2: 1093–1112.

Mandl, G., and Crans, W. 1981. Gravitational gliding in deltas. pp.41–54. In: Thrust and Nappe Tectonics. Geol. Soc. London.

Mann, K.H. 1975. Relationship between morphometry and biological functioning in three coastal inlets of Nova Scotia. pp. 634–644. In: L.E. Cronin (ed.) Estuarine Research, Vol. 1. Academic Press, New York.

———. 1982. Ecology of Coastal Waters. Univ. Calif. Press, Berkeley, 322 p.

Mantoura, R.F.C., and Woodward, E.M.S. 1983. Conservative behaviour of riverine dissolved organic carbon in the Severn Estuary: Chemical and geochemical implications. Geochim. Cosmochim. Acta 47: 1293.

Mantoura, R.F.C., Dickson, A., and Riley, J.P. 1978. The complexation of metals with humic materials in natural waters. Est. Coast. Mar. Sci. 6: 387–408.

Marcotte, B.M. 1980. The meiobenthos of fjords: A review and prospectus. pp. 557–568. In: H.J. Freeland, D.M. Farmer, and C.D. Levings (eds.) Fjord Oceanography. Plenum Press, New York.

Martin, J.H., and Knauer, G.A. 1973. The elemental composition of plankton. Geochim. Cosmochim. Acta 37: 1639–1653.

———. 1980. Manganese cycling in the northeast Pacific waters. Earth Planet. Sci. Lett. 51: 266–274.

Martin, J.M., and Whitfield, M. 1983. The significance of the river input of chemical elements to the ocean. pp. 265–296. In: C.S. Wong, E. Boyle, K.W. Bruland, J.D. Burton, and E.D. Goldberg (eds.) Trace Metals in Seawater. Plenum Press, New York.

Matisoff, G. 1982. Mathematical models of bioturbation. pp. 289–330. In: P.L. McCall and M.J.S. Tevesz (eds.) Animal-Sediment Relations. Plenum Press, New York.

Mathews, W.H. 1967. Profiles of Late Pleistocene glaciers in New Zealand. N.Z. J. Geol. Geophys. 10: 146–163.

Mathews, W.H., Fyles, J.G., and Nasmith, H.W. 1970. Postglacial crustal movements in southwestern British Columbia and adjacent Washington state. Can. J. Earth Sci. 7: 690–702.

Matthews, J.B. 1981. The seasonal circulation of the Glacier Bay, Alaska fjord system. Est. Coast. Shelf Sci. 12: 679–700.

———. 1983. Some aspects of circulation along the Alaska Beaufort Sea coast. pp. 475–497. In: H.G. Gade, A. Edwards, and H. Svendsen (eds.) Coastal Oceanography. Plenum Press, New York.

Matthews, J.B., and Quinlan, A.V. 1975. Seasonal characteristics of water masses in Muir Inlet, a fjord with tidewater glaciers. J Fish. Res. Bd. Can. 32: 1693–1703.

Matthews, J.B.L., and Heimdal, B.R. 1980. Pelagic productivity and food chains in fjord systems. pp. 377–398. In: H.J. Freeland, D.M. Farmer, and C.D. Levings (eds.) Fjord Oceanography. Plenum Press, New York.

McAlister, W.B., Rattray, M., and Barnes, C.A. 1959. The dynamics of a fjord estuary: Silver Bay, Alaska. Dept. of Oceanography Tech. Rep. 62. Univ. of Washington, Seattle, Washington.

McCall, P.L. 1977. Community patterns and adaptive strategies of the infaunal benthos of the Long Island Sound. J. Mar. Res. 35: 221–266.

———. 1978. Spatial-temporal distributions of Long Island Sound infauna: The role of bottom disturbance in a nearshore marine habitat. pp. 191–219. In: M.L. Wiley (ed.) Estuarine Interactions. Academic Press, New York.

McCall, P.L., and Tevesz, M.J.S. (eds.) 1982. Animal-Sediment Relations. Plenum Press, New York. 336 p.

McCann, S.B. (ed.) 1980a. The Coastline of Canada G.S.C. paper 80-10, 439 p.

———. (ed.) 1980b. Sedimentary Processes and Animal-Sediment Relationships in Tidal Environments. G.A.C. Short Course No. 1, Halifax, Canada, 232 p.

McCann, S.B., Dale, J.E., and Hale, P.B. 1981. Subarctic tidal flats in areas of large tidal range, southern Baffin Island, eastern Canada. Geographie physique et Quaternaire XXXV: 183–204.

McCave, I.N. 1973. Mud in the North Sea. pp. 75–100. In: Goldberg, E.D. (ed.) North Sea Science. M.I.T. Press, Cambridge.

———. 1975. Vertical flux of particles in the ocean. Deep-Sea Res. 22: 491–502.

———. (ed.) 1976. The Benthic Boundary Layer. Plenum Press, New York.

McClimans, T.A. 1978a. Fronts in fjords. Geophys. Astrophys. Fluid Dynamics 11: 23–34.

———. 1978b. On the energetics of tidal inlets to landlocked fjords. Mar. Sci. Comm. 4: 121–137.

———. 1979. On the energetics of river plume entrainment. Geophys. Astrophys. Fluid Dynamics 13: 67–82.

McDaniels, N.G. 1973. A survey of the benthic macro-invertebrate fauna and solid pollutants in Howe Sound. Fish. Res. Bd. Can. Tech. Rep. 385: 64 p.
McLaren, I.A. 1967. Physical and chemical characteristics of Ogac Lake, a landlocked fjord on Baffin Island. J. Fish. Res. Bd. Canada 24: 981–1015.
McMahon, T.G., and Patching, J.W. 1984. Fluxes of organic carbon in a fjord on the west coast of Ireland. Est. Coast. Shelf Sci. 19: 205–215.
Meadows, P.S., and Campbell, J.L. 1972. Habitat selection by aquatic invertebrates. Adv. Mar. Biol. 10: 271–382.
Measures, C.I., and Burton, J.D. 1980. The vertical distribution and oxidation states of dissolved selenium in the northeast Atlantic Ocean and their relationship to biological processes. Earth. Planet. Sci. Lett. 46: 385–396.
Mehta, A.J., Parchure, T.M., Dixit, J.G., and Ariathurai, R. 1982. Resuspension potential of deposited cohesive sediment beds. pp. 591–609. In: V.S. Kennedy (ed.) Estuarine Comparisons. Academic Press, New York.
Meier, M.F., and Post, A. 1969. What are glacier surges? Can. J. Earth Sci. 6: 807–817.
Menard, H.W. 1964. Turbidity currents. In: Marine Geology of the Pacific. McGraw-Hill, New York. pp. 191–254.
Menzie, C.A. 1984. Diminishment of recruitment: A hypothesis concerning impacts on benthic communities. Mar. Poll. Bull. 15: 127–128.
Menzel, D.W. 1977. Summary of experimental results: Controlled ecosystem pollution experiment. Bull. Mar. Sci. 27: 142–145.
Mercer, F.H. 1956. Geomorphology and glacial history of southernmost Baffin Island. G.S.A. Bull. 67: 553–570.
———. 1961. The response of a fjord glacier to changes in the firn limit. J. Glaciology 3: 850–858.
Meyer-Peter, E., and Muller, R. 1948. Formulas for Bed-load Transport. Intern. Assoc. Hydr. Res. 2nd meeting, Stockholm.
Middleton, G.V. 1970. The generation of log-normal size frequency distribution in sediments. pp. 34–42. In: M.A. Romanova and O.V. Sarmand (eds.) Topics in Mathematical Geology. Consultants Bureau, New York, N.Y.
Middleton, G.V., and Hampton, M.A. 1976. Subaqueous sediment transport and deposition by sediment gravity flows. pp. 197–218. In: D.J. Stanley and D.J.W. Swift (eds.) Marine Sediment Transport and Environmental Management. John Wiley & Sons, New York.
Middleton, G.V., and Southard, J.B. 1977. Mechanics of sediment movement. Soc. Econ. Paleont. Mineral. Short Course 3, Tulsa, Oklahoma.
Miller, A.A.L., Mudie, P.S., and Scott, D.B. 1982. Holocene history of Bedford Basin, Nova Scotia: Foraminifera, dinoflagellate and pollen records. Can. J. Earth Sci. 19: 2342–2367.
Miller, D.J. 1960. Giant waves in Lituya Bay, Alaska. U.S. Geol. Surv. Prof. Paper 354 C: 51–86.
Miller, G.H. 1980. Late Foxe glaciation of southern Baffin Island, N.W.T., Canada. Geol. Soc. Am. Bull. 91: 399–405.
Milliman, J.D. 1980. Sedimentation in the Fraser River and its estuary, southwestern British Columbia (Canada): Est. Coast. Mar. Sci. 10: 609–633.
Milliman, J.D., and Meade, R.H. 1983. World-wide delivery of river sediment to the oceans. J. Geol. 91: 1–21.
Mills, C.E. 1982. Patterns and mechanisms of vertical distribution of medusae and ctenophores. Unpublished Ph.D. thesis, Univ. Victoria, Victoria, B.C. 384 p.
Mills, G.L., and Quinn, J.G. 1981. Isolation of dissolved organic matter and copper-organic complexes from estuarine waters using reverse-phase liquid chromatography. Mar. Chem. 10: 93–102.
———. 1984. Dissolved copper and copper-organic complexes in the Narragansett Bay estuary. Mar. Chem. 15: 151–172.
Milne, P.H. 1972. Hydrography of Scottish west coast sea lochs. Marine Research Report No. 3, Dept. Agricult. Fish. Scotland, Edinburgh, 40 p.
Mirza, F.B., and Gray, J.S. 1981. The fauna of benthic sediments from the organically enriched Oslofjord, Norway. J. Exp. Mar. Biol. Ecol. 54: 181–207.
Misar, Z. 1968. A contribution to the geomorphological history of the Sermilik Fjord area, Frederikshab district, S.W. Greenland. Acta Universitatis Carolinae Geographica 1: 51–66.
Molnia, B.F. 1979. Sedimentation in coastal embayments, northeastern Gulf of Alaska. 11th Offshore Tech. Conf. 665–676.
———. 1983. Subarctic glacial-marine sedimentation: A model. pp.95–114. In: B.F. Molnia (ed.) Glacial-Marine Sedimentation. Plenum Press, New York.
Molnia, B.F., and Sangrey, D.A. 1979. Glacially derived sediments in the northern Gulf of Alaska—geology and engineering characteristics. 11th Offshore Tech. Conf. 647–655.
Molvaer, J. 1980. Deep-water renewals in the Frierfjord—An intermittently anoxic basin. pp. 531–539. In: H.J. Freeland, D.M. Farmer, and C.D. Levings (eds.) Fjord Oceanography. Plenum Press, New York.
Möller, J.S. 1984. Hydrodynamics of an arctic fjord. Field study. Affarlikassaa, West Greenland. Resonance of internal seiches and buoyancy driven circulation. Inst. of hydro-dynamics and hydraulic engineering, Lyngby, Denmark. Series Paper 34, 197 p.
Möller, P., and Rosenberg, R. 1983. Recruitment, abundance and production of *Mya arenaria* and

Cardium edula in marine shallow waters, western Sweden. Ophelia 22: 33–55.
Moore, D.G. 1969. Reflection profiling studies of the California continental borderland: Structure and Quaternary turbidite basins. Geol. Soc. Am. Spec. Paper 107, 142 p.
Moore, R.M., and Burton, J.D. 1978. Dissolved copper in the Zaire estuary. Neth. J. Sea Res. 12: 355–363.
Moore, R.M., Burton, J.D., Williams, P.J.B., and Young, M.L. 1979. The behavior of dissolved organic material, iron and manganese in estuarine mixing. Geochim. Cosmochim. Acta 43: 919–926.
Moore, W.L., and Masch, F.D. 1962. Experiments on the scour resistance of cohesive sediments. J. Geophys. Res. 67: 1437–1449.
Morel, F., and Morgan, J.J. 1972. A numerical method for computing equilibria in aqueous chemical systems. Environ. Sci. Tech. 6: 58–67.
Morgan, J.J. 1967a. Applications and limitations of chemical thermodynamics in water systems. pp. 1–29. In: R.F. Gould (ed.) Equilibrium Concepts in Natural Water Systems. Amer. Chem. Soc., Washington, D.C.
———. 1967b. Chemical equilibria and kinetic properties of manganese in natural water. pp. 561–622. In: S.D. Faust and J.V. Hunter (eds.) Principles and Applications of Water Chemistry. John Wiley & Sons, New York.
Morgenstern, N.R. 1967. Submarine slumping and the initiation of turbidity currents. pp. 189–220. In: A.F. Richards (ed.) Marine Geotechnique, Univ. of Illinois Press.
Mörner, N-A. 1971. Eustatic changes during the last 20,000 years and a method of separating the isostatic and eustatic factors in an uplifted area. Palaeogeog. Palaeoclim. Palaeoecol. 9: 153–181.
———. 1980. Eustasy and geoid changes as a function of Core/Mantle changes. pp. 535–553. In: N-A. Mörner (ed.) Earth Rheology, Isostasy and Eustasy. John Wiley & Sons, London.
Morris, A.W., Mantoura, R.F.C., Bate, A.J., and Howland, R.J.M. 1978. Very low salinity regions of estuaries: Important sites for chemical and biological reactions. Nature 274: 678–680.
Morris, S., and Leaney, A.J. 1980. The Somass River Estuary, Status of Environmental Knowledge to 1980. Special Estuary Series. Environment Canada.
Muench, R.D., and Heggie, D.T. 1978. Deep water exchange in Alaskan subarctic fjords. pp. 239–268. In: B. Kjerfve (ed.) Estuarine Transport Processes. Univ. S. Carolina Press, Columbia.
Muench, R.D., and Nebert, D.L. 1973. Physical oceanography. pp. 103–149. In: D.W. Hood, W.E. Shiels, and E.J. Kelley (eds.) Environmental Studies of Port Valdez. Occ. Publ. No. 3, Inst. Mar. Sci., Univ. Alaska, Fairbanks.
Muller, F. 1959. Beobachtungen über Pingos. Meddel om Grønland 153: 1–128.
Müller, G., Grimmer, G., and Böhnke, H. 1977. Sedimentary record of heavy metal and polycyclic aromatic hydrocarbons in Lake Constance. Naturwissenschaften 64: 427–431.
Müller, P.J., and Mangani, A. 1980. Organic carbon decomposition rates in the sediments of the Pacific manganese nodule belt dated by ^{234}Th and ^{231}Pa. Earth Planet. Sci. Lett. 51: 94–114.
Mullin, M.M. 1980. Interactions between marine zooplankton and suspended particles. pp. 233–241. In: M.C. Kavanaugh and J.O. Leckie (eds.) Particulates in Water. Advances in Chemistry Series No. 189. Amer. Chem. Soc., Washington, D.C.
Murray, J. 1886. Temperature observations on the Firth and Lochs of Clyde, made by the Scottish Marine Station, from March to November 1886. J. Scott. Met. Soc. Ser. 3: 7(3), 313–351.
Murray, J.W. 1975. The interaction of metal ions at the manganese dioxide-solution interface. Geochim. Cosmochim. Acta 39: 505–519.
Murray, J.W., and Gill, G. 1978. The geochemistry of iron in Puget Sound. Geochim. Cosmochim. Acta 42: 9–19.
Murray, J.W., and Grundmanis, V. 1980. Oxygen consumption in pelagic marine sediments. Science 209: 1527–1530.
Murray, J.W., Grundmanis, V., and Smethie, W.M. 1978. Interstitial water chemistry in the sediments of Saanich Inlet. Geochim. Cosmochim. Acta 42: 1011–1026.
Murty, T.S. 1979. Submarine slide-generated water waves in Kitimat Inlet, British Columbia. J. Geophys. Res. 84: 7777–7779.
Myers, A.C. 1978. Sediment processing in a marine subtidal sandy bottom community: I. Physical aspects. J. Mar. Res. 35: 609–632.
Naiman, R.J., and Sibert, J.R. 1979. Detritus and juvenile salmon production in the Nanaimo River estuary. III. Importance of detrital carbon to the estuarine ecosystem. J. Fish. Res. Bd. Can. 36: 504–520.
Nardin, T.R., Hein, F.J., Gorsline, D.S., and Edwards, B.D. 1979. A review of mass movement processess, sediment and acoustic characteristics and contrasts in slope and base-of-slope systems versus canyon-fan-basin floor systems. Soc. Econ. Paleont. Mineral. Spec. Publ.27: 61–73.
NAS. 1975. Petroleum in the Marine Environment. Nat. Acad. Science, Washington, D.C., 107 p.
Nealson, K.H., and Ford, J. 1980. Surface enhancement of bacterial manganese oxidation: Implications for aquatic environment. Geomicrobiol. J. 2: 21–37.
Nedwell, B.B., Hull, S.E., Andersson, A., Hagström, Å.F., and Lindström, E.B. 1983. Seasonal changes

in the distribution and exchange of inorganic nitrogen between sediment and water in the Northern Baltic (Gulf of Bothnia). Est. Coast. Shelf Sci. 17: 169–179.

Neihof, R.A., and G.I. Loeb. 1974. Dissolved organic matter in seawater and the electric charge of immersed surfaces. J. Mar. Res. 32: 5–12.

Nelson, C.H., Carlson, P.R., Byrne, J.V., and Alpha, T.R. 1970. Development of the Astoria Canyon-fan physiography and comparison with similar systems. Mar. Geol. 8: 259–291.

Nemoto, T., and Harrison, G. 1981. High latitude ecosystems. pp. 95–126. In: A.R. Longhurst (ed.) Analysis of Marine Ecosystems. Academic Press, London.

Newman, H.N. 1974. Microbial films in nature. Microbios. 9: 247–257.

Newman, W.S., Thurber, D.H., Zeiss, H.S., Rokach, A., and Musich, L. 1969. Late Quaternary geology of the Hudson River Estuary: A preliminary report. Trans. New York Acad. Sci. 31: 548–570.

Nichols, R.L. 1960. Geomorphology of Marguerite Bay area Palmer Peninsula, Antarctica. Bull. Geol. Soc. Amer. 71(10): 1421–1450.

Niebauer, H.J. 1980. A numerical model of circulation in a continental shelf-silled fjord coupled system. Est. Coast. Mar. Sci. 10: 507–521.

Nihoul, J.C.J. (ed.). 1978. Hydrodynamics of Estuaries and Fjords. Elsevier Scientific Publishing Company, Amsterdam, 546 p.

Nissenbaum, A., and Swaine, D.J. 1976. Organic matter-metal interactions in recent sediments: The role of humic substances. Geochim. Cosmochim. Acta 40: 809–816.

Nordseth, K. 1976. Suspended and bed material load in Norwegian Rivers. pp. 33–42. In: S. Skreslet, R. Leinebø, J.B.L. Matthews, and E. Sakshaug (eds.) Freshwater on the Sea. Assoc. Norw. Oceanogr., Oslo.

Northcote, T.G., Wilson, M.S., and Hurn, D.R. 1964. Some characteristics of Nitinat Lake, an inlet on Vancouver Island, British Columbia. J. Fish. Res. Bd. Can. 21: 1069–1081.

Nowell, A.R.M., Jumars, P.A., and Eckman, J.E. 1981. Effects of biological activity on the entrainment of marine sediments. Mar. Geol. 42: 133–153.

Nozaki, Y. 1977. Distributions of natural radionuclides in sediments influenced by bioturbation. J. Geol. Soc. Japan 8: 699–706.

Nutt, D.C. 1963. Fjords and Marine basins of Labrador. Polar notes, Occasional Publication of the Stefansson Collection 5: 9–23. Dartmouth College, Dartmouth.

Nye, J.F. 1976. Water flow in glaciers: Jökulhlaups, tunnels and veins. J. Glaciol. 17: 181–207.

Nyffeler, V.P., Li, Y.H., and Santschi, P.H. 1984. A kinetic approach to describe trace-element distribution between particles and solution in natural aquatic systems. Geochim. Cosmochim. Acta 48: 1513–1522.

Nyholm, N., Nielsen, T.K., and Pedersen, K. 1983. Modeling heavy metals transport in an arctic fjord system polluted from mine tailings. ISEM Conference. Modelling the rate and effects of toxic substances in the environment, Copenhagen.

Oakley, S.M. 1981. The geochemical partitioning and bioavailability of trace metals in marine sediments. Unpublished Ph.D. thesis. Oregon State Univ., Corvallis, Oregon, 92 p.

O'Brien, N.R., and Burrell, D.C. 1970. Mineralogy and distribution of clay size sediment in Glacier Bay, Alaska. J. Sed. Pet. 40: 650–655.

Odum, H.T. 1965. An energy circuit language for ecological and social system. pp. 1–35. In: B. Patton (ed.) Systems Analysis and Simulation in Ecology. Academic Press, New York.

Officer, C.B. 1982. Mixing, sedimentation rates and age dating for sediment cores. Mar. Geol. 46: 261–278.

Officer, C.B. 1983. Physics of estuarine circulation. pp. 15–42. In: B.H. Ketchum (ed.) Estuaries and Enclosed Seas. Elsevier Sc. Publ. Co., Amsterdam.

Officer, C.B., and Ryther, J.H. 1977. Secondary sewage treatment versus ocean outfalls: An assessment. Science 197: 1056–1060.

Officer, C.B., and J.H. Ryther. 1980. The possible importance of silicon in marine eutrophication. Mar. Ecol. Prog. Ser. 3:83.

Oftstad, E.B., Lunde, G., Martinsen, K., and Rygg, B. 1978. Chlorinated aromatic hydrocarbons in fish from an area polluted by industrial effluents. Sci. Total Environ. 10: 219–230.

Olausson, E. 1973. The Byfjord: Sediments, sedimentation and geochemistry. Medd. Maringeologiska Lab., Göteborg., No. 5, 6 p.

Olsen, O., and Reeh, N. 1969. Preliminary report on glacier observations in Nordwestfjord, east Greenland. Rapp. Grønland. Geol. Unders. 21: 41–53.

Olsson, I., Rosenberg, R., and Olundh, E. 1973. Benthic fauna and zooplankton in some polluted Swedish estuaries. Ambio 2: 158–163.

Østrem, G. 1975. Sediment transport in glacial meltwater streams. In: A.V. Jopling and B.C. McDonald (eds.) Glaciofluvial and Glaciolacustrine sedimentation. Soc. Ec. Paleont. Mineral. Spec. Publ. 23: 101–122.

Østrem, G., Bridge, C.W., and Rannie, W.F. 1967. Glacio-hydrology, discharge and sediment transport in the Decade Glacier Area, Baffin Island, N.W.T. Geog. Annaler 49A: 268–282.

Østrem, G., Riegler, T., and Ekman, S.R. 1970. Slamtransportundersøkelser i norske bre-elver 1969. Rep. 6/70, Hydrologisk avdeling, Norges vassdrags- og elektrisitetsvesen, 68 p.

Otsuka, K. 1976. Regional distribution of clay minerals of Sagami Bay, Japan. Rep. Fac. Sci. Shizuoku Univ. 11: 179–190.

Ovenshine, A.T. 1970. Observations of iceberg rafting in Glacier Bay, Alaska and the identification of ancient ice-rafted deposits. Geol. Soc. of Am. Bull. 81: 891–894.

Ovenshine, A.T., and Bartsch-Winkler, S. 1978. Portage, Alaska: Case history of an earthquake's impact on an estuarine system. pp. 275–284. In: M.L. Wiley, (ed.) Estuarine Interactions. Academic Press, New York.

Ovenshine, A.T., and Kachadoorian, R. 1976. Estimate of the time required for natural restoration of the effects of the 1964 earthquake at Portage. In: E.H. Cobb (ed.) The United States Geological Survey in Alaska: Accomplishments during 1975. U.S.G.S. circular 773: 53–54.

Ovenshine, A.T., Bartsch-Winkler, S., O'Brien, N.R., and Lawson, D.E. 1976. Sediment of the high tidal range environment of upper Turnagain Arm, Alaska. M1-M26. In: T.P. Miller (ed.) Recent and Ancient Sedimentary Environments in Alaska. Alaska Geol. Soc. Sym. Proc.

Owen, M.W. 1977. Problems in the modeling of transport, erosion, and deposition of cohesive sediments. pp. 515–537. In: E.D. Goldberg, I.N. McCave, J.J. O'Brien, and J.H. Steele (eds.) The Sea, Vol. 6. John Wiley & Sons, New York.

Owens, T.L., Burrell, D.C., and Weiss, H.V. 1980. Reaction and flux of manganese within the oxic sediment and basin water of an Alaskan fjord. pp. 607–675. In: H.J. Freeland, D.M. Farmer, and C.D. Levings (eds.) Fjord Oceanography. Plenum Press, New York.

Paasche, E., and Østergren, I. 1980. The annual cycle of plankton diatom growth and silica production in the inner Oslofjord. Limnol. Oceanogr. 25: 481–494.

Pantin, H.M. 1964. Sedimentation in Milford Sound. In: T.M. Skerman (ed.) Studies of a southern fiord. N. Zeal. Ocean. Inst. Mem. 17: 35–48.

Parfitt, R.L., Atkinson, R.J., and Smart, R. 1975. The mechanism of phosphate fixation by iron oxides. Soil Sci. Soc. Am. Proc. 39: 837–841.

Parsons, T.R. Ecological relations. In: D.W. Hood and S. Zimmerman (eds.) The Gulf of Alaska. NOAA. U.S. Gov. Print Office, Washington, D.C. (In press).

Parsons, T.R., LeBrasseur, R.J., and Barraclough, W.E. 1970. Levels of production in the pelagic environment of the Strait of Georgia, British Columbia: A review. J. Fish. Res. Bd. Can. 27: 1251–1264.

Parsons, T.R., Stronach, J., Borstad, G.A., Louttit, G., and Perry, R.I. 1981. Biological fronts in the Strait of Georgia, British Columbia, and their relation to recent measurements of primary production. Mar. Ecol. Prog. Ser. 6: 237–242.

Parsons, T.R., Perry, R.I., Nutbrown, E.D., Hsieh, W., and Lalli, C.M. 1983. Frontal zone analysis at the mouth of Saanich Inlet, British Columbia, Canada. Mar.Biol. 73: 1–5.

Parsons, T.R., Dovey, H.M., Cochlan, W.P., Perry, R.I., and Crean, P.B. 1984. Frontal zone analysis at the mouth of a fjord—Jarvis Inlet, British Columbia. Sarsia 69: 133–137.

Partheniades, E. 1972. Recent investigations in stratified flows related to estuarial hydraulics. Geol. Soc. Am. Mem. 133: 29–70.

Paxeus, N., Hall, P.D.J., and Iverfeldt, Å. Humic substances and sediment-water interactions in a coastal environment (in prep.).

Peake, E. 1978. Organic constituents of sediments from Nancy Sound. pp. 63–71. In: G.P. Glasby (ed.) Fiord Studies: Caswell and Nancy Sounds, New Zealand. Memoir No. 79, New Zealand Oceanographic Institute, D.S.I.R., Auckland, New Zealand.

Pearson, T.H. 1970. The benthic ecology of Loch Linnhe and Loch Eil, a sea-loch system on the west coast of Scotland. I. The physical environment and distribution of the macrobenthic fauna. J. Exp. Mar. Biol. Ecol. 5: 1–34.

———. 1971. Studies on the ecology of the macrobenthic fauna of Lochs Linnhe and Eil, west coast of Scotland. II. Analysis of the macrobenthic fauna by comparison of feeding groups. Vie et Milieu, Suppl. 22: 53–91.

———. 1975. The benthic ecology of Loch Linnhe and Loch Eil, a sea-loch system on the west coast of Scotland. IV. Changes in the benthic fauna attributable to organic enrichment. J. Exp. Mar. Biol. Ecol. 20: 1–41.

———. 1980a. The macrobenthos of fjords. pp. 569–602. In: H.J.Freeland, D.M. Farmer, and D.C. Levings (eds.) Fjord Oceanography. Plenum Press, New York.

———. 1980b. Marine pollution effects of pulp and paper industry wastes. In: O. Kinne and H.P. Bulnheim (eds.) 14th European Biology Symposium. Protection of life in the sea. Helgoländer Meeresuntersuchungen 33: 340–360.

———. 1981. Stress and catastrophe in marine benthic ecosystems. pp. 201–214. In: G.W. Barrett and R. Rosenberg (eds.) Stress Effects on Natural Ecosystems. John Wiley & Sons, Chichester.

———. 1982. The Loch Eil project: Assessment and synthesis with a discussion of certain biological questions arising from a study of the organic pollution of sediments. J. Exp. Mar. Biol. Ecol. 57: 92–124.

Pearson, T.H., and Eleftheriou, A. 1981. The benthic ecology of Sullom Voe. Proc. R. Soc. Edinb. 80B: 241–269.

Pearson, T.H., and Rosenberg, R. 1976. A comparative study of the effects on the marine environment of water from cellulose industries in Scotland and Sweden. Ambio 5: 77–79.

———. 1978. Macrobenthic succession in relation to organic enrichment and pollution of the marine environment. Oceanogr. Mar. Biol. Ann. Rev. 16: 229–311.

Pearson, T.H., Duncan, G., and Nuttall, J. 1982. The Loch Eil project: Population fluctuations in the macrobenthos. J. Exp. Mar. Biol. Ecol. 56: 305–321.

Pearson, T.H., Gray, J.S., and Johannessen, P.J. 1983. Objective selection of sensitive species indicative of pollution-induced change in benthic communities. 2. Data analyses. Mar. Ecol. Prog. 12: 237–255.

Pearson, T.H., Burrell, D.C., and Feder, H.M. Fjords. In: R.E. Turner and W.J. Wolff (eds.) Coastal Ecology Source Book (in press).

Pedersen, T.F. 1984. Interstitial water metabolic chemistry in a marine mine tailings deposit, Rupert Inlet, British Columbia. Can. J. Earth Sci. 21: 1–9.

Pedersen, T.F., and Price, N.B. 1982. The geochemistry of manganese carbonate in Panama Basin sediments. Geochim. Cosmochim. Acta 46: 59–68.

Peltier, W.R. 1974. The impulse response of a Maxwell earth. Revs. Geophys. Space Phys. 12: 649–669.

———. 1976. Glacial isostatic adjustment-II. The inverse problem. Geophys. J. Roy. Astr. Soc. 46: 669–705.

Peltier, W.R., and Andrews, J.T. 1976. Glacial isostatic adjustment-I. The forward problem. Geophys. J. Roy. Astr. Soc. 46: 605–646.

Perdue, E.M. 1979. Acid-base equilibrium of river water humic substances. pp. 99–114. In: E.A. Jenne (ed.) Chemical Modeling in Aqueous Systems. Amer. Chem. Soc. Washington, D.C.

Perkin, R.G., and Lewis, E.L. 1978. Mixing in an Arctic fjord. J. Phys. Oceanogr. 8: 873–880.

Petersen, C.G.J. 1913. Evaluation of the sea. II. The animal communities of the sea bottom and their importance for marine zoogeography. Rep. Dan. Biol. Stn. 23: 3–28.

———. 1915. Om Havbundens Dyresamfund i Skagerak, Kristianiafjord og de Danske Farvande. Beretn. Minist. Landbr. Fisk. dan biol. Stan. 23: 3–26.

Petersen, C.G.J., and Boysen-Jensen, P. 1911. Valuation of the sea. I. Animal life of the sea bottom, its food and quantity. Rep. Dan. Biol. Stn. 20: 3–81.

Petersen, G.H. 1977. Biological effects of sea-ice and icebergs in Greenland. pp. 319–329. In: M.J. Dunbar (ed.) Polar Oceans. Arctic Institute of North America, Calgary.

———, G.H. 1978. Life cycles and population dynamics of marine benthic bivalves from the Disko Bugt area of West Greenland. Ophelia 17: 95–120.

Petersen, G.H., and Curtis, M.A. 1980. Differences in energy flow through major components of subarctic, temperate and marine shelf ecosystems. Dana 1: 53–64.

Pewe, T.L. 1959. Sand-wedge polygons in the McMurdo Sound region, Antarctica—A progress report. Am. J. Sci. 257: 545–552.

———. 1960. Multiple glaciation in the McMurdo Sound region, Antarctica. A progress report. J. Geol. 68: 498–514.

Picard, M.D., and High, L.R., Jr. 1972. Criteria for recognizing lacustrine rocks. pp. 108–145. In: J.K. Rigby, and W.K. Hamblin (eds.) Recognition of ancient sedimentary environments. Soc. Econ. Paleon. Mineral. Spec. Publ. 16. Tulsa.

Pickard, G.L. 1961. Oceanographic features of inlets in the British Columbia mainland coast. J. Fish. Res. Bd. Can. 18: 907–984.

———. 1963. Oceanographic characteristics of inlets of Vancouver Island, British Columbia. J. Fish. Res. Bd. Canada 20: 1109–1144.

———. 1967. Some oceanographic characteristics of the larger inlets of southeast Alaska. J. Fish. Res. Bd. Can. 24: 1475–1506.

———. 1971. Some physical oceanographic features of inlets of Chile. J. Fish. Res. Bd. Can. 28: 1077–1106.

———. 1973. Water structure in Chilean fjords. pp. 95–104. In: R. Fraser (ed.) Oceanography of the South Pacific. N. Zeal. Nat. Com. UNESCO, Wellington.

———. 1975. Annual and longer term variations of deepwater properties in the coastal waters of southern British Columbia. J. Fish. Res. Bd. Can. 32: 1561–1587.

Pickard, G.L., and Giovando, L.F. 1960. Some observations of turbidity in British Columbia inlets. Limnol. Oceanogr. 5: 162–170.

Pickard, G.L., and Rodgers, K. 1959. Current measurements in Knight Inlet, British Columbia. J. Fish. Res. Bd. Can. 16: 635–678.

Pickard, G.L., and Stanton, B.R. 1980. Pacific fjords—A review of their water characteristics, pp. 1–51. In: H.J. Freeland, D.M. Farmer, and C.D. Levings (eds.) Fjord Oceanography. Plenum Press, New York.

Pickrill, R.A., and Irwin, J. 1983. Sedimentation in a deep glacier-fed lake—Lake Tekapo, New Zealand. Sedimentology 30: 63–75.

Pickrill, R.A., Irwin, J., and Shakespeare, B.S. 1981. Circulation and sedimentation in a tidal-influenced fjord lake: Lake McKerrow, New Zealand. Est. Coast. Shelf Sci. 12: 23–37.

Pielou, E.C. 1975. Ecological Diversity. John Wiley & Sons, New York. 165 p.

Pihl, L., and Rosenberg, R. 1982. Production, abundance, and biomass of mobile epibenthic marine fauna in shallow waters, western Sweden. J. Exp. Mar. Bio. Ecol. 57: 273–301.

Pingree, R.D., Holligan, P.M., and Mardell, G.T. 1978. The effects of vertical stability on phytoplankton distributions in the summer on the Northwest European shelf. Deep-Sea Res. 25: 1011–1028.

Piper, D.J.W., and Keen, M.J. 1976. Geological Studies in St. Margarets Bay Nova Scotia. Geol. Surv. Canada Paper 78-18, 18 p.

Piper, D.J.W., and Porritt, C.J. 1965. Some pingos in Spitsbergen. Norsk. Polarinstitutt Arbok, Oslo 81–84.

Piper, D.J.W., Wrightman, D.M., Lewis, J.F., and Dweyer, G.J.T. 1975. Late Quaternary geology of Nain Bay, Labrador. Maritime Sediments 11: 53–54.

Piper, D.J.W., Letson, J.R.J., Delure, A.M., and Barrie, C.Q. 1983. Sediment accumulation in low-sedimentation, wave dominated, glacial inlets. Sed. Geol. 36: 195–215.

Platt, T. 1975. Analysis of the importance of spatial and temporal heterogeneity in the estimation of annual production by phytoplankton in a small enriched basin. J. Exp. Mar. Biol. Ecol. 18: 99-109.

Platt, T., and Conover, R.J. 1971. Variability and its effect on the 24 h chlorophyll budget of a small marine basin. Mar. Biol. 10: 52–65.

Platt, T., and Subba Rao, D.V. 1970. Primary production measurements on a natural plankton bloom. J. Fish. Res. Bd. Can. 27: 887–899.

Pocklington, R. 1976. Terrigenous organic matter in surface sediments for the Gulf of St. Lawrence. J. Fish. Res. Bd. Can. 33: 93–97.

Pocklington, R., and Leonard, J.D. 1979. Terrigenous organic matter in sediments of the St. Lawrence Estuary and the Saguenay Fjord. J. Fish. Res. Bd. Can. 36: 1250–1255.

Poling, G.W. 1982. The characteristics of mill tailings and their behavior in marine environments. pp. 63–84. In: D.V. Ellis (ed.) Marine Tailings Disposal. Ann Arbor Science, Ann Arbor, Michigan.

Poole, A.L. 1951. New Zealand—American Fiordland Expedition. N.Z. Dep. Sci. Industr. Res. Bull. 103, 99 p.

Post, A. 1974. Subaqueous deposits in Lituya Bay, Alaska. U.S. Geol. Surv. Prof. Paper 900, 151 p.

Poulet, S.A. 1976. Feeding of *Pseudocalanus minutus* on living and non-living particles. Mar. Biol. 34: 117–125.

Powell, R.D. 1980. Holocene glacimarine sediment deposition by tidewater glaciers in Glacier Bay, Alaska. Thesis, Ohio State Univ. Columbus, Ohio, 420 p. Unpubl.

———. 1981a. A model for sedimentation by tidewater glaciers. Annals of Glaciology 2: 129–134.

———. 1981b. Sedimentation conditions in Taylor Valley, Antarctica, inferred from textural analysis of DRDP cores. In: L.D. McGinnis (ed.) Am. Geophy. Union, Antarctic Res. Ser. 33: 331–349.

———. 1983. Glacial-marine sedimentation processes and lithofacies of temperate tidewater glaciers, Glacier Bay, Alaska. pp. 185–232. In: B.F. Molnia (ed.) Glacial-Marine Sedimentation. Plenum Press, New York.

Presley, B.J., Kolodny, Y., Nissenbaum, A., and Kaplan, I.R. 1972. Early diagenesis in a reducing fjord, Saanich Inlet, British Columbia. II. Trace element distribution in interstitial water and sediments. Geochim. Cosmochim. Acta 36: 1073–1090.

Price, R.J. 1965. The changing proglacial environment of the Casement Glacier, Glacier Bay, Alaska. Trans. Inst. Br. Geog. 36: 107–116.

———. 1980. Rates of geomorphological changes in proglacial areas. pp. 79–93. In: R.A. Cullingford, D.A. Davidson, and J. Lewin (eds.) Timescales in Geomorphology. John Wiley & Sons, London.

Price, N.B., and Skei, J.M. 1975. Aereal and seasonal variations in the chemistry of particulate matter in a deep water fjord. Est. Coast. Mar. Sci. 3: 349–369.

Prior, D.B., and Coleman, J.M. 1979. Submarine landslides—Geometry and nomenclature. Zeitschrift Geomorph. 23: 415–426.

Prior, D.B., Wiseman, W.J., Jr., and Gilbert, R. 1981. Submarine slope processes on a fan delta, Howe Sound, British Columbia. Geo-Marine Lett. 1: 85–90.

Prior, D.B., Wiseman, Jr., W.J., and Bryant, W.R. 1981. Submarine chutes of the slopes of fjord deltas. Nature 209: 326–328.

Prior, D.B., Bornhold, B.D., Coleman, J.M., and Bryant, W.R. 1982a. Morphology of a submarine slide, Kitimat Arm, British Columbia: Geology 10: 588–592.

Prior, D.B., Coleman, J.M., and Bornhold, B.D. 1982b. Results of a known sea floor instability event. Geo-Marine Lett. 2: 117–122.

Prior, D.B., Bornhold, B.D., and Coleman, J.M. 1983. Geomorphology of a submarine landslide, Kitimat Arm, British Columbia: Geol. Surv. Can. Open File Rep. 961.

Pritchard, D.W. 1967. Observations of circulation in coastal plain estuaries. pp. 37–44. In: G.H. Lauff (ed.) Estuaries. Am. Assoc. Adv. Sci., Publ. 83, Washington, D.C.

Propp, M.V., Denisov, V.A., Pogreloov, V.M., and Ryabushko, V.I. 1975. The ecological system of a fiord of the Barents Sea. The Soviet J. Mar. Biol. 1: 201–212.

Pye, M.J.A. 1980. Studies of burrows in recent sublittoral fine sediments off the west coast of Scotland. Unpublished Ph.D. thesis, Univ. Glasgow, Glasgow, 253 p.

Quigley, R.M. 1980. Geology, mineralogy and geochemistry of Canadian soft soils: A geotechnical perspective. Can. Geotechnical J. 17: 261–285.

Quinlan, G., and Beaumont, C. 1981. A comparison of observed and theoretical post glacial relative sea level in Atlantic Canada. Can. J. Earth Sci. 18: 1146–1163.

Rasmussen, E. 1973. Systematics and ecology of the Isefjord marine fauna (Denmark). Ophelia 11: 1–507.

Redfield, A.C. 1934. On the proportion of organic derivatives in seawater and their relation to the composition of plankton. pp. 176–192. In: James Johnston Memorial Volume, University Press, Liverpool.

Reeburgh, W.S. 1980. Anaerobic methane oxidation: Rate depth distributions in Skan Bay sediments. Earth Planet. Sci. Letters 47: 345–352.

Reeburgh, W.S., Muench, R.D., and Cooney, R.T. 1976. Oceanographic conditions during 1973 in Russel Fjord, Alaska. Est. Coast. Mar. Sci. 4: 129–145.

Reid, S.J., Trites, R.W., Lawrence, D.J., Louck, S.R.H., and Seibert, G.H. 1976. Deep water exchange processes in the Saguenay Fjord. Bedford Institute of Oceanography report.

Reimers, C.E. 1982. Organic matter in anoxic sediments off central Peru: Relations of porosity, microbial decomposition and deformation properties. Mar. Geol. 46: 175–197.

Reimnitz, E., Von Huene, R., and Wright, F.F. 1970. Detrital gold and sediments in Nuka Bay, Alaska. U.S. Geol. Surv. Prof. Paper 700-C: 35–42.

Relling, O., and Nordseth, K. 1979. Sedimentation of a river suspension into a fjord basin, Gaupnefjord in Western Norway. Norsk Geografisk Tidsskrift 33: 187–203.

Reuter, J.H., and Perdue, E.M. 1977. Importance of heavy metal-organic matter interactions in natural waters. Geochim. Cosmochim. Acta 41: 325–334.

Revsbech, N.P., Jørgensen, B.B., and Blackburn, T.H. 1980. Oxygen in the sea bottom measured with a microelectrode. Science 207: 1355–1356.

Rhoads, D.C. 1974. Organism-sediment relations on the muddy sea floor. Oceanogr. Mar. Biol. Ann. Rev. 12: 263–300.

Rhoads, D.C., and Boyer, L.F. 1982. The effects of marine benthos on the physical properties of sediments: A successional perspective. pp.3–52. In: P.L. McCall, and M.J.S. Tevesz (eds.) Animal-Sediment Relations. Plenum Press, New York.

Rhoads, D.C., and Morse, P.W. 1971. Evolutionary and ecologic significance of oxygen deficient marine basins. Letharia 4: 413–428.

Rhoads, D.C., and Young, D.K. 1970. The influence of deposit-feeding organisms on sediment stability and community trophic structure. J. Mar. Res. 28: 150–178.

Rhoads, D.C., Aller, R.C., and Goldhaber, M.B. 1977. The influence of colonizing benthos on physical properties and chemical diagenesis of the estuarine seafloor. pp. 113–138. In: B.C. Coull (ed.) Ecology of Marine Benthos. Univ. S. Carolina Press, Columbia.

Rhoads, D.C., McCall, P.L., and Yingst, J.Y. 1978a. Disturbance and production on the estuarine sea floor. Amer. Sci. 66: 577–586.

Rhoads, D.C., Yingst, J.Y., and Ullman, U.J. 1978b. Seafloor stability in central Long Island Sound. Part I. Temporal changes in erodibility of fine-grained sediment. pp. 221–244. In: M.L. Wiley (ed.) Estuarine Interactions. Academic Press, New York.

Richards, A.F. 1976. Marine geotechnics of the Oslofjorden region. pp. 41–63. In: Contributions to Soil Mechanics, Laurits Bjerrum Memorial Volume. Oslo, Norweg. Geotech. Instit.

Richards, A.F., and Parks, J.M. 1976. Marine geotechnology. pp. 157–181. In: I.N. McCave (ed.) The Benthic Boundary Layer. Plenum Press, New York.

Richards, F.A. 1965. Anoxic basins and fjords. pp. 611–645. In: J.P. Riley, and G. Skirrow (eds.) Chemical Oceanography, Vol. 1. Academic Press, London.

Riddihough, R. 1983. Contemporary vertical movements and tectonics on Canada's west coast. Geological Assoc. of Can., abstracts Vol. 8: A57.

Robb, M.S. 1981. Composition and manganese association of suspended particulate matter at the head of a southeast Alaska fjord. Unpublished M.S. thesis, Univ. Alaska, Fairbanks.

Rosen, P.S. 1979. Boulder barricades in central Labrador. J. Sed. Pet. 49: 1113–1123.

Rosenberg, R. 1972. Benthic faunal recovery in a swedish fjord following the closure of a sulphite pulp mill. Oikos 23: 92–108.

———. 1973. Succession in benthic macrofauna in a Swedish fjord subsequent to the closure of a sulphite pulp mill. Oikos 24: 244–258.

———. 1974. Spatial dispersion of an estuarine benthic faunal community. J. Exp. Mar. Biol. Ecol. 15: 69–80.

———. 1976. Benthic faunal dynamics during succession following pollution abatement in a Swedish estuary. Oikos 27: 414–427.

———. 1977. Benthic macrofaunal dynamics, production, and dispersion in an oxygen-deficient estuary of west Sweden. J. Exp. Mar. Biol. Ecol. 26: 107–133.

———. 1980. Effect of oxygen deficiency on benthic macrofauna in fjords. pp. 499–514. In: H.J. Freeland, D.M. Farmer, and C.D. Levings (eds.) Fjord Oceanography. Plenum Press, New York.

Rosenberg, R., and Möller, P. 1979. Salinity stratified benthic macrofaunal communities and long-term monitoring along the west coast of Sweden. J. Exp. Mar. Biol. Ecol. 37: 175–203.

Rosenberg, R., Olsson, I., and Olundh, E. 1977. Energy flow model of an oxygen-deficient estuary on the Swedish west coast. Mar. Biol. 42: 99–107.

Rosenfeld, J.K. 1979. Ammonium adsorption in nearshore anoxic sediments. Limnol. Oceanogr. 24: 356–364.

———. 1981. Nitrogen diagenesis in Long Island Sound sediments. Amer. J. Sci. 281: 436–462.

Royer, T.C. 1975. Seasonal variations of waters in the northern Gulf of Alaska. Deep-Sea Res. 22: 403–416.

———. 1979. On the effects of precipitation and runoff on coastal circulation in the Gulf of Alaska. J. Phys. Oceanog. 9: 555–563.

———. 1983. Observations on the Alaska coastal current. pp. 9–30. In: H.G. Gade, A. Edwards, and H. Svendsen (eds.) Coastal Oceanography. Plenum Press, New York.

Russell, S.O. 1972. Behaviour of steep creeks in a large flood. B.C. Geography series 14: 223–227.

Rucker, T.L. 1983. The life history of the intertidal barnacle *Balanus balanoides* L. in Port Valdez, Alaska. Unpublished M.S. thesis, Univ. Alaska, Fairbanks, 251 p.

Rust, B.R. 1978. A classification of alluvial channel systems. In: A.D. Miall (ed.) Fluvial Sedimentology. Can. Soc. Petrol. Geol. Mem. 5: 187–198.

Rygg, B., and Skei, J. 1984. Correlation between pollutant load and the diversity of marine soft-bottom fauna communities. pp. 153–183. In: Proceedings of the International Workshop in Biological Testing of Effluents and Related Receiving Waters. OECD/U.S. EPA/Environ. Canada.

Sadler, H.E., and Serson, H.V. 1980. An unusual polynya in an arctic fjord. pp. 299–305. In: H.J. Freeland, D.M. Farmer, and C.D. Levings (eds.) Fjord Oceanography. Plenum Press, New York.

Saelen, O.H. 1967. Some features on the hydrography of Norwegian fjords. pp. 63–71. In: G.H. Lauff (ed.) Estuaries. Am. Assoc. Adv. Sci., Pub. 83, Washington, D.C.

———. 1976. General hydrography of fjords. pp. 43-49. In: S. Skreslet, R., Leinbø, J.B.L. Matthews, and E. Sakshaug (eds.) Freshwater on the Sea. Assoc. Norw. Oceanogr., Oslo.

Sakamoto, W. 1972. Study on the process of river suspension from flocculation to accumulation in estuary. Bull. Ocean Research Inst. University Tokyo 5, 46 p.

Sakshaug, E. 1978. The influence of environmental factors on the chemical composition of cultivated and natural populations of marine phytoplankton. Unpublished Ph.D. thesis, University Trondheim, Trondheim, 83 p.

Sakshaug, E., and Myklestad, S. 1973. Studies on the phytoplankton ecology of the Trondheimsfjord. III. Dynamics of phytoplankton blooms in relation to environmental factors, bioassay experiments and parameters for the physiological state of the populations. J. Exp. Mar. Biol. Ecol. 11: 157–188.

Salomons, W., and Eysink, W. 1981. Pathways of mud and particulate trace metals from rivers to the Southern North Sea. In: S.D. Nio, R.T.E. Schuttenhelm, and T.C.E. Weering (eds.) Holocene Marine Sedimentation in the North Sea Basin. Spec. Publ. Int. Assoc. Sedimentol. 5: 429–450.

Salomons, W., and Förstner, U. 1984. Metals in the hydrocycle. Springer-Verlag, Berlin, 349 p.

Sanders, H.L. 1968. Marine benthic diversity: A comparative study. Amer. Nat. 102: 243–282.

Sandnes, O.K., and Gulliksen, B. 1980. Monitoring and manipulation of a sublittoral hard bottom biocoenosis in Balsfjord, northern Norway. Helgolander Meeresunters. 33: 467–472.

Sangrey, D.A., Cornell, U., Clukey, E.C., and Molnia, B.F. 1979. Geotechnical engineering analysis of underconsolidated sediments from Alaska coastal waters. 11th Offshore Technology Conf., Houston 1: 677–682.

Santos, S.L., and Simon, J.L. 1980. Marine soft-bottom community establishment following annual defaunation: Larval or adult recruitment? Mar. Ecol. Prog. Ser. 2: 235–241.

Santschi, P.H., Adler, D.M., and Amdurer, M. 1983. The fate of particles and particle-reactive trace metals in coastal waters: Radioisotope studies in microcosms. pp. 331–349. In: C.S. Wong, E. Boyle, K.W. Bruland, J.D. Burton, and E.D. Goldberg (eds.) Trace Metals in Seawater. Plenum Press, New York.

Savdra, C.E., Bottjer, D.J., and Gorsline, D.S. 1984. Development of a comprehensive oxygen-deficient marine biofacies model: Evidence from Santa Monica, San Pedro, and Santa Barbara basins, California Continental Borderland. Am. Assoc. Petroleum Geol. Bull. 68: 1179–1193.

Schafer, C.T., and Blakeney, C.P. 1984. Baffin Island fjords. Sea Frontiers 30: 94–105.

Schafer, C.T., Smith, J.N., and Loring, D.H. 1980. Recent sedimentation events at the head of Saguenay Fjord, Canada. Environmental Geology 3: 139–150.

Schafer, C.T., Smith, J.N., and Seibert, G. 1983. Significance of natural and anthropogenic sediment inputs to the Saguenay Fjord, Quebec. Sed. Geol. 36: 177–195.

Scheltema, R.S. 1974. Relationship of dispersal to geographical distribution and morphological variation in the polychaete family *Chaetopteridae*. Thalassia Jugosl. 10: 297–312.

Schink, D.R., and Guinasso, N.L. 1977. Effects of bioturbation on sediment-seawater interaction. Mar. Biol. 23: 133–154.

Schnitzer, M., and Khan, S.U. 1972. Humic Substances in the Environment. Marcel Dekker, New York.

Schoklitsch, A. 1950. Handbuch des Wasserbaues. Springer-Verlag, Vienna.

Scholl, D.W., von Huene, R., and Ridlon, J.B. 1968. Spreading of the ocean floor: Undeformed sediments in the Peru-Chile Trench. Science 159: 869–971.

Schrader, H.-J. 1971. Fecal pellets: Role in sedimentation of pelagic diatoms. Nature 174: 55–57.

Schubel, J.R. 1971. The estuarine environment. Estuaries and estuarine sedimentation. Short course lecture notes American Geological Institute, Washington D.C. 324 p.

Schubel, J.R., and Hirschberg, P.J. 1978. Estuarine graveyards, climatic change and the importance of estuarine environments. pp. 285–303. In: M.L. Wiley (ed.) Estuarine Interactions. Academic Press, New York.

Schwarz, H.-U. 1982. Subaqueous Slope Failures—Experiments and Modern Occurences. Contributions to Sedimentology 11. Elsevier, Stuttgart.

Schwinghammer, P. 1981. Characteristic size distributions of integral benthic communities. Can. J. Fish. Aquat. Sci. 38: 1255–1263.

Sclater, F.R., Boyle, E.A., and Edmond, J.M. 1976. On the marine geochemistry of nickel. Earth Planet. Sci. Lett. 31: 119–128.

Seibert, G.H., Trites, R.W., and Reid, S.J. 1979. Deep water exchange processes in the Saguenay Fjord. J. Fish. Res. Bd. Can. 36: 42–53.

Seki, H. 1982. Organic Material in Aquatic Ecosystems. CRC Press, Boca Raton, 201 p.

Semb, A. 1978. Deposition of trace elements from the atmosphere in Norway. Research Report 13, Acid Precipitation—Effects on Forest and Fish. 28 p.

Shanks, A.L., and Trent, J.D. 1980. Marine snow: Sinking rates and potential role in vertical flux. Deep Sea Res. 27A: 137–143.

Sharma, G.D. 1979. The Alaskan Shelf: Hydrographic, Sedimentary, and Geochemical Environment. Springer-Verlag, New York, 498 p.

Sharma, G.D., and Burbank, D.C. 1973. Geological oceanography, pp. 15–99. In: D.W. Hood, W.E. Shiels, and E.J. Kelley (eds.) Environmental Studies of Port Valdez. Occas. Publ. No. 3, Inst. Mar. Sci., Univ. Alaska, Fairbanks.

Sharma, G.D., and Burrell, D.C. 1970. Sedimentary environment and sediments of Cook Inlet, Alaska. Amer. Assoc. Pet. Geol. Bull. 54: 647–654.

Sheehan, P.J. 1984. Effects on community and ecosystem structure and dynamics. pp. 51–99. In: P.J. Sheehan, D.R. Miller, G.C. Butler, and Ph. Bourdeau (eds.) Effects of Pollutants at the Ecosystem Level. John Wiley & Sons, Chichester.

Shepard, F.P. 1973. Submarine Geology, 3rd. ed. Harper & Row, New York. 517 p.

Shick, J.M. 1976. Physiological and behavioral responses to hypoxia and hydrogen sulfide in the infaunal asteroid *Ctenodiscus crispatus*. Mar. Biol. 37: 279–289.

Shields, A. 1936. Anwendung der Ähnlichkeitsmechanik und Turbulenzforschung auf die Geschiebebewegung. Mitteil. Preuss. Versuchsanst. Wasser, Erd, Schiffsbau, Berlin, No. 26.

Sholkovitz, E.R. 1976. Flocculation of dissolved organic and inorganic matter during the mixing of river water and seawater. Geochim. Cosmochim. Acta 40: 831–845.

Sholkovitz, E.R., Boyle, E.A., and Price, N.B. 1978. The removal of dissolved humic acids and iron during estuarine mixing. Earth Planet. Sci. Lett. 40: 130–136.

Sholkovitz, E.R., Cochran, J.K., and Carey, A.E. 1983. Laboratory studies of the diagenesis and mobility of ^{239}, ^{240}Pu and ^{137}Cs in nearshore sediments. Geochim. Cosmochim. Acta 47: 1369–1379.

Sieburth, J. McN. 1969. Studies on algal substances in the sea. III. The production of extra-cellular organic matter by littoral marine algae. J. Exp. Biol. Ecol. 3: 290–309.

Skadsheim, A. 1983. The ecology of intertidal amphiods in the Oslofjord. Distribution and response to physical factors. Crustaceana 44: 225–244.

Skaven-Haug, S. 1955. Submarine slides in Trondheim harbour. Teknisk Ukeblad 102: 133–144.

Skei, J.M. 1975. The marine chemistry of Sørfjorden, West Norway. Unpubl. Ph.D. thesis, Univ. of Edinburgh, 207 p.

———. 1978. Serious mercury contamination of sediments in a Norwegian semi-enclosed bay. Mar. Pollut. Bull. 9: 191–193.

———. 1980. The chemistry of suspended particulate

matter from two oxygen deficient Norwegian fjords—with special reference to manganese. pp. 693–697. In: H.J. Freeland, D.M. Farmer, and C.D. Levings (eds.) Fjord Oceanography. Plenum Press, New York.

———. 1981a. The entrapment of pollutants in Norwegian fjord sediments—A beneficial situation for the North Sea. Spec. Publs. Int. Ass. Sediment. 5: 461–468.

———. 1981b. Et biogeokjemisk studium av en permanent anoksisk fjord-Framvaren ved Farsund. Norsk Institutt for Vannforskning (internal report in Norwegian) 108 p.

———. 1981c. Dispersal and retention of pollutants in Norwegian fjords. Rapp. P.-V. Réun. Cons. int. Explor. Mer. 181: 78–86.

———. 1982. Fluxes of pollutants to Norwegian fjord sediments based on lead-210. Proc. Symp. Mar. Chem. into the Eighties, Sidney, B.C. 29–47.

———. 1983a. Permanently anoxic marine basins-exchange of substances across boundaries. In: R. Hallberg (ed.) Environmental Biogeochemistry. Ecol. Bull. 35: 419–424.

———. 1983b. Why sedimentologists are interested in fjords. Sedim. Geol. 36: 75–80.

———. 1983c. Geochemical and sedimentological considerations of a permanently anoxic fjord—Framvaren, South Norway. Sedim. Geol. 36: 131–145.

Skei, J.M., and Melsom, S. 1982. Seasonal and vertical variations in the chemical composition of suspended particulate matter in an oxygen-deficient fjord. Est. Coast. Shelf Sci. 14: 61–78.

Skei, J.M., and Paus, P.E. 1979. Surface metal enrichment and partitioning of metals in a dated sediment core from a Norwegian fjord. Geochim. Cosmochim. Acta 43: 239–246.

Skei, J.M., Price, N.B., and Calvert, S.E., and Holtedahl, H. 1972. The distribution of heavy metals in sediments of Sørfjord, Norway. Water, Air, and Soil Pollution 1: 452–461.

Skei, J.M., Saunders, M., and Price, N.B. 1976. Mercury in plankton from a polluted Norwegian fjord. Mar. Pollut. Bull. 7: 34–35.

Skerman, T.M. 1964. Studies of a southern fjord. New Zealand Oceanographic Institute Mem. No. 17, 102 p.

Skjoldal, H.R., and Lannergren, C. 1978. The spring phytoplankton bloom in Lindåspollene, a land-locked Norwegian fjord. II. Biomass and activity of net and nanoplankton. Mar. Biol. 47: 313–323.

Skreslet, S., and Loeng, H. 1977. Deep water renewal and associated processes in Skjomen, a fjord in north Norway. Est. Coast. Mar. Sci. 5: 383–398.

Slatt, R.M. 1974. Formation of palimpsest sediments, Conception Bay, southeastern Newfoundland. Geol. Soc. Am. Bull. 85: 821–826.

———. 1975. Dispersal and geochemistry of surface sediments in Halls Bay, north-central Newfoundland: Application to mineral exploration. Can. J. Earth Sci. 12: 1346–1361.

Slatt, R.M., and Gardiner, W.W. 1976. Comparative petrology and source of sediments in Newfoundland fjords. Can. J. Earth Sci. 13: 1460–1465.

Slatt, R.M., and Hoskin, C.M. 1968. Water and sediment in the Norris Glacier outwash area upper Taku Inlet, South eastern Alaska. J. Sed. Pet. 38: 434–456.

Smayda, T.J. 1969. Some measurements of the sinking rate of fecal pellets. Limnol. Oceanogr. 14: 621–625.

———. 1970. The suspension and sinking of phytoplankton in the sea. Oceanogr. Mar. Biol. Ann. Rev. 8: 353–414.

Smetacek, V. 1980. Annual cycle of sedimentation in relation to plankton ecology—Western Kiel Bight. Ophelia Suppl. 1: 65–76.

Smethie, W.M. 1980. Estimation of vertical mixing rates in fjords using naturally occurring radon-222 and salinity as tracers. pp. 241–249. In: H.J. Freeland, D.M. Farmer, and C.D. Levings (eds.) Fjord Oceanography. Plenum Press, New York.

———. 1981. Vertical mixing rates in fjords determined using radon and salinity as tracers. Est. Coast. Shelf Sci. 12: 131–152.

Smethie, W.M., Nittrauer, C.A., and Self, R.F.L. 1981. The use of radon-222 as a tracer of sediment irrigation and mixing on the Washington continental shelf. Mar. Geol. 42: 173–200.

Smith, J.D., and Farmer, D.M. 1980. Mixing induced by internal hydraulic disturbances in the vicinity of sills. pp. 251–257. In: H.J. Freeland, D.M. Farmer, and C.D. Levings (eds.) Fjord Oceanography. Plenum Press, New York.

Smith, J.N., and Ellis, K.K. 1982. Transport mechanism for Pb-210, Cs-137 and Pu fallout radionuclides through fluvial-marine systems. Geochim. Cosmochim. Acta 46: 941–954.

Smith, J.N., and Loring, D.H. 1981. Geochronology for mercury pollutions in the sediments of the Saguenay Fjord, Quebec. Environ. Sci. Technol. 15: 944–951.

Smith, J.N., and Schafer, C.T. 1985. A 20th century record of seasonally-modulated sediment accumulation rates in a Canadian fjord based on Pb-210 measurements. J. Quaternary Res. (in press).

Smith, J.N., and Walton, A. 1980. Sediment accumulation rates and geochronologies measured in the Saguenay Fjord using Pb-210 dating method. Geochim. Cosmochim. Acta 44: 225–240.

Smith, J.N., Schafer, C.T., and Loring, D.H. 1980.

Depositional processes in an anoxic, high sedimentation regime in the Saguenay Fjord. pp.625–631. In: H.J. Freeland, D.M. Farmer, and C.D. Levings (eds.) Fjord Oceanography. Plenum Press, New York.

Smith, N., and Syvitski, J.P.M. 1982. Sedimentation in a glacier-fed lake: The role of pelletization on deposition of fine-grained suspensates. J. Sed. Pet. 52: 503–513.

Solorzano, L., and Grantham, B. 1975. Surface nutrients, chlorophyll α and phaeopigments in some Scottish sea lochs. J. Exp. Mar. Biol. Ecol. 20: 63–76.

Soot-Ryen, T. 1924. Faunistische untersuchungen im Ranafjorden. Tromsø Mus. Årb. 45: 1–106.

Sousa, W.P. 1979. Disturbance in marine intertidal boulder fields: The nonequilibrium maintenance of species diversity. Ecology 60: 1225–1239.

Spencer, D.W., and Brewer, P.G. 1971. Vertical advection, diffusion and redox potentials as controls on the distribution of manganese and other trace metals dissolved in waters of the Black Sea. J. Geophys. Res. 6: 5877–5892.

Spencer, D.W., Brewer, P.G., and Sachs, P.L. 1972. Aspects of the distribution and trace element composition of suspended particulate matter in the Black Sea. Geochim. Cosmochim. Acta 36: 71–86.

Sporstöl, S., Gjøs, N., Lichtenthaler, R.G., Gustavsen, K.O., Urdal, K., Oreld, F., and Skei, J. 1983. Source identification of aromatic hydrocarbons in sediments using GC/MS. Environ. Sci. Technol. 17: 283–286.

Squamish Estuary Management Plan. 1981. Air and Water Quality Work Group Final Report. Publ. Environment Canada & Prov. British Columbia.

Stanley, D.J. 1968. Reworking of glacial sediments in the N.W. Arm, a fjord-like inlet on the southeast coast of Nova Scotia. J. Sed. Pet. 38: 1224–1241.

———. 1972. The Mediterranean Sea: A natural sedimentation laboratory. Dowden, Hutchison & Ross, Stroudsburg, Pa. 765 p.

Stanley, S.O., Pearson, T.H., and Brown, C.M. 1978. Marine microbial ecosystems and the degradation of organic pollutants. pp. 60–79. In: W.W.A. Chater and H.J. Sommerville (eds.) The Oil Industry and Microbial Ecosystems. Heyden, London.

Stanley, S.O., Leftley, J.W., Lightfoot, A., Robertson, N., Stanley, I.M., and Vance, I. 1981. The Loch Eil project: Sediment chemistry, sedimentation and the chemistry of the overlying water in Loch Eil. J. Exp. Mar. Biol. Ecol. 55: 299–313.

Stanton, B.R. 1978. Hydrology of Caswell and Nancy Sounds. pp. 73–82. In: G.P. Glasby (ed.) Fiord Studies: Caswell and Nancy Sounds, New Zealand Memoir N.Z. Oceanographic Inst. No. 79.

———. 1984. Some oceanographic observations in the New Zealand fjords. Est. Coast. Mar. Sci. 19: 89–104.

Stanton, B.R., and Pickard, G.L. 1981. Physical Oceanography of the New Zealand Fiords. New Zealand Oceanographic Inst. Memoir 88: 1–36.

Steele, J.H., and Baird, I.E. 1968. Production ecology of a sandy beach. Limnol. Oceanogr. 13: 14–25.

Steele, J.H., and Baird, I.E. 1972. Sedimentation of organic matter in a Scottish sea loch. Me. Ist. Ital. Idrobiol. 29: 73–88.

Steeman Nielsen, E. 1958. A survey of recent Danish measurements of the organic productivity in the sea. Rapp. P.V. Cons. Perm. Int. Explor. Mer. 144: 92–95.

Stehman, C.F. 1976. Pleistocene and Recent sediments of northern Placentia Bay, Newfoundland. Can. J. Earth Sci. 13: 1386–1392.

Stephens, K., and Sibert, J. 1976. Primary production in Rupert Inlet, B.C. Fisheries and Marine Service, Nanaimo, British Columbia, Unpublished manuscript.

Stephens, K., Sheldon, R.W., and Parsons, T.R. 1967. Seasonal variations in the availability of food for benthos in a coastal environment. Ecology 48: 852–855.

Stewart, T.G.. and England, J. 1983. Holocene sea-ice variations and paleoenvironmental change, northernmost Ellesmere Island, N.W.T., Canada. Arctic Alp. Res. 15: 1–17.

Stigebrandt, A. 1976. Vertical diffusion driven by internal waves in a sill fjord. J. Phys. Oceanogr. 6: 486–495.

———. 1979. Observational evidence for vertical diffusion driven by internal waves of tidal origin in the Oslofjord. J. Phys. Oceanogr. 9: 435–441.

———. 1980. Some aspects of tidal interaction with fjord constrictions. Est. Coast. Mar. Sci. 11: 151–166.

———. 1981. A mechanism governing the estuarine circulation in deep, strongly stratified fjords. Est. Coast. Shelf Sci. 13: 197–211.

Stockner, J.G., and Cliff, D.D. 1979. Phytoplankton ecology of Vancouver Harbour. J. Fish. Res. Bd. Can. 36: 1–10.

Stockner, J.G., Cliff, D.D., and Buchanan, D.B. 1977. Phytoplankton production and distribution in Howe Sound, British Columbia: A coastal marine embayment-fjord under stress. J. Fish. Res. Bd. Can. 34: 907–917.

Stockner, J.G., Cliff, D.D., and Shortreed, K.R.S. 1979. Phytoplankton ecology of the Strait of Georgia, British Columbia. J. Fish. Res. Bd. Can. 36: 657–666.

Strickland, R.M. 1983. The Fertile Fjord: Plankton

in Puget Sound. Univ. of Washington Press, Seattle, 145 p.

Strøm, K.M. 1936. Land-locked water. Hydrography and bottom deposits in badly ventilated Norwegian fjords with remarks upon sedimentation under anaerobic conditions. Skrift. Norsk. Videnskap. Akad., Oslo, No. 7, 85 p.

———. 1955. Land-locked waters and the deposition of black muds. pp. 356–372. In: P.D. Trask (ed.) Recent Marine Sediments. Amer. Assoc. Pet. Geol., Tulsa.

———. 1957. A lake with trapped sea water? Nature 180: 982–983.

———. 1961. A second lake with old sea water at its bottom. Nature 189: 913.

Strömgren, T. 1974. The use of weighted arithmetic mean for describing the sediments of a landlocked basin (Borgenfjorden, Western Norway). Deep-Sea Res. 21: 155–160.

Stucchi, D.J. 1980. The tidal jet in Rupert-Holberg Inlet. pp. 491–497. In: H.J. Freeland, D.M. Farmer, and C.D. Levings (eds.) Fjord Oceanography. Plenum Press, New York.

Stucchi, D., and Farmer, D.M. 1976. Deep-water exchange in Rupert-Holberg Inlet. Pacific Mar. Sci. Rep. No. 76-10, Inst. Ocean Sci., Sidney, British Columbia, 31 p.

Stumm, W., and Brauner, D.A. 1975. Chemical speciation. pp. 173–234. In: J.P. Riley, and G. Skirrow (eds.) Chemical Oceanography, 2nd ed., Vol. 1. Academic Press, London.

Stumm, W., and Lee, G.F. 1960. The chemistry of aqueous iron. Schweiz. Z. Hydrol. 22: 295–319.

Stumm, W., and Morgan, J.J. 1981. Aquatic Chemistry, 2nd ed. John Wiley & Sons, New York. 780 p.

Suess, E. 1976. Nutrients near the depositional interface. pp. 57–79. In: I.N. McCave (ed.) The Benthic Boundary Layer. Plenum Press, New York.

———. 1979. Mineral phases found in anoxic sediments by microbial decomposition of organic matter. Geochim. Cosmochim. Acta 43: 339–352.

———. 1980. Particulate organic carbon flux in the oceans—Surface productivity and oxygen utilization. Nature 288: 260–263.

Sugai, S.F. 1985. Processes controlling trace metal and nutrient geochemistry in two southeast Alaskan fjords. Unpublished Ph.D. thesis. Univ. Alaska, Fairbanks, 139 p.

Sugai, S.F., and Burrell, D.C. 1984a. Transport of dissolved organic carbon, nutrients, and trace metals from the Wilson and Blossom Rivers to Smeaton Bay, southeast Alaska. Can. J. Fish. Aquat. Sci. 41: 180–190.

———. 1984b. Riverine influences upon geochronologies and trace metal distributions in sediment from two southeastern Alaskan fjords. EOS 65: 1075.

Sugai, S.F., and Healy, M.L. 1978. Voltammetric studies of the organic association of copper and lead in two Canadian inlets. Mar. Chem. 6: 291–308.

Sundborg, A. 1956. The river Klaralven, a study of fluvial processes. Geogr. Annaler 38: 127–316.

Sundby, B., and Loring, D.H. 1978. Geochemistry of suspended particulate matter in the Saguenay Fjord. Can. J. Earth Sci. 15: 1002–1011.

Sundby, B., Anderson, L., Hall, P., Iverfeldt, Å., van der Loeff, R., and Westerlund, S. The effect of oxygen on release and uptake of cobalt, manganese, iron and phosphate at the sediment-water interface. Limnol. Oceanogr. (in press).

Surlyk, F. 1978. Submarine fan sedimentation along fault scarps on tilted fault blocks (Jurassic-Cretaceous boundary, East Greenland). Bull. Grønlands Geol. Unders. 128, 108 p.

Svendsen, H. 1977. A study of the circulation in a sill fjord on the west coast of Norway. Mar. Sci. Comm. 3: 151–209.

Svensson, T. 1980. Water exchange and mixing in fjords. Dept. of Hydraulics, Chalmers Univ. of Tech. Gøteborg, Sweden, Rept. Series A:7: 266 p.

Sverdrup, H.V. 1931. The transport of material by pack ice. Geographical J. 77: 399–400.

———. 1953. On conditions for the vernal blooming of phytoplankton. J. Cons. Perm. Int. Explor. Mer. 18: 287–295.

Syvitski, J.P.M. 1978. Sedimentological advances concerning the flocculation and zooplankton pelletization of suspended sedimentation in Howe Sound, B.C. A fjord receiving glacial melt water. Ph.D. thesis, Univ. of British Columbia, 291 p.

———. 1980. Flocculation, agglomeration and zooplankton pelletization of suspended sediment in a fjord receiving glacial meltwater. pp. 615–623. In: H.J. Freeland, D.M. Farmer, and C.D. Levings (eds.) Fjord Oceanography. Plenum Press, New York.

———. 1984a. Sedimentology of Arctic Fjords Experiment: HU83-028 data report, Vol. 2. Can. Data Rep. Hydrog. Ocean Sci. 28: 1130 p.

———. 1984b. SAFE: 1983 Geophysical investigations. pp. 16-1 to 16-32. In: J.P.M. Syvitski (compiler) Sedimentology of Arctic Fjords Experiment: HU83-028 Data Report, Vol. 2. Can. Data Rep. Hydrog. Oc. Sci. 28.

Syvitski, J.P.M., and Blakeney, C.P. 1983. Sedimentology of Artic Fjords Experiment: HU82-031 data report, Vol. 1. Can. Data Rep. Hydrog. Oc. Sciences 12: 935 p.

Syvitski, J.P.M., and van Everdingen, D.A. 1981. A revaluation of the geologic phenomenon of sand flotation: A field and experimental approach. J. Sed. Pet. 51: 1315–1322.

Syvitski, J.P.M., and Farrow, G.E. 1983. Structures and processes in bayhead deltas: Knight and Bute Inlet, British Columbia. Sed. Geol. 36: 217–244.

Syvitski, J.P.M., and Lewis, A.G. 1980. Sediment ingestion by *Tigriopus californicus* and other zooplankton: Mineral transformation and sedimentological considerations. J. Sed. Pet. 50: 869–880.

Syvitski, J.P.M., and Macdonald, R. 1982. Sediment character and provenance in a complex fjord: Howe Sound, British Columbia. Can. J. Earth Sci. 19: 1025–1044.

Syvitski, J.P.M., and Murray, J.W. 1977. A discussion on grain size distributions using log-probability plots. Can. Soc. Pet. Bull. 25: 683–694.

———. 1981. Particle interaction in fjord suspended sediment. Mar. Geol. 39: 215–242.

Syvitski, J.P.M., and Schaefer, C.T. 1985. Sedimentology of Artic Fjords Experiment (SAFE): 1. Project Introduction. Arctic 38: 264–270.

Syvitski, J.P.M., and Swinbanks, D.D. 1980. VSA: A new fast size analysis technique for low sample weight based on Stokes' settling velocity. Can. Geotech. J. 17: 304–312.

Syvitski, J.P.M., Asprey, K.W., Blakeney, C.P., and Clattenburg, D. 1983a. SAFE: 1982 delta report. pp. 18-1 to 18-41. In: J.P.M. Syvitski and C.P. Blakeney (Comp.). Sedimentology of Artic Fjords Experiment: HU82-031 Data Report, Vol. 1. Can. Data Rep. Hydrogr. Oc. Sci. 12.

Syvitski, J.P.M., Silverberg, N., Ouellet, G., and Asprey, K.W. 1983b. First observations of benthos and seston from a submersible in the lower St. Lawrence estuary. Geog. Phys. Quat. 37: 227–240.

Syvitski, J.P.M., Fader, G.B., Josenhans, H.W., MacLean, B., and Piper, D.J.W. 1983c. Seabed investigations of the Canadian east coast and arctic using Pisces IV. Geoscience Canada 10: 59–68.

Syvitski, J.P.M., Blakeney, C.P., and Hay, A.E. 1983d. SAFE: Sidescan Sonar and Sounder Profiles. pp. 16-1 to 16-49. In: J.P.M. Syvitski and C.P. Blakeney (Comp.) Sedimentology of Arctic Fjords Experiment: HU82-031 data report, Vol. 1. Can. Data Rep. Hydrogr. Oc. Sci. 12.

Syvitski, J.P.M., Farrow, G.E., Taylor, R., Gilbert, R., and Emory-Moore, M. 1984a. SAFE: 1983 delta report. pp. 18-1–18-91. In: J.P.M. Syvitski (Comp.) Sedimentology of Arctic Fjords Experiment: HU83-028 Data Report Vol. 2. Can. Data Rep. Hydrogr. Ocean Sci. 28.

Syvitski, J.P.M., Lamplugh, M., and Kelly, B. 1984b. Fjord morphology. In: J.P.M. Syvitski (Comp.) Sedimentology of Arctic Fjords Experiment: HU83-028 Data Report, Vol. 2. Can. Data Rep. Hydrogr. Oc. Sci. 28: 20-1 to 20-27.

Syvitski, J.P.M., Asprey, K.W., Clattenburg, D.A., and Hodge, G.D. 1985. The prodelta environment of a fjord: Suspended particle dynamics. Sedimentology 32: 83–107.

Takahashi, M., D.L. Seibert, and W.H. Thomas. 1977. Occasional blooms of phytoplankton during summer in Saanich Inlet, B.C., Canada. Deep-Sea Res. 24: 775.

Tan, F.C., and Strain, P.M. 1979. Organic carbon isotope ratios in recent sediments in the St. Lawrence estuary and the Gulf of St. Lawrence. Est. Coast. Mar. Sci. 8: 213–225.

Tanner, V. 1939. Om de blockrika strandgördlana vid Subarktiska ocean-kuster, förekomstsätt och uppkomst. Terra 60: 157–165.

Taylor, G.B. 1975. Saguenay River sections from fifteen cruises 1961–1974. Bedford Inst. of Oceanogr. Data Series BI-D-75-2 38 p.

Taylor, J.H., and Price, N.B. 1983. The geochemistry of iron and manganese in the waters and sediments of Bolstadfjord, S.W. Norway. Est. Coast. Shelf Sci. 17: 1–19.

Taylor, R.B., and McCann, S.B. 1983. Coastal depositional landforms in northern Canada. pp. 53–75. In: D.E. Smith, and A.G. Dawson (eds.) Shorelines and Isostasy. Academic Press, London.

Tebo, B.M., Nealson, K.H., Emerson, S., and Jacobs, L. 1984. Microbial mediation of Mn (II) and Co (II) precipitation at the O_2/H_2S interfaces in two anoxic fjords. Limnol. Oceanogr. 29: 1247–1258.

Terzaghi, K. 1956. Varieties of submarine slope failures. Proceedings of the Eighth Texas Conference on Soil Mechanics and Foundation Engineering, 41 p.

Tesaker, E. 1978. Sedimentation in recipients. Disposal of particulate mine waste. VHL-Report No. STF Go A78105, 72 p.

Tett, P., and Wallis, A. 1978. The general annual cycle of chlorophyll standing crop in Loch Creran. J. Ecol. 66: 227–239.

Thayer, C.W. 1975. Morphologic adaptations of benthic invertebrates to soft substrata. J. Mar. Res. 33: 177–189.

Theede, H. 1973. Comparative studies on the influence of oxygen deficiency and hydrogen sulfide on marine bottom invertebrates. Neth. J. Sea Res. 7: 244–252.

Theede, H. 1981. Studies on the role of benthic animals of the Western Baltic and the flow of energy and organic material. Kieler Meeresforsch. Sonderh. 5: 634.

Theede, H., Ponat, A., Hiruki, K., and Schlieper, C. 1969. Studies on the resistance of marine bottom invertebrates to oxygen-deficiency and hydrogen sulphide. Mar. Biol. 2: 325–337.

Therriault, J.C., and Lacroix, G. 1975. Penetration of the deep layer of the Saguenay Fjord by surface waters of the St. Lawrence Estuary. J. Fish. Res. Board. Can. 32: 2372–2377.

Therriault, J.C., de Ladurantaye, R., and Ingram, R.G. 1980. Particulate matter exchange processes between the St. Lawrence estuary and the Saguenay Fjord. pp. 363–373. In: H.J. Freeland, D.M. Farmer, and D.C. Levings (eds.) Fjord Oceanography. Plenum Press, New York.

Therriault, J.C., de Ladurantaye, R., and Ingram, R.G. 1984. Particulate matter exchange across a fjord sill. Est. Coast Shelf Sci. 18: 51–64.

Thistle, D. 1981. Natural physical disturbances and communities of marine soft bottoms. Mar. Ecol. Prog. Ser. 6: 223.

Thomas, W.H., and Seibert, D.L.R. 1977. Effects of copper on the dominance and the diversity of algae: Controlled ecosystem pollution experiment. Bull. Mar. Sci. 27: 23–33.

Thompson, D.E. 1980. Surging glaciers as an oscillatory stable flow. pp.389–400. In: H.E. Holt and E.C. Kosters (eds.) Reports of the Planetary Geology Program. NASA, Washington.

Thompson, J.A.J., and McComas, F.T. 1974. Copper and zinc levels in submerged mine tailings at Britannia Beach, B.C. Fish. Res. Bd. Can. Tech. Rep. 437: 33 p.

Thompson, J.A.J., Macdonald, R.W., and Wong, C.S. 1980. Mercury geochemistry in sediments of a contaminated fjord of coastal British Columbia. Geochem. J. 14: 71–82.

Thomson, R.E. 1981. Oceanography of the British Columbia Coast. In: Canadian Special Publication. Fisheries and Aquatic Sciences 56: 1–291.

Thorson, G. 1950. Reproduction and larval ecology of marine bottom invertebrates. Biol. Reviews 25: 1–45.

Tiltze, W. 1973. Comments on relief metamorphosis in the Pleistocene. Geoforum 15: 62–63.

Tipping, E. 1981. The adsorption of aquatic humic substances by iron oxides. Geochim. Cosmochim. Acta 45: 191–199.

Tipping, E., and Cooke, D. 1982. The effect of adsorbed humic substances in the surface charge of goethite (α-FeOOH) in freshwaters. Geochim. Cosmochim. Acta 46: 75–80.

Tollan, A. 1972. Variability of seasonal and annual runoff in Norway. Nordic Hydrology 3: 73–79.

Topping, G. 1976. Sewage and the sea. pp. 303–351. In: R. Johnston (ed.) Marine Pollution. Academic Press, London.

Torrance, J.K. 1983. Towards a general model of quick clay development. Sedimentology 30: 547–556.

Toth, D.J., and Lerman, A. 1977. Organic matter reactivity and sedimentation rates in the ocean. Am. J. Sci. 277: 465–485.

Trites, R.W., Petrie, W.M., Hay, A.E., and DeYoung, B. 1983. Synoptic Oceanography. pp. 2-1 to 2-129. In: J.P.M. Syvitski and C.P. Blakeney (Comp.) Sedimentology of Arctic Fjords Experiment: HU82-031 data report Vol. 1. Can. Data Rep. Hydrogr. Ocean Sci. 12: 935 p.

Tulkki, P. 1965. Disappearance of the benthic fauna from the basin of Bornholm (southern Baltic) due to oxygen deficiency. Cah. Biol. Mar. 6: 455–463.

Tulkki, P. 1968. Effect of pollutions on the benthos of Gothenberg. Helgo. wiss Meeresunters. 17: 209–215.

Tunnicliffe, V. 1981. High species diversity and abundance of the epibenthic community in an oxygen-deficient basin. Nature 294: 354–356.

Tunnicliffe, V., and Syvitski, J.P.M. 1983. Corals assist boulder movement: An unusual mechanism of sediment movement. Limnol. Oceanogr. 8: 564–568.

Turner, D.R., Dickson, A.G., and Whitfield, M. 1980. Water-rock partition coefficients and the composition of natural waters—a reassessment. Mar. Chem. 9: 211–218.

Ullman, W.J., and Aller, R.C. 1982. Diffusion coefficients in nearshore marine sediments. Limnol. Oceanogr. 27: 552–556.

Vance, I., Stanley, S.O., and Brown, C.M. 1982. The Loch Eil project: Cellulose degrading bacteria in the sediments of Loch Eil and the Lynn of Lorne. J. Exp. Mar. Biol. Ecol. 56: 267–278.

Vancouver, G. 1798. A Voyage of Discovery to the North Pacific Ocean and Round the World. 3 vols. G.G. and J. Robinson and J. Edwards, London.

van den Berg, C.M.G., and Dharmvanij, S. 1984. Organic complexation of zinc in estuarine interstitial and surface water samples. Limnol. Oceanogr. 29: 1025–1036.

Vanderborght, J.P., and Billen, G. 1975. Vertical distribution of nitrate concentration in interstitial water of marine sediments with nitrification and denitrification. Limnol. Oceanogr. 20: 953–961.

Vanderborght, J.P., Wollast, R., and Billen, G. 1977. Kinetic models of diagenesis in disturbed sediments. I. Mass transfer properties and silica diagenesis. Limnol. Oceanogr. 22: 787–793.

van der Loeff, M.M.R., Anderson, L.G., Hall, P.O.J., Iverfeldt, Å., Josefson, A.B., Sundby, B., and Westerlund, S.F.G. 1984. The asphyxiation technique: An approach to distinguishing between molecular diffusion and biologically mediated transport at the sediment-water interface. Limnol. Oceanogr. 29: 675–686.

Verwey, E.J.W., and Overbeck, J.T.G. 1948. Theory of the Stability of Hydrophobic Colloids. Elsevier, New York.

Vernberg, W.B. 1983. Responses to estuarine stress. pp. 53–63. In: B.H. Ketchum (ed.) Estuaries and Enclosed Seas. Elsevier Scientific, Amsterdam.

Vilks, G., and Mudie, P.J. 1983. Evidence for postglacial paleoceanographic and paleoclimatic changes in Lake Melville, Labrador, Canada. Arctic Alpine Res. 15: 307–320.

Vilks, G., Deonarine, B., Rashid, M., and Winters, G. 1984. Late glacial-postglacial land-sea interaction in Lake Melville, southeastern Labrador. pp. 56–57. 13th annual Arctic Workshop, Boulder Colorado, Abstract Volume.

VKI. 1979. Laboratory assays on metal release from tailings material from Marmorilik. Marmorilik duplication. 1978. Project report prepared for Greenex A/S.

von Heune, R. 1966. Glacial-marine geology of Nuka Bay, Alaska, and the adjacent continental shelf. Mar. Geol. 4: 291–304.

von Heune, R, and Shor, G.G. Jr. 1969. The structure and tectonic history of the eastern Aleutian Trench. Geol. Soc. Am. Bull. 80: 1889–1902.

von Heune, R., Shor, G.G. Jr., and Reimnitz, E. 1967. Geological interpretation of seismic profiles in Prince William Sound, Alaska. Geol. Soc. Am. Bull. 78: 259–268.

Vorndran, V.G., and Sommerhoff, G. 1974. Glaziologisch-glazialmorphologische Untersuchungen im Gebiet des Qorqup-Austafsyle & Schers (Südwest-Grönland). Polarforschung 44: 137–147.

VTN. 1984. Integrative data analysis: Coastal and marine biology program. Unpublished report to U.S. Borax & Chemical Corporation.

Vuceta, J., and Morgan, J.J. 1978. Chemical modelling of trace metals in fresh waters: Role of complexation and adsorption. Environ. Sci. Tech. 12: 1302–1309.

Wacasey, J.W. 1975. Biological productivity of the southern Beaufort Sea: zoobenthic studies. Beaufort Sea project, Technical Report No. 126, Environment Canada, Victoria, 39 p.

Walcott, R.I. 1970. Isostatic response to loading of the crust in Canada. Can. J. Earth. Sci. 7: 716–727.

Waldichuck, M. 1960. Industrial pollution. I. Effects of non-metallic contaminants. Fish. Res. Bd. Can. MS Rep. 1115: 52 p.

———. 1980. Advective vertical processes in the upward transport of particulate materials and dissolved metals from mine tailings discharge into two different fjord systems. Am. Geophys. Union Trans. 61, 275 p.

Walsh, J.J., Rowe, G.T., Iverson, R.L., and McRoy, C.P. 1981. Biological export of shelf carbon is a neglected sink of the global CO_2 cycles. Nature 291: 196–201.

Wang, D.P., and D.W. Kravitz. 1980. A semi-implicit two-dimensional model of estuarine circulation. J. Phys. Oceanog. 10: 441–454.

Wassmann, P. 1983. Sedimentation of organic and inorganic particulate material in Lindåspollene, a stratified, land-locked fjord in western Norway. Mar. Ecol. Prog. Ser. 13: 237–248.

———. 1984. Sedimentation and benthic mineralization of organic detritus in a Norwegian fjord. Mar. Biol. 83: 83–94.

Wassmann, P., and Aadnesen, A. 1984. Hydrography, nutrients, suspended organic matter, and primary production in a shallow fjord system on the west coast of Norway. Sarsia 69: 139–153.

Watanabe, Y., and Tsunogai, S. 1984. Adsorption-desorption control of phosphate in anoxic sediment of a coastal sea, Funka Bay, Japan. Mar. Chem. 15: 71–83.

Water Survey of Canada. 1969. Surface Runoff Data. Atlantic. Environment Canada.

———. 1980. Surface Runoff Data. Atlantic. Environment Canada.

———. 1981. Surface Runoff Data. British Columbia. Environment Canada.

———. 1982. Surface Runoff Data. British Columbia. Environment Canada.

Watling, L. 1975. Analysis of structural variations in a shallow estuarine deposit-feeding community. J. Exp. Mar. Biol. Ecol. 19: 275–313.

Weidick, A. 1968. Observations on some Holocene glacier fluctuations in West Greenland. Medd. om Grønland 163: 202 p.

Welschmeyer, N.A., and Lorenzen, C.J. 1985. Chlorophyll budgets: Zooplankton grazing and phytoplankton growth in a temperate fjord and the central pacific gyres. Limnol. Oceanogr. 30: 1–21.

Westerlund, S., Anderson, L., Hall, P., Iverfeldt, Å, van der Loeff, M.R., and Sundby, B. Benthic fluxes (in prep.).

Westrich, J.T., and Berner, R.A. 1984. The role of sedimentary organic matter in bacterial sulfate reduction: The G model tested. Limnol. Oceanogr. 29: 236–249.

Whitehouse, V.G., Jeffrey, L.M., and Debrecht, J.D. 1960. Differential settling tendencies of clay minerals in saline waters. pp. 1–79. In: A. Swineford (ed.) Clays and Clay Minerals proc. 7th Natl. Conf. London. Pergamon Press.

Whitfield, M. 1979. The mean oceanic residence time (MORT) concept—A rationalization. Mar. Chem. 8: 101–123.

Whitfield, M., and Turner, D.R. 1979. Water-rock partition coefficients and the composition of seawater and river water. Nature 278: 132–137.

Wildish, D.J. 1977. Factors controlling marine and estuarine sublittoral macrofauna. Helgol. wiss. Meeresunters. 30: 454–465.

Wildish, D.J., and Peer, D. 1983. Tidal current speed and production of benthic macrofauna in the lower Bay of Fundy. Can. J. Fish. Aquat. Sci. 40: 309–321.

Wilke, R.J., and Daval, R. 1982. The behavior of iron, manganese and silicon in the Peconic River estuary, New York. Est. Coast. Shelf. Sci. 15: 577–586.

Williams, P.M., Mathews, W.H., and Pickard, G.L. 1961. A lake in British Columbia containing old seawater. Nature, 191: 830–832.

Willumsen, P. 1975. En undersögelse af foraminiferene i to grönlandske fjorde. Inst. Historisk Geol. Paleontologi, Univ. of Copenhagen.

Wilson, D.E. 1980. Surface and complexation effects on the rate of Mn (II) oxidation in natural water. Geochim. Cosmochim. Acta 44: 1311–1317.

Wilson, W.H. 1981. Sediment mediated interactions in a densely populated infaunal assemblage: the effects of the polychaete, *Abarenicola pacifica*. J. Mar. Res. 39: 735–748.

Windom, H., Wallace, G., Smith, R., Dudek, N., Maeda, M., Dulmage, R., and Storti, F. 1983. Behavior of copper in southeastern United States estuaries. Mar. Chem. 12: 183–193.

Winiecki, C.I., and Burrell, D.C. 1985. Benthic community development and seasonal variations in an Alaskan fjord. pp. 299–309. In: P.E. Gibbs (ed.). Proceedings of the 19th European Marine Biology Symposium. Cambridge University Press, Cambridge.

Winter, D.F. 1973. A similarity solution for steady state gravitational circulation in fjords. Est. Coast. Mar. Sci. 1: 387–400.

Winter, D.F., Banse, K., and Anderson, G.C. 1975. The dynamics of phytoplankton blooms in Puget Sound, a fjord in the northwestern United States. Mar. Biol. 29: 139.

Winters, G.V., and Buckley, D.E. 1980. *In situ* determination of suspended particulate matter and dissolved organic matter concentrations in an estuarine environment by means of an optical beam attenuance meter. Est. Coast. Mar. Sci. 10: 455–466.

Winters, G.V., Syvitski, J.P.M., and Maillet, L. 1985. Distribution and dynamics of suspended particulate matter in Baffin Island fjords pp. 73–77. In: G. Vilks and J.P.M. Syvitski (Comp.) 14th Arctic Workshop: Arctic Land-Sea Interaction Abstracts, Dartmouth.

Wittmann, G. 1979. Toxic Metals. pp. 3–70. In: U. Førstner and G.T.W. Wittman (eds.) Metal Pollution in the aquatic environment. Springer-Verlag, Berlin-Heidelberg-New York.

Wolff, W.J. 1983. Estuarine benthos. pp. 151–182. In: B.H. Ketchum (ed.) Estuaries and Enclosed Seas. Elsevier Scientific, Amsterdam.

Wood, B.J.B., Teet, P.B. and Edwards, A. 1973. An introduction to the phytoplankton, primary production and relevant hydrography of Loch Etive. J. Ecol. 61: 569–585.

Wood, J.M. 1974. Biological cycles for toxic elements in the environment. Science 183: 1049–1052.

Woodin, S.A. 1983. Biotic interactions in recent marine sedimentary environments. pp. 3–38. In: M.J.S. Tevesz and P.L. McCall (eds.) Biotic Interactions in Recent and Fossil Benthic Communities. Plenum Press, New York.

Woodin, S.A., and Jackson, J.B.C. 1979. Interphyletic competition among marine benthos. Amer. Zool. 19: 1029–1043.

Wright, F.F., Sharma, G.D., and Burbank, D.C. 1973. ERTS-1 observations of sea surface circulation and sediment transport, Cook Inlet, Alaska. pp. 1315–1322. Proc. Symp. on Significant Results obtained from ERTS-1. NASA, Washington, D.C.

Wyatt, T. 1980. The growth season in the sea. J. Plank. Res. 2: 81–96.

Xiong, Q., and Royer, T.C. 1984. Coastal temperature and salinity in the northern Gulf of Alaska, 1970–1983. J. Geophys. Res. 89: 8061–8068.

Yeats, P.A., and Bewers, J.M. 1976. Trace metals in the waters of the Saguenay fjord. Can. J. Earth Sci. 13: 1319–1327.

Yoshida, S. 1980. Mixing mechanisms of density current system at a river mouth. In: T. Carstens and T. McClimans (eds.) 2nd International Symposium on Stratified Flows.

Young, D.K. 1971. Effects of infauna on the sediment and seston of a subtidal environment. Vie et Milieu Supplement 22: 557–571.

Young, R.A., Clarke, T.L., Mann, R., and Swift, D.J.P. 1981. Temporal variability of suspended particulate concentrations in the New York Bight. J. Sed. Pet. 51: 293–305.

Zajac, R.N., and Whitlatch, R.B. 1982. Responses of estuarine infauna to disturbance. II. Spatial and temporal variation of succession. Mar. Ecol. Prog. Ser. 10: 15–27.

Ziegler, T. 1973. Materialtransportundersökelser i norske bre-elver 1971. Rept. 4/73, Hydrologisk avdeling Norges vassdrags-og electrisitetsvesen, 91 p.

Subject Index

Fjord Index